Permutation Statistical Methods

Kenneth J. Berry • Paul W. Mielke, Jr. •
Janis E. Johnston

Permutation Statistical Methods

An Integrated Approach

Springer

Kenneth J. Berry
Department of Sociology
Colorado State University
Fort Collins
Colorado, USA

Paul W. Mielke, Jr.
Department of Statistics
Colorado State University
Fort Collins
Colorado, USA

Janis E. Johnston
U.S. Government
Alexandria
Virginia, USA

ISBN 978-3-319-28768-3 ISBN 978-3-319-28770-6 (eBook)
DOI 10.1007/978-3-319-28770-6

Library of Congress Control Number: 2016938914

Springer Cham Heidelberg New York Dordrecht London

Printed on acid-free paper

Springer International Publishing AG Switzerland is part of Springer Science+Business Media (www.springer.com)

For our families: Nancy T. Berry, Ellen E. Berry, Laura B. Berry, Roberta R. Mielke, William W. Mielke, Emily (Mielke) Spear, Lynn (Mielke) Basila, Lindsay A. Johnston, James B. Johnston, and Tayla, Malia, Ollie, Cami, and Brian.

Preface

Permutation Statistical Methods: An Integrated Approach provides a synthesis of a number of statistical tests and measures, which, at first consideration, appear disjointed and unrelated. No attempt is made to synthesize all of statistics—a daunting undertaking—but a wide variety of commonly-used statistics illustrate an underlying commonality. Many years ago the authors realized that much of statistical analysis could be integrated and condensed into a small set of methods that unified many conventional tests and measures under a common rubric. Since our joint specialty is permutation methods, it was only natural that the organizing rubric be the permutation model, as contrasted with the more popular population model, although the two are compared and contrasted throughout the book.

Permutation statistical methods possess several advantages over classical statistical methods in that they are optimal for small samples, can be utilized to analyze nonrandom samples, are completely data-dependent, are free of distributional assumptions, and yield exact probability values. Today, permutation statistical tests are considered by many to be a gold standard against which conventional statistical tests should be evaluated and validated. An obvious drawback to permutation statistical methods is the amount of computation required. While it took the advent of high-speed computing to make permutation methods feasible for many problems, today powerful computational algorithms and modern computers make permutation analyses practical for many research applications.

This book begins with a description of a generalized Minkowski distance function, from which a five-dimensional model is constructed, each cell of which contains a conventional statistic, a permutation analogue of a conventional statistic, or the mathematical formulation for a new statistic. Originally, the authors thought that most of the cells would describe existing statistical tests and measures, but as the writing of the book progressed, it became apparent that a majority of the cells contained entirely new and previously unknown statistics, many of which appear to be quite useful.

The first of the five dimensions simply divides statistical models into the analysis of two data types: completely randomized data and randomized-block data; for example, completely randomized one-way or between-subjects analysis of variance, on the one hand, and randomized-block analysis of variance, sometimes called repeated-measures, or within-subjects analysis of variance, on the other.

The second dimension divides data into three levels of measurement: nominal, ordinal, and interval. Examples for nominal-level (categorical) data include statistical tests such as the chi-squared goodness-of-fit test and the chi-squared test of independence, Goodman and Kruskal's t_a and t_b asymmetric measures of nominal association, and Cohen's unweighted κ measure of agreement. Ordinal-level (rank) statistical tests include the Wilcoxon–Mann–Whitney two-sample rank-sum test, Goodman and Kruskal's γ measure of ordinal association, and the Kruskal–Wallis multi-sample rank-sum test. Interval-level statistical tests include Student's t test, the F test for the analysis of variance, and the Pearson product-moment correlation coefficient.

The third dimension divides the analysis of data into two entirely different approaches. One approach utilizes squared Euclidean distances between observations, as is customary with conventional statistical tests. The other approach utilizes ordinary Euclidean (absolute) distances between observations. Examples of these two approaches include ordinary least squares (OLS) regression and least absolute deviation (LAD) regression.

The fourth dimension divides the focus of the statistical analysis into tests of differences and measures of relationship, recognizing that one can often be transformed into the other. For example, Student's t test for differences between means and one-way analysis of variance, on the one hand, and the Pearson product-moment correlation between two variables and Spearman's rank-order correlation coefficient, on the other.

Finally, the fifth dimension divides data into univariate and multivariate response measurements. For example, analysis of variance (ANOVA) and simple linear regression and correlation are appropriate for univariate data, and multivariate analysis of variance (MANOVA) and multiple regression and correlation are appropriate for multivariate data.

Altogether, 48 five-dimensional cells are identified and explored using a generalized Minkowski distance function and two permutation-based derivatives. One derivative, denoted as δ, provides for tests of differences, and the other, denoted as $\Re$, provides for measures of relationships. The two permutation statistics are seminal constructs for integrating a variety of statistical tests and measures. Figure 1 graphically displays the 24 analysis cells for completely randomized experimental designs, shaded in gray, and Fig. 2 graphically displays the 24 analysis cells for randomized-block experimental designs, also shaded in gray.

The foundation of the synthesizing model is a generalized Minkowski distance function. Derived from the generalized Minkowski distance function are two permutation approaches: multi-response permutation procedures (MRPP), designed for analyzing completely-randomized data, and multivariate randomized-block permutation (MRBP) procedures, designed for analyzing randomized-block data. The generalized Minkowski distance function, together with MRPP and MRBP, provide, for the analysis of completely randomized and randomized-block data, both univariate and multivariate, at the nominal, ordinal, and interval levels of measurement, utilizing either squared Euclidean distances or ordinary Euclidean distances.

Completely-Randomized Experimental Designs							
Interval-Level Data Analysis							
Squared Euclidean Distance				Ordinary Euclidean Distance			
Tests of Differences		Measures of Relationships		Tests of Differences		Measures of Relationships	
Univariate analysis	Multivariate analysis	Univariate analysis	Multivariate analysis	Univariate analysis	Multivariate analysis	Univariate analysis	Multivariate analysis
Ordinal-Level Data Analysis							
Squared Euclidean Distance				Ordinary Euclidean Distance			
Tests of Differences		Measures of Relationships		Tests of Differences		Measures of Relationships	
Univariate analysis	Multivariate analysis	Univariate analysis	Multivariate analysis	Univariate analysis	Multivariate analysis	Univariate analysis	Multivariate analysis
Nominal-Level Data Analysis							
Squared Euclidean Distance				Ordinary Euclidean Distance			
Tests of Differences		Measures of Relationships		Tests of Differences		Measures of Relationships	
Univariate analysis	Multivariate analysis	Univariate analysis	Multivariate analysis	Univariate analysis	Multivariate analysis	Univariate analysis	Multivariate analysis

Fig. 1 Diagram for completely randomized experimental designs with analysis cells shaded in *gray*

Randomized-Block Experimental Designs							
Interval-Level Data Analysis							
Squared Euclidean Distance				Ordinary Euclidean Distance			
Tests of Differences		Measures of Relationships		Tests of Differences		Measures of Relationships	
Univariate analysis	Multivariate analysis	Univariate analysis	Multivariate analysis	Univariate analysis	Multivariate analysis	Univariate analysis	Multivariate analysis
Ordinal-Level Data Analysis							
Squared Euclidean Distance				Ordinary Euclidean Distance			
Tests of Differences		Measures of Relationships		Tests of Differences		Measures of Relationships	
Univariate analysis	Multivariate analysis	Univariate analysis	Multivariate analysis	Univariate analysis	Multivariate analysis	Univariate analysis	Multivariate analysis
Nominal-Level Data Analysis							
Squared Euclidean Distance				Ordinary Euclidean Distance			
Tests of Differences		Measures of Relationships		Tests of Differences		Measures of Relationships	
Univariate analysis	Multivariate analysis	Univariate analysis	Multivariate analysis	Univariate analysis	Multivariate analysis	Univariate analysis	Multivariate analysis

Fig. 2 Diagram for randomized-block experimental designs with analysis cells shaded in *gray*

Both MRPP, illustrated in Fig. 1, and MRBP, illustrated in Fig. 2, generate two test statistics, δ and $\Re$, providing for a number of statistical tests of differences and measures of association. For MRPP, test statistic δ is the weighted mean of the average distance-function values for all distinct pairs of objects in all treatment groups, and $\Re$ is a chance-corrected within-group measure of effect size. For MRBP, δ is the balanced mean of the distance-function values for all distinct pairs of objects in all treatment groups, and $\Re$ is a chance-corrected within-blocks measure of effect size. Finally, test statistics δ and $\Re$ are applied to three levels of measurement that are commonly encountered in statistical analyses: interval, ordinal, and nominal. Taken together, the five-dimensional structure contains 48 distinct analysis cells, many of which contain new statistical tests and measures. Most of the new statistics are based on ordinary Euclidean distances as most conventional statistics are based on squared Euclidean distances. However, other new statistics result from generalizing conventional statistics designed for univariate data to statistics designed for multivariate data.

The book is comprised of 11 chapters, each of which is designed to stand alone. Thus, each chapter can be read independently of the other chapters without any loss of understanding. Chapters 1–7 focus on MRPP and the analysis of completely randomized data, and Chaps. 8–11 focus on MRBP and the analysis of randomized-block data.

Chapter 1 of *Permutation Statistical Methods* provides an introduction to the remaining 10 chapters, compares the population and permutation models of statistical analysis, and presents the three main approaches to permutation statistical methods: exact, moment-approximation, and resampling-approximation permutation tests.

Chapter 2 develops a general set of synthesizing multi-response permutation procedures (MRPP) for permutation statistical tests and measures, designed for completely randomized data sets. The two MRPP test statistics, δ and $\mathfrak{R}$, are introduced and derived from a generalized Minkowski distance function. The two test statistics provide the mathematical foundation for the permutation tests and measures presented in Chaps. 3–7.

Chapter 3 applies the multi-response permutation procedures for completely randomized data developed in Chap. 2 to permutation statistical tests and measures designed to analyze univariate and multivariate response measurements at the interval level of measurement. Permutation analogues of Student's two-sample t test, Hotelling's two-sample T^2 test, one-way fixed-effects analysis of variance, and one-way multivariate analysis of variance illustrate the application of MRPP statistics δ and $\mathfrak{R}$ to interval-level response measurements.

Chapter 4 continues the analysis of interval-level response measurements presented in Chap. 3, analyzing the response measurements with appropriate regression models, both ordinary least squares (OLS) and least absolute deviation (LAD) models. Included in Chap. 4 are permutation regression analyses of one-way randomized designs, with and without a covariate, one-way randomized block, factorial, Latin square, and nested analysis of variance designs.

Chapter 5 applies the multi-response permutation methods developed in Chap. 2 to univariate ordinal-level response measurements. Permutation analogues of the Wilcoxon two-sample rank-sum test, the Kruskal–Wallis multi-sample rank-sum test, the Ansari–Bradley and Mood rank-sum tests for dispersion, the Brown–Mood median test, the Mielke power-of-rank function tests, and the Whitfield two-sample rank sum test illustrate the application of MRPP statistics δ and $\mathfrak{R}$ to ordinal-level response measurements.

Chapter 6 continues the analysis of ordinal-level response measurements, generalizing the univariate permutation procedures developed in Chap. 5 to multivariate response measurements. As in Chap. 5, example statistical tests and measures include permutation versions of two-sample rank-sum tests, multiple sample rank-sum tests, rank-sum tests for dispersion, sum-of-squared-rank tests, median tests, and power-of-rank function tests.

Chapter 7 uses the multi-response permutation methods developed in Chap. 2 to analyze nominal-level (categorical) response measurements. Permutation versions of Goodman and Kruskal's t_a and t_b asymmetric measures of nominal association, Light and Margolin's categorical analysis of variance, tests to analyze multiple binary choices, and various multivariate measures of association for a nominal-level independent variable and nominal-, ordinal-, and interval-level dependent variables illustrate the application of statistics δ and $\Re$ to categorical response measurements.

Chapter 8 develops multivariate randomized-block permutation (MRBP) procedures for analyzing randomized-block data, generates MRBP statistics δ and $\Re$ from a generalized Minkowski distance function, and provides the mathematical foundation for the permutation tests and measures presented in Chaps. 9–11.

Chapter 9 applies the multivariate randomized-block permutation procedures developed in Chap. 8 to interval-level response measurements. Permutation analogues of Student's matched-pairs t test, Hotelling's matched-pairs T^2 test, one-way randomized-block analysis of variance with univariate response measurements, and one-way randomized-block analysis of variance with multivariate response measurements illustrate the application of MRPP statistics δ and $\Re$ to interval-level response measurements.

Chapter 10 applies the multivariate randomized-block methods developed in Chap. 8 to ordinal-level response measurements. Permutation analogues of a variety of statistical tests illustrate the application of statistics δ and $\Re$ to ordinal-level response measurements, including the Wilcoxon signed-rank test, the sign test, Spearman's rank-order and footrule measures of correlation, Friedman's analysis of variance for ranks, Kendall's coefficient of concordance, Cohen's weighted kappa measure of agreement, Kendall's t_a and t_b measures of ordinal association, Stuart's t_c statistic, Goodman and Kruskal's γ measure of ordinal association, Yule's Q, and Somers' d_{yx} and d_{xy} asymmetric measures of ordinal association.

Chapter 11 applies the multivariate randomized-block methods developed in Chap. 8 to nominal-level response measurements. Permutation analogues of a number of statistical tests and measures illustrate the application of statistics δ and $\Re$ to nominal-level response measurements, including Cohen's unweighted kappa measure of chance-corrected agreement, McNemar's and Cochran's Q tests for change, Kendall's t_a and Yule's Q measures of association, the odds ratio, Somers' d_{xy} and d_{yx} asymmetric measures of association, Pearson's product-moment correlation coefficient, percentage differences, and chi-squared. Finally, the book closes with a brief Epilogue.

Acknowledgments The authors wish to thank the editors and staff at Springer-Verlag. Very special thanks to Federica Corradi Dell'Acqua, Associate Editor, Statistics and Natural Language Processing, who guided the project through from beginning to end. We are grateful to Roberta Mielke who read the entire manuscript. Finally, we wish to thank Steve and Linda Jones, proprietors of the Rainbow Restaurant, 212 West Laurel Street, Fort Collins, Colorado, for their gracious hospitality. Like our previous books, much of this book was written at Table 22 in their restaurant adjacent to the campus of Colorado State University.

Fort Collins, CO, USA — Kenneth J. Berry
Fort Collins, CO, USA — Paul W. Mielke, Jr.
Alexandria, VA, USA — Janis E. Johnston
August 2015

Contents

Introduction 1

Commencing with the seminal contributions of R.A. Fisher, E.J.G. Pitman, and other mathematicians and scientists in the 1920s and 1930s, permutation statistical methods were initially developed to validate the normality and homogeneity assumptions of classical statistical methods, a point made repeatedly by Fisher in his second book on statistics, *The Design of Experiments* [119, Chaps. 20 and 21]. Over the subsequent eighty or so years, permutation methods have emerged as a statistical approach to hypothesis testing in their own right. Permutation statistical methods possess several advantages over classical statistical methods in that they are optimal for small data sets, can be utilized to analyze non-random samples, are completely data-dependent, are free of distributional assumptions, and yield exact probability values. These attributes make permutation statistical methods ideal for research areas that often have to deal with small non-random samples; e.g., atmospheric science, biology, ecology, medical research, and psychology.

This book presents a synthesis of permutation statistical methods that unifies many previously described tests and measures, defines a continuous methodological spectrum, and weaves together what are usually considered to be disjoint families of statistical tests and measures. The incorporation of a large family of statistics into a unifying statistical approach under a common rubric provides a new perspective on traditional statistics composed of seemingly unrelated tests and measures.

While permutation tests have been developed as counterparts to a number of conventional parametric tests, permutation tests are not limited to parametric analogues. This book describes and illustrates a large number of new permutation tests with no parametric complements. When available, a permutation test is compared with its parametric alternative; otherwise, new permutation tests are presented as solutions to statistical problems for which no corresponding parametric tests are currently available.

The typical first course in statistics is often seen as an unorganized and confusing maze of unconnected chapters because, frequently, the material is presented without a synthesizing model with which to link and understand the disparate chapters.

K.J. Berry et al., *Permutation Statistical Methods*,
DOI 10.1007/978-3-319-28770-6_1

Many first-year, non-mathematical courses in statistics, especially in the social and behavioral sciences, are presented as a variety of ostensibly unrelated statistical tests and measures, making those tests appear independent and disjointed, with little or no discernible segue among topics. These various tests and measures often include *t* tests, both independent and matched-pairs; simple correlation and regression, with Spearman's rank-order correlation coefficient sometimes included; chapters on the analysis of variance covering completely randomized, randomized-block, and factorial designs, with Latin squares, split-plot, and nested designs sometimes included; and (usually) a final chapter on chi-squared containing tests of goodness-of-fit and independence, which often includes such chi-squared-based measures of association as Pearson's ϕ^2, Tschuprov's (Čhuprov's) T^2, and Cramér's V^2. Consequently, students often do not see the important functional relationships between, for example, the t test for two independent samples and the F test for a one-way analysis of variance, the chi-squared test of independence and the product-moment correlation coefficient, the analysis of variance and linear regression, or the percentage difference and the unstandardized slope of a regression line.

The Argentine fabulist, Jorge Luis Borges, in a 1941 review of the movie *Citizen Kane* quoted G.K. Chesterton as saying, "There is nothing more frightening than a labyrinth that has no center."[1] In this book the authors hope to provide a center to a piece of the statistical maze that often confronts and confounds beginning students of statistics.

1.1 Models of Statistical Inference

Essentially, two models of statistical inference coexist: the population model and the permutation model; see, for example, discussions by Curran-Everett [85], Hubbard [186], Kempthorne [204], Kennedy [212], Lachin [226], Ludbrook [247, 248], and Ludbrook and Dudley [252]. The population model, formally proposed by Jerzy Neyman and Egon Pearson in a seminal two-part article on statistical inference in *Biometrika* in 1928, assumes random sampling from one or more specified populations [319, 320]. Under the population model, the level of statistical significance that results from applying a statistical test to the results of an experiment or survey corresponds to the frequency with which the null hypothesis would be rejected in repeated random samplings from the same specified population(s). Because repeated sampling of the specified population(s) is usually impractical, it is assumed that the sampling distribution of the test statistics generated under repeated random sampling conforms to an approximating theoretical distribution, such as the normal distribution. The size of the statistical test, e.g., 0.05, is the probability under a specified null hypothesis that repeated outcomes based on random samples of the same size are equal to or more extreme than the observed outcome.

[1]The actual quote, from the Father Brown mystery "The Head of Caesar" by G. K Chesterton is "What we all dread most is a maze with no centre" [68, p. 229].

The permutation model was introduced by R.A. Fisher in 1925 [118], further developed by R.C. Geary in 1927 [134] and T. Eden and F. Yates in 1933 [103], and made explicit in three seminal articles by E.J.G. Pitman in 1937 and 1938 [340–342]. In a permutation statistical test the only assumption is that experimental variability has caused the observed result. That assumption, or null hypothesis, is then tested as follows. A test statistic is computed for the observed data, then the observations are permuted over all possible arrangements of the data and the selected test statistic is computed for each equally-likely arrangement of the observed data. The proportion of arrangements with test statistic values equal to or more extreme than the observed test statistic yields the exact probability of the observed test statistic value.

1.2 Permutation Statistical Tests

Permutation statistical tests are considered by many to be a gold standard against which conventional statistical tests should be evaluated and validated. In 1940 Friedman, comparing tests of significance for multiple rankings, referred to an exact permutation test as "the correct one" [129, p. 88]. In 1973 Feinstein remarked that conventional statistical tests "yield reasonably reliable approximations of the more exact results provided by permutation procedures" [113, p. 912]. In 1992 Good noted that Fisher himself regarded randomization as a technique for validating tests of significance, i.e., ensuring that conventional probability values were accurate [144, p. 263].[2] Bakeman, Robinson, and Quera remarked in 1996 that "like Read and Cressie . . . we think permutation tests represent the standard against which asymptotic tests must be judged" [18, p. 6]. And in 2007 Edgington and Onghena observed that "randomization tests . . . have come to be recognized by many in the field of medicine as the 'gold standard' of statistical tests for randomized experiments" [109, p. 9].

The value of permutation statistical tests was recognized by early statisticians, even during periods in which the computationally intensive nature of permutation tests made them impractical. In 1955 Kempthorne wrote that "tests of significance in the randomized experiment have frequently been presented by way of normal law theory, whereas their validity stems from randomization theory" [202, p. 947] and "there seems little point in the present state of knowledge in using [a] method of inference other than randomization analysis" [202, p. 966]. Similarly, in 1959 Scheffé stated that the conventional analysis of variance F-ratio "can often be regarded as a good approximation to a permutation test, which is an exact test under a less restrictive model" [365, p. 313]. In 1966, Kempthorne re-emphasized that "the proper way to make tests of significance in the simple randomized experiments [sic] is by way of the randomization (or permutation) test" [203, p. 20] and "in the randomized experiment one should, logically, make tests of significance by way of

[2] The terms "permutation test" and "randomization test" are often used interchangeably.

the randomization test" [203, p. 21]. Later, in 1968, Bradley observed that "eminent statisticians have stated that the randomization test is the truly correct one and that the corresponding parametric test is valid only to the extent that it results in the same statistical decision" [52, p. 85].

Because permutation statistical methods are inherently computationally intensive, it took the development of high-speed computers for permutation methods to achieve their potential. Computers, as we know them, did not exist in the 1920s and 1930s, although mechanical calculators such as the Millionaire calculator used by R.A. Fisher in the Statistical Laboratory at the Rothamsted Experimental Station or the Brunsviga calculator used by K. Pearson in the Biometric Laboratory at University College, London, were commonplace in large research centers. These early mechanical calculators were eventually replaced by electro-mechanical calculators such as those produced by the Burroughs, Victor, Monroe, Marchant, and Sundstrand companies [156]. In turn, electro-mechanical calculators were largely supplanted by early computers in the 1940s and 1950s.

The few computers that became available to researchers in the 1940s and 1950s were large, slow, inefficient, very expensive to use, and located at only a few computing centers. Moreover, in large part their use was restricted to military and industrial applications and thus were not generally available to those involved in the development of permutation statistical methods [192]. Today, a small netbook computer outperforms even the largest mainframe computers of previous decades. Consequently, in the 21st century permutation statistical methods have become both feasible and practical and have found applications in diverse fields of research ranging from agriculture to zoology. Fields of research that examine small non-random samples, such as atmospheric science, psychology, ecology, biology, and medicine, have been especially receptive to permutation methods. This is due in part to strong advocates of permutation methods in these fields, including Hugh Dudley [252–255], Eugene Edgington [104–109], Alvan Feinstein [113, 114], Phillip Good [145–148], Oscar Kempthorne [201–204], John Ludbrook [247–250], Bryan Manly [258–261], and John Tukey [57, 403–405].

Three types of permutation tests are common in the statistical literature: exact, moment-approximation, and resampling-approximation permutation tests. To this taxonomy might be added network-algorithm permutation tests. Although the three types of permutation tests are methodologically quite different, all three types are based on the same specified null hypothesis.

1.2.1 Exact Permutation Tests

An exact permutation test exhaustively enumerates all equally-likely arrangements of the observed data. Then, for each arrangement, the desired test statistic is calculated. The observed data yield the observed value of the test statistic. The probability of obtaining the observed value of the test statistic, or one more extreme, is the proportion of the enumerated test statistics with values equal to or more extreme than the value of the observed test statistic. For large samples the total number of possible arrangements can be considerable and exact permutation methods are quickly rendered impractical. For example, permuting two small samples of sizes $n_1 = n_2 = 25$ yields

$$M = \frac{(n_1 + n_2)!}{n_1!\, n_2!} = \frac{(25 + 25)!}{(25!)^2} = 126{,}410{,}606{,}437{,}752$$

different arrangements of the observed data set—far too many statistical values to compute in a reasonable amount of time.

The origin of exact permutation statistical methods is often traced back to the early work of R.A. Fisher, but R.C. Geary, T. Eden, F. Yates, and E.J.G. Pitman also played substantial roles in the early development of exact permutation methods. For descriptions of their contributions, see *A Chronicle of Permutation Statistical Methods* by Berry, Johnston, and Mielke [41, pp. 31–33, 37–41, 78–82]. The following three analyses by R.A. Fisher illustrate early exact permutation statistical analyses.

Example 1

On 18 December 1934, R.A. Fisher presented an invited paper describing the logic of permutation statistical tests to the Royal Statistical Society, a paper that was subsequently published in *Journal of the Royal Statistical Society* the following year [120].[3] Fisher described data on 30 criminal same-sex twins from a study originally conducted by Dr. Johannes Lange, Chief Physician at the Munich-Schwabing Hospital in Schwabing, a northern suburb of Munich.

The Lange data analyzed by Fisher consisted of 13 pairs of monozygotic (identical) twins and 17 pairs of dizygotic (fraternal) twins [229]. For each of the 30 pairs of twins, one twin was known to be a convict. The study considered whether the twin brother of the known convict was himself "convicted" or "not convicted," thus forming a 2×2 contingency table with 12 "convicted" and 18 "not convicted" twins cross-classified by the 13 "monozygotic" and 17 "dizygotic" twins. The 2×2 contingency table, as analyzed by Fisher, is presented in Fig. 1.1.

Fisher determined all possible arrangements of the four cell frequencies, given the observed marginal frequency totals; in this case, 13 different arrangements of the cell frequencies. Fisher then calculated the hypergeometric probability value for each of the 13 cell arrangements, summing those probability values that were equal to or less than the hypergeometric probability value of the observed cell frequency arrangement. Fisher concluded, "The test of significance is therefore direct, and

[3] As was customary in scientific societies at the time, these special papers were printed in advance and circulated to the membership of the Society. Then, only a brief summary was made by the author at the meeting and the remaining time was devoted to a discussion of the paper. By tradition, the "proposer of the vote of thanks" advanced what he thought was commendable about the paper, and the seconder put forward what he thought was not so worthy. Subsequently, there was a general discussion by the Fellows of the Society and often a number of prominent statisticians offered comments, suggestions, or criticisms, all of which were subsequently printed along with the published paper in the journal of the Society [50, p. 41].

Fig. 1.1 Convictions of like-sex twins of criminals; data from Lange [229]

Twin type	Convicted	Not convicted	Total
Monozygotic	10	3	13
Dizygotic	2	15	17
Total	12	18	30

Table 1.1 Listing of the 13 possible 2×2 contingency tables from Lange's data [229], with associated hypergeometric probability values

Table 1		Probability	Table 2		Probability
0	13	7.1543×10^{-5}	1	12	1.8601×10^{-3}
12	5		11	6	
Table 3		**Probability**	**Table 4**		**Probability**
2	11	1.7538×10^{-2}	3	10	8.0384×10^{-2}
10	7		9	8	
Table 5		**Probability**	**Table 6**		**Probability**
4	9	2.0096×10^{-1}	5	8	2.8938×10^{-1}
8	9		7	10	
Table 7		**Probability**	**Table 8**		**Probability**
6	7	2.4554×10^{-1}	7	6	1.2277×10^{-1}
6	11		5	12	
Table 9		**Probability**	**Table 10**		**Probability**
8	5	3.5414×10^{-2}	9	4	5.6212×10^{-3}
4	13		3	14	
Table 11		**Probability**	**Table 12**		**Probability**
10	3	4.4970×10^{-4}	11	2	1.5331×10^{-5}
2	15		1	16	
Table 13		**Probability**			
12	1	1.5030×10^{-7}			
0	17				

exact for small samples. No process of estimation is involved" [120, p. 50]. The 13 arrangements of cell frequencies and the associated hypergeometric probability values are listed in Table 1.1. Fisher observed, given that each table has only one degree of freedom, it was only necessary to compute the probability of one of the four cells; he chose the convicted dizygotic twins, the lower-left cell in Fig. 1.1 with a frequency of 2.

For a 2×2 contingency table, such as depicted in Fig. 1.2, the hypergeometric point probability of any specified cell, say cell (2,1), is given by

$$P(n_{21}|n_{2.},n_{.1},N) = \frac{n_{1.}!\,n_{2.}!\,n_{.1}!\,n_{.2}!}{N!\,n_{11}!\,n_{12}!\,n_{21}!\,n_{22}!}\,,$$

where $n_{1.}$ and $n_{2.}$ denote the marginal frequency totals for rows 1 and 2, $n_{.1}$ and $n_{.2}$ denote the marginal frequency totals for columns 1 and 2, n_{ij} denotes the cell frequency for $i,j = 1, \ldots, 2$, and $N = n_{11} + n_{12} + n_{21} + n_{22}$.

Fig. 1.2 Conventional notation for a 2×2 contingency table

n_{11}	n_{12}	$n_{1.}$
n_{21}	n_{22}	$n_{2.}$
$n_{.1}$	$n_{.2}$	N

Computing the discrepancies from proportionality as great or greater than the observed cell frequency configuration in Fig. 1.1, Fisher computed for 2, 1, and 0 convicted dizygotic twins, a one-tailed hypergeometric probability value of

$$P\{2|17,12,30\}+P\{1|17,12,30\}+P\{0|17,12,30\}$$
$$=\frac{13!\;17!\;12!\;18!}{30!\;10!\;3!\;2!\;15!}+\frac{13!\;17!\;12!\;18!}{30!\;11!\;2!\;1!\;16!}+\frac{13!\;17!\;12!\;18!}{30!\;12!\;1!\;0!\;17!}$$
$$=4.4970\times10^{-4}+1.5331\times10^{-5}+1.5030\times10^{-7}$$
$$=4.6518\times10^{-4}\;.$$

For the frequency data given in Fig. 1.1, a two-tailed hypergeometric probability includes all probability values less than or equal to the probability of the observed table, i.e., $P = 4.4970\times10^{-4}$. In this case, the additional probability value associated with Table 1 in Table 1.1 with 12 dizygotic convicts, i.e., $P = 7.1543\times10^{-5}$. Thus, the two-tailed probability value is computed as

$$P\{2|17,12,30\}+P\{1|17,12,30\}+P\{0|17,12,30\}+P\{12|17,12,30\}$$
$$=\frac{13!\;17!\;12!\;18!}{30!\;10!\;3!\;2!\;15!}+\frac{13!\;17!\;12!\;18!}{30!\;11!\;2!\;1!\;16!}+\frac{13!\;17!\;12!\;18!}{30!\;12!\;1!\;0!\;17!}+\frac{13!\;17!\;12!\;18!}{30!\;0!\;13!\;12!\;5!}$$
$$=4.4970\times10^{-4}+1.5331\times10^{-5}+1.5030\times10^{-7}+7.1543\times10^{-5}$$
$$=5.3672\times10^{-4}\;.$$

The point of the twin analysis—that exact tests are possible for small samples, eliminating the need for estimation—indicates an early understanding of the superiority of exact probability values computed from discrete permutation distributions, over approximations based on assumed theoretical distributions. It should also be noted, however, that the exact solution proposed by Fisher was not without controversy; see, for example, a 1992 article by Routledge in *Canadian Journal of Statistics* [356]. Stephen Senn wryly observed in 2012 that "statisticians have caused the destruction of whole forests to provide paper to print their disputes regarding the analysis of 2×2 tables" [370, p. 33].

Example 2

In 1935 Fisher described a hypothetical experiment in his second book on statistics, *The Design of Experiments*, in which a woman claimed to be able to tell the difference between tea with milk added to the cup first and tea with milk added to the cup second [119, Chap. 2]. He designed an experiment whereby the woman sampled eight cups of tea, four of each type, and identified the point at which the milk had been added—before the tea, or after.[4] Again Fisher constructed a 2×2 contingency table in which there were five possible arrangements of cell frequencies, given the observed marginal frequency totals. The five possible arrangements of cell frequencies tables are presented in Table 1.2. Fisher then calculated a hypergeometric probability value for each of the five possible cell frequency arrangements, summing those probability values equal to or less than the hypergeometric probability value of the observed arrangement.

The null hypothesis in the lady-tasting-tea experiment was that the judgments of the lady were in no way influenced by the order in which the ingredients were added. Fisher explained that the probability of correctly classifying all eight cups of tea was one in 70, i.e., the hypergeometric point probability value for the cell arrangement in Table 1 in Table 1.2 given by

$$P\{0|4,4,8\} = \frac{4!\ 4!\ 4!\ 4!}{8!\ 0!\ 4!\ 4!\ 0!} = \frac{24}{1{,}680} = \frac{1}{70}\ .$$

Fisher went on to note that only if every cup was correctly classified would the lady be judged successful; a single mistake would reduce her performance below the level of significance, in this case $\alpha = 0.05$. For example, with one misclassification the one-tailed hypergeometric probability value for the cell arrangements in

Table 1.2 Five possible arrangements of cell frequencies with $N = 8$ and identical marginal frequency totals of 4, 4, 4, and 4

Table 1		Table 2		Table 3		Table 4		Table 5	
0	4	1	3	2	2	3	1	4	0
4	0	3	1	2	2	1	3	0	4

[4]The experiment was obviously inspired by an actual tea-tasting experiment at the Rothamsted Experimental Station some dozen years prior, where Fisher was employed as a statistician from 1919 to 1933. The woman tasting the tea was Dr. B. Muriel Bristol, an algologist at the Station. For descriptions of the tea-tasting experiment at the Rothamsted Experimental Station, see discussions by Agresti, [2, pp. 91–97], Berry, Johnston, and Mielke [41, pp. 58–61, 429–432], Box [48], Box [49, pp. 134–135], Fisher [119, pp. 11–29], Fisher [121, Chap. 6], Gridgeman [155], Hall [165], Lehmann [236, pp. 63–64], Okamoto [324], Salsburg [361, pp. 1–2], Senn [369–371], and Springate [384].

Tables 1 and 2 in Table 1.2 is given by

$$P\{1|4,4,8\} + P\{0|4,4,8\} = \frac{4!\ 4!\ 4!\ 4!}{8!\ 1!\ 3!\ 3!\ 1!} + \frac{4!\ 4!\ 4!\ 4!}{8!\ 0!\ 4!\ 4!\ 0!} = \frac{16}{70} + \frac{1}{70} = \frac{17}{70}$$

and $17/70 = 0.2429$ is much greater than $\alpha = 0.05$, whereas $1/70 = 0.0143$ is considerably less than $\alpha = 0.05$.

This procedure became widely known as the Fisher exact probability, or FEP, test. It should be noted, however, that the test was independently developed by Frank Yates in 1934 [433] and by Joseph Irwin in 1935 [191]. Thus, the test is sometimes referred to as the Fisher–Yates exact test or the Fisher–Irwin exact test. Today, the Fisher–Yates–Irwin test remains the iconic data-dependent, distribution-free, exact permutation test.

Example 3

Fisher provided a second discussion of permutation statistical tests in *The Design of Experiments*, describing a way to compare the arithmetic means of randomized pairs of observations by permutation [119, Sect. 21]. For this more ambitious permutation analysis, Fisher analyzed original data collected by Charles Darwin on $N = 15$ pairs of planters containing *Zea mays* ("maize" in the United States) seeds in similar soils and locations, with heights to be measured when the plants reached a predetermined age [89]. The data from the experiment are given in the first two columns of Fig. 1.3 and are adapted from Table XCVII in Darwin's 1876 book on *The Effects of Cross and Self Fertilisation in the Vegetable Kingdom* [89, p. 234].

Zea mays is monoecious, so half of the plants were allowed to fertilize spontaneously, while the other (matching) half were crossed with pollen taken from a separate plant. Astonishingly, Fisher calculated the sums of the differences for all the $2^N = 2^{15} = 32{,}768$ possible arrangements of the observed data. An exact probability value was computed as the proportion of differences in height as extreme, or more extreme, than the observed differences in the heights of the observed plants.[5] For the analysis of Darwin's *Zea mays* data, Fisher also calculated a conventional matched-pairs t test and compared the results of the two analyses. After a correction for continuity was administered to the observed t value, Fisher noted that the *Zea mays* example analysis served to demonstrate that an "independent check" existed for the "more expeditious methods" that were typically in use, such as Student's matched-pairs t test [119, pp. 45–46]. In this regard, Fisher was fond of referring to a 1931 article in *The Journal of Agricultural Science* by Olaf Tedin [395] in which Tedin convincingly demonstrated that when the assumptions of the classical

[5]For a concise summary of the *Zea mays* experiment, see an informative discussion by Erich Lehmann in his posthumously published 2011 book on *Fisher, Neyman, and the Creation of Classical Statistics* [236, pp. 65–66].

Fig. 1.3 Heights of cross- and self-fertilized *Zea mays* plants in inches; data from Darwin [89, p. 234]

Pot	Cross-fertilized	Self-fertilized	Difference (eighths)
I	$23\frac{4}{8}$	$17\frac{3}{8}$	+49
	12	$20\frac{3}{8}$	−67
	21	20	+8
II	22	20	+16
	$19\frac{1}{8}$	$18\frac{3}{8}$	+6
	$21\frac{4}{8}$	$18\frac{5}{8}$	+23
III	$22\frac{1}{8}$	$18\frac{5}{8}$	+28
	$20\frac{3}{8}$	$15\frac{2}{8}$	+41
	$18\frac{2}{8}$	$16\frac{4}{8}$	+14
	$21\frac{5}{8}$	18	+29
	$23\frac{2}{8}$	$16\frac{2}{8}$	+56
IV	21	18	+24
	$22\frac{1}{8}$	$12\frac{6}{8}$	+75
	23	$15\frac{4}{8}$	+60
	12	18	−48

analysis of variance F test are met in practice, the classical test and the corresponding permutation test yield essentially identical probability values [338].[6]

Specifically, using the data in the last column of Fig. 1.3 where the differences between the heights of the crossed- and self-fertilized plants were recorded in eighths of an inch, Fisher first calculated a matched-pairs t test. He found the mean difference (d) between the crossed- and self-fertilized *Zea mays* plants to be

$$\bar{d} = \frac{1}{N}\sum_{i=1}^{N} d_i = \frac{314}{15} = 20.933$$

and the estimated standard error to be

$$s_{\bar{d}} = \sqrt{\frac{\sum_{i=1}^{N} d_i^2 - \bar{d}\sum_{i=1}^{N} d_i}{N(N-1)}} = \sqrt{\frac{26{,}518 - (20.933)(314)}{15(15-1)}} = 9.746\ .$$

[6]Olaf Tedin (1898–1966) was a Swedish geneticist who spent most of his professional career as a plant breeder with the Swedish Seed Association, Svalöf, where he was in charge of breeding barley and fodder roots in the Weibullsholm Plant Breeding Station, Landskrona.

Then, Student's matched-pairs t test yielded an observed statistic of

$$t = \frac{\bar{d}}{s_{\bar{d}}} = \frac{20.933}{9.746} = +2.148 \; .$$

Fisher pointed out that the 5 % critical value with 14 degrees of freedom was $t = \pm 2.145$ and concluded that since $+2.148$ just exceeded $+2.145$, the result was "significant" at the 5 % level.[7]

Fisher then turned his attention to an exact permutation test, calculating sums of the differences for the $2^{15} = 32{,}768$ possible arrangements of the observed measurements, based on the null hypothesis of no difference between self- and cross-fertilized *Zea mays* plants. The exact probability value was calculated as the proportion of values with differences as extreme, or more extreme, than the observed value. Fisher found that in 835 out of 32,768 cases the deviations were greater than the observed value of 314; in an equal number of cases, less than 314; and in 28 cases, exactly equal to 314. Fisher explained that in just $835 + 28 = 863$ out of a possible 32,768 cases, the total deviation would have a positive value as great or greater than the observed value of 314, and in an equal number of cases it would have as great a negative value. The two groups added together ($863 + 863 = 1{,}726$) constituted $1{,}726/32{,}768 = 5.267$ %, or slightly more than 5 % of the possibilities available, a result very nearly equivalent to that obtained using Student's t test, where the two-tailed probability value for $t = +2.148$ under the null hypothesis with 14 degrees of freedom is 4.970 % (slightly less than 5 %) [122, p. 47].

Finally, Fisher argued that, because the t distribution is continuous and the permutation distribution is discrete, the t distribution was counting only half of the 28 cases that corresponded exactly with the observed total of 314. He went on to show that making an adjustment corresponding to a correction for continuity provided a probability value more in line with the exact probability value. The corrected value was $t = +2.139$, yielding a value of 5.054 % which is closer to the exact value of 5.267 % than the unadjusted value of 4.970 %.

1.2.2 Moment-Approximation Permutation Tests

The moment-approximation of a test statistic requires computation of the exact moments of the test statistic, assuming equally-likely arrangements of the observed response measurements. The moments are then used to fit a specified distribution that approximates the underlying discrete permutation distribution and provide an approximate, but often highly accurate, probability value. Historically, the beta distribution was commonly used for the approximating distribution, but in recent years

[7] For a brief history of R.A. Fisher and the origins of $\alpha = 0.05$, see a 2011 book by Erich Lehmann on *Fisher, Neyman, and the Creation of Classical Statistics* [236].

the Pearson type III distribution has largely replaced the beta distribution.[8] For many years moment-approximation permutation tests provided an important intermediary approximation when computers lacked the speed for calculating exact permutation tests. In recent years, with the advent of high-speed computers, resampling-approximation permutation tests have largely replaced moment-approximation permutation procedures.

Moment-approximation permutation tests were popular from the early days of permutation statistical methods. For example, E.J.G. Pitman used a moment approach to obtain approximate probability values in each of his three seminal papers published in 1937 and 1938 [340–342]. In these three papers on permutation versions of two-sample tests, bivariate correlation, and randomized-block analysis of variance, moments based on the observed data were equated to the moments of the beta distribution to obtain the correspondence between the probabilities obtained from observed response measurements and probabilities from the associated beta distribution. A drawback to this approach was that use of the beta distribution required standardization of the test statistic to ensure that the statistic varied between 0 and 1, the limits of the beta distribution. For example, in his 1938 paper on randomized-block analysis of variance, Pitman defined statistic

$$W = \frac{SS_{\text{Between}}}{SS_{\text{Between}} + SS_{\text{Within}}} ,$$

which is a monotonic increasing function of $F = SS_{\text{Between}}/SS_{\text{Within}}$ that is bounded by 0 and 1. Other early researchers who utilized moments of the permutation distribution to compare results to asymptotic distributions were Welch [418] and Friedman [128] in 1937; Olds [325] and Kendall [205] in 1938; and Kendall and Babington Smith [209], Kendall, Kendall, and Babington Smith [211], and McCarthy in 1939 [269].

In a 1943 paper on "Statistical inference in the non-parametric case" in *Annals of Mathematical Statistics*, Henry Scheffé sharply criticized the use of moments to approximate discrete permutation distributions, stating that in his opinion the justification for moment approximations had never been mathematically satisfactory [364, p. 311]. Although Scheffé did not specifically mention the beta distribution, it was so widely used at the time that it can be assumed with some confidence that Scheffé included the beta distribution in his criticism of moment approximation procedures.

From 1976 through 1980, P.W. Mielke and his many collaborators utilized the beta distribution in a number of publications; however, in 1981 the beta distribution was replaced with the Pearson type III distribution due to the difficulty of making simple associations between the parameters of the beta distribution and the moments of the discrete permutation distribution, even after reparameterization

[8]It was the Pearson type III distribution that Student (W.S. Gosset) used to fit the distribution of sample variances in his classic 1908 article on "The probable error of a mean" [390, p. 4].

[283, 301].[9] The first published paper by Mielke in which a Pearson type III distribution was used was a 1981 article by Mielke, Berry, and Brier on "Application of multi-response permutation procedures for examining seasonal changes in monthly mean sea-level pressure patterns," in *Monthly Weather Review* [301].

The Pearson type III distribution, as a three-parameter gamma distribution, has the advantage of being totally characterized by the exact mean, variance, and skewness, in the same manner that the normal distribution, as a two-parameter distribution, is fully characterized by the exact mean and variance—a property not possessed by the beta distribution.[10] An added advantage of the Pearson type III distribution is when the skewness parameter is zero, the distribution is normal.[11] In describing the Pearson type III distribution, Pearson noted "This generalized probability curve fits with a high degree of accuracy a number of measurements and observations hitherto not reduced to theoretical treatment" [332, p. 331]. With the advent of high-speed computers, moment-approximation permutation tests have largely been replaced by resampling-approximation permutation tests.

1.2.3 Resampling-Approximation Permutation Tests

Resampling-approximation (hereafter, resampling) permutation tests generate and examine a Monte Carlo random subset of all possible, equally-likely arrangements of the observed response measurements.[12] For each resampled arrangement of the observed data, the desired test statistic is calculated. The probability of obtaining the observed value of the test statistic, or one more extreme, is the proportion of the resampled test statistics with values equal to or more extreme than the value of the observed test statistic. With a sufficient number of resamplings, a probability value can be computed to any reasonable accuracy. The current recommended practice is to use $L = 1{,}000{,}000$ resampled arrangements of the observed data to ensure a probability value with three decimal places of accuracy [195].

[9] The Pearson type III distribution was one of four distributions introduced by Karl Pearson in 1895 [333], although the type III distribution had previously been presented without discussion by Pearson in 1893 [332, p. 331]. The type V distribution introduced by Pearson in 1895 was simply the normal distribution and the Pearson type I distribution was a generalized beta distribution.

[10] Mielke, Berry, and Brier were not the first to adopt the Pearson type III distribution to approximate a discrete permutation distribution. For example, B.L. Welch utilized the Pearson type III distribution in a 1936 paper on the specification of rules for rejecting too variable a product [417] and used it again in a 1938 paper on testing the significance of differences between the means of two independent samples when the population variances were unequal [419].

[11] For a one-way analysis of variance utilizing a moment-approximation approach, see a 1983 article by Berry and Mielke [23].

[12] It is generally accepted that the term "Monte Carlo" method was coined by Stanislaw Ulam, John von Neumann, and Nicholas Metropolis in 1946 while they were working on nuclear weapon projects at the Los Alamos National Laboratory [278, 415]. However, in a 2012 book on *Turing's Cathedral*, George Dyson attributes the coining of the term "Monte Carlo" solely to Nicholas Metropolis [102, p. 192].

Meyer Dwass is usually credited with the formal development of resampling permutation tests, first presented in an article on "Modified randomization tests for nonparametric hypotheses" published in *The Annals of Mathematical Statistics* in 1957 [100]. While researchers prior to 1957 certainly utilized resampling permutation methods to provide approximate probability values—witness Eden and Yates in their 1933 investigation into height measurements of Yeoman II wheat shoots in which they analyzed a random sample of 1,000 out of a possible 4,586,471,424 arrangements—Dwass provided the first rigorous investigation into the precision of resampling probability approximations.[13]

Presently, resampling permutation tests are the method of choice for most researchers, with exact permutation tests reserved for smaller data sets. There are three notable advantages to resampling permutation tests. First, resampling permutation tests are highly efficient given the ready availability of high-speed computers and the recent development of rapid pseudorandom number generators such as the Mersenne Twister, on which resampling permutation tests are highly dependent. Second, in some applications a resampling permutation test is much more efficient than an exact permutation test, even for small samples. For example, in the permutation analysis of contingency tables an exact permutation test must calculate a hypergeometric probability value for each of, potentially, thousands of cell frequency arrangements, while a resampling permutation test need only count the number of cell arrangements as extreme or more extreme than the observed cell arrangement. Third, algorithms for exact permutation tests are non-existent or completely impractical for analyzing certain problems, such as multi-way contingency tables, while an efficient resampling algorithm is presently available for multi-way tables; see, for example, a 2007 article by Mielke, Berry, and Johnston in *Psychological Reports* [307].

1.2.4 Mehta–Patel Network Algorithm

Finally, mention should be made of the Mehta–Patel network enumeration algorithm for $r \times c$ contingency tables, a computer algorithm that cleverly circumvents the need to completely enumerate all possible arrangements of the observed cell frequencies, given the observed marginal frequency totals, yet still provides an exact probability value; see, for example, a paper by Mehta, Patel, and Gray in 1985 [277] and two papers by Mehta and Patel in 1986 [275, 276].[14]

[13]It should be noted that the 1957 Dwass article on modified randomization tests for non-parametric hypotheses relied heavily on the theoretical contributions of an article titled "On the theory of some non-parametric hypotheses" by Erich Lehmann and Charles Stein published in *The Annals of Mathematical Statistics* in 1949 [237].

[14]The Mehta–Patel network algorithm was subsequently applied to many more statistical analyses than the highly limited analysis of $r \times c$ contingency tables.

The Mehta–Patel network algorithm is a directed non-cyclic network consisting of nodes in a sequence of stages, corresponding to the reference set of $r \times c$ contingency tables. Distances between the nodes, called arcs, are defined so that the total distance of a path through the network corresponds to the value of the test statistic. At each intermediary node, the network algorithm computes the longest and shortest path for all paths passing through that node. The value of the test statistic is compared with the longest and shortest paths to determine (1) if all paths through the node contribute to the probability value, (2) if none of the paths through the node contributes to the probability value, or (3) if neither of these situations occurs. The Mehta–Patel network algorithm has greatly increased the range of exact permutation tests and, having been incorporated into various computer packages, is widely available to researchers in a convenient and usable format.[15]

1.3 Permutation and Parametric Statistical Tests

Permutation statistical tests, based on the permutation model, differ from traditional parametric tests, based on the population model, in several ways. First, permutation tests are data-dependent in that all the information required for analysis is contained within the observed data set.[16] Implicit in this data-dependency is the understanding that statistical inferences are limited to the actual experiment or survey that has been performed. Second, permutation tests are appropriate for non-random samples, such as are common in many fields of research. Third, permutation tests are distribution-free in that they do not depend on the assumptions associated with traditional parametric tests, such as normality and homogeneity of variance. Fourth, permutation tests provide exact probability values based on the discrete permutation distribution of equally-likely test statistic values, rather than approximate probability values based on a theoretical approximating distribution, such as a normal, χ^2, t, or F distribution. Fifth, permutation tests are ideal for very small data sets, whereas distribution functions often provide very poor fits.

1.3.1 Permutation Tests and Normality

The assumption of normality is so basic to classical statistics that it deserves special attention. Two points should be emphasized. First, permutation tests make no distributional assumptions and, therefore, do not depend on the assumption of normality. Second, the assumption of normality by conventional tests is always unrealistic and never justified in practice.

[15]For a detailed description of the Mehta–Patel network enumeration algorithm, see Berry, Johnston, and Mielke [41, pp. 288–293].

[16]For the importance of data-dependent analysis, see a 1988 article by Biondini, Mielke, and Berry on "Data-dependent permutation techniques for the analysis of ecological data" [44] and a 2002 article by Mielke and Berry on "Data-dependent analyses in psychological research" [296].

In 1957 R.C. Geary famously proclaimed: "Normality is a myth; there never has, and never will be, a normal distribution" [135, p. 241]. The French physicist and Nobel laureate in physics, Gabriel Lippmann, once wrote in a letter to Henri Poincaré à propos the normal curve:

> Les expérimentateurs s'imaginent que c'est un théorèm de mathématiques, et les mathématiciens d'être un fait expérimental.
> Experimentalists think that it is a mathematical theorem, while mathematicians believe it to be an experimental effect.

(Lippman, quoted in D'Arcy Wentworth Thompson's *On Growth and Form* [396, p. 121]). And in 1954 Bross pointed out that statistical methods "are based on certain assumptions—assumptions which not only can be wrong, but in many situations *are* wrong" [58, p. 815].[17,18] Others have empirically demonstrated the prevalence of highly skewed and heavy-tailed distributions in a variety of academic disciplines; see, for example, discussions by Schmidt and Johnson [366], Bradley [53], Saal, Downey, and Lahey [359], Bernardin and Beatty [22], Micceri, and Murphy and Cleveland [314], the best known of which is Micceri's widely quoted 1989 article on "The unicorn, the normal curve, and other improbable creatures" in *Psychological Bulletin* [280].

O'Boyle and Aguinis cautioned that "assuming normality... can lead to misspecified theories and misleading practices" [323, p. 116], noting that the assumption of normality, like random sampling, belongs to the class of "received doctrines" that are

> taught in undergraduate and graduate classes, enforced by gatekeepers (e.g., grant panels, reviewers, editors, dissertation committee members), discussed among colleagues, and otherwise passed along among pliers of the trade far and wide and from generation to generation [228, p. 281].

The development of a cohesive methodology of basic tests by R.A. Fisher was under the assumption of normality [235, p. 45]. Egon Pearson, in reviewing the second edition of Fisher's *Statistical Methods for Research Workers* in 1929 wrote:

> There is one criticism, however, which must be made from the statistical point of view. A large number of the tests developed are based upon the assumption that the population sampled is of 'normal' form. That this is the case may be gathered from very careful reading of the text, but the point is not sufficiently emphasized. It does not appear reasonable to lay stress on the 'exactness' of tests, when no means whatever are given of appreciating how rapidly they become inexact as the population sampled diverges from normality. That the tests, for example, connected with the analysis of variance are far more dependent on normality than those involving 'Student's' z (or t) distribution is almost certain, but no clear indication of the need for caution in their application is given... [335, pp. 866–867].

[17] Emphasis in the original.

[18] See also a short but comprehensive 2010 article on this topic by Tom Siegfried in *Science News* [377].

An obvious drawback to permutation statistical tests is the amount of computation required, with exact permutation tests being impractical for many statistical analyses. Even resampling permutation tests often require the enumeration of tens of millions of random permutations in order to guarantee sufficient accuracy. Two features of permutation methods mitigate this problem: first, mathematical recursion with an arbitrary initial value and, second, calculation of only the variable portion of the selected test statistic.

1.3.2 Mathematical Recursion

Mathematical recursion, in a statistical context, is a process in which an initial probability value of a test statistic is calculated, then successive probability values are generated from the initial value by a recursive process.[19] The initial value need not be an actual probability value, but can be a completely arbitrary positive value by which the resultant relative probability values are adjusted for the initializing value at the conclusion of the recursion process.

In 1934 Frank Yates used recursion with an arbitrary initial value to calculate the Fisher–Yates exact test for 2×2 contingency tables [433]. Here, Yates was able to generate all possible probability values without evaluating even a single factorial expression, a process that was extremely efficient given that, under the usual method, there are nine factorial expressions to be computed for each possible arrangement of the observed response measurements. Maurice Kendall, in 1938, was another early statistician who utilized a recursive process in the calculation of exact probability values for his new measure of rank correlation, τ [205]. It is also true, however, that recursion methods were not new in the 1930s, having been used historically by Blaise Pascal, Christiaan Huygens, James Bernoulli, Willem 'sGravesande, Pierre Rémond de Montmort, and Adolphe Quetelet, among others [162, 163]. Presently, computer algorithms employing recursion methods are powerful tools for the efficient generation of exact probability values.

Mathematical recursion is so fundamental to permutation methods that a detailed example of a recursion process is important to illustrate the procedure. Perhaps no better example of the statistical recursion procedure exists than that provided by Frank Yates. In 1934 Yates published an article on the analysis of contingency tables containing small cell frequencies in *Supplement to the Journal of the Royal Statistical Society* [433]. The stated purpose of the article was threefold: first, to introduce statisticians to Fisher's exact probability test, which was very new at the time; second, to use Fisher's exact probability test as a gold standard against which

[19] A recursive process is one in which items are defined in terms of items of similar kind. Using a recursive relation, a class of items can be constructed from one or a few initial values (a base) and a small number of relationships (rules). For example, given the base, $F_0 = 0$ and $F_1 = F_2 = 1$, the Fibonacci series $\{0, 1, 1, 2, 3, 5, 8, 13, 21, \ldots\}$ can be constructed by the recursive rule $F_n = F_{n-1} + F_{n-2}$ for $n > 2$.

a	b	$N-n$
c	d	n
$N-n'$	n'	N

Fig. 1.4 Notation for a 2×2 contingency table as defined by Yates [433]

the small-sample performance of the Pearson chi-squared test might be judged; and third, to present a correction for continuity to the chi-squared test of independence, resulting in a better approximation to Fisher's exact probability test [177]. Yates succinctly described the recursion process:

> In cases where N is not too large the distribution with any particular numerical values of the marginal totals can be computed quite quickly, using a table of factorials to determine some convenient term, and working out the rest of the distribution term by term, by simple multiplications and divisions. If a table of factorials is not available we may start with any convenient term as unity, and divide by the sum of the terms so obtained [433, p. 219],

where N in this context denotes the total number of observations. Yates defined a 2×2 contingency table using the notation in Fig. 1.4, where $n \leq n' \leq N/2$.

Giving due credit to Fisher, Yates showed that the probability value corresponding to any set of cell frequencies, a, b, c, d, was the hypergeometric point-probability value given by

$$P = \frac{n!\, n'!\, (N-n)!\, (N-n')!}{N!\, a!\, b!\, c!\, d!} \, .$$

Since the exact probability value of a 2×2 contingency table with fixed marginal frequency totals and one degree of freedom is equivalent to the probability value of any one cell, determining the probability value of cell a is sufficient. If

$$P\{a+1|N-n, N-n', N\} = P\{a|N-n, N-n', N\} \times f(a) \, ,$$

then solving for $f(a)$ produces

$$f(a) = \frac{P\{a+1|N-n, N-n', N\}}{P\{a|N-n, N-n', N\}} = \frac{a!\, b!\, c!\, d!}{(a+1)!\, (b-1)!\, (c-1)!\, (d+1)!}$$

and, after cancelling, yields

$$f(a) = \frac{bc}{(a+1)(d+1)} \, . \tag{1.1}$$

Yates provided an example analysis based on data from Milo Hellman on bottle feeding and malocclusion that had been published in *Dental Cosmos* in 1914 [172]. The data are summarized in Fig. 1.5 and the six exhaustive 2×2 contingency tables from Hellman's data given in Fig. 1.5 are listed in Table 1.3. Yates generated

Fig. 1.5 Hellman's data on breastfeeding and malocclusion [172]

	Teeth		
Feeding type	Normal	Malocclusion	Total
Breast-fed	4	16	20
Bottle-fed	1	21	22
Total	5	37	42

Table 1.3 Six possible arrangements of cell frequencies with $N = 42$ and marginal frequency totals of 20, 22, 5, and 37

Table 1		Table 2		Table 3		Table 4		Table 5		Table 6	
0	20	1	19	2	18	3	17	4	16	5	15
5	17	4	18	3	19	2	20	1	21	0	22

the entire exact probability distribution as follows.[20] The probability of obtaining zero normal breast-fed babies for the cell arrangement in Table 1 in Table 1.3 was given by

$$P\{a = 0|20, 5, 42\} = \frac{5!\ 37!\ 20!\ 22!}{42!\ 0!\ 20!\ 5!\ 17!} = 0.030957$$

and calculated utilizing a table of factorials. Then, the probability values for $a = 1, 2, 3, 4$, and 5 in Table 1.3 were generated recursively utilizing Eq. (1.1). Thus,

$$P\{a = 1|20, 5, 42\} = 0.030957 \times \frac{(20)(5)}{(1)(18)} = 0.171982\ ,$$

$$P\{a = 2|20, 5, 42\} = 0.171982 \times \frac{(19)(4)}{(2)(19)} = 0.343965\ ,$$

$$P\{a = 3|20, 5, 42\} = 0.343964 \times \frac{(18)(3)}{(3)(20)} = 0.309568\ ,$$

$$P\{a = 4|20, 5, 42\} = 0.309568 \times \frac{(17)(2)}{(4)(21)} = 0.125301\ ,$$

and

$$P\{a = 5|20, 5, 42\} = 0.125301 \times \frac{(16)(1)}{(5)(22)} = 0.018226\ ,$$

[20]Exact probability values in this example are given to six places to demonstrate the accuracy of recursion processes with an arbitrary initial value.

respectively. In this manner, Yates was able to generate the entire discrete permutation distribution from $\min(a) = \max(0, N - n - n') = \max(0, -17) = 0$ to $\max(a) = \min(N - n, N - n') = \min(20, 5) = 5$.

1.3.3 Calculation with an Arbitrary Initial Value

To illustrate Yates' use of an arbitrary origin in a recursion procedure, consider Table 1 in Table 1.3 and set $C\{a = 0|20, 5, 42\}$ to a small arbitrarily chosen value, say 5.00; thus, $C\{a = 0|20, 5, 42\} = 5.00$. Then, a recursion procedure produces

$$C\{a = 1|20, 5, 42\} = 5.000000 \times \frac{(20)(5)}{(1)(18)} = 27.777778 ,$$

$$C\{a = 2|20, 5, 42\} = 27.777778 \times \frac{(19)(4)}{(2)(19)} = 55.555556 ,$$

$$C\{a = 3|20, 5, 42\} = 55.555556 \times \frac{(18)(3)}{(3)(20)} = 50.000000 ,$$

$$C\{a = 4|20, 5, 42\} = 50.000000 \times \frac{(17)(2)}{(4)(21)} = 20.238095 ,$$

and

$$C\{a = 5|20, 5, 42\} = 20.238095 \times \frac{(16)(1)}{(5)(22)} = 2.943723 ,$$

for a total of $C\{0, \ldots, 5|20, 5, 42\} = 5.00 + 27.777778 + \cdots + 2.943723 = 161.515152$. The desired exact probability values are then obtained by dividing each relative probability value by the recursively obtained total; for example,

$$P\{a = 0|20, 5, 42\} = \frac{5.000000}{161.515152} = 0.030957 ,$$

$$P\{a = 1|20, 5, 42\} = \frac{27.777778}{161.515152} = 0.171982 ,$$

$$P\{a = 2|20, 5, 42\} = \frac{55.555556}{161.515152} = 0.343965 ,$$

$$P\{a = 3|20, 5, 42\} = \frac{50.000000}{161.515152} = 0.309568 ,$$

$$P\{a = 4|20, 5, 42\} = \frac{20.238095}{161.515152} = 0.125301 ,$$

and

$$P\{a = 5|20, 5, 42\} = \frac{2.943723}{161.515152} = 0.018226 .$$

In this manner, the entire analysis could be conducted utilizing an arbitrary initial value and a recursion procedure, thereby eliminating all factorial expressions. When the number of potential contingency tables given by $\max(a) - \min(a) + 1$ is large, the computational savings can be substantial.

1.3.4 Variable Portion of a Test Statistic

Under permutation, only the variable portion of the test statistic need be computed for each arrangement of the observed data. As this is often only a very small portion of the desired test statistic, calculations can often be reduced by several factors; see, for example, a discussion by Scheffé in 1959 [365, pp. 314–317]. For example, in computing the permutation probability value of Student's two-sample t test, only the sum of the response measurements in the smaller of the two samples need be calculated for each arrangement of the observed response measurements, thus eliminating a great deal of calculation for each random arrangement of the observed data, a technique utilized by Pitman in his 1937 permutation analysis of two independent samples [340].

For another example, in 1933 Thomas Eden and Frank Yates substantially reduced calculations in their randomized-block analysis of Yeoman II wheat shoots by recognizing that the block and total sums of squares would be constant for all of their 1,000 random samples and, consequently, the value of z for each sample would be uniquely defined by the treatment (between) sum of squares, i.e., the treatment sum of squares was sufficient for a permutation analysis of variance test [103].[21] Also in 1937, Bernard Welch, in a permutation analysis of randomized-block, considered a monotonically increasing function of z that contained only the portion of z that varied under permutation [418]. In this case, as with Eden and Yates, Welch calculated only the treatment sum of squares.

Furthermore, Maurice Kendall and Bernard Babington Smith, in their discussion of the problem of m rankings, substantially reduced their calculations by recognizing that the number of rankings (m) and the number of ranks (N) were invariant over permutation of the observed data and, therefore, calculated only the sum of squared deviations from the mean of the ranks in their permutation analysis of m rankings [209]. Likewise, Kendall, Kendall, and Babington Smith, in their permutation analysis of Spearman's rank-order correlation coefficient, considered only the sum of the squared differences between ranks, which reduced computation considerably for each of the $N!$ arrangements of the observed rank-order statistics [211].

[21] The letter F for the analysis of variance (variance-ratio) test statistic was introduced in 1934 by George Snedecor at Iowa State University, much to the displeasure of R.A. Fisher [378, p. 15]. Prior to 1934 the test statistic was indicated by z, the letter originally assigned to it by Fisher.

A few brief examples of analyzing test statistics using only the variable portion illustrate the efficiency of the procedure. First, consider Spearman's rank-order correlation coefficient given by

$$\rho = 1 - \frac{6\sum_{i=1}^{N}(x_i - y_i)^2}{N(N^2 - 1)} , \tag{1.2}$$

where x_i and y_i for $1 = 1, \ldots, N$ objects represent ranks on two ordinal variables [381]. In this case, N, 1, and 6 are constants in Eq. (1.2) so it is only necessary to calculate $\sum_{i=1}^{N}(x_i - y_i)^2$ for each arrangement of the ranks. Moreover, for a permutation analysis it suffices to shuffle only the x or the y ranks, holding the other set of ranks constant.

Second, consider the Pearson product-moment correlation coefficient for two variables, x and y, given by

$$r_{xy} = \frac{N\sum_{i=1}^{N}x_i y_i - \sum_{i=1}^{N}x_i \sum_{i=1}^{N}y_i}{\left\{\left[N\sum_{i=1}^{N}x_i^2 - \left(\sum_{i=1}^{N}x_i\right)^2\right]\left[N\sum_{i=1}^{N}y_i^2 - \left(\sum_{i=1}^{N}y_i\right)^2\right]\right\}^{1/2}} , \tag{1.3}$$

where x_i and y_i for $i = 1, \ldots, N$ objects represent bivariate response measurements on two interval-level variables. Here, N and all the summations, with the exception of $\sum_{i=1}^{N} x_i y_i$ in Eq. (1.3), are constants under permutation, so for any arrangement of the observed data it is only necessary to calculate the sum of the products of x_i and y_i for $i = 1, \ldots, N$. Like Spearman's rank-order correlation coefficient, for efficiency only the response measurement scores in variable x or y need be shuffled, while the other variable is held constant.

Finally, consider Cohen's unweighted kappa measure of inter-rater agreement given by

$$\hat{\kappa} = \frac{\sum_{i=1}^{r} O_{ii} - \sum_{i=1}^{r} E_{ii}}{N - \sum_{i=1}^{r} E_{ii}} , \tag{1.4}$$

where O_{ii} and E_{ii} for $i = 1, \ldots, r$ denote the observed and expected cell frequencies, respectively, on the principal diagonal of an $r{\times}r$ contingency (agreement) table [70]. Since the E_{ii}, $i = 1, \ldots, r$, are based on N and the row and column marginal frequency totals, the variable portion of $\hat{\kappa}$ in Eq. (1.4) is simply the sum of the observed cell frequencies, $\sum_{i=1}^{r} O_{ii}$, on the principal diagonal for each arrangement of the cell frequencies, given the fixed marginal frequency totals.

These two features, mathematical recursion with an arbitrary initial value and computation of only the variable portion under permutation, combined with powerful resampling algorithms and high-speed computers, produce a highly efficient permutation statistical approach that, today, makes permutation analyses both feasible and practical for many research applications.

1.4 Overviews of Chaps. 2–11

The book is organized into 11 chapters. This first chapter was an introduction to the organization of the remaining ten chapters and presented the three main approaches to permutation statistical methods: exact, moment-approximation, and resampling-approximation permutation tests. In addition, recursion with an arbitrary initial value and calculation of only the variable portion of test statistics under permutation were shown to have distinct advantages in both exact and resampling permutation methods.

Chapter 2 introduces a generalized Minkowski distance function and describes the synthesizing algorithm under which various permutation statistical tests and measures designed for completely randomized data are derived. Chapters 3 and 4 examine tests and measures designed for completely randomized response measurements at the interval level of measurement. Chapters 5 and 6 examine tests and measures for completely randomized response measurements at the ordinal (ranked) level of measurement. Chapter 7 examines tests and measures for completely randomized response measurements at the nominal (categorical) level of measurement. Both univariate and multivariate response measurements are considered in Chaps. 3–7.

Chapter 8 utilizes the generalized Minkowski distance function described in Chap. 2 to develop a synthesizing algorithm under which various permutation statistical tests and measures designed for randomized-block data are derived. Chapters 9, 10, and 11 examine permutation statistical tests and measures designed for randomized-block response measurements at the interval, ordinal, and nominal levels of measurement, respectively. Like the completely randomized data analyzed in Chaps. 3–7, both univariate and multivariate response measurements are considered in Chaps. 8–11.

1.4.1 Chapter 2: Completely Randomized Data

Chapter 2 develops a general set of synthesizing Multi-Response Permutation Procedures (MRPP) for permutation statistical tests and measures designed for completely randomized data sets. Included within MRPP is a generalized chance-corrected measure of effect size, $\Re$. The multi-response permutation procedures for completely randomized data are grounded in a generalized Minkowski distance function and are sufficiently general to accommodate interval-, ordinal-, and nominal-level response measurements, both univariate and multivariate. Chapter 2 provides an introduction and mathematical foundation for the permutation statistical tests and measures for completely randomized data that are further developed in Chaps. 3–7.

1.4.2 Chapter 3: Randomized Interval-Level Data

Chapter 3 applies the multi-response permutation procedures for completely randomized data developed in Chap. 2 to permutation statistical tests and measures designed to analyze univariate and multivariate responses at the interval level of measurement. Example statistical tests and measures presented and illustrated in Chap. 3 include permutation tests corresponding to Student's t test for two independent samples, Hotelling's generalized T^2 test for two independent samples, the one-way analysis of variance F test (ANOVA), the one-way multivariate analysis of variance F test (MANOVA), the Bartlett–Nanda–Pillai trace test, and the unbiased correlation ratio. Also included in Chap. 3 are discussions of measures of effect size and applications to multiple regression.

1.4.3 Chapter 4: Regression Analysis of Interval Data

Chapter 4 continues and expands the analyses described in Chap. 3, applying multi-response permutation procedures to permutation statistical tests and measures designed to analyze univariate and multivariate responses at the interval level of measurement. In contrast to Chap. 3, Chap. 4 utilizes multi-response permutation procedures to analyze regression residuals from ordinary least squares (OLS) and least absolute deviation (LAD) regression models. Example designs presented and analyzed in Chap. 4 include one-way completely randomized, one-way randomized with a covariate, one-way randomized block, two-way randomized-block, two-way factorial, Latin square, split-plot, and two-factor nested analysis-of-variance designs.

1.4.4 Chapter 5: Randomized Ordinal-Level Data—I

Chapter 5 applies the multi-response permutation procedures for completely randomized data developed in Chap. 2 to permutation statistical tests and measures designed to analyze univariate and multivariate responses at the ordinal (ranked) level of measurement. Example statistical tests and measures presented and illustrated in Chap. 5 include permutation tests corresponding to the Wilcoxon two-sample rank-sum test, the Kruskal–Wallis multi-sample rank-sum test, the Ansari–Bradley rank-sum test for dispersion, the Taha sum-of-squared-ranks test, the Mood rank-sum test for dispersion, the Brown–Mood median test, the Mielke power-of-rank function tests, the Whitfield two-sample rank-sum test, and the Cureton rank-biserial test.

1.4.5 Chapter 6: Randomized Ordinal-Level Data—II

Chapter 6 generalizes the analyses described in Chap. 5 to multivariate responses at the ordinal level of measurement. As in Chap. 5, example statistical tests and measures presented and illustrated in Chap. 6 include permutation tests corresponding to the Wilcoxon two-sample rank-sum test, the Kruskal–Wallis multiple-sample rank-sum test, the Ansari–Bradley rank-sum test for dispersion, the Taha sum-of-squared-ranks test, the Mood rank-sum test for dispersion, the Brown–Mood median test, the Mielke power-of-rank function tests, the Whitfield two-sample rank-sum test, and the Cureton rank-biserial test.

1.4.6 Chapter 7: Randomized Nominal-Level Data

Chapter 7 applies the multi-response permutation procedures for completely randomized data developed in Chap. 2 to permutation statistical tests and measures designed to analyze univariate and multivariate responses at the nominal (categorical) level of measurement. Example statistical tests and measures presented and illustrated in Chap. 7 include permutation tests corresponding to the Goodman and Kruskal's t_a and t_b measures of categorical association, Light and Margolin's categorical analysis of variance, tests to analyze multiple binary choices, and various multivariate measures of association for a nominal-level independent variable and nominal-, ordinal-, and interval-level dependent variables.

1.4.7 Chapter 8: Randomized Block Data

Chapter 8 develops a general set of synthesizing Multivariate Randomized-Block Permutation (MRBP) procedures for permutation statistical tests and measures designed for randomized-block data sets. Included within MRBP is a generalized chance-corrected within-block measure of effect size, $\Re$. The multivariate randomized-block permutation procedures are grounded in a generalized Minkowski distance function and are sufficiently general to accommodate interval-, ordinal-, and nominal-level response measurements, both univariate and multivariate. Chapter 8 provides an introduction and mathematical foundation for the permutation statistical tests and measures that are further developed in Chaps. 9, 10, and 11.

1.4.8 Chapter 9: Blocked Interval-Level Data

Chapter 9 applies the multivariate randomized-block permutation procedures developed in Chap. 8 to permutation statistical tests and measures designed to analyze univariate and multivariate responses at the interval level of measurement. Example statistical tests and measures presented and illustrated in Chap. 9 include permutation tests corresponding to Student's matched-pairs t test, Hotelling's generalized T^2 test for two dependent samples, the randomized-block F test, the multivariate randomized-block test, and Pearson's product-moment correlation coefficient.

1.4.9 Chapter 10: Blocked Ordinal-Level Data

Chapter 10 applies the multivariate randomized-block permutation procedures developed in Chap. 8 to permutation statistical tests and measures designed to analyze univariate and multivariate responses at the ordinal (rank) level of measurement. Example statistical tests and measures presented and illustrated in Chap. 10 include permutation tests corresponding to the Wilcoxon signed-ranks test, the Friedman analysis of variance for ranks, Spearman's rank-order and footrule measures of correlation, Kendall's coefficient of concordance, Cohen's weighted kappa measure of chance-corrected agreement, Kendall's t_a and t_b measures of ordinal association, Stuart's t_c statistic, Goodman and Kruskal's γ measure of ordinal association, and Somers' d_{xy} and d_{yx} measures of ordinal association.

1.4.10 Chapter 11: Blocked Nominal-Level Data

Chapter 11 applies the multivariate randomized-block permutation procedures developed in Chap. 8 to permutation statistical tests and measures designed to analyze univariate and multivariate responses at the nominal (categorical) level of measurement. Example statistical tests and measures presented and illustrated in Chap. 11 include permutation tests corresponding to McNemar's and Cochran's Q tests, Cohen's unweighted kappa measure of chance-corrected agreement, Yule's Q and Y measures of association, percentage differences, the odds ratio, and chi-squared.

1.5 Coda

Chapter 1 provided an introduction to the next ten chapters, compared the population and permutation models of statistical analysis, and presented the three main approaches to permutation statistical methods: exact, moment-approximation, and resampling-approximation permutation tests. Chapters 2 and 8 introduce permutation procedures for completely randomized and randomized-block data, respectively. The substantive material in Chaps. 3–7 for completely randomized data, and Chaps. 9–11 for randomized-block data contain only an illustrative sample of possible applications of permutation statistical methods.

It is not the intent of the authors to provide a synthesis of all statistical methods, but rather to derive and illustrate a common model under which many statistical tests and measures can be understood. Interestingly, because of the organization of the permutation model it was inevitable that some new statistical tests and measures were uncovered that were previously unknown and for which applications might prove interesting. Finally, an emphasis on permutation-based statistical methods throughout the book promotes permutation methods as a data-dependent, distribution-free approach to statistical analysis that does not require random sampling from a specified, well-defined population, and yields exact probability values.

Chapter 2

Chapter 2 introduces Multi-Response Permutation Procedures (MRPP) for univariate and multivariate completely randomized response measurement data and establishes the relationships between the MRPP test statistics, δ and $\mathfrak{R}$ developed in Chap. 2, and selected conventional tests and measures designed for the analysis of completely randomized data at the interval level of measurement in Chaps. 3 and 4, the ordinal level of measurement in Chaps. 5 and 6, and the nominal level of measurement in Chap. 7.

2 Completely Randomized Data

This second chapter of *Permutation Statistical Methods* introduces a generalized distance function that provides the foundation for a set of multi-response permutation procedures specifically designed for univariate and multivariate completely randomized data. Multi-Response Permutation Procedures (MRPP) were introduced by Mielke, Berry, and Johnson in 1976 and constitute a class of permutation methods for one or more response measurements on each object that were initially developed to distinguish possible differences among two or more groups of objects [300].[1] The multi-response permutation procedures presented here are based on a generalized Minkowski distance function and provide a synthesizing foundation for a variety of statistical tests and measures for completely randomized data that are further developed in Chaps. 3–7.

2.1 Minkowski Distance Function

Hermann Minkowski (1864–1909), German mathematician and creator of the geometry of numbers, utilized geometrical methods to solve problems in number theory, mathematical physics, and the theory of relativity. Minkowski was a close friend of David Hilbert while teaching at Königsberg University and taught Albert Einstein while employed at Eidgenössische Polytechnikum in Zürich (now, ETH Zürich). In 1891 Minkowski introduced a measure of metric distance between

[1]The 1976 paper by Mielke, Berry, and Johnson was the first published account of MRPP [300]. Previously, Mielke utilized MRPP in a study sponsored by the National Communicable Disease Center that involved comparisons of proportional contributions of five plague organism protein bands based on electrophoresis measurements obtained from samples of organisms associated with distinct geographical regions.

K.J. Berry et al., *Permutation Statistical Methods*,
DOI 10.1007/978-3-319-28770-6_2

two points in *Crelle's Journal* [310].[2] The Minkowski metric distance of order p between two points in an r-dimensional Euclidean space, $x' = (x_1, x_2, \ldots, x_r)$ and $y' = (y_1, y_2, \ldots, y_r) \in \mathbb{R}^r$, is given by

$$d(x, y) = \left(\sum_{i=1}^{r} |x_i - y_i|^p \right)^{1/p},$$

where $p \geq 1$.

The Minkowski distance function is typically used with $p = 1, 2$, or ∞. When $p = 1$, the distance is a first-order Minkowski metric, often called a city-block, Manhattan [231], rectilinear [54], or taxicab [222] metric, the latter named for the distance between two points that a car or taxicab would drive in a city laid out in square blocks. When $p = 2$, the distance is a second-order Minkowski metric and is the ordinary Euclidean distance between points, a generalization of the Pythagorean theorem to more than two coordinates. When $p = \infty$, the Minkowski metric is known as the Tchebycheff (Chebyshev), von Neumann, or, in the two-dimensional case, the chess-board Minkowski distance [167].

Conventional statistical tests and measures, such as t tests, F tests, and ordinary least-squares (OLS) regression and correlation, are based on squared Euclidean distances between response measurement scores, which are not metric. The Minkowski distance function, however, is limited to metric distances and, under its standard definition, cannot accommodate most conventional statistical tests. Therefore, consider a generalized Minkowski distance function given by

$$\Delta(x, y) = \left(\sum_{i=1}^{r} |x_i - y_i|^p \right)^{v/p}, \tag{2.1}$$

where $p \geq 1$ and $v > 0$ [297, p. 5]. When $r \geq 2$, $p = 2$, and $v = 1$, $\Delta(x, y)$ is rotationally invariant in an $r \geq 2$ dimensional space. When $v = p = 1$, $\Delta(x, y)$ is a city-block metric, which is not rotationally invariant. When $v = 1$ and $p = 2$, $\Delta(x, y)$ is an ordinary Euclidean distance metric. And when $v = p = 2$, $\Delta(x, y)$ is a squared Euclidean distance, which is not a metric distance function since the triangle inequality is not satisfied.[3]

[2]The *Journal für die Reine und Angewandte Mathematik* was founded by August Leopold Crelle in 1826. It continues today, although it is more popularly known as *Crelle's Journal*.

[3]A distance function is a metric if it satisfies three properties given by (1) $\Delta(x, y) \geq 0$ and $\Delta(x, x) = 0$, i.e., the distance is positive between two different points and is equal to zero from any point to itself; (2) the distance is symmetric: $\Delta(x, y) = \Delta(y, x)$, i.e., the distance between points x and y is the same in either direction; and (3) the triangle inequality is satisfied: $\Delta(x, y) \leq \Delta(x, z) + \Delta(z, y)$, i.e., the distance between any two points is the shortest distance along any path.

2.2 Multi-response Permutation Procedures

Multi-Response Permutation Procedures (MRPP) were originally designed to statistically determine possible differences among one or more response measurement scores among two or more groups of objects or subjects [300]. Let $\Omega = \{\omega_1, \ldots, \omega_N\}$ denote a finite sample of N objects that represents a target population, let $x_i' = (x_{1i}, \ldots, x_{ri})$ be a transposed vector of r commensurate response measurement scores for object ω_i, $i = 1, \ldots, N$, and let $S_1, \ldots, S_g$ designate an exhaustive partitioning of the N objects into g disjoint treatment groups.[4] The MRPP test statistic is a weighted mean given by

$$\delta = \sum_{i=1}^{g} C_i \xi_i \,, \tag{2.2}$$

where $C_i > 0$ is a positive weight for treatment group S_i, $i = 1, \ldots, g$, $\sum_{i=1}^{g} C_i = 1$,

$$\xi_i = \binom{n_i}{2}^{-1} \sum_{j<k} \Delta(j,k)\, \Psi_i(\omega_j)\, \Psi_i(\omega_k) \tag{2.3}$$

is the average distance-function value for all distinct pairs of objects in treatment group S_i, $i = 1, \ldots, g$, $n_i \geq 2$ is the number of a priori objects classified into treatment group S_i, $i = 1, \ldots, g$,

$$N = \sum_{i=1}^{g} n_i,$$

$\sum_{j<k}$ is the sum over all j and k such that $1 \leq j < k \leq N$, and $\Psi(\cdot)$ is an indicator function given by

$$\Psi_i(\omega_j) = \begin{cases} 1 & \text{if } \omega_j \in S_i \,, \\ 0 & \text{otherwise} \,. \end{cases}$$

The choice of the treatment-group weights, $C_1, \ldots, C_g$, and the generalized Minkowski distance function given in Eq. (2.1) on p. 30 specify the structure of

---

[4]Multi-response permutation procedures also provide for a group of unclassified response measurement scores such as might result from a survey with question choices that include "none of the above" or "not applicable." See, for example, a 1983 article on lead concentrations in inner-city soils by Mielke, Anderson, Berry, Mielke, Chaney, and Leech [302] and a discussion by Mielke and Berry in 2007 [297, pp. 35–40].

MRPP. The original choice of C_i given by Mielke, Berry, and Johnson in 1976 was

$$C_i = \frac{n_i(n_i - 1)}{\sum_{j=1}^{g} n_j(n_j - 1)}$$

for $i = 1, \ldots, g$ [300]. However, a variety of other treatment-group weights can be considered; for example,

$$C_i = \frac{n_i}{N}\,, \quad C_i = \frac{n_i - 1}{N - g}\,, \quad \text{or} \quad C_i = \frac{1}{g}$$

for $i = 1, \ldots, g$. The efficient choice of $C_i = n_i/N$, $i = 1, \ldots, g$, forces the population variance, σ_x^2, to be proportional to N^{-2} and eliminates all terms of order $1/N$ in the variance of δ [297, pp. 26, 30].

The null hypothesis (H_0) states that equal probabilities are assigned to each of the

$$M = \frac{N!}{\prod_{i=1}^{g} n_i!}$$

possible, equally-likely allocations of the N objects to the g treatment groups, $S_1, \ldots, S_g$. Under H_0 the N multi-response measurements are exchangeable multivariate random variables.[5] The probability value associated with an observed value of δ, δ_o, is the probability under the null hypothesis (H_0) of observing a value of δ as extreme or more extreme than δ_o. Thus, an exact probability value for δ_o may be expressed as

$$P(\delta \leq \delta_o | H_0) = \frac{\text{number of } \delta \text{ values} \leq \delta_o}{M}\,.$$

When M is very large, an approximate probability value for δ may be obtained from a resampling procedure, where

$$P(\delta \leq \delta_o | H_0) = \frac{\text{number of } \delta \text{ values} \leq \delta_o}{L}$$

[5] A sufficient condition for a permutation statistical test is the exchangeability of the random variables. Sequences that are independent and identically distributed (i.i.d.) are always exchangeable, but so is sampling without replacement from a finite population. However, while i.i.d. implies exchangeability, exchangeability does not imply i.i.d. [150, 168, 217].

and L denotes the number of randomly sampled test statistic values. Typically, L is set to a large number to ensure accuracy.

Number of Resamplings Necessary

Exact permutation tests are restricted to relatively small samples, given the large number of possible permutations. On the other hand, resampling permutation tests are not limited by the size of the samples. Resampling permutation tests also have been shown to provide good approximations to exact probability values as a function of the number of resamplings considered. An early concern regarding the systematic use of resampling permutation tests was the speed of the computers used for calculating the probability values. Given modern high-speed computers, the question of computational speed is moot when probability values are not too small. The remaining question is: how many resamplings are required for a specified accuracy?

The number of resamplings suggested in books and articles on permutation methods is varied and likely dated due to previous limitations of computer speed and memory. Some authors have proposed as few as 100 resamplings to as many as 5,000; for example, see discussions by Dwass in 1957 [100]; Hope in 1968 [180]; Edwards in 1985 [110]; Jockel in 1986 [193]; Keller-McNulty and Higgins in 1987 [199]; Bailer in 1989 [16]; Kim, Nelson, and Startz in 1991 [216]; Manly in 1991 [258, pp. 32–35]; McQueen in 1992 [274]; Rickerts and Berry in 1994 [347]; Kennedy in 1995 [212]; Maxim in 1999 [265, p. 356]; Lunneborg in 2000 [256, pp. 210–213]; Good in 2001 [149, p. 47]; Higgins in 2004 [176]; and Edgington and Onghena in 2007 [109, pp. 40–41]. On the other hand, examples provided by Howell as recently as 2007 utilized as many as 10,000 resamplings [184, pp. 642–646]. Resampling computing packages such as Resampling Stats [14] and StatXact [15] typically use 10,000 resamplings as the default value.

The accuracy of a resampling probability value depends on both the probability value (P) and the number of resamplings (L). Confidence limits on the probability value can be obtained from the binomial distribution when L is large. The $1 - \alpha$ confidence limits of the binomial distribution are given by

$$\hat{P} \pm Z_{\alpha/2} \sqrt{\frac{P(1-P)}{L}} \,, \tag{2.4}$$

where P is the probability value in question and $\hat{P}$ denotes the estimated value of P. Define

$$x_i = \begin{cases} 1 & \text{if } \hat{P} \leq \hat{P}_{\text{o}} \,, \\ 0 & \text{otherwise} \,, \end{cases}$$

for $i = 1, \ldots, L$, where $\hat{P}_o$ denotes the observed value of $\hat{P}$. Then $\hat{P}$, the expected value of $\hat{P}$, the variance of $\hat{P}$, and the skewness of $\hat{P}$ are given by

$$\hat{P} = \frac{1}{L}\sum_{i=1}^{L} x_i \,,$$

$$\mathrm{E}[\hat{P}] = P \,,$$

$$\sigma_{\hat{P}}^2 = \frac{P(1-P)}{L} \,,$$

and

$$\gamma_{\hat{P}} = \frac{1-2P}{\sqrt{LP(1-P)}} \,,$$

respectively [195, p. 916]. If L is small and P is close to either 0 or 1, the skewness term $\gamma_{\hat{P}}$ becomes large and Eq. (2.4) may not be appropriate. For example, if $L = 100$ and $P = 0.01$,

$$\gamma_{\hat{P}} = \frac{1-2P}{\sqrt{LP(1-P)}} = \frac{1-2(0.01)}{\sqrt{100(0.01)(1-0.01)}} = 0.9849 \,.$$

Table 2.1 lists a selected number of probability values ($P = 0.50$, 0.25, 0.10, 0.05, and 0.01), a variety of resamplings ($L = 100$, 1000, 10,000, 1,000,000, and 100,000,000), computed skewness values, errors on the 95 % confidence limits determined from Eq. (2.4), and the simulated lower and upper errors on the 95 % confidence limits based on L resamplings and determined from the smallest value for which the cumulative binomial distribution is equal to or less than 0.025 and equal to or greater than 0.975, respectively. In general, as can be seen from Table 2.1, two additional orders of magnitude are required to increase accuracy by just one decimal place.

To illustrate the number of resamplings required to yield a predetermined number of decimal places of accuracy, given a known probability value, consider the interval-level data listed in Fig. 2.1.

The data listed in Fig. 2.1 are adapted from Berry, Mielke, and Mielke [38] and represent soil lead (Pb) quantities from two school districts in metropolitan New Orleans. Elevated Pb levels have been linked to a number of physiological, neurological, and endocrine effects in children, including difficulties in learning, perception, social behavior, and fine motor skills. The $n_1 = 20$ soil lead samples collected in District 1 yielded a mean value of $\bar{x}_1 = 203.9350$ mg/kg and the $n_2 = 20$ soil lead samples collected in District 2 yielded a mean value of $\bar{x}_2 = 1{,}661.7800$ mg/kg. There are

$$M = \frac{(n_1 + n_2)!}{n_1!\; n_2!} = \frac{(20+20)!}{20!\; 20!} = 137{,}846{,}528{,}820$$

Table 2.1 Five probability (P) values, four levels of resampling (L), skewness ($\gamma_{\hat{P}}$), and asymptotic and simulated errors on 95 % confidence limits; table adapted from Johnston, Berry, and Mielke [195, p. 917]

			Error on 95 % confidence limits		
P	L	$\gamma_{\hat{P}}$	Asymptotic	Lower	Upper
0.50	100	0.00	0.10	0.10	0.10
	10,000	0.00	0.010	0.010	0.010
	1,000,000	0.00	0.0010	0.0010	0.0010
	100,000,000	0.00	0.00010	0.00010	0.00010
0.25	100	0.11547	0.09	0.08	0.09
	10,000	0.01155	0.009	0.009	0.009
	1,000,000	0.00115	0.0009	0.0008	0.0008
	100,000,000	0.00012	0.00009	0.00008	0.00008
0.10	100	0.26667	0.06	0.05	0.06
	10,000	0.02667	0.006	0.006	0.006
	1,000,000	0.00267	0.0006	0.0006	0.0006
	100,000,000	0.00027	0.00006	0.00006	0.00006
0.05	100	0.41295	0.04	0.04	0.05
	10,000	0.04129	0.004	0.004	0.004
	1,000,000	0.00413	0.0004	0.0004	0.0004
	100,000,000	0.00041	0.00004	0.00004	0.00004
0.01	100	0.98494	0.02	0.01	0.02
	10,000	0.09849	0.002	0.002	0.002
	1,000,000	0.00985	0.0002	0.0002	0.0002
	100,000,000	0.00098	0.00002	0.00002	0.00002

possible permutations of the soil lead data listed in Fig. 2.1 to be considered. Under the null hypothesis of no difference between the two group means in the population, a Fisher–Pitman permutation F test [38] yields an exact two-sided probability value of

$$P(F \geq F_{\text{o}}|H_0) = \frac{\text{number of } F \text{ values} \geq F_{\text{o}}}{M}$$

$$= \frac{2{,}056{,}423{,}782}{137{,}846{,}528{,}820} = 0.0149182123$$

for the soil lead data listed in Fig. 2.1. Figure 2.2 summarizes the results for eight different resamplings of the data listed in Fig. 2.1 and the associated two-sided resampling probability values with $\alpha = 0.05$. Each of the probability values was generated using a common seed and the same pseudorandom number generator [197]. The last row of Fig. 2.2 contains the exact probability value based on all $M = 137{,}846{,}528{,}820$ possible permutations of the soil lead data listed in Fig. 2.1.

Fig. 2.1 Ordered soil Pb data in mg/kg from two school attendance districts in metropolitan New Orleans

n	District 1	District 2
1	16.0	4.7
2	34.3	10.8
3	34.6	35.7
4	57.6	53.1
5	63.1	75.6
6	88.2	105.5
7	94.2	200.4
8	111.8	212.8
9	112.1	212.9
10	139.0	215.2
11	165.6	257.6
12	176.7	347.4
13	216.2	461.9
14	221.1	566.0
15	276.7	984.0
16	362.8	1,040.0
17	373.4	1,306.0
18	387.1	1,908.0
19	442.2	3,559.0
20	706.0	21,679.0

Fig. 2.2 Comparison of eight resampled probability values with the exact probability value given in the last row, based on the soil lead data listed in Fig. 2.1

Resampling (L)	Probability ($\hat{P}$)
10	0.06
1,000	0.020
10,000	0.0110
100,000	0.01556
1,000,000	0.014946
10,000,000	0.0149302
100,000,000	0.01488510
1,000,000,000	0.014917218
Exact P value	0.0149182123

Given the results of the resampling probability analyses listed in Fig. 2.2, $L =$ 1,000,000 is recommended whenever three decimal places of accuracy are required. There are four reasons for promoting $L =$ 1,000,000 resamplings: accuracy, practicality, error, and consistency. First, inspection of Fig. 2.2 indicates that with an exact probability value of $P = 0.0149182123$ and $\alpha = 0.05$, $L =$ 1,000,000 resamplings is the minimum number of resamplings necessary to ensure three decimal places of accuracy. Second, given the speed of modern computers and the efficiency of resampling algorithms, such as the Mersenne Twister, $L =$ 1,000,000 resamplings can be used on a routine basis. Third, there is the potential for additional type I error, the magnitude of which is of concern when the number of resamplings (L) is very small. Fourth, some researchers object to the use of resampling statistics because different pseudorandom number generators and different seeds can produce widely varying results. This is certainly true when L is very small. For example, in Fig. 2.2, $L = 100$ yields a probability value of $P = 0.06$. Varying the seed with

$L = 100$ and the same pseudorandom number generator produced observed probability values ranging from $P = 0.01$ to $P = 0.11$. However, with $L = 1{,}000{,}000$, varying the seed produced no differences in the third decimal place.

When the number of possible arrangements (M) is very large and the exact probability value (P) is exceedingly small, a resampling permutation procedure may produce no δ values equal to or less than δ_o, even with $L = 1{,}000{,}000$, yielding an approximate resampling probability value of $P = 0.00$. In such cases, moment-approximation permutation procedures based on fitting the first three exact moments of the discrete permutation distribution to a Pearson type III distribution provide approximate probability values, as detailed in Chap. 1, Sect. 1.2.2; see also references [284] and [300].

An Index of Agreement

It is oftentimes desirable to have an index of the amount of agreement among response measurement scores within g treatment groups. A useful measure for this purpose is a chance-corrected within-group coefficient of agreement given by

$$\Re = 1 - \frac{\delta}{\mu_\delta} \,, \tag{2.5}$$

where μ_δ is the arithmetic average of the M δ values calculated on all possible arrangements of the observed response measurement scores, given by

$$\mu_\delta = \frac{1}{M} \sum_{i=1}^{M} \delta_i \,. \tag{2.6}$$

$\Re$ is a chance-corrected measure of agreement since $\mathrm{E}[\Re | H_0] = 0$.[6] Because μ_δ is a constant under H_0, the permutation distributions of δ and $\Re$ are equivalent, viz.,

$$P\big(\delta \leq \delta_o | H_0\big) = P\left(\Re \geq \Re_o | H_0\right) \,,$$

where

$$\Re_o = 1 - \frac{\delta_o}{\mu_\delta}$$

and δ_o and $\Re_o$ denote the observed values of δ and $\Re$, respectively. Possible values of $\Re$ range from slightly negative values to a maximum of $\Re = +1$ for the extreme

[6] As will be shown in Chap. 3, $\Re$ may also be interpreted as a chance-corrected measure of effect size.

case when all response measurements on objects within each of the g classified treatment groups are identical, i.e., $\delta = 0$.

The generalized Minkowski distance function, $\Delta(x, y)$, as defined in Eq. (2.1) on p. 30, determines the analysis space of the MRPP test statistic, δ. The data space in question for almost all statistical analyses is an ordinary Euclidean distance space. If the distance function of the MRPP test statistic is based on $p = 2$ and $v = 1$, then the data and analysis spaces are congruent, so that the resulting statistical analyses represent the data in question. Unfortunately, commonly used statistical analyses based on the arithmetic mean, such as Student's two-sample t test and Fisher's one-way analysis of variance, are based on $p = v = 2$, yielding a non-metric squared-distance analysis space that is not congruent with the data space. The difference between the data and analysis spaces associated with the most popular statistical analyses is a reason that problems occur with what should be routine analyses. Examples illustrating this problem are given elsewhere; see, for example, references [41, pp. 404–410] and [297, pp. 50–53]. Any statistical analysis is questionable when the data and analysis spaces are not congruent.

2.2.1 Chance-Corrected Agreement Measures

Chance-corrected measures yield values that are interpreted as a proportion above that expected by chance alone. Chance-corrected agreement measures provide clear and meaningful interpretations of the amount of, or lack of, agreement present in the data. In general, chance-corrected measures of agreement, such as $\Re$, are equal to $+1$ when perfect agreement among the response measurement scores occurs, 0 when agreement is equal to that expected under independence, and negative when agreement among the response measurement scores is less than that expected by chance. For example, define a chance-corrected measure such that

$$A_i = 100 \left(\frac{O_i - E_i}{N - E_i} \right) ,$$

where O_i and E_i denote the Observed (earned) and Expected (chance) score from purely guessing, respectively, on a multiple-choice examination with N questions for the ith student in a class of m students [175, p. 912].

Thus, on a 50-question multiple-choice examination with five choices per question, chance would indicate that a student could answer $50 \times 0.20 = 10$ questions correctly simply by guessing. If a student answered only eight questions correctly, then a chance-corrected measure of agreement would yield a grade of

$$A = 100 \left(\frac{8 - 10}{50 - 10} \right) = 100 \left(\frac{-2}{40} \right) = -5 ,$$

since the score was less than expected by chance, i.e., only eight of 50 questions were answered correctly. The lowest grade would occur when a student answered all 50 questions incorrectly, yielding a score of

$$A = 100\left(\frac{0-10}{50-10}\right) = 100\left(\frac{-10}{40}\right) = -25\ .$$

Note that while a student with the highest possible score of 50 correct answers would score

$$A = 100\left(\frac{50-10}{50-10}\right) = 100\left(\frac{40}{40}\right) = 100\ ,$$

the lowest possible score is −25, not −100. Thus, the distributions of chance-corrected measures are usually asymmetric.

Since the mean value of $\Re$ under H_0 is 0, homogeneity of within-classified-group response measurements is associated with $\Re > 0$, and heterogeneity of within-classified-group response measurements is associated with $\Re \leq 0$ [28]. The distribution of $\Re$ is usually asymmetric and the upper and lower bounds depend on both the nature of the data and the structure of δ. The degree of homogeneity or heterogeneity depends on the discrete permutation distribution of $\Re$. If large values of $n_1, \ldots, n_g$ and N are involved, a very small value of $P(\delta \leq \delta_o|H_0)$ may be associated with a small positive observed value of $\Re$, say $\Re_o$. Conversely, with small values of $n_1, \ldots, n_g$ and N, a large value of $\Re_o$ may be associated with a relatively large value of $P(\delta \leq \delta_o|H_0)$.

2.2.2 Example Univariate MRPP Analysis with $v = 2$

Although multi-response permutation procedures were originally designed for analyzing multivariate response measurement scores, they can also be used for analyzing univariate data. Consider a comparison between two mutually exclusive groups of objects, S_1 and S_2, where a single response measurement, x, has been obtained from each object. For this example, there is $r = 1$ response measurement score for each object, $g = 2$ disjoint groups, and a total of $N = 6$ objects with $n_1 = 2$ and $n_2 = 4$ in treatment groups S_1 and S_2, respectively. Suppose that the $n_1 = 2$ observed response measurement scores for treatment group S_1 are $\{5, 4\}$ and the $n_2 = 4$ response measurement scores for treatment group S_2 are $\{2, 3, 7, 9\}$. The treatment-group sizes and the response measurement scores are deliberately kept small to simplify the example analysis. The treatment-group sizes and the univariate response measurement scores are listed in Fig. 2.3.

For this example analysis, let $v = 2, p = 2, r = 1$,

$$C_1 = \frac{n_1}{N} = \frac{2}{6}\ , \quad \text{and} \quad C_2 = \frac{n_2}{N} = \frac{4}{6}\ ,$$

Fig. 2.3 Example data with $g = 2$, $r = 1$, $n_1 = 2$, $n_2 = 4$, and $N = n_1 + n_2 = 6$

Group	Object	Value
S_1	ω_1	5
	ω_2	4
S_2	ω_3	2
	ω_4	3
	ω_5	7
	ω_6	9

so that the S_1 and S_2 treatment groups are weighted proportional to their group sizes of $n_1 = 2$ and $n_2 = 4$, respectively. For univariate response measurement scores with $r = 1$, Eq. (2.1) on p. 30 reduces to

$$\Delta(j,k) = \left(|x_j - x_k|^p\right)^{v/p} . \tag{2.7}$$

Thus, for treatment group S_1 with $n_1 = 2$ objects, $p = 2$, and $v = 2$, the generalized Minkowski distance function yields

$$\Delta(1,2) = \left(|5-4|^2\right)^{2/2} = 1.00 ,$$

and for treatment group S_2 with $n = 4$ objects, the generalized Minkowski distance function yields

$$\Delta(3,4) = \left(|2-3|^2\right)^{2/2} = 1.00 ,$$

$$\Delta(3,5) = \left(|2-7|^2\right)^{2/2} = 25.00 ,$$

$$\Delta(3,6) = \left(|2-9|^2\right)^{2/2} = 49.00 ,$$

$$\Delta(4,5) = \left(|3-7|^2\right)^{2/2} = 16.00 ,$$

$$\Delta(4,6) = \left(|3-9|^2\right)^{2/2} = 36.00 ,$$

and

$$\Delta(5,6) = \left(|7-9|^2\right)^{2/2} = 4.00 .$$

Then following Eq. (2.3) on p. 31, the average distance-function values for all distinct pairs of objects in treatment groups S_i, $i = 1, 2$, are

$$\xi_1 = \binom{n_1}{2}^{-1} \left[\Delta(1,2) \right] = \binom{2}{2}^{-1} (1.00) = 1.00$$

and

$$\xi_2 = \binom{n_2}{2}^{-1} \left[\Delta(3,4) + \Delta(3,5) + \Delta(3,6) + \Delta(4,5) + \Delta(4,6) + \Delta(5,6) \right]$$
$$= \binom{4}{2}^{-1} (1.00 + 25.00 + 49.00 + 16.00 + 36.00 + 4.00) = 21.8333 \,.$$

Following Eq. (2.2) on p. 31, the observed weighted mean of the ξ_1 and ξ_2 values, based on $v = 2$ and $C_i = n_i/N$ for $i = 1, 2$ is

$$\delta_o = C_1\xi_1 + C_2\xi_2 = \left(\frac{2}{6}\right)(1.00) + \left(\frac{4}{6}\right)(21.8333) = 14.8889 \,.$$

Smaller values of δ_o indicate a concentration of response measurement scores within the g treatment groups, whereas larger values of δ_o indicate a lack of concentration between response measurement scores among the g treatment groups [301]. The $N = 6$ objects can be partitioned into $g = 2$ treatment groups, S_1 and S_2, respectively, with $n_1 = 2$ and $n_2 = 4$ response measurement scores preserved in

$$M = \frac{N!}{n_1!\, n_2!} = \frac{6!}{2!\, 4!} = 15$$

possible, equally-likely ways. The $M = 15$ possible arrangements of the observed data in Fig. 2.3, along with the corresponding ξ_1, ξ_2, and δ values, are listed in Table 2.2 and ordered by the δ values from lowest to highest. The observed MRPP test statistic, $\delta_o = 14.8889$, obtained from the realized arrangement,

$$\{5, 4\} \quad \{2, 3, 7, 9\} \,,$$

(Order 9 in Table 2.2) is not unusual since five of the remaining δ values (δ_{11} to δ_{15}) exceed the observed value of $\delta_o = 14.8889$ and 10 values of δ (δ_1 to δ_{10}) are equal to or less than the observed value. If all arrangements of the $N = 6$ observed response measurement scores listed in Fig. 2.3 occur with equal chance, the exact probability value of $\delta_o = 14.8889$ computed on the $M = 15$ possible arrangements of the observed data with $n_1 = 2$ and $n_2 = 4$ response measurement scores preserved for

Table 2.2 Permutations of the observed data in Fig. 2.3 for treatment groups S_1 and S_2 with values for ξ_1, ξ_2, and δ based on $v = 2$, ordered by values of δ from lowest to highest

Order	S_1	S_2	ξ_1	ξ_2	δ
1	{7, 9}	{2, 5, 3, 4}	4.0000	3.3333	3.5556
2	{2, 3}	{5, 4, 7, 9}	1.0000	9.8333	6.8889
3	{2, 4}	{5, 3, 7, 9}	4.0000	13.3333	10.2222
4	{5, 9}	{2, 3, 4, 7}	16.0000	9.3333	11.5556
5	{3, 4}	{2, 5, 7, 9}	1.0000	17.8333	12.2222
6	{2, 5}	{3, 4, 7, 9}	9.0000	15.1667	13.1111
7	{5, 3}	{2, 4, 7, 9}	4.0000	19.3333	14.2222
8	{5, 7}	{2, 3, 4, 9}	4.0000	19.3333	14.2222
9	{5, 4}	{2, 3, 7, 9}	1.0000	21.8333	14.8889
10	{4, 9}	{2, 5, 3, 7}	25.0000	9.8333	14.8889
11	{4, 7}	{2, 5, 3, 9}	9.0000	19.1667	15.7778
12	{3, 7}	{2, 5, 4, 9}	16.0000	17.3333	16.8889
13	{2, 7}	{5, 3, 4, 9}	25.0000	13.8333	17.5556
14	{3, 9}	{2, 5, 4, 7}	36.0000	8.6667	17.7778
15	{2, 9}	{5, 3, 4, 7}	49.0000	5.8333	20.2222

each arrangement is

$$P(\delta \le \delta_o | H_0) = \frac{\text{number of } \delta \text{ values} \le \delta_o}{M} = \frac{10}{15} = 0.6667 \,.$$

For comparison, a conventional Student two-sample pooled t test calculated on the $N = 6$ response measurement scores listed in Fig. 2.3 yields an observed value of $t_o = -0.3004$. Assuming independence, normality, and homogeneity of variance, t is approximately distributed as Student's t under the null hypothesis with $N - 2 = 6 - 2 = 4$ degrees of freedom. Under the null hypothesis, the observed value of $t_o = -0.3004$ yields an approximate two-sided probability value of $P = 0.7789$.

Following Eq. (2.6) on p. 37, the exact average value of the $M = 15$ δ values listed in Table 2.2 is $\mu_\delta = 13.60$. Thus, the observed chance-corrected coefficient of agreement, following Eq. (2.5) on p. 37, is

$$\Re_o = 1 - \frac{\delta_o}{\mu_\delta} = 1 - \frac{14.8889}{13.60} = -0.0948 \,,$$

indicating that within-group agreement is well below that expected by chance.

2.2.3 Example Univariate MRPP Analysis with $v = 1$

Permutation statistical tests and measures are data-dependent, distribution-free, and non-parametric; consequently, they require no distributional assumptions and make no estimates of population parameters. Thus, it is not necessary to set $v = 2$ and to square the response-measurement differences between objects. While conventional tests and measures that assume normality must estimate the mean and variance, μ_x and σ_x^2, of the normal distribution, both of which are based on squared deviations from the mean, permutation tests and measures do not assume normality and are not restricted to $v = 2$, which is not a metric distance function. A distance function based on $v = 1$ is an attractive alternative to $v = 2$ as it is a metric distance function, satisfies the triangle inequality, is robust to extreme values, provides an easy-to-understand ordinary Euclidean distance between objects, and ensures that the data and analysis spaces are congruent [284–287, 289, 295]. In addition, choosing $v = 1$ over $v = 2$ can make a substantial difference in the results of an MRPP analysis; see, for example, a discussion by Mielke and Berry in 2007 [297, pp. 45–50].

To illustrate the computation of δ with $v = 1$, consider the same finite sample of $N = 6$ objects listed in Fig. 2.3 on p. 40 and let S_1 and S_2 denote an exhaustive partitioning of the $N = 6$ objects into $g = 2$ disjoint treatment groups. As previously, let S_1 consist of $n_1 = 2$ objects, each with a single response measurement, and let S_2 consist of $n_2 = 4$ objects, each with a single response measurement.

Given the univariate data listed in Fig. 2.3, let $r = 1, p = 2$,

$$C_1 = \frac{n_1}{N} = \frac{2}{6} \,, \quad \text{and} \quad C_2 = \frac{n_2}{N} = \frac{4}{6} \,,$$

but in this case set $v = 1$ instead of $v = 2$, employing ordinary Euclidean distance instead of squared Euclidean distance between objects. Following Eq. (2.7) on p. 40 for treatment group S_1 with $n_1 = 2$ objects, $p = 2$, and $v = 1$, the generalized Minkowski distance function yields

$$\Delta(1,2) = \left(|(5-4|^2\right)^{1/2} = 1.00 \,,$$

and for treatment group S_2 with $n = 4$ objects, the generalized Minkowski distance function yields

$$\Delta(3,4) = \left(|2-3|^2\right)^{1/2} = 1.00 \,,$$

$$\Delta(3,5) = \left(|2-7|^2\right)^{1/2} = 5.00 \,,$$

$$\Delta(3,6) = \left(|2-9|^2\right)^{1/2} = 7.00 \,,$$

$$\Delta(4,5) = \left(|3-7|^2\right)^{1/2} = 4.00 \,,$$

$$\Delta(4,6) = \left(|3-9|^2\right)^{1/2} = 6.00 \,,$$

and

$$\Delta(5,6) = \left(|7-9|^2\right)^{1/2} = 2.00\,.$$

Then following Eq. (2.3) on p. 31, the average distance-function values for all distinct pairs of objects in treatment group S_i, $i = 1, 2$, are

$$\xi_1 = \binom{n_1}{2}^{-1} \left[\Delta(1,2)\right] = \binom{2}{2}^{-1} (1.00) = 1.00$$

and

$$\xi_2 = \binom{n_2}{2}^{-1} \left[\Delta(3,4) + \Delta(3,5) + \Delta(3,6) + \Delta(4,5) + \Delta(4,6) + \Delta(5,6)\right]$$

$$= \binom{4}{2}^{-1} (1.00 + 5.00 + 7.00 + 4.00 + 6.00 + 2.00) = 4.1667\,.$$

Following Eq. (2.2) on p. 31, the observed weighted mean of the ξ_1 and ξ_2 values, based on $C_i = n_i/N$ for $i = 1, 2$ is

$$\delta_o = C_1\xi_1 + C_2\xi_2 = \left(\frac{2}{6}\right)(1.00) + \left(\frac{4}{6}\right)(4.1667) = 3.1111\,.$$

As in the previous MRPP example with $v = 2$, the $N = 6$ objects can be partitioned into $g = 2$ treatment groups, S_1 and S_2, with $n_1 = 2$ and $n_2 = 4$ response measurement scores preserved for each arrangement of the observed data in

$$M = \frac{N!}{n_1!\,n_2!} = \frac{6!}{2!\,4!} = 15$$

possible, equally-likely ways. The $M = 15$ possible arrangements of the observed data in Fig. 2.3, along with the corresponding ξ_1, ξ_2, and δ values, are listed in Table 2.3 and ordered by the δ values from lowest to highest. The observed MRPP test statistic, $\delta_o = 3.1111$, obtained from the realized arrangement,

$$\{5,4\} \quad \{2,3,7,9\}\,,$$

(Order 5 in Table 2.3) is not unusual since eight of the remaining δ values (δ_8 to δ_{15}) exceed the observed value of $\delta_o = 3.1111$ and seven values of δ (δ_1 to δ_7) are equal to or less than the observed value. If all arrangements of the $N = 6$ observed

Table 2.3 Permutations of the observed data in Fig. 2.3 for treatment groups S_1 and S_2 with values for ξ_1, ξ_2, and δ based on $v = 1$, ordered by values of δ from lowest to highest

Order	S_1	S_2	ξ_1	ξ_2	δ
1	{7, 9}	{2, 5, 3, 4}	2.0000	1.6667	1.7778
2	{2, 3}	{5, 4, 7, 9}	1.0000	2.8333	2.2222
3	{2, 4}	{5, 3, 7, 9}	2.0000	3.3333	2.8889
4	{3, 4}	{2, 5, 7, 9}	1.0000	3.8333	2.8889
5	{5, 4}	{2, 3, 7, 9}	1.0000	4.1667	3.1111
6	{5, 7}	{2, 3, 4, 9}	2.0000	3.6667	3.1111
7	{5, 9}	{2, 3, 4, 7}	4.0000	2.6667	3.1111
8	{2, 5}	{3, 4, 7, 9}	3.0000	3.5000	3.3333
9	{5, 3}	{2, 4, 7, 9}	2.0000	4.0000	3.3333
10	{4, 7}	{2, 5, 3, 9}	3.0000	3.8333	3.5556
11	{4, 9}	{2, 5, 3, 7}	5.0000	2.8333	3.5556
12	{2, 7}	{5, 3, 4, 9}	5.0000	3.1667	3.7778
13	{3, 7}	{2, 5, 4, 9}	4.0000	3.6667	3.7778
14	{2, 9}	{5, 3, 4, 7}	7.0000	2.1667	3.7778
15	{3, 9}	{2, 5, 4, 7}	6.0000	2.6667	3.7778

response measurement scores listed in Fig. 2.3 occur with equal chance, the exact probability value of $\delta_o = 3.1111$ computed on the $M = 15$ possible arrangements of the observed data with $n_1 = 2$ and $n_2 = 4$ response measurement scores preserved for each arrangement is

$$P(\delta \leq \delta_o | H_0) = \frac{\text{number of } \delta \text{ values} \leq \delta_o}{M} = \frac{7}{15} = 0.4667 \ .$$

For comparison, for the univariate data listed in Fig. 2.3 the exact probability value based on $v = 2$, $M = 15$, and $C_i = n_i/N$ for $i = 1, 2$ in the previous example is $P = 0.6667$. No comparison is made with the conventional Student two-sample t test as Student's t test is undefined for $v = 1$.

Following Eq. (2.6) on p. 37, the exact average value of the $M = 15$ δ values listed in Table 2.3 is $\mu_\delta = 3.20$. Thus, the observed chance-corrected coefficient of agreement, following Eq. (2.5) on p. 37, is

$$\Re_o = 1 - \frac{\delta_o}{\mu_\delta} = 1 - \frac{14.8889}{3.20} = +0.0278 \ ,$$

indicating very little within-group agreement above that expected by chance.

2.2.4 Example Bivariate MRPP Analysis with $v = 2$

In this second example, bivariate response measurement scores are used for simplicity to demonstrate a multivariate MRPP analysis. To illustrate the computation of MRPP with bivariate response measurement scores for each object, consider a finite

Fig. 2.4 Example data with $g = 2$, $r = 2$, $n_1 = 4$, $n_2 = 3$, and $N = n_1 + n_2 = 7$

Group	Object	Values	
		x_1	x_2
S_1	ω_1	5	1
S_1	ω_2	4	6
S_1	ω_3	5	2
S_1	ω_4	6	3
S_2	ω_5	2	3
S_2	ω_6	3	4
S_2	ω_7	2	4

sample of $N = 7$ objects and let S_1 and S_2 denote an exhaustive partitioning of the N objects into $g = 2$ disjoint treatment groups. Further, let S_1 consist of $n_1 = 4$ objects with $r = 2$ commensurate response measurement scores (x_{1i} and x_{2i}) on each object for $i = 1, \ldots, 4$, with $x_1' = (5,\ 1)$, $x_2' = (4,\ 6)$, $x_3' = (5,\ 2)$, and $x_4' = (6,\ 3)$, and let S_2 consist of $n_2 = 3$ objects with $r = 2$ commensurate response measurement scores (x_{1i} and x_{2i}) on each object for $i = 1, 2, 3$ with $x_5' = (2,\ 3)$, $x_6' = (3,\ 4)$, and $x_7' = (2,\ 4)$. The treatment group sizes and the response measurement scores are deliberately kept small to simplify the example analysis. The bivariate response measurement scores for the $N = 7$ objects are listed in Fig. 2.4.

For this example analysis, let $v = 2, p = 2, r = 2$,

$$C_1 = \frac{n_1}{N} = \frac{4}{7}\,, \quad \text{and} \quad C_2 = \frac{n_2}{N} = \frac{3}{7}\,,$$

so that the S_1 and S_2 treatment groups are weighted proportional to their group sizes of $n_1 = 4$ and $n_2 = 3$, respectively. Following Eq. (2.1) on p. 30 for treatment group S_1 with $n_1 = 4$ objects, $p = 2$, and $v = 2$, the generalized Minkowski distance function yields

$$\Delta(1,2) = \left(|5-4|^2 + |1-6|^2\right)^{2/2} = 26.00\,,$$

$$\Delta(1,3) = \left(|5-5|^2 + |1-2|^2\right)^{2/2} = 1.00\,,$$

$$\Delta(1,4) = \left(|5-6|^2 + |1-3|^2\right)^{2/2} = 5.00\,,$$

$$\Delta(2,3) = \left(|4-5|^2 + |6-2|^2\right)^{2/2} = 17.00\,,$$

$$\Delta(2,4) = \left(|4-6|^2 + |6-3|^2\right)^{2/2} = 13.00\,,$$

and

$$\Delta(3,4) = \left(|5-6|^2 + |2-3|^2\right)^{2/2} = 2.00\ ,$$

and for treatment group S_2 with $n_2 = 3$ objects, the generalized Minkowski distance function yields

$$\Delta(5,6) = \left(|2-3|^2 + |3-4|^2\right)^{2/2} = 2.00\ ,$$

$$\Delta(5,7) = \left(|2-2|^2 + |3-4|^2\right)^{2/2} = 1.00\ ,$$

and

$$\Delta(6,7) = \left(|3-2|^2 + |4-4|^2\right)^{2/2} = 1.00\ .$$

Then following Eq. (2.3) on p. 31, the average distance-function values for all distinct pairs of objects in treatment group S_i, $i = 1, 2$, are

$$\xi_1 = \binom{n_1}{2}^{-1} \left[\Delta(1,2) + \Delta(1,3) + \Delta(1,4) + \Delta(2,3) + \Delta(2,4) + \Delta(3,4)\right]$$

$$= \binom{4}{2}^{-1} (26.00 + 1.00 + 5.00 + 17.00 + 13.00 + 2.00) = 10.6667$$

and

$$\xi_2 = \binom{n_2}{2}^{-1} \left[\Delta(5,6) + \Delta(5,7) + \Delta(6,7)\right]$$

$$= \binom{3}{2}^{-1} (2.00 + 1.00 + 1.00) = 1.3333\ .$$

Following Eq. (2.2) on p. 31, the observed weighted mean of the ξ_1 and ξ_2 values, based on $v = 2$ and $C_i = n_i/N$ for $i = 1, 2$ is

$$\delta_o = C_1\xi_1 + C_2\xi_2 = \left(\frac{4}{7}\right)(10.6667) + \left(\frac{3}{7}\right)(1.3333) = 6.6667\ .$$

The $N = 7$ objects can be partitioned into $g = 2$ treatment groups, S_1 and S_2, with $n_1 = 4$ and $n_2 = 3$ response measurement scores preserved for each arrangement of the observed data in

$$M = \frac{N!}{n_1!\, n_2!} = \frac{7!}{4!\, 3!} = 35$$

possible, equally-likely ways. The $M = 35$ possible arrangements of the observed bivariate data in Fig. 2.4, along with the corresponding ξ_1, ξ_2, and δ values, are listed in Table 2.4 and ordered by the δ values from lowest to highest. The observed MRPP test statistic, $\delta_o = 6.6667$, obtained from the realized arrangement,

$$\{(5,1)(4,6)(5,2)(6,3)\} \quad \{(2,3)(3,4)(2,4)\} ,$$

(Order 3 in Table 2.4) is unusual since 32 of the remaining δ values (δ_4 to δ_{35}) exceed the observed value of $\delta_o = 6.6667$ and only two values of δ are less than the observed value: $\delta_1 = 4.0000$ and $\delta_2 = 6.4762$. If all arrangements of the $N = 7$ observed bivariate response measurement scores listed in Fig. 2.4 occur with equal chance, the exact probability value of $\delta_o = 6.6667$ computed on the $M = 35$ possible arrangements of the observed data with $n_1 = 4$ and $n_2 = 3$ response measurement scores preserved for each arrangement is

$$P(\delta \leq \delta_o | H_0) = \frac{\text{number of } \delta \text{ values} \leq \delta_o}{M} = \frac{3}{35} = 0.0857 .$$

A conventional Hotelling two-sample T^2 test is given by

$$T^2 = \frac{n_1 n_2}{N} (\bar{\mathbf{y}}_1 - \bar{\mathbf{y}}_2)' \, \mathbf{S}^{-1} (\bar{\mathbf{y}}_1 - \bar{\mathbf{y}}_2) , \tag{2.8}$$

where $\bar{\mathbf{y}}_1$ and $\bar{\mathbf{y}}_2$ denote vectors of mean differences between treatment groups S_1 and S_2, n_1 and n_2 are the number of interval-level multivariate response measurement scores in treatment groups S_1 and S_2, and $\mathbf{S}$ is a pooled variance–covariance matrix.

For the example data listed in Fig. 2.4, $\bar{y}_{11} = 5.00$, $s_{11}^2 = 0.6167$, $\bar{y}_{12} = 3.00$, $s_{12}^2 = 4.6667$, $\text{cov}(1,2)_1 = -1.00$, $\bar{y}_{21} = 2.3333$, $s_{21}^2 = 0.3333$, $\bar{y}_{22} = 3.6667$, $s_{22}^2 = 0.3333$, and $\text{cov}(1,2)_2 = +0.1667$. Then, $\bar{\mathbf{y}}_1 = \bar{y}_{11} - \bar{y}_{21} = 5.00 - 2.3333 = +2.6667$ and $\bar{\mathbf{y}}_2 = \bar{y}_{12} - \bar{y}_{22} = 3.00 - 3.6667 = -0.6667$.

The variance–covariance matrices for treatment groups S_1 and S_2 in Fig. 2.4 are

$$\hat{\boldsymbol{\Sigma}}_1 = \begin{bmatrix} 0.6667 & -1.0000 \\ -1.0000 & 4.6667 \end{bmatrix} \quad \text{and} \quad \hat{\boldsymbol{\Sigma}}_2 = \begin{bmatrix} 0.3333 & +0.1667 \\ +0.1667 & 0.3333 \end{bmatrix} ,$$

Table 2.4 Permutations of the observed data set in Fig. 2.4 for treatment groups S_1 and S_2 with values for ξ_1, ξ_2, and δ based on $v = 2$, ordered by values of δ from lowest to highest

Order	S_1	S_2	ξ_1	ξ_2	δ
1	{(4, 6) (2, 3) (3, 4) (2, 4)}	{(5, 1) (5, 2) (6, 3)}	5.0000	2.6667	4.0000
2	{(5, 1) (5, 2) (6, 3) (2, 3)}	{(4, 6) (3, 4) (2, 4)}	7.8333	4.6667	6.4762
3	{(5, 1) (4, 6) (5, 2) (6, 3)}	{(2, 3) (3, 4) (2, 4)}	10.6667	1.3333	6.6667
4	{(5, 1) (5, 2) (6, 3) (3, 4)}	{(4, 6) (2, 3) (2, 4)}	6.5000	7.3333	6.8571
5	{(5, 1) (5, 2) (6, 3) (2, 4)}	{(4, 6) (2, 3) (3, 4)}	9.3333	6.6667	8.1905
6	{(4, 6) (6, 3) (3, 4) (2, 4)}	{(5, 1) (5, 2) (2, 3)}	9.0000	8.0000	8.5714
7	{(5, 1) (2, 3) (3, 4) (2, 4)}	{(4, 6) (5, 2) (6, 3)}	8.0000	10.6667	9.1429
8	{(5, 1) (5, 2) (2, 3) (2, 4)}	{(4, 6) (6, 3) (3, 4)}	9.3333	9.3333	9.3333
9	{(5, 2) (2, 3) (3, 4) (2, 4)}	{(5, 1) (4, 6) (6, 3)}	5.8333	14.6667	9.6190
10	{(4, 6) (6, 3) (2, 3) (2, 4)}	{(5, 1) (5, 2) (3, 4)}	11.3333	7.3333	9.6190
11	{(4, 6) (5, 2) (6, 3) (3, 5)}	{(5, 1) (2, 3) (2, 4)}	9.1667	10.6667	9.8095
12	{(4, 6) (5, 2) (3, 4) (2, 4)}	{(5, 1) (6, 3) (2, 3)}	8.6667	11.3333	9.8095
13	{(5, 1) (5, 2) (2, 3) (3, 4)}	{(4, 6) (6, 3) (2, 4)}	7.8333	12.6667	9.9048
14	{(4, 6) (5, 2) (2, 3) (2, 4)}	{(5, 1) (6, 3) (3, 4)}	10.3333	9.3333	9.9048
15	{(4, 6) (6, 3) (2, 3) (3, 4)}	{(5, 1) (5, 2) (2, 4)}	9.8333	10.6667	10.1905
16	{(5, 1) (4, 6) (6, 3) (3, 4)}	{(5, 2) (2, 3) (2, 4)}	12.0000	8.0000	10.2857
17	{(5, 1) (4, 6) (2, 3) (2, 4)}	{(5, 2) (6, 3) (3, 4)}	13.1667	6.6667	10.3810
18	{(4, 6) (5, 2) (6, 3) (2, 4)}	{(5, 1) (2, 3) (3, 4)}	11.6667	9.3333	10.6667
19	{(6, 3) (2, 3) (3, 4) (2, 4)}	{(5, 1) (4, 6) (5, 2)}	7.8333	14.6667	10.7619
20	{(5, 1) (4, 6) (3, 4) (2, 4)}	{(5, 2) (6, 3) (2, 3)}	11.8333	9.3333	10.7619
21	{(5, 1) (6, 3) (2, 3) (2, 4)}	{(4, 6) (5, 2) (3, 4)}	11.6667	10.0000	10.9524
22	{(4, 6) (5, 2) (2, 3) (3, 4)}	{(5, 1) (6, 3) (2, 4)}	9.1667	13.3333	10.9524
23	{(5, 1) (6, 3) (2, 3) (3, 4)}	{(4, 6) (5, 2) (2, 4)}	9.8333	12.6667	11.0476
24	{(5, 1) (5, 2) (3, 4) (2, 4)}	{(2, 6) (6, 3) (2, 3)}	9.0000	14.0000	11.1429
25	{(5, 1) (4, 6) (6, 3) (2, 4)}	{(5, 2) (2, 3) (3, 4)}	14.5000	6.6667	11.1429
26	{(4, 6) (5, 2) (6, 3) (2, 3)}	{(5, 1) (3, 4) (2, 4)}	11.8333	10.6667	11.3333
27	{(5, 1) (4, 6) (6, 3) (2, 3)}	{(5, 2) (3, 4) (2, 4)}	14.3333	7.3333	11.3333
28	{(5, 1) (4, 6) (2, 3) (3, 4)}	{(5, 2) (6, 3) (2, 4)}	12.0000	10.6667	11.4286
29	{(5, 1) (4, 6) (5, 2) (3, 4)}	{(6, 3) (2, 3) (2, 4)}	11.6667	11.3333	11.5238
30	{(5, 1) (4, 6) (5, 2) (2, 3)}	{(6, 3) (3, 4) (2, 4)}	13.3333	9.3333	11.6190
31	{(5, 1) (6, 3) (3, 4) (2, 4)}	{(4, 6) (5, 2) (2, 3)}	10.6667	13.3333	11.8095
32	{(5, 2) (6, 3) (2, 3) (2, 4)}	{(5, 1) (4, 6) (3, 4)}	9.8333	14.6667	11.9048
33	{(5, 1) (4, 6) (5, 2) (2, 4)}	{(6, 3) (2, 3) (3, 4)}	13.8333	9.3333	11.9048
34	{(5, 2) (6, 3) (2, 3) (3, 4)}	{(5, 1) (4, 6) (2, 4)}	8.0000	17.3333	12.0000
35	{(5, 2) (6, 3) (3, 4) (2, 4)}	{(5, 1) (4, 6) (2, 3)}	8.5000	17.3333	12.2857

respectively, and the pooled variance–covariance matrix and its inverse are

$$\mathbf{S} = \begin{bmatrix} 0.5333 & -0.5333 \\ -0.5333 & 2.9333 \end{bmatrix} \quad \text{and} \quad \mathbf{S}^{-1} = \begin{bmatrix} +2.9167 & +0.4167 \\ +0.4167 & +0.4167 \end{bmatrix} ,$$

respectively.[7]

Following Eq. (2.8), the observed value of Hotelling's T^2 is

$$T_o^2 = \frac{n_1 n_2}{N} (\bar{\mathbf{y}}_1 - \bar{\mathbf{y}}_2)' \, \mathbf{S}^{-1} \, (\bar{\mathbf{y}}_1 - \bar{\mathbf{y}}_2)$$

$$= \frac{(4)(3)}{7} \begin{bmatrix} +2.6667 & -0.6667 \end{bmatrix} \begin{bmatrix} +2.2917 & +0.4167 \\ +0.4167 & +0.4167 \end{bmatrix} \begin{bmatrix} +2.6667 \\ -0.6667 \end{bmatrix}$$

$$= (1.7143)(15.00) = 25.7143$$

and the observed F-ratio for Hotelling's T^2 is

$$F_o = \frac{N - r - 1}{r(N - r)} T_o^2 = \frac{7 - 2 - 1}{2(7 - 2)} (25.7145) = 10.2858 \, .$$

Assuming independence, normality, and homogeneity of variance, F is approximately distributed as Snedecor's F under the null hypothesis with $\nu_1 = r = 2$ and $\nu_2 = N - r - 1 = 7 - 2 - 1 = 4$ degrees of freedom. Under the null hypothesis, the observed value of $F_o = 10.2858$ yields an approximate probability value of $P = 0.0265$. While there is a considerable difference between the exact probability value of $P = 0.0857$ and the approximate probability value of $P = 0.0265$, it is not surprising, as Hotelling's T^2 test was not designed for samples as small as $n_1 = 4$ and $n_2 = 3$.

Following Eq. (2.6) on p. 37, the exact average value of the $M = 35$ δ values listed in Table 2.4 is $\mu_\delta = 10.0952$. Thus, the observed chance-corrected coefficient of agreement, following Eq. (2.5) on p. 37, is

$$\Re_o = 1 - \frac{\delta_o}{\mu_\delta} = 1 - \frac{6.6667}{10.0952} = +0.3396 \, ,$$

indicating approximately 34 % within-group agreement above that expected by chance.

[7] Each element of the **S** matrix is constructed from two corresponding elements in the $\hat{\boldsymbol{\Sigma}}$ matrices, weighted by the degrees of freedom, i.e., $n - 1$. For example, the first element of the **S** matrix is $0.5333 = [(4 - 1)(0.6667) + (3 - 1)(0.3333)]/(4 + 3 - 2)$.

Fig. 2.5 Example data with $g = 2$, $r = 2$, $n_1 = 4$, $n_2 = 3$, and $N = n_1 + n_2 = 7$

Group	Object	Values	
		x_1	x_2
S_1	ω_1	5	1
S_1	ω_2	4	6
S_1	ω_3	5	2
S_1	ω_4	6	3
S_2	ω_5	2	3
S_2	ω_6	3	4
S_2	ω_7	2	4

2.2.5 Example Bivariate MRPP Analysis with $v = 1$

As mentioned in the univariate example on p. 43, the choice of v can make a substantial difference in the results of an MRPP analysis. To illustrate the computation of MRPP with bivariate data and $v = 1$, consider the same finite sample of $N = 7$ objects listed in Fig. 2.4 on p. 46 and let S_1 and S_2 denote an exhaustive partitioning of the N objects into $g = 2$ disjoint treatment groups. As previously, let S_1 consist of $n_1 = 4$ objects with $r = 2$ commensurate response measurement scores (x_{1i} and x_{2i}) on each object for $i = 1, \ldots, 4$, with $x_1' = (5,\ 1)$, $x_2' = (4,\ 6)$, $x_3' = (5,\ 2)$, and $x_4' = (6,\ 3)$, and let S_2 consist of $n_2 = 3$ objects with $r = 2$ commensurate response measurement scores (x_{1i} and x_{2i}) on each object for $i = 1, 2, 3$ with $x_5' = (2,\ 3)$, $x_6' = (3,\ 4)$, and $x_7' = (2,\ 4)$.

The bivariate response measurement scores for the $N = 7$ objects are listed in Fig. 2.4 on p. 46 and are replicated in Fig. 2.5 for convenience.

For this example analysis, let $r = 2$, $C_1 = n_1/N = 4/7$, $C_2 = n_2/N = 3/7$, and $p = 2$, but in this case set $v = 1$ instead of $v = 2$, employing ordinary Euclidean distance between objects. Following Eq. (2.1) on p. 30 for treatment group S_1 with $n_1 = 4$ objects, $p = 2$, and $v = 1$, the generalized Minkowski distance function yields

$$\Delta(1,2) = \left(|5-4|^2 + |1-6|^2\right)^{1/2} = 5.0990\ ,$$

$$\Delta(1,3) = \left(|5-5|^2 + |1-2|^2\right)^{1/2} = 1.0000\ ,$$

$$\Delta(1,4) = \left(|5-6|^2 + |1-3|^2\right)^{1/2} = 2.2361\ ,$$

$$\Delta(2,3) = \left(|4-5|^2 + |6-2|^2\right)^{1/2} = 4.1231\ ,$$

$$\Delta(2,4) = \left(|4-6|^2 + |6-3|^2\right)^{1/2} = 3.6056\ ,$$

and

$$\Delta(3,4) = \left(|5-6|^2 + |2-3|^2\right)^{1/2} = 1.4142\,,$$

and for treatment group S_2 with $n_2 = 3$ objects, the generalized Minkowski distance function yields

$$\Delta(5,6) = \left(|2-3|^2 + |3-4|^2\right)^{1/2} = 1.4142\,,$$

$$\Delta(5,7) = \left(|2-2|^2 + |3-4|^2\right)^{1/2} = 1.0000\,,$$

and

$$\Delta(6,7) = \left(|3-2|^2 + |4-4|^2\right)^{1/2} = 1.0000\,.$$

Then, following Eq. (2.3) on p. 31, the average distance-function values for all distinct pairs of objects in treatment group S_i, $i = 1, 2$, are

$$\xi_1 = \binom{n_1}{2}^{-1} \Big[\Delta(1,2) + \Delta(1,3) + \Delta(1,4) + \Delta(2,3) + \Delta(2,4) + \Delta(3,4)\Big]$$

$$= \binom{4}{2}^{-1} (5.0990 + 1.0000 + 2.2361 + 4.1231 + 3.6056 + 1.4142)$$

$$= 2.9130$$

and

$$\xi_2 = \binom{n_2}{2}^{-1} \Big[\Delta(5,6) + \Delta(5,7) + \Delta(6,7)\Big]$$

$$= \binom{3}{2}^{-1} (1.4142 + 1.0000 + 1.0000) = 1.1381\,.$$

Following Eq. (2.2) on p. 31, the observed weighted mean of the ξ_1 and ξ_2 values, based on $v = 1$ and $C_i = n_i/N$ for $i = 1, 2$ is

$$\delta_o = C_1\xi_1 + C_2\xi_2 = \left(\frac{4}{7}\right)(2.9130) + \left(\frac{3}{7}\right)(1.1381) = 2.1523\,.$$

The $N = 7$ objects listed in Fig. 2.5 can be partitioned into $g = 2$ treatment groups, S_1 and S_2, with $n_1 = 4$ and $n_2 = 3$ response measurement scores preserved for each arrangement of the observed data in

$$M = \frac{N!}{n_1!\ n_2!} = \frac{7!}{4!\ 3!} = 35$$

possible, equally-likely ways. The $M = 35$ possible arrangements of the observed data in Fig. 2.5, along with the corresponding ξ_1, ξ_2, and δ values, are listed in Table 2.5 and ordered by the δ values from lowest to highest. The observed MRPP test statistic, $\delta_o = 2.1523$, obtained from the realized arrangement,

$$\{(5,1)(4,6)(5,2)(6,3)\} \quad \{(2,3)(3,4)(2,4)\} ,$$

(Order 2 in Table 2.5) is unusual since 33 of the remaining δ values (δ_3 to δ_{35}) exceed the observed value of $\delta_o = 2.1523$ and only one value is less than the observed value: $\delta_1 = 1.8152$. If all arrangements of the $N = 7$ observed bivariate response measurement scores listed in Fig. 2.5 occur with equal chance, the exact probability value of $\delta_o = 2.1523$ computed on the $M = 35$ possible arrangements of the observed data with $n_1 = 4$ and $n_2 = 3$ response measurement scores preserved for each arrangement is

$$P(\delta \le \delta_o | H_0) = \frac{\text{number of } \delta \text{ values} \le \delta_o}{M} = \frac{2}{35} = 0.0571 .$$

For comparison, for the bivariate response measurement scores listed in Fig. 2.5 the exact probability value based on $v = 2$ and $C_i = n_i/N$ for $i = 1, 2$ in the first example is $P = 0.0857$. No comparison is made with the conventional Hotelling T^2 test as Hotelling's T^2 is undefined for $v = 1$.

Following Eq. (2.6) on p. 37, the exact average value of the $M = 35$ δ values listed in Table 2.5 is $\mu_\delta = 2.9475$. Thus, the observed chance-corrected coefficient of agreement, following Eq. (2.5) on p. 37, is

$$\Re_o = 1 - \frac{\delta_o}{\mu_\delta} = 1 - \frac{2.1523}{2.9475} = +0.2698 ,$$

indicating approximately 27 % within-group agreement above that expected by chance.

2.3 Coda

Chapter 2 provided the foundation for Multi-Response Permutation Procedures (MRPP), with special emphasis on the generalized Minkowski distance function, $\Delta(x, y)$, as defined in Eq. (2.1) on p. 30; δ, the weighted mean of the specified distance function values for all distinct pairs of objects in treatment group S_i for

Table 2.5 Permutations of the observed data set in Fig. 2.5 for treatment groups S_1 and S_2 with values for ξ_1, ξ_2, and δ based on $v = 1$, ordered by values of δ from lowest to highest

Order	S_1	S_2	ξ_1	ξ_2	δ
1	{(4,6) (2,3) (3,4) (2,4)}	{(5,1) (5,2) (6,3)}	2.0140	1.5501	1.8152
2	{(5,1) (4,6) (5,2) (6,3)}	{(2,3) (3,4) (2,4)}	2.9130	1.1381	2.1523
3	{(5,1) (5,2) (6,3) (2,3)}	{(4,6) (3,4) (2,4)}	2.5697	2.0215	2.3347
4	{(5,1) (5,2) (6,3) (3,4)}	{(4,6) (2,3) (2,4)}	2.3744	2.4780	2.4188
5	{(5,1) (5,2) (6,3) (2,4)}	{(4,6) (2,3) (3,4)}	2.7703	2.4186	2.6196
6	{(5,1) (2,3) (3,4) (2,4)}	{(4,6) (5,2) (6,3)}	2.4780	3.0476	2.7221
7	{(4,6) (6,3) (3,4) (2,4)}	{(5,1) (5,2) (2,3)}	2.8259	2.5893	2.7245
8	{(5,2) (2,3) (3,4) (2,4)}	{(5,1) (4,6) (6,3)}	2.1684	3.6469	2.8020
9	{(6,3) (2,3) (3,4) (2,4)}	{(5,1) (4,6) (5,2)}	2.4499	3.4074	2.8603
10	{(5,1) (5,2) (2,3) (2,4)}	{(4,6) (6,3) (3,4)}	2.7693	3.0013	2.8687
11	{(4,6) (6,3) (2,3) (2,4)}	{(5,1) (5,2) (3,4)}	3.1938	2.4780	2.8870
12	{(4,6) (5,2) (6,5) (3,4)}	{(5,1) (2,3) (2,4)}	2.8949	2.9494	2.9183
13	{(4,6) (6,3) (2,3) (3,4)}	{(5,1) (5,2) (2,4)}	3.0039	2.9494	2.9806
14	{(4,6) (5,2) (3,4) (2,4)}	{(5,1) (6,3) (2,3)}	2.7703	3.2805	2.9890
15	{(5,1) (5,2) (2,3) (3,4)}	{(4,6) (6,3) (2,4)}	2.6027	3.5190	2.9954
16	{(5,1) (4,6) (2,3) (2,4)}	{(5,2) (6,3) (3,4)}	3.3969	2.4683	2.9989
17	{(5,1) (4,6) (6,3) (3,4)}	{(5,2) (2,3) (2,4)}	3.3241	2.5893	3.0092
18	{(4,6) (5,2) (2,3) (2,4)}	{(5,1) (6,3) (3,4)}	3.0542	3.0013	3.0315
19	{(5,1) (4,6) (3,4) (2,4)}	{(5,2) (6,3) (2,3)}	3.1686	2.8588	3.0359
20	{(5,1) (4,6) (5,2) (3,4)}	{(6,3) (2,3) (2,4)}	3.1487	3.0410	3.1026
21	{(4,6) (5,2) (6,3) (2,4)}	{(5,1) (2,3) (3,4)}	3.2833	2.8751	3.1084
22	{(5,1) (6,3) (2,3) (2,4)}	{(4,6) (5,2) (3,4)}	3.2012	3.0625	3.1418
23	{(5,1) (4,6) (5,2) (2,3)}	{(6,3) (3,4) (2,4)}	3.4326	2.7618	3.1451
24	{(5,1) (5,2) (3,4) (2,4)}	{(2,6) (6,3) (2,3)}	2.7137	3.7370	3.1523
25	{(4,6) (5,2) (6,3) (2,3)}	{(5,1) (3,4) (2,4)}	3.3184	2.9494	3.1603
26	{(5,1) (4,6) (6,3) (2,4)}	{(5,2) (2,3) (3,4)}	3.6891	2.4683	3.1659
27	{(4,6) (5,2) (2,3) (3,4)}	{(5,1) (6,3) (2,4)}	2.8949	3.5339	3.1688
28	{(5,1) (4,6) (2,3) (3,4)}	{(5,2) (6,3) (2,4)}	3.2610	3.0476	3.1695
29	{(5,1) (4,6) (6,3) (2,3)}	{(5,2) (3,4) (2,4)}	3.6920	2.4780	3.1717
30	{(5,2) (6,3) (2,3) (2,4)}	{(5,1) (4,6) (3,4)}	2.8842	3.6469	3.2111
31	{(5,1) (4,6) (5,2) (2,4)}	{(6,3) (2,3) (3,4)}	3.4831	2.8588	3.2156
32	{(5,1) (6,3) (2,3) (3,4)}	{(4,6) (5,2) (2,4)}	3.0039	3.5190	3.2247
33	{(5,2) (6,3) (2,3) (3,4)}	{(5,1) (4,6) (2,4)}	2.6636	4.0567	3.2606
34	{(5,2) (6,3) (3,4) (2,4)}	{(5,1) (4,6) (2,3)}	2.6889	4.1034	3.2951
35	{(5,1) (6,3) (3,4) (2,4)}	{(4,6) (5,2) (2,3)}	3.0616	3.6303	3.3053

$i = 1, \ldots, g$, as defined in Eq. (2.2) on p. 31; and $\Re$, the chance-corrected within-group coefficient of agreement, as defined in Eq. (2.4) on p. 33. Chapters 3 and 4 provide applications of MRPP for completely randomized data at the interval level of measurement, Chaps. 5 and 6 provide applications of MRPP for completely randomized data at the ordinal (ranked) level of measurement, and Chap. 7 provides applications of MRPP for completely randomized data at the nominal (categorical) level of measurement.

Chapter 3

Chapter 3 establishes the relationship between the MRPP test statistics, δ and $\Re$, and selected conventional tests and measures designed for the analysis of completely randomized data at the interval level of measurement. Considered in Chap. 3 are Student's two-sample t test with interval-level univariate response measurement scores, Hotelling's two-sample T^2 test with interval-level multivariate response measurement scores, one-way fixed-effects analysis of variance (ANOVA) with interval-level univariate response measurement scores, and one-way multivariate analysis of variance (MANOVA) with interval-level multivariate response measurement scores.

Randomized Designs: Interval Data

3

This third chapter of *Permutation Statistical Methods* utilizes the Multi-Response Permutation Procedures (MRPP) presented in Chap. 2 to develop the analysis of completely randomized data at the interval level of measurement. As detailed in Chap. 2, the structure of the MRPP test statistic, δ, depends on the value of v in the generalized Minkowski distance function given by

$$\Delta(x, y) = \left(\sum_{i=1}^{r} \left| x_i - y_i \right|^p \right)^{v/p} ,$$

where $p \geq 1$, $v > 0$, and the treatment-group weights are given by

$$C_i = \frac{n_i - 1}{N - g} \quad \text{or} \quad C_i = \frac{n_i}{N} ,$$

for $i = 1, \ldots, g$ treatment groups. The choices of v and C_i for $i = 1, \ldots, g$ permit the MRPP test statistic, δ, to be transformed into a wide variety of tests and measures and provide the flexibility for δ to analyze univariate and multivariate data at the interval, ordinal, and nominal levels of measurement.

The genesis for $C_i = (n_i - 1)/(N - g)$, $i = 1, \ldots, g$, as a treatment-group weight is the assumption of normality that requires fitting estimates of population means for each of the g treatment groups. Consequently, one degree of freedom is lost for each estimate of a population parameter; here, the population means. Because the assumption of normality is never satisfied in practice, $C_i = n_i/N$, $i = 1, \ldots, g$, simply weighting each treatment group proportional to its size, is a more appropriate choice for weighting treatment-groups in a permutation analysis, as permutation tests negate the need for estimating population parameters entirely. The weighting function, $C_i = (n_i - 1)/(N - g)$ for $i = 1, \ldots, g$, is useful, however, when making direct comparisons of corresponding conventional and permutation tests, such as the F test for a fully randomized analysis of variance, on

K.J. Berry et al., *Permutation Statistical Methods*,
DOI 10.1007/978-3-319-28770-6_3

the one hand, and the Fisher–Pitman permutation test for g treatment groups, on the other.

Because multi-response permutation procedures are distribution-free, data-dependent, and non-parametric, there is no reason to square differences between response measurements, nor to weight treatment groups by degrees of freedom. Therefore, $v = 1$ and $C_i = n_i/N$ for $i = 1, \ldots, g$ are preferred for all applications of MRPP [32, 293, 297].

Permutation analogues of four selected tests are examined in this chapter: (1) Student's two-sample t test with interval-level univariate response measurement scores, (2) Hotelling's two-sample T^2 test with interval-level multivariate response measurement scores, (3) one-way fixed-effects analysis of variance (ANOVA) with interval-level univariate response measurement scores, and (4) one-way multivariate analysis of variance (MANOVA) with interval-level multivariate response measurement scores. The four tests are illustrated with examples analyzed with $v = 2$ and $C_i = (n_i - 1)/(N - g)$ for $i = 1, \ldots, g$, $v = 1$ and $C_i = (n_i - 1)/(N - g)$ for $i = 1, \ldots, g$, and $v = 1$ and $C_i = n_i/N$ for $i = 1, \ldots, g$.

As developed more completely in Chap. 2, let $\Omega = \{\omega_1, \ldots, \omega_N\}$ denote a finite sample of N objects and let $S_1, \ldots, S_g$ designate an exhaustive partitioning of the N objects into g disjoint treatment groups. The MRPP test statistic is a weighted mean given by

$$\delta = \sum_{i=1}^{g} C_i \xi_i \,, \tag{3.1}$$

where $C_i > 0$ is a positive treatment-group weight for $i = 1, \ldots, g$,

$$\xi_i = \binom{n_i}{2}^{-1} \sum_{j<k} \Delta(j,k)\, \Psi_i(\omega_j)\, \Psi_i(\omega_k) \tag{3.2}$$

is the average distance-function value for all distinct pairs of objects in treatment group S_i for $i = 1, \ldots, g$, $n_i \geq 2$ is the number of a priori objects classified into treatment group S_i for $i = 1, \ldots, g$,

$$N = \sum_{i=1}^{g} n_i ,$$

$\sum_{j<k}$ is the sum over all j and k such that $1 \leq j < k \leq N$, and $\Psi_i(\cdot)$ is an indicator function given by

$$\Psi_i(\omega_j) = \begin{cases} 1 & \text{if } \omega_j \in S_i \,, \\ 0 & \text{otherwise} \,. \end{cases}$$

The null hypothesis (H_0) states that equal probabilities are assigned to each of the

$$M = \frac{N!}{\prod_{i=1}^{g} n_i!}$$

possible, equally-likely allocations of the N objects to the g treatment groups, $S_1, \ldots, S_g$. The probability value associated with an observed value of δ, δ_o, is the probability under the null hypothesis (H_0) of observing a value of δ as extreme or more extreme than δ_o. Thus, an exact probability value for δ_o may be expressed as

$$P\big(\delta \leq \delta_o | H_0\big) = \frac{\text{number of } \delta \text{ values} \leq \delta_o}{M} \,.$$

When M is large, an approximate probability value for δ may be obtained from a resampling procedure, where

$$P\big(\delta \leq \delta_o | H_0\big) = \frac{\text{number of } \delta \text{ values} \leq \delta_o}{L}$$

and L denotes the number of randomly sampled test statistic values. Typically, L is set to a large number to ensure accuracy, e.g., $L = 1{,}000{,}000$. Also, when M is very large and P is exceedingly small, a resampling-approximation permutation procedure may produce no δ values equal to or less than δ_o, even with $L = 1{,}000{,}000$, yielding an approximate resampling probability value of $P = 0.00$. In such cases, moment-approximation permutation procedures based on fitting the first three exact moments of the discrete permutation distribution to a Pearson type III distribution provide approximate probability values, as detailed in Chap. 1, Sect. 1.2.2 [284, 300].

A chance-corrected measure of agreement among response measurement scores is given by

$$\mathfrak{R} = 1 - \frac{\delta}{\mu_\delta} \,, \tag{3.3}$$

where μ_δ is the arithmetic average of the M δ values calculated on all possible arrangements of the observed response measurement scores, i.e.,

$$\mu_\delta = \frac{1}{M} \sum_{i=1}^{M} \delta_i \,. \tag{3.4}$$

3.1 Permutation Analogue of Student's *t* Test

A common research design calls for a test of difference between $g = 2$ independent treatment groups when univariate ($r = 1$) response measurement scores have been obtained for each object. The conventional approach to such research situations is Student's t test for two independent samples (groups of objects) given by

$$t = \frac{\bar{x}_1 - \bar{x}_2}{\left[s_p^2 \left(\dfrac{1}{n_1} + \dfrac{1}{n_2} \right) \right]^{1/2}} ,$$

where the pooled estimate of the population variance is given by

$$s_p^2 = \frac{s_1^2(n_1 - 1) + s_2^2(n_2 - 1)}{N - 2} , \tag{3.5}$$

the sample estimate of the population variance for the ith treatment group is given by

$$s_i^2 = \frac{1}{n_i - 1} \sum_{j=1}^{n_i} \left(x_{ij} - \bar{x}_i \right)^2 , \qquad i = 1, \ldots, g ,$$

n_i is the number of objects in the ith of $g = 2$ treatment groups,

$$N = \sum_{i=1}^{g} n_i$$

is the total number of objects in the g treatment groups, $\bar{x}_i$ is the arithmetic mean of the response measurement scores for the ith of g treatment groups, given by

$$\bar{x}_i = \frac{1}{n_i} \sum_{j=1}^{n_i} x_{ij} , \qquad i = 1, \ldots, g ,$$

and x_{ij} is a univariate response measurement score for the jth object in the ith treatment group. Assuming independence, normality, and homogeneity of variance, t is approximately distributed as Student's t under the null hypothesis with $N - 2$ degrees of freedom.

When $r = 1$, $v = 2$, and the treatment-group weights are given by

$$C_i = \frac{n_i - 1}{N - g} , \qquad i = 1, \ldots, g ,$$

it can easily be shown that δ, as defined in Eqs. (3.1) and (3.2) on p. 58, is the permutation analogue of Student's two-sample t test. The functional relationships between test statistic δ and Student's t statistic for two independent samples are given by

$$\delta = \frac{2\,(NT - S^2)}{N(N-2+t^2)} \quad \text{and} \quad t = \left[\frac{2(NT-S^2)}{N\delta} - N + 2\right]^{1/2} ,$$

where

$$S = \sum_{i=1}^{N} x_i \,, \quad T = \sum_{i=1}^{N} x_i^2 \,,$$

and x_i is a univariate response measurement score for the ith of N objects. Also, given $r = 1$, $g = 2$, $v = 2$, and treatment-group weights

$$C_i = \frac{n_i - 1}{N - g} \,, \qquad i = 1, \ldots, g \,,$$

the average distance-function values are related to the sample estimates of the population variance by

$$\xi_i = 2s_i^2 \,, \qquad i = 1, \ldots, g \,, \tag{3.6}$$

the MRPP test statistic is related to the pooled estimate of the population variance by

$$\delta = 2s_p^2 \,, \tag{3.7}$$

and the arithmetic mean of the M δ values is related to SS_{Total} by

$$\mu_\delta = \frac{2SS_{\text{Total}}}{N-1} \,, \tag{3.8}$$

where $SS_{\text{Total}} = T - S^2/N$.

3.2 Measures of Effect Size

The fact that statistical tests of null hypotheses, such as Student's two-sample t test, produce low probability values indicates only that there are differences among the response measurement scores in the $g = 2$ treatment groups that (possibly) cannot be attributed to error. The obtained probability values do not indicate whether these differences are of any practical value.[1]

Statisticians have raised a number of issues and concerns with null hypothesis statistical testing (NHST). There are literally hundreds of articles and chapters dealing with the problems of NHST, far too many to be summarized here. However, a brief overview of the limitations of null hypothesis statistical testing will suffice for these purposes.[2]

First, the null hypothesis is almost never literally true, so rejection of the null hypothesis is relatively uninformative; see, for example, articles by Baken [17], Carver [66, 67], Levine, Weber, Hullett, Park, and Massi Lindsey [240], Levine, Weber, Park, and Hullett [241], McLean and Ernest [272], and Nix and Barnette [321,322]. Second, tests of significance are highly dependent on sample sizes. When sample sizes are small, important effects can be non-significant, and when sample sizes are large, even trivial effects can produce very small probability values; see, for example, articles by Daniel [86] and Levine and Hullett [239]. Third, the requirement of obtaining a random sample from a well-defined population is seldom met in practice; see, for example, articles by Altman and Bland [6], Bradbury [51], Feinstein [113], Frick [127], LaFleur and Greevy [227], Ludbrook [247], Ludbrook and Dudley [254], and Still and White [388]. Fourth, the assumptions of normality and homogeneity of variance are rarely satisfied in real-data situations; see, for example, articles by Bernardin and Beatty [22], Bross [58], Feinstein [113], Geary [134], Micceri [280], Murphy and Cleveland [314], Saal, Downey, and Lahey [359], and Schmidt and Johnson [366].[3]

In February 2015, the Editor, David Trafimow, and Associate Editor, Michael Marks, of *Basic and Applied Social Psychology* formally banned NHST procedures from its pages, including probability values, t values, F values and other statements about significant differences [401, p. 1]; the ban had been announced previously with a one-year grace period [400]. The editors argued that NHST is logically invalid and disallowed all null hypothesis statistical testing as well as the use of confidence intervals as an alternative to reject null hypotheses [401, p. 1]. Instead, the editors stated that the journal would henceforth favor "strong descriptive statistics, including effect sizes," and the use of larger sample sizes "because as the sample size increases, descriptive statistics become increasingly stable and sampling error is less of a problem" [401, p. 1].

[1] In the literature, "practical value" is often referred to as "practical significance" as contrasted with "statistical significance" [219].

[2] A comprehensive bibliography for the limitations of null hypothesis statistical testing has been compiled by William Thompson [397].

[3] William Thompson has compiled an extensive list of quotes from various authors detailing the limits of null hypothesis statistical testing [398].

Moreover, as Roger Kirk observed in 1968, test statistics such as t and F and their associated probability values provide no information as to the size of treatment effects, only whether they are statistically significant [218, p. 135]. As Kirk explained in 1996 [219, p. 747], the one individual most responsible for bringing the shortcomings of hypothesis testing to the attention of researchers was the psychologist Jacob Cohen with two articles with unconventional titles in *American Psychologist*: "Things I have learned (so far)" in 1990 and "The earth is round ($p < .05$)" in 1994 [73, 74]. As a result of the identified challenges with NHST and the reporting of probability values, various measures of effect size have been designed to reflect the substantive importance and practical value of treatment differences; see, for example, a 2000 book by Rosenthal, Rosnow, and Rubin on *Contrasts and Effect Sizes in Behavioral Research* [353] and a 2005 book by Grissom and Kim on *Effect Sizes for Research* [157].

Recent trends in the literature have stressed the importance of reporting a measure of effect size along with a test of significance when analyzing experimental data [70, 174, 219, 428]. For example, as far back as 1994 the 4th edition of the *Publication Manual of the American Psychological Association* strongly encouraged reporting measures of effect size in conjunction with probability values. In 1999 the American Psychological Association Task Force on Statistical Inference, under the direction of Leland Wilkinson noted that "reporting and interpreting effect sizes in the context of previously reported effects is essential to good research" [430, p. 599]. Today many journals require authors to provide measures of effect size in addition to tests of significance [63, 64]. As a result of increased attention to measures of effect size, introductory statistics textbooks often include discussions of such measures as Cohen's $\hat{d}$, Pearson's r^2, Kelley's ϵ^2, and Hays' $\hat{\omega}^2$; see, for example, references [93, 153, 185, 239, 331, 421, 427].

The appropriate use of effect-size measures such as $\hat{d}$, r^2, ϵ^2, and $\hat{\omega}^2$ is restricted to data with homogeneous variances [140].[4] Micceri [280] and Wilcox [426] have both argued that assuming normality for measures of effect size may not be realistic.[5] However, the permutation-based chance-corrected measure of agreement, $\Re$, as defined in Eq. (3.3) on p. 59, is a universal measure of effect size and is appropriate for homogeneous or heterogeneous, normal or non-normal, data sets [194]. In addition, $\Re$ is suitable for any number of treatment groups. The various measures of effect size and their relationships are described in this section and illustrated in the following section with univariate response measurement scores for a two-sample test of differences.

---

[4] As noted by Olejnik and Algina, if the variance equality assumption is not met, then the standard deviation for one of the g treatment groups should be used as the standardizer [326, p. 246]. In the context of comparing an experimental and a control group, Glass, McGraw, and Smith recommended using the standard deviation for the control group [142].

[5] Scheffé noted that the usual measures of effect size do not assume normality [365]; see also, a 1969 article by Vaughan and Corballis [411].

Currently, the most popular measures of effect size are Cohen's $\hat{d}$, Pearson's r^2, Kelley's ϵ^2, and Hays' $\hat{\omega}^2$. There are, of course, many other useful measures of effect size, such as Cohen's f [72], Glass's Δ [138], and Kirk's $\hat{f}$ [219]. For a comprehensive list of effect-size measures, see an article by Kirk on "Practical significance: A concept whose time has come" in 1996 and another article by Kirk on "Effect magnitude: A different focus" in 2006 [219, 220], a book by Rosenthal, Rosnow, and Rubin on *Contrasts and Effect Sizes in Behavioral Research* published in 2000 [353], and a book by Grissom and Kim on *Effect Sizes for Research* published in 2005 [157] with a second edition published in 2012 [158]. Assume for purposes of exposition that the problem is to choose an appropriate measure of effect size for Student's two-sample t test.

3.2.1 Cohen's $\hat{d}$

In 1969 Jacob Cohen [72] defined a new measure of effect size, $\hat{d}$, based on the difference between two treatment-group means divided by the pooled estimate of the population standard deviation, s_p; consequently, Cohen's standardized measure of mean differences is only appropriate for populations with homogeneous variances. While other measures of effect size preceded Cohen's $\hat{d}$, $\hat{d}$ was the first measure of effect size that was explicitly labeled as such [219, p. 749]. Cohen's $\hat{d}$ is given by

$$\hat{d} = \frac{|\bar{x}_1 - \bar{x}_2|}{s_p} ,$$

where $\bar{x}_1$ and $\bar{x}_2$ are the sample means for treatment groups 1 and 2, respectively, and s_p is the pooled estimate of the population standard deviation given by

$$s_p = \left[\frac{s_1^2(n_1 - 1) + s_2^2(n_2 - 1)}{N - 2} \right]^{1/2} , \tag{3.9}$$

where n_1 and n_2 are the sample sizes for treatment groups 1 and 2, respectively, $N = n_1 + n_2$ is the total size of the two treatment groups combined, and s_1^2 and s_2^2 are the sample estimates of the population variance for treatment groups 1 and 2 given by

$$s_i^2 = \frac{1}{n_i - 1} \sum_{j=1}^{n_i} \left(x_{ij} - \bar{x}_i \right)^2 , \qquad i = 1, 2 .$$

Cohen's $\hat{d}$ is expressed in standard deviation units and, in this context, measures the effect size for Student's t test for two independent samples.

In 1969 Cohen [72] provided crude estimates with which to evaluate and interpret $\hat{d}$ values. Cohen proposed that if $\hat{d} \leq 0.20$, the effect should be considered "small"; if $0.20 < \hat{d} \leq 0.80$, the effect size should be considered "medium" or "moderate"; and if $\hat{d} > 0.80$, the effect size should be considered "large."[6,7]

3.2.2 Hedges' *g*

Cohen's $\hat{d}$ was originally defined as

$$\delta_c = \frac{|\mu_1 - \mu_2|}{\sigma_x} , \tag{3.10}$$

where μ_1 and μ_2 denote the two population means and σ_x is the common population standard deviation.[8] In 1981 and 1982, Hedges proposed

$$g = \frac{|\bar{x}_1 - \bar{x}_2|}{\sqrt{s_p^2}} ,$$

where $\bar{x}_1$ and $\bar{x}_2$ denote the two sample means and s_p^2 is the pooled estimate of the population variance, as given in Eq. (3.5) on p. 60 [170]. Hedges argued that g could be made an unbiased estimator of δ_c, showing that $\mathrm{E}[g] = \delta_c / c(m)$, where

$$c(m) = \frac{\Gamma(m/2)}{\sqrt{m/2} \times \Gamma\left[(m-1)/2\right]}$$

and $m = n_1 + n_2 - 2$. Hedges further showed that $c(m)$ could be closely approximated by $c(m) \simeq 1 - 3/(4m - 1)$.

If m is even, then

$$\Gamma(m/2) = \left(\frac{m-2}{2}\right)! \quad \text{and} \quad \Gamma\left[(m-1)/2\right] = \frac{(m-3)!\sqrt{\pi}}{2^{m-3}\left(\frac{m-4}{2}\right)!} .$$

---

[6]In a 2006 article, McGrath and Meyer took issue with these values and suggested slightly higher values for "medium" and "large" effect sizes [270].

[7]Cohen did not select his effect sizes capriciously. Effect size 0.20 was chosen to correspond to 15 % overlap (85 % non-overlap) between the sampling distributions of the two sample means, $\bar{x}_1$ and $\bar{x}_2$, and effect size 0.80 was chosen to correspond to 50 % overlap (50 % non-overlap) between the two sampling distributions.

[8]δ_c is used in Eq. (3.10) so as not to confuse Cohen's δ with the MRPP test statistic δ defined in Eq. (3.9) on p. 64.

Substituting and simplifying yields

$$c(m) = \frac{\left(\frac{m-2}{2}\right)!\left(\frac{m-4}{2}\right)!\,2^{m-3}}{(m-3)!\,\sqrt{\frac{m\pi}{2}}}\,.$$

If m is odd, then

$$\Gamma\,(m/2) = \frac{(m-2)!\,\sqrt{\pi}}{2^{m-2}\left(\frac{m-3}{2}\right)!} \quad \text{and} \quad \Gamma\,[(m-1)/2] = \left(\frac{m-3}{2}\right)!\,.$$

Substituting and simplifying yields

$$c(m) = \frac{(m-2)!\,\sqrt{\frac{2\pi}{m}}}{2^{m-2}\left[\left(\frac{m-3}{2}\right)!\right]^2}\,.$$

To illustrate, if m is even, say 8, then

$$c(8) = \frac{\left(\frac{8-2}{2}\right)!\left(\frac{8-4}{2}\right)!\,2^{8-3}}{(8-3)!\,\sqrt{\frac{(8)(3.1416)}{2}}} = \frac{(3!)(2!)(32)}{5!\sqrt{12.5664}} = \frac{(6)(2)(32)}{(120)(3.5449)} = 0.9027$$

and

$$c(8) \simeq 1 - \frac{3}{4m-1} = 1 - \frac{3}{(4)(8)-1} = 1 - 0.0968 = 0.9032\,.$$

And, if m is odd, say 7, then

$$c(7) = \frac{(7-2)!\,\sqrt{\frac{(2)(3.1416)}{7}}}{2^{7-2}\left[\left(\frac{7-3}{2}\right)!\right]^2} = \frac{5!\sqrt{0.8976}}{2^5\,[2!]^2} = \frac{(120)(0.9474)}{(32)(4)} = 0.8882$$

and

$$c(7) \simeq 1 - \frac{3}{4m-1} = 1 - \frac{3}{(4)(7)-1} = 1 - 0.1111 = 0.8889\,.$$

3.2.3 Pearson's r^2

The second measure of effect size, r^2, is the familiar squared Pearson product-moment correlation coefficient. For Student's two-sample t test, r^2 may be expressed as

$$r^2 = \frac{t^2}{t^2 + N - 2} . \tag{3.11}$$

It is not uncommon for r^2 in Eq. (3.11) to be labeled as r_{pb}^2, indicating that this measure of effect size is the point-biserial correlation between the response measurement scores and a dummy-coded variable representing the two treatment groups, i.e., the correlation between the response measurement scores and group membership; see, for example, discussions by Friedman [130] and Howell [185, pp. 307–309]. In other applications, especially in the analysis of variance, r^2 is designated as η^2, where it is known as the "correlation ratio" and defined simply as

$$\eta^2 = \frac{SS_{\text{Between}}}{SS_{\text{Total}}} , \tag{3.12}$$

i.e., the proportion of the total variability attributable to the treatment or intervention. The measure of effect size, r^2 (η^2), however, has been criticized repeatedly in the literature for its positive bias, especially for small sample sizes; see, for example, articles by Levine and Hullett [239] and Maxwell, Camp, and Arvey [266].

3.2.4 Kelley's ϵ^2

The third measure of effect size is Kelley's ϵ^2 [200] and, defined for Student's two-sample t test, is given by

$$\epsilon^2 = \frac{t^2 - 1}{t^2 + N - 2} . \tag{3.13}$$

Oftentimes in the literature ϵ^2 is designated as $\hat{\eta}^2$, i.e., η^2 adjusted for degrees of freedom, and it is typically termed the "unbiased correlation ratio." It has been well established and is widely recognized that ϵ^2 is not, in fact, unbiased, but since the title of Truman Kelley's article was "An unbiased correlation ratio measure" [200], the label has survived for over 80 years.

3.2.5 Hays' $\hat{\omega}^2$

The fourth measure of effect size is Hays' $\hat{\omega}^2$ [169, pp. 323–332]. Hay's $\hat{\omega}^2$ estimates the proportion of total variance attributable to treatment. Thus $\hat{\omega}^2$ is a ratio of variance estimates given by

$$\hat{\omega}^2 = \frac{\hat{\sigma}_t^2}{\hat{\sigma}_t^2 + \hat{\sigma}_x^2} ,$$

where $\hat{\sigma}_t^2$ is an estimate of the treatment variance and $\hat{\sigma}_x^2$ is an estimate of the population variance. For Student's two-sample t test, Hays' $\hat{\omega}^2$ is given by

$$\hat{\omega}^2 = \frac{t^2 - 1}{t^2 + N - 1} . \tag{3.14}$$

Hays defined $\hat{\omega}^2$ as the proportion of variance in the observations attributable to group membership and, alternatively, as the relative reduction in uncertainty about the observations given by knowledge of group membership [169, p. 325]. Note the high degree of similarity between ϵ^2 in Eq. (3.13) and $\hat{\omega}^2$ in Eq. (3.14). It has been shown empirically by Carroll and Nordholm that ϵ^2 and $\hat{\omega}^2$ will ordinarily differ very little for a given set of response measurement scores [65]. In fact, as sample sizes increase, Kelley's ϵ^2 and Hays' $\hat{\omega}^2$ both converge to the same value [266, p. 527]. There are actually two quite different $\hat{\omega}^2$ measures of effect size: one for a fixed-effects analysis-of-variance model and another for a random-effects analysis-of-variance model. However, when $g = 2$, both measures yield the same result.[9]

3.2.6 Mielke and Berry's $\mathfrak{R}$

Finally, a chance-corrected measure of effect size, $\mathfrak{R}$, is defined as

$$\mathfrak{R} = 1 - \frac{\delta}{\mu_\delta} ,$$

as given in Eq. (3.3) on p. 59, where δ is the weighted mean of the observed response measurement scores, as defined in Eq. (3.1) on p. 58, and μ_δ is the arithmetic average of the δ values calculated on all possible, equally-likely arrangements of the observed response measurement scores, as defined in Eq. (3.4) on p. 59.

The five measures of effect size, $\hat{d}$, r^2 (η^2), ϵ^2 ($\hat{\eta}^2$), $\hat{\omega}^2$, and $\mathfrak{R}$ usually produce similar results when $r = 1$, $v = 2$, and $C_i = (n_i - 1)/(N - g)$ for $i = 1, \ldots, g$ and are directly related to each other and to Student's t test for two independent samples, as detailed in Table 3.1 [194] . While r^2 and η^2 are equivalent measures of effect size, convention dictates that r^2 is used when $g = 2$ and η^2 is used when $g > 2$. Also, ϵ^2

[9]Actually, there exist a large number of $\hat{\omega}^2$ measures of effect size designed for a wide variety of experimental designs; see, for example, articles by Dodd and Schultz [98], Dwyer [101], Fleiss [123], Friedman [130], Gaebelein, Soderquist, and Powers [131], Golding [143], Hays [169], and Vaughan and Corballis [411].

Table 3.1 Equivalencies among the pooled t test statistic for two independent samples, $\Re$, Cohen's $\hat{d}$, Pearson's r^2, and Hays' $\hat{\omega}^2$ when $r = 1$, $v = 2$ and $C_i = (n_i - 1)/(N - g)$ for $i = 1, \ldots, g$ [194]

Relationship	Equivalencies	
$\Re$ and t	$\Re = \dfrac{t^2 - 1}{t^2 + N - 2}$	$t = \left[\dfrac{\Re(N-2)+1}{1-\Re}\right]^{1/2}$
$\Re$ and r^2	$\Re = r^2 - (t^2 + N - 2)^{-1}$	$r^2 = \Re + (t^2 + N - 2)^{-1}$
$\Re$ and $\hat{\omega}^2$	$\Re = \hat{\omega}^2\left(\dfrac{t^2+N-1}{t^2+N-2}\right)$	$\hat{\omega}^2 = \Re\left(\dfrac{t^2+N-2}{t^2+N-1}\right)$
$\Re$ and $\hat{d}$	$\Re = \dfrac{n_1 n_2 \hat{d}^2 - N}{n_1 n_2 \hat{d}^2 + (N)(N-2)}$	$\hat{d} = \left[\dfrac{\Re(N)(N-2)+N}{(1-\Re)(n_1 n_2)}\right]^{1/2}$
t and $\hat{d}$	$t = \hat{d}\left[\dfrac{n_1 n_2}{N}\right]^{1/2}$	$\hat{d} = t\left[\dfrac{N}{n_1 n_2}\right]^{1/2}$
t and r^2	$t = \left[\dfrac{r^2(N-2)}{1-r^2}\right]^{1/2}$	$r^2 = \dfrac{t^2}{t^2+N-2}$
$\hat{d}$ and r^2	$\hat{d} = \left[\dfrac{r^2(N)(N-2)}{(1-r^2)(n_1 n_2)}\right]^{1/2}$	$r^2 = \dfrac{n_1 n_2 \hat{d}^2}{n_1 n_2 \hat{d}^2 + (N)(N-2)}$
$\hat{\omega}^2$ and r^2	$\hat{\omega}^2 = \dfrac{r^2(N-1)-1}{N-(1+r^2)}$	$r^2 = \dfrac{\hat{\omega}^2(N-1)+1}{\hat{\omega}^2+N-1}$
$\hat{\omega}^2$ and t	$\hat{\omega}^2 = \dfrac{t^2-1}{t^2+N-1}$	$t = \left[\dfrac{\hat{\omega}^2(N-1)+1}{1-\hat{\omega}^2}\right]^{1/2}$
$\hat{\omega}^2$ and $\hat{d}$	$\hat{\omega}^2 = \dfrac{n_1 n_2 \hat{d}^2 - N}{n_1 n_2 \hat{d}^2 + N(N-1)}$	$\hat{d} = \left[\dfrac{N[\hat{\omega}^2(N-1)+1]}{(1-\hat{\omega}^2)n_1 n_2}\right]^{1/2}$

and $\hat{\eta}^2$ are equivalent measures, but the ϵ^2 notation appears to be the choice of most authors, especially in recent publications [239].

Note that the relationships between t and $\hat{d}$, t and r^2, $\hat{d}$ and r^2, $\hat{\omega}^2$ and r^2, $\hat{\omega}^2$ and t, and $\hat{\omega}^2$ and $\hat{d}$ described in Table 3.1 hold only for Student's *pooled* two-sample t test. The measures of effect size, $\hat{d}$, r^2, and $\hat{\omega}^2$, all require homogeneity of variance and the relationships given in Table 3.1 do not hold for Student's non-pooled two-sample t test. On the other hand, $\Re$ does not require homogeneity of variance and is appropriate for both pooled and non-pooled two-sample t tests.

It is widely recognized that r^2 is a positively biased estimate of the squared Pearson population correlation coefficient, ρ^2. An adjusted r^2 coefficient that compensates for degrees of freedom was introduced by M.J.B. Ezekiel in 1930 [112][10]; see also discussions by Larson [230] and Wherry in 1931 [422]. An adjusted r^2 value is produced by most statistical computer programs and is given by

$$\hat{r}^2 = 1 - \frac{(1-r^2)(N-1)}{N-2}$$

[10]The formula for an adjusted r^2 was actually first presented by Ezekiel at the December 1928 meeting of the American Mathematical Society in Chicago, Illinois.

for $g = 2$ treatment groups.[11] It can easily be shown that $\Re = \hat{\eta}^2 = \epsilon^2 = \hat{r}^2$ when $r = 1$, $v = 2$, and $C_i = (n_i - 1)/(N - g)$, $i = 1, \ldots, g$; see, for example, discussions by Cohen and Cohen [75, p. 188] and Maxwell, Camp, and Arvey [266]. Thus, since $\Re$ is a chance-corrected measure, $\hat{\eta}^2$, ϵ^2, and $\hat{r}^2$ are also chance-corrected measures of effect size. To clarify the relationship and emphasize that the adjustment is for the degrees of freedom, $\hat{\eta}^2$, ϵ^2, $\hat{r}^2$, and $\Re$ can be redefined in an analysis of variance context as

$$\Re = \hat{\eta}^2 = \epsilon^2 = \hat{r}^2 = 1 - \left(\frac{N-1}{N-g}\right)\frac{SS_{\text{Within}}}{SS_{\text{Total}}} \tag{3.15}$$

and expressed in terms of the conventional F-ratio as

$$\Re = \hat{\eta}^2 = \epsilon^2 = \hat{r}^2 = \frac{(F-1)(g-1)}{F(g-1)+N-g} . \tag{3.16}$$

As is evident in Eq. (3.16), when $F < 1$, $\Re$, $\hat{\eta}^2$, ϵ^2, and $\hat{r}^2$ are all negative. It is disconcerting, to say the least, to try to interpret squared coefficients with negative values, as a negative value does not constitute a valid estimate of the population variance [379, p. 344]. It is also important to note that negative estimates of effect size cannot be simply excluded on theoretical grounds [271, p. 1000]. In 1968 Friedman noted that ϵ^2 could sometimes be negative [130]. In 1981 Maxwell, Camp, and Arvey also observed that $\hat{r}^2$ could be negative and suggested that negative values of $\hat{r}^2$, $\hat{\omega}^2$, and ϵ^2 be treated as zero [266], failing to recognize that negative values of ϵ^2 represent effect sizes less than expected by chance.[12] As can be seen in Eq. (3.15), when $SS_{\text{Within}} = 0$, $\Re = \epsilon^2 = \hat{r}^2 = 1$; and when $SS_{\text{Within}} = SS_{\text{Total}}$, then

$$\Re = \hat{\eta}^2 = \epsilon^2 = \hat{r}^2 = 1 - \frac{N-1}{N-g} = -\left(\frac{g-1}{N-g}\right) ,$$

i.e., the negated ratio of the numerator and denominator degrees of freedom, which is the most extreme negative value that can be obtained for these equivalent chance-corrected measures of effect size; and when $\delta = \mu_\delta$, i.e., the observed result is expected only by chance, $\Re = \hat{\eta}^2 = \epsilon^2 = \hat{r}^2 = 0$. Thus, positive reported values of $\Re$, $\hat{\eta}^2$, ϵ^2 and $\hat{r}^2$ are to be interpreted as effect sizes greater than expected by chance, and negative values are to be interpreted as effect sizes less than expected by chance, i.e., the treatment group means are closer together than expected under randomization of the N objects.

[11] In the literature, $\hat{r}^2$ is variously termed "adjusted" or "shrunken" r^2.

[12] As noted by Scheffé [365, pp. 112–119] and by Vaughan and Corballis [411, p. 212], replacing a negative estimate by zero introduces a positive bias and both advise reporting the negative value. See also a 2001 article by Fidler and Thompson [117].

While $\hat{r}^2$, $\hat{\eta}^2$, and ϵ^2 are often reported as measures of effect size for two-sample t tests and one-way analysis of variance, and although they all are estimates of the population effect size, they are widely recognized as being difficult to interpret. Cast in the light of agreement theory, $\hat{r}^2$, $\hat{\eta}^2$, and ϵ^2 are revealed as chance-corrected measures of effect size. This previously undocumented feature provides a new and improved interpretation of these three measures. The fact that $\hat{r}^2$, $\hat{\eta}^2$, and ϵ^2 can yield negative values is recast as a favorable attribute and places the three measures of effect size into the family of chance-corrected measures, which includes such well-known members as Scott's coefficient of inter-coder agreement [368], Cohen's coefficient of weighted agreement [71], Spearman's footrule [382], and when the two variables consist of ranks from 1 to N with no tied ranks, Spearman's rank-order correlation coefficient [381].

Hays' $\hat{\omega}^2$ also produces negative values—again, not appropriate for a squared coefficient of effect size. The value of $\hat{\omega}^2$ will be negative whenever the value of the computed F is less than 1. Defining $\hat{\omega}^2$ in terms of F makes this clear. For a fixed-effects one-way analysis of variance,

$$\hat{\omega}^2 = \frac{(F-1)(g-1)}{(F-1)(g-1)+N} . \tag{3.17}$$

If $F < 1$, then the numerator of Eq. (3.17) will be negative and $\hat{\omega}^2$ will ipso facto be negative. For a random-effects one-way analysis of variance,

$$\hat{\omega}^2 = \frac{F-1}{F+n-1} , \tag{3.18}$$

where n denotes the common number of objects in each of g treatments. Again, if $F < 1$, then the numerator of Eq. (3.18) will be negative and $\hat{\omega}^2$ will also be negative.

Negative values of $\hat{\omega}^2$ have led many researchers to advocate treating negative values as zero, including Hays [169, pp. 327, 383]; see also Kenny [213, p. 234]. Although $\hat{\omega}^2$ does not norm properly between 0 and 1, i.e., its minimum value is given by

$$-\left(\frac{g-1}{N-g+1}\right) ,$$

it is in fact a chance-corrected measure of effect size like $\Re$, $\hat{\eta}^2$, ϵ^2, and $\hat{r}^2$. The relationships between the chance-corrected measure of effect size, $\Re$, and Hays' $\hat{\omega}^2$ in terms of F, for a fixed-effects one-way analysis of variance, are given by

$$\Re = \hat{\omega}^2\left(\frac{F+N-1}{F+N-2}\right) \quad \text{and} \quad \hat{\omega}^2 = \Re\left(\frac{F+N-2}{F+N-1}\right) .$$

Finally, it should be noted that since $\Re$ is completely data-dependent, it is irrelevant whether the model is fixed or random.

3.2.7 Biased Estimators

In general, statisticians prefer sample estimates of population parameters that are unbiased, e.g., the sample mean, $\bar{x}$, is an unbiased estimator of the population mean, μ_x, and the sample variance, s_x^2, is an unbiased estimator of the population variance, σ_x^2.[13] It is well known that, under the population model of inference whereby repeated random samples are hypothetically drawn from a normal population, measures of effect size such as r^2, $\hat{r}^2$, η^2, $\hat{\eta}^2$, ϵ^2, and $\hat{\omega}^2$ are biased estimators of their respective population parameters [219,348,379]. The terms "biased" and "unbiased" possess quite different meanings under the permutation model of inference, as there is no population parameter to be estimated. For the permutation model, an unbiased measure simply means that the average value of the measure of effect size obtained from all M possible arrangements of the observed response measurement scores is zero. In the case of $\hat{\eta}^2 = \epsilon^2 = \hat{r}^2 = \Re$, the expected values are indeed zero and all four chance-corrected measures of effect size are unbiased under the permutation model. On the other hand, while $\hat{\omega}^2$ is a chance-corrected measure of effect size like $\hat{\eta}^2$, ϵ^2,$\hat{r}^2$, and $\Re$, it is not an unbiased estimator under either the permutation or population models of inference. That said, however, the positive bias of $\hat{\omega}^2$ is typically quite small, within the context of a fixed-effects one-way analysis of variance. Under the permutation model, it can be shown that while the expected value of $\hat{\omega}^2$ is not zero, it is given by

$$\mathrm{E}\left[\hat{\omega}^2\right] = \frac{1}{M}\sum_{i=1}^{M}\left(\frac{N\delta_i}{\mu_\delta(N-1)+\delta_i}\right) ,$$

where

$$M = \frac{\left(\displaystyle\sum_{i=1}^{g} n_i\right)!}{\displaystyle\prod_{i=1}^{g} n_i!}$$

and n_i denotes the number of objects in the ith of g treatment groups.

It is important to note that conventional measures of effect size such as $\hat{d}$, $\hat{r}^2$, $\hat{\eta}^2$, ϵ^2, and $\hat{\omega}^2$ depend on homogeneity of variance [311, p. 96]. In this regard, Mitchell and Hartmann documented this dependency and a number of additional weaknesses

[13]It should be noted that the sample standard deviation, s_x, is not an unbiased estimator of the population standard deviation, σ_x.

of measures of effect size, leading them to conclude that

> the *uncritical* use of magnitude of effects statistics as a cure for the problems of conventional hypothesis testing methods of assessing treatment effectiveness may very well represent a remedy as troublesome as the original problems [311, p. 99].[14]

In addition, these traditional measures require that $v = 2$ and the weighting factor be

$$C_i = \frac{n_i - 1}{N - g} , \qquad i = 1, \ldots, g .$$

Researchers have repeatedly called for an index of effect size that can quantify substantive importance in such a way that the index can be meaningfully interpreted when population variances differ [175, p. 910]. To this end, $\Re$ does not depend on the assumption of homogeneity; moreover, $\Re$ is sufficiently flexible to accommodate any $v > 0$ and any C_i, $i = 1, \ldots, g$.

3.3 Example Univariate MRPP Analyses with $g = 2$

In this section, three example analyses illustrate the permutation approach to typical two-sample problems. The first example is designed to correspond to the conventional Student's two-sample t test using a small set of univariate response measurement scores with $v = 2$ and treatment-group weights given by $C_i = (n_i - 1)/(N - g)$ for $i = 1, \ldots, g$; the second example analyzes the same small set of univariate response measurement scores, but with $v = 1$ and treatment-group weights given by $C_i = (n_i - 1)/(N - g)$ for $i = 1, \ldots, g$; and the third example analyzes the same set of response measurement scores using $v = 1$, but adopts a simple proportional treatment-group weighting function given by $C_i = n_i/N$ for $i = 1, \ldots, g$.

3.3.1 Example 1

Consider the univariate response measurement scores listed in Fig. 3.1 where $r = 1$, $g = 2$, $n_1 = n_2 = 10$, and $N = n_1 + n_2 = 20$. For this analysis let $v = 2$, employing squared Euclidean distance between response measurement scores, and let the treatment-group weights be given by

$$C_i = \frac{n_i - 1}{N - g} , \qquad i = 1, \ldots, g ,$$

to correspond to Student's two-sample t test.

[14]Emphasis in the original.

Fig. 3.1 Example univariate response measurement scores with $r = 1$, $g = 2$, $n_1 = n_2 = 10$, and $N = n_1 + n_2 = 20$

Treatment			
1		2	
99	94	98	90
99	95	98	86
97	89	97	86
98	96	92	85
95	94	92	60

An exact solution is feasible for the univariate response measurement scores listed in Fig. 3.1 since there are only

$$M = \frac{N!}{\prod_{i=1}^{g} n_i!} = \frac{20!}{10!\ 10!} = 184{,}756$$

possible, equally-likely arrangements of the $N = 20$ observed response measurement scores listed in Fig. 3.1. Following Eq. (3.2) on p. 58, the univariate response measurement scores listed in Fig. 3.1 yield $g = 2$ average distance-function values of

$$\xi_1 = 17.8667 \quad \text{and} \quad \xi_2 = 248.0889 \ .$$

Following Eq. (3.1) on p. 58, the observed value of the MRPP test statistic based on $v = 2$ and treatment-group weights

$$C_i = \frac{n_i - 1}{N - g} \ , \qquad i = 1, 2 \ ,$$

is

$$\delta_o = \sum_{i=1}^{g} C_i \xi_i = \frac{10 - 1}{20 - 2}\big(17.8667 + 248.0889\big) = 132.9778 \ .$$

If all arrangements of the $N = 20$ observed response measurement scores listed in Fig. 3.1 occur with equal chance, the exact probability value of $\delta_o = 132.9778$ computed on the $M = 184{,}756$ possible arrangements of the observed data with $n_1 = n_2 = 10$ univariate response measurement scores preserved for each arrangement is

$$P\big(\delta \leq \delta_o | H_0\big) = \frac{\text{number of } \delta \text{ values} \leq \delta_o}{M} = \frac{5{,}648}{184{,}756} = 0.0306 \ .$$

For comparison, a conventional pooled two-sample t test calculated on the univariate response measurement scores listed in Fig. 3.1 yields $\bar{x}_1 = 95.60$, $\bar{x}_2 = 88.40$, $s_1^2 = 8.9333$, $s_2^2 = 124.0444$, $s_p^2 = 66.4889$, and an observed value of $t_o = +1.9744$. Assuming independence, normality, and homogeneity of variance, t is approximately distributed as Student's t under the null hypothesis with $N - 2 = 20 - 2 = 18$ degrees of freedom. Under the null hypothesis, the observed value of $t_o = +1.9744$ yields an approximate two-sided probability value of $P = 0.0639$.

Following Eq. (3.4) on p. 59, the exact expected value of the $M = 184{,}756$ δ values is $\mu_\delta = 153.2632$ and, following Eq. (3.3) on p. 59, the observed chance-corrected measure of effect size is

$$\Re_o = 1 - \frac{\delta_o}{\mu_\delta} = 1 - \frac{132.9778}{153.2632} = +0.1324 ,$$

indicating approximately 13 % within-group agreement above that expected by chance. For comparison, the conventional measures of effect size are $\hat{d} = 0.8830$, $r^2 = \eta^2 = 0.1780$, $\hat{\omega}^2 = 0.1266$, and $\epsilon^2 = \hat{\eta}^2 = \hat{r}^2 = 0.1324$.

Note also that when $v = 2$ and the treatment-group weights are given by

$$C_i = \frac{n_i - 1}{N - g} , \qquad i = 1, 2 ,$$

$\xi_1 = 2s_1^2 = 2(8.9333) = 17.8667$, $\xi_2 = 2s_2^2 = 2(124.0444) = 248.0889$, $\delta_o = 2s_p^2 = 2(66.4889) = 132.9778$, and $\mu_\delta = 2SS_{\text{Total}}/(N-1) = 1{,}456/(20-1) = 153.2632$, as shown in Eqs. (3.6) on p. 61, (3.7) on p. 61, and (3.8) on p. 61, where $SS_{\text{Total}} = T - S^2/N = 170{,}736 - (1{,}840)^2/20 = 1{,}456$.

Given the univariate response measurement scores listed in Fig. 3.1 on p. 74, the observed values of S and T are

$$S_o = \sum_{i=1}^{N} x_i = 99 + 99 + 97 + \cdots + 60 = 1{,}840$$

and

$$T_o = \sum_{i=1}^{N} x_i^2 = 99^2 + 99^2 + 97^2 + \cdots + 60^2 = 170{,}736 ,$$

and the identities relating Student's two-sample t test and the MRPP test statistic are

$$t_o = \left[\frac{2(NT_o - S_o^2)}{N\delta_o} - N + 2 \right]^{1/2}$$

$$= \left\{ \frac{2[(20)(170{,}736) - 1{,}840^2]}{(20)(132.9778)} - 20 + 2 \right\}^{1/2} = 1.9744$$

and

$$\delta_o = \frac{2(NT_o - S_o^2)}{N(N-2+t_o^2)} = \frac{2[(20)(170{,}736) - 1{,}840^2]}{20(20-2+1.9744^2)} = 132.9778 .$$

Thus, Student's t test for two independent samples may be considered a special case of the MRPP test statistic, δ, with $v = 2$ and $C_i = (n_i - 1)/(N - g)$ for $i = 1, 2$.

While the above analysis is based on Student's pooled two-sample t test, it is readily apparent that, given the univariate response measurement scores listed in Fig. 3.1, the variances should not be pooled, as $s_1^2 = 8.9333$ and $s_2^2 = 124.0444$ are markedly different, with a ratio of 13.8856. This discrepancy is also reflected in the ξ_i values with $\xi_1 = 17.8667$ and $\xi_2 = 248.0889$, also with a ratio of 13.8856 since $\xi_i = 2s_i^2$ for $i = 1, 2$. It is generally recognized that for a two-sample t test, minor deviations from population normality are less important than inequality of population variances. In general, if the larger of two samples has the greater variance, there is increased risk of a type II or β error: failure to reject a false null hypothesis. However, if the smaller sample has the greater variance, there is increased risk of a type I or α error: rejection of a true null hypothesis [141]. In this instance, however, the point is moot as the two samples are of equal size with $n_1 = n_2 = 10$.

A non-pooled t test yields $t_o = 1.9744$, the same as the pooled t test since $n_1 = n_2$ in this example. However, the probability values of the two tests differ due to different degrees of freedom. In the case of the non-pooled t test, t is approximately distributed as Student's t with an estimated 10.2896 degrees of freedom. Under the null hypothesis, the observed value of $t_o = 1.9744$ yields an approximate two-sided probability value of $P = 0.0758$, which is slightly larger than the pooled t test probability value of $P = 0.0639$ and considerably greater than the exact probability value of $P = 0.0306$. The estimated degrees of freedom is based on a solution by Satterthwaite [363] that provides an approximate degrees of freedom given by

$$\min(n_1 - 1, n_2 - 1) \le \frac{\left(\dfrac{s_1^2}{n_1} + \dfrac{s_2^2}{n_2}\right)^2}{\dfrac{\left(\dfrac{s_1^2}{n_1}\right)^2}{n_1 - 1} + \dfrac{\left(\dfrac{s_2^2}{n_2}\right)^2}{n_2 - 1}} \le n_1 + n_2 - 2 .$$

Thus, for the data listed in Fig. 3.1, $\min(n_1 - 1, n_2 - 1) = \min(10 - 1, 10 - 1) = 9$, $n_1 + n_2 - 2 = 10 + 10 - 2 = 18$, and $9 \le 10.2826 \le 18$.[15]

[15] An alternative approximation of the degrees of freedom was proposed by B.L. Welch in 1938 [419]. For the data listed in Fig. 3.1, the Welch procedure yields approximately 10.5762 degrees of freedom, compared with the Satterthwaite procedure of approximately 10.2826 degrees of freedom, yielding an approximate two-sided probability value of $P = 0.0750$.

Note that it is inconsequential to the permutation test whether the population variances are equal or unequal, as the permutation test is strictly a data-dependent test. Therefore, the MRPP analysis does not change. This is not to say that heterogeneity of variances does not affect the value of the MRPP test statistic, but only that, unlike Student's t distribution, the discrete permutation distribution and the associated exact probability value are not dependent on the assumption of homogeneity of variance.

3.3.2 Example 2

For a second example analysis of the univariate response measurement scores listed in Fig. 3.1, let the treatment-group weights be given by

$$C_i = \frac{n_i - 1}{N - g} , \qquad i = 1, \ldots, g ,$$

as in Example 1, but set $v = 1$ instead of $v = 2$, employing ordinary Euclidean distance between response measurement scores. The univariate response measurement scores listed in Fig. 3.1 on p. 74 contain one extreme value of $x_{2,10} = 60$, relative to the other values in Treatment 2. Permutation tests based on $v = 1$ are quite robust to extreme values, while permutation tests (and conventional tests) based on $v = 2$ can be highly influenced by even a single extreme value due to squaring of the differences between the response measurement scores [295, pp. 13–15].

Following Eq. (3.2) on p. 58, the univariate response measurement scores listed in Fig. 3.1 on p. 74 yield $g = 2$ average distance-function values of

$$\xi_1 = 3.3778 \quad \text{and} \quad \xi_2 = 11.2889 .$$

Following Eq. (3.1) on p. 58, the observed value of the MRPP test statistic based on $v = 1$ and treatment-group weights

$$C_i = \frac{n_i - 1}{N - g} , \qquad i = 1, 2 ,$$

is

$$\delta_o = \sum_{i=1}^{g} C_i \xi_i = \frac{10 - 1}{20 - 2}(3.3778 + 11.2889) = 7.3333 .$$

If all arrangements of the $N = 20$ observed response measurement scores listed in Fig. 3.1 on p. 74 occur with equal chance, the exact probability value of $\delta_o = 7.3333$ computed on the $M = 184{,}756$ possible arrangements of the observed data with $n_1 = n_2 = 10$ univariate response measurement scores preserved for each arrangement is

$$P(\delta \leq \delta_o | H_0) = \frac{\text{number of } \delta \text{ values} \leq \delta_o}{M} = \frac{5{,}288}{184{,}756} = 0.0286 \; .$$

For comparison, the exact probability value based on $v = 2$, $M = 184{,}756$, and $C_i = (n_i - 1)/(N - g)$ for $i = 1, 2$ in Example 1 is $P = 0.0306$. No comparison is made with Student's two-sample t test as Student's t test is undefined for $v = 1$, as are the conventional measures of effect size for two-sample tests: $\hat{d}$, r^2, $\hat{r}^2$, η^2, $\hat{\eta}^2$, ϵ^2, and $\hat{\omega}^2$.

Following Eq. (3.4) on p. 59, the exact expected value of the $M = 184{,}756$ δ values is $\mu_\delta = 8.0842$ and, following Eq. (3.3) on p. 59, the observed chance-corrected measure of effect size is

$$\Re_o = 1 - \frac{\delta_o}{\mu_\delta} = 1 - \frac{7.3333}{8.0842} = +0.0929 \; ,$$

indicating approximately 9 % within-group agreement above that expected by chance.

To demonstrate the robustness of statistical analyses based on $v = 1$, consider again the single extreme value of $x_{2,10} = 60$ located in Treatment 2 in Fig. 3.1 on p. 74, replicated in Fig. 3.2 for convenience. Successively diminishing the value of $x_{2,10} = 60$ to 50, 40, 30, 20, 10, and finally to 0, does not change the exact permutation probability value of $P = 0.0286$. For comparison, Student's two-sample pooled t test with $x_{2,10} = 0$ yields an observed value of $t_o = +1.4139$ with an approximate two-sided probability value of $P = 0.1745$, instead of $P = 0.0639$ with $x_{2,10} = 60$, and a two-sample non-pooled t test with $x_{2,10} = 0$ yields an observed value of $t_o = +1.4139$ with an approximate two-sided probability value of $P = 0.1904$, instead of $P = 0.0758$ with $x_{2,10} = 60$. The probability values for the exact two-sample analysis with $v = 1$, Student's pooled t test, and the non-pooled t test, based on Satterthwaite's approximation, are listed in Fig. 3.3.

Fig. 3.2 Example univariate response measurement scores with $r = 1$, $g = 2$, $n_1 = n_2 = 10$, and $N = n_1 + n_2 = 20$

Treatment			
1		2	
99	94	98	90
99	95	98	86
97	89	97	86
98	96	92	85
95	94	92	60

Fig. 3.3 Exact, pooled, and non-pooled probability values for $x_{2,10}$ values of 60, 50, 40, 30, 20, 10, and 0

	Probability		
$x_{2,10}$	Exact	Pooled	Non-pooled
60	0.0286	0.0639	0.0758
50	0.0286	0.0876	0.1016
40	0.0286	0.1099	0.1249
30	0.0286	0.1296	0.1452
20	0.0286	0.1468	0.1626
10	0.0286	0.1616	0.1775
0	0.0286	0.1745	0.1904

3.3.3 Example 3

The treatment-group weighting function given by

$$C_i = \frac{n_i - 1}{N - g} \,, \qquad i = 1, \ldots, g \,,$$

is essential for the classical t test, but is not required for a permutation test, as degrees of freedom are irrelevant for distribution-free permutation methods. Thus, for this third analysis of the univariate response measurement scores listed in Fig. 3.2, the treatment-group weighting function is set to

$$C_i = \frac{n_i}{N} \,, \qquad i = 1, \ldots, g \,,$$

simply weighting each treatment group proportional to the number of observations in the group, and setting $v = 1$, employing ordinary Euclidean distance between response measurement scores, as in Example 2. Following Eq. (3.2) on p. 58, the univariate response measurement scores listed in Fig. 3.2 yield $g = 2$ average distance-function values of

$$\xi_1 = 3.3778 \quad \text{and} \quad \xi_2 = 11.2889 \,.$$

Following Eq. (3.1) on p. 58, the observed value of the MRPP test statistic based on $v = 1$ and treatment-group weights

$$C_i = \frac{n_i}{N} \,, \qquad i = 1, 2 \,,$$

is

$$\delta_o = \sum_{i=1}^{g} C_i \xi_i = \frac{10}{20}(3.3778 + 11.2889) = 7.3333 \,.$$

If all arrangements of the $N = 20$ observed response measurement scores listed in Fig. 3.2 occur with equal chance, the exact probability value of $\delta_o = 7.3333$ computed on the $M = 184{,}756$ possible arrangements of the observed data with $n_1 = n_2 = 10$ univariate response measurement scores preserved for each arrangement is

$$P(\delta \leq \delta_o|H_0) = \frac{\text{number of } \delta \text{ values} \leq \delta_o}{M} = \frac{5{,}288}{184{,}756} = 0.0286\,.$$

For comparison, the exact probability values based on $v = 2$, $M = 184{,}756$, and $C_i = (n_i - 1)/(N - g)$ for $i = 1, 2$ in Example 1 and $v = 1$, $M = 184{,}756$, and $C_i = (n_i - 1)/(N - g)$ for $i = 1, 2$ in Example 2 are $P = 0.0306$ and $P = 0.0286$, respectively.[16] No comparison is made with Student's two-sample t test as Student's t test is undefined for both $v = 1$ and $C_i = n_i/N$ for $i = 1, \ldots, g$, as are the conventional measures of effect size for two-sample tests: $\hat{d}$, r^2, $\hat{r}^2$, η^2, $\hat{\eta}^2$, ϵ^2, and $\hat{\omega}^2$.

Following Eq. (3.4) on p. 59, the exact expected value of the $M = 184{,}756$ δ values is $\mu_\delta = 8.0842$ and, following Eq. (3.3) on p. 59, the observed chance-corrected measure of effect size is

$$\Re_o = 1 - \frac{\delta_o}{\mu_\delta} = 1 - \frac{7.3333}{8.0842} = +0.0929\,,$$

indicating approximately 9 % within-group agreement above that expected by chance.

3.4 Permutation Analogue of Hotelling's T^2 Test

It is sometimes necessary to test for the difference between $g = 2$ independent treatment-groups when $r \geq 2$ response measurements have been obtained for each object. The usual approach to such research applications is Hotelling's T^2 test for two independent samples given by

$$T^2 = \frac{n_1 n_2}{N} (\bar{\mathbf{y}}_1 - \bar{\mathbf{y}}_2)' \, \mathbf{S}^{-1} (\bar{\mathbf{y}}_1 - \bar{\mathbf{y}}_2)\,, \tag{3.19}$$

where $\bar{\mathbf{y}}_1$ and $\bar{\mathbf{y}}_2$ denote vectors of mean differences between treatment groups 1 and 2, respectively, n_1 and n_2 are the number of interval-level multivariate response measurement scores in treatment groups 1 and 2, respectively, $N = n_1 + n_2$, and $\mathbf{S}$

[16]When $n_1 = n_2$, as in this case with $n_1 = n_2 = 10$, $C_i = n_i/N$ and $C_i = (n_i - 1)/(N - g)$ for $i = 1, \ldots, g$ are equivalent, yielding the same δ and P values.

is a variance–covariance matrix given by

$$\mathbf{S} = \begin{bmatrix} \frac{1}{N}\sum_{I=1}^{N}(y_{1I}-\bar{y}_1)^2 & \cdots & \frac{1}{N}\sum_{I=1}^{N}(y_{1I}-\bar{y}_1)(y_{rI}-\bar{y}_r) \\ \vdots & & \vdots \\ \frac{1}{N}\sum_{I=1}^{N}(y_{rI}-\bar{y}_r)(y_{1I}-\bar{y}_1) & \cdots & \frac{1}{N}\sum_{I=1}^{N}(y_{rI}-\bar{y}_r)^2 \end{bmatrix} \tag{3.20}$$

[291, p. 228].[17] The observed value of Hotelling's T^2, T^2_{o}, is conventionally transformed into an observed F-ratio by

$$F_{\text{o}} = \frac{N-r-1}{r(N-2)}\,T^2_{\text{o}}\,,$$

which is approximately distributed as Snedecor's F under the null hypothesis with $\nu_1 = r$ and $\nu_2 = N-r-1$ degrees of freedom.

Whenever the data consist of $r \geq 2$ response measurement scores for each object, the response measurement scores may be expressed in entirely different units of measurement, such as height in inches and weight in pounds. The disparate response measurement scores must be made commensurate (i.e., standardized) prior to analysis. Let $y_i' = (y_{1i}, \ldots, y_{ri})$ for $i = 1, \ldots, N$ denote N non-commensurate r-dimensional values $(r \geq 2)$. The corresponding N commensurate r-dimensional values denoted by $x_i' = (x_{1i}, \ldots, x_{ri})$ for $i = 1, \ldots, N$ are given by $x_{ij} = y_{ij}/\phi_j$, where

$$\phi_j = \left[\sum_{I<J}\left|y_{jI} - y_{jJ}\right|^v\right]^{1/v}$$

for $j = 1, \ldots, r$. The commensurated response measurement scores have the desired property that

$$\sum_{I<J}\left|x_{jI} - x_{jJ}\right|^v = 1$$

for $j = 1, \ldots, r$ and $v > 0$. The commensuration procedure is based on the distance between the r response measurements of objects ω_I and ω_J and is given by the

[17] As noted by Anderson [8, p. 1], Hotelling's original notation for the variance–covariance matrix was "A," while the current convention is "S."

generalized Minkowski distance function

$$\Delta(I,J) = \left[\sum_{j=1}^{r} \left(x_{jI} - x_{jJ} \right)^2 \right]^{v/2} ,$$

where $v > 0$. The commensuration is termed Euclidean commensuration when $v = 1$ and Hotelling commensuration when $v = 2$ [297, pp. 53–57]. Hotelling commensuration with $v = 2$ is based on the distance function

$$\Delta(I,J) = \left[(y_I - y_J)' \, \mathbf{S}^{-1} \, (y_I - y_J) \right]^{v/2} ,$$

where $\mathbf{S}$ is the $r \times r$ variance–covariance matrix given in Eq. (3.20).

3.5 Example Bivariate MRPP Analyses with $g = 2$

In this section, three example analyses with bivariate response measurement scores illustrate the permutation approach to two-sample problems with multivariate response measurement scores. The first example is designed to correspond to the conventional Hotelling two-sample T^2 test using a small set of bivariate response measurement scores with $v = 2$, Hotelling commensuration, and treatment-group weights given by $C_i = (n_i - 1)/(N - g)$ for $i = 1, \ldots, g$; the second example analyzes the same small set of bivariate response measurement scores, but with $v = 1$, Euclidean commensuration, and treatment-group weights given by $C_i = (n_i - 1)/(N - g)$ for $i = 1, \ldots, g$; and the third example analyzes the same set of bivariate response measurement scores using $v = 1$ and Euclidean commensuration, but adopts a simple proportional treatment-group weighting given by $C_i = n_i/N$ for $i = 1, \ldots, g$.

3.5.1 Example 1

Consider the bivariate response measurement scores listed in Fig. 3.4, where $r = 2$, $g = 2$, $n_1 = 4$, $n_2 = 6$, and $N = n_1 + n_2 = 10$. For this first analysis, let $v = 2$, employing squared Euclidean distance between response measurement scores, and let the treatment-group weights be given by

$$C_i = \frac{n_i - 1}{N - g} , \qquad i = 1, \ldots, g ,$$

to correspond to Hotelling's two-sample T^2 test [181]. An exact permutation solution is feasible for the bivariate response measurement scores listed in Fig. 3.4 since

Fig. 3.4 Example bivariate response measurement scores with $r = 2$, $g = 2$, $n_1 = 4$, $n_2 = 6$, and $N = n_1 + n_2 = 10$

Treatment	
1	2
(1.2, 3.1)	(3.7, 6.1)
(2.9, 6.8)	(6.1, 8.3)
(1.8, 2.1)	(6.2, 7.9)
(5.2, 6.1)	(4.8, 9.7)
	(5.1, 9.9)
	(4.2, 7.8)

there are only

$$M = \frac{N!}{\prod_{i=1}^{g} n_i!} = \frac{10!}{4!\;6!} = 210$$

possible, equally-likely arrangements of the $N = 10$ observed bivariate response measurement scores listed in Fig. 3.4. Following Eq. (3.2) on p. 58, the bivariate response measurement scores listed in Fig. 3.4 yield $g = 2$ average distance-function values of

$$\xi_1 = 0.4862 \quad \text{and} \quad \xi_2 = 0.2737 \ .$$

Following Eq. (3.1) on p. 58, the observed value of the MRPP test statistic based on $v = 2$ and treatment-group weights

$$C_i = \frac{n_i - 1}{N - g} \ , \qquad i = 1, 2 \ ,$$

is

$$\delta_\text{o} = \sum_{i=1}^{g} C_i \xi_i = \frac{1}{10-2}\big[(4-1)(0.4862) + (6-1)(0.2737)\big] = 0.3534 \ .$$

If all arrangements of the $N = 10$ observed bivariate response measurement scores listed in Fig. 3.4 occur with equal chance, the exact probability value of $\delta_\text{o} = 0.3534$ computed on the $M = 210$ possible arrangements of the observed data with $n_1 = 4$ and $n_2 = 6$ bivariate response measurement scores preserved for each arrangement is

$$P\big(\delta \leq \delta_\text{o} | H_0\big) = \frac{\text{number of } \delta \text{ values} \leq \delta_\text{o}}{M} = \frac{12}{210} = 0.0571 \ .$$

For comparison, a conventional two-sample Hotelling T^2 test of the $N = 10$ bivariate response measurement scores listed in Fig. 3.4 yields $\bar{y}_{11} = 2.7750$, $s_{11}^2 = 3.1092$, $\bar{y}_{12} = 4.5250$, $s_{12}^2 = 5.1892$, $\text{cov}(1,2)_1 = +2.9042$, $\bar{y}_{21} = 5.0167$, $s_{21}^2 = 1.0057$, $\bar{y}_{22} = 8.2833$, $s_{22}^2 = 1.9537$, and $\text{cov}(1,2)_2 = +0.5323$. Then, $\bar{\mathbf{y}}_1$ and $\bar{\mathbf{y}}_2$ in Eq. (3.19) are

$$\bar{\mathbf{y}}_1 = \bar{y}_{11} - \bar{y}_{21} = 2.7550 - 5.0167 = -2.2417$$

and

$$\bar{\mathbf{y}}_2 = \bar{y}_{12} - \bar{y}_{22} = 4.5250 - 8.2833 = -3.7583\ .$$

The variance–covariance matrices for Treatments 1 and 2 in Fig. 3.4 are

$$\hat{\mathbf{\Sigma}}_1 = \begin{bmatrix} 3.1092 & +2.9042 \\ +2.9042 & 5.1892 \end{bmatrix} \quad \text{and} \quad \hat{\mathbf{\Sigma}}_2 = \begin{bmatrix} 1.0057 & +0.5323 \\ +0.5323 & 1.9537 \end{bmatrix},$$

respectively, and the pooled variance–covariance matrix and its inverse are

$$\mathbf{S} = \begin{bmatrix} 1.7945 & +1.4218 \\ +1.4218 & 3.1670 \end{bmatrix} \quad \text{and} \quad \mathbf{S}^{-1} = \begin{bmatrix} +0.8649 & -0.3883 \\ -0.3883 & +0.4901 \end{bmatrix},$$

respectively.[18]

Following Eq. (3.19) on p. 80, the observed value of Hotelling's T^2 is

$$\begin{aligned} T_o^2 &= \frac{n_1 n_2}{N} (\bar{\mathbf{y}}_1 - \bar{\mathbf{y}}_2)' \, \mathbf{S}^{-1} \, (\bar{\mathbf{y}}_1 - \bar{\mathbf{y}}_2) \\ &= \frac{(4)(6)}{10} \begin{bmatrix} -2.2417 & -3.7583 \end{bmatrix} \begin{bmatrix} +0.8649 & -0.3883 \\ -0.3883 & +0.4901 \end{bmatrix} \begin{bmatrix} -2.2417 \\ -3.7583 \end{bmatrix} \\ &= (2.40)(4.7260) = 11.3423 \end{aligned}$$

and the observed F-ratio for Hotelling's T^2 is

$$F_o = \frac{N - r - 1}{r(N-2)} T_o^2 = \frac{10 - 2 - 1}{2(10-2)} (11.3423) = 4.9623\ .$$

Assuming independence, normality, and homogeneity of variance, F is approximately distributed as Snedecor's F under the null hypothesis with $\nu_1 = r = 2$ and

[18]Each element of the **S** matrix is constructed from two corresponding elements in the $\hat{\mathbf{\Sigma}}$ matrices, weighted by the degrees of freedom, i.e., $n - 1$. For example, the first element of the **S** matrix is $1.7945 = [(4-1)(3.1092) + (6-1)(1.0057)]/(4+6-2)$.

$\nu_2 = N - r - 1 = 10 - 2 - 1 = 7$ degrees of freedom. Under the null hypothesis, the observed value of $F_o = 4.9623$ yields an approximate probability value of $P = 0.0455$.

Following Eq. (3.4) on p. 59, the exact expected value of the $M = 210$ δ values is $\mu_\delta = 0.4444$ and, following Eq. (3.3) on p. 59, the observed chance-corrected measure of effect size is

$$\Re_o = 1 - \frac{\delta_o}{\mu_\delta} = 1 - \frac{0.3534}{0.4444} = +0.2049 \, ,$$

indicating approximately 20 % within-group agreement above that expected by chance.

The identity relating Hotelling's two-sample T^2 test and the MRPP test statistic is given by

$$\delta = \frac{2(r - V^{(s)})}{N - g} \, , \tag{3.21}$$

where

$$V^{(s)} = \frac{T^2}{T^2 + N - g} \tag{3.22}$$

and $s = \min(g-1, r)$; in this case with $g - 1 = 2 - 1 = 1$ and $r = 2$, $s = \min(2 - 1, 2) = 1$. Thus, following Eqs. (3.21) and (3.22), the observed value of $V_o^{(1)}$ is

$$V_o^{(1)} = \frac{11.3423}{11.3423 + 10 - 2} = \frac{11.3423}{19.3423} = 0.5864$$

and the observed value of δ is

$$\delta_o = \frac{2(2 - 0.5864)}{10 - 2} = \frac{2.8272}{8} = 0.3534 \, .$$

It is obvious from an inspection of Eq. (3.22) that $V^{(1)}$ is simply the squared multiple correlation coefficient, $R^2_{y_1,y_2.x}$, where the $n_1 = 4$ objects in treatment-group 1 are dummy-coded by some numerical value, say 0, and the $n_2 = 6$ objects in treatment-group 2 are coded by some other numerical value, say 1.[19] Figure 3.5 displays the multiple correlation data where variable x is the dummy-coded independent variable, variable y_1 is a dependent variable containing the first of the $r = 2$ response measurement scores for each object, and variable y_2 is a second dependent variable containing the second of the response measurement scores for each object.

[19] Actually, any two different numerical values will suffice for dummy coding, but 0 and 1 values are conventional for two treatment groups.

Fig. 3.5 Example bivariate regression response measurement scores with $N = 10$ cases, independent variable x, and dependent variables y_1 and y_2

Variable		
x	y_1	y_2
0	1.2	3.1
0	2.9	6.8
0	1.8	2.1
0	5.2	6.1
1	3.7	6.1
1	6.1	8.3
1	6.2	7.9
1	4.8	9.7
1	5.1	9.9
1	4.2	2.8

For the bivariate response measurement scores listed in Fig. 3.5, $R^2_{y_1,y_2.x} = 0.5864$. More simply,

$$\eta^2 = \frac{V^{(s)}}{s} = \frac{V^{(s)}}{\min(g-1,r)} = \frac{V^{(1)}}{1} = \frac{0.5864}{1} = 0.5864 \; .$$

This example analysis demonstrates that the Hotelling two-sample T^2 test may simply be considered a special version of the MRPP test statistic, δ, with $v = 2$ and $C_i = (n_i - 1)/(N - g)$ for $i = 1, 2$. Considering the distributional problems under the multivariate normal assumptions that are rarely satisfied in practice [7], the exact permutation analogue of Hotelling's T^2 test offers a vast improvement over any approach in the current literature. In addition, the conventional Hotelling T^2 test fails if $r > g$, while δ processes such cases without any problems; see, for example, a 1996 paper by Mielke, Berry, and Neidt in *Psychological Reports* [304].

3.5.2 Example 2

As with Student's two-sample t test, it is not necessary to set $v = 2$, thereby squaring the response-measurement differences between objects. For a second example analysis of the bivariate response measurement scores listed in Fig. 3.4 on p. 83, let the treatment-group weights be given by

$$C_i = \frac{n_i - 1}{N - g}, \qquad i = 1, \ldots, g \; ,$$

as in Example 1, but set $v = 1$ instead of $v = 2$, employing ordinary Euclidean distance between response measurement scores. Note that with $v = 1$, Euclidean commensuration is selected. Following Eq. (3.2) on p. 58, the bivariate response measurement scores listed in Fig. 3.4 yield $g = 2$ average distance-function values of

$$\xi_1 = 3.7865 \quad \text{and} \quad \xi_2 = 2.2200 \, .$$

Following Eq. (3.1) on p. 58, the observed value of the MRPP test statistic based on $v = 1$ and treatment-group weights

$$C_i = \frac{n_i - 1}{N - g} \, , \qquad i = 1, 2 \, ,$$

is

$$\delta_o = \sum_{i=1}^{g} C_i \xi_i = \frac{1}{10 - 2} \big[(4 - 1)(3.7865) + (6 - 1)(2.2200) \big] = 2.8074 \, .$$

If all arrangements of the $N = 10$ observed bivariate response measurement scores listed in Fig. 3.4 occur with equal chance, the exact probability value of $\delta_o = 2.8074$ computed on the $M = 210$ possible arrangements of the observed data with $n_1 = 4$ and $n_2 = 6$ bivariate response measurement scores preserved for each arrangement is

$$P(\delta \leq \delta_o | H_0) = \frac{\text{number of } \delta \text{ values} \leq \delta_o}{M} = \frac{4}{210} = 0.0190 \, .$$

For comparison, the exact probability value based on $v = 2$, $M = 210$, and $C_i = (n_i - 1)/(N - g)$ for $i = 1, 2$ in Example 1 is $P = 0.0571$. No comparison is made with Hotelling's two-sample T^2 test as Hotelling's T^2 is undefined for $v = 1$, as are the conventional measures of effect size for two-sample tests: $\hat{d}$, r^2, $\hat{r}^2$, η^2, $\hat{\eta}^2$, ϵ^2, and $\hat{\omega}^2$.

Following Eq. (3.4) on p. 59, the exact expected value of the $M = 210$ δ values is $\mu_\delta = 3.7628$ and, following Eq. (3.3) on p. 59, the observed chance-corrected measure of effect size is

$$\Re_o = 1 - \frac{\delta_o}{\mu_\delta} = 1 - \frac{2.8074}{3.7628} = +0.2539 \, ,$$

indicating approximately 25 % within-group agreement above that expected by chance.

3.5.3 Example 3

As with Student's two-sample t test, the treatment-group weighting function given by

$$C_i = \frac{n_i - 1}{N - g}\,, \qquad i = 1,\, \ldots,\, g\,,$$

is, from a permutation perspective, a relic from the classical t test and is not suitable for a distribution-free permutation test, as degrees of freedom are never appropriate for permutation methods, except when validating a corresponding conventional test. Thus, for this third analysis of the bivariate response measurement scores listed in Fig. 3.4, the treatment-group weighting function is set to

$$C_i = \frac{n_i}{N}\,, \qquad i = 1,\, \ldots,\, g\,,$$

weighting each treatment group proportional to the number of observations in the group, and the distance function is set to $v = 1$ as in Example 2, again selecting Euclidean commensuration.

Following Eq. (3.2) on p. 58, the bivariate response measurement scores listed in Fig. 3.4 yield $g = 2$ average distance-function values of

$$\xi_1 = 3.7865 \quad \text{and} \quad \xi_2 = 2.2200\,.$$

Following Eq. (3.1) on p. 58, the observed value of the MRPP test statistic based on $v = 1$ and treatment-group weights

$$C_i = \frac{n_i}{N}\,, \qquad i = 1, 2\,,$$

is

$$\delta_o = \sum_{i=1}^{g} C_i \xi_i = \frac{1}{10}\big[(4)(3.7865) + (6)(2.2200)\big] = 2.8466\,.$$

If all arrangements of the $N = 10$ observed bivariate response measurement scores listed in Fig. 3.4 occur with equal chance, the exact probability value of $\delta_o = 2.8466$ computed on the $M = 210$ possible arrangements of the observed data with $n_1 = 4$ and $n_2 = 6$ bivariate response measurement scores preserved for each arrangement is

$$P\big(\delta \le \delta_o | H_0\big) = \frac{\text{number of } \delta \text{ values} \le \delta_o}{M} = \frac{4}{210} = 0.0190\,.$$

For comparison, the exact probability values based on $v = 2$, $M = 210$, and $C_i = (n_i - 1)/(N - g)$ for $i = 1, 2$ in Example 1 and $v = 1$, $M = 210$, and $C_i = (n_i - 1)/(N - g)$ for $i = 1, 2$ in Example 2 are $P = 0.0571$ and $P = 0.0190$, respectively. No comparison is made with the Hotelling two-sample T^2 test as T^2 is undefined for both $v = 1$ and $C_i = n_i/N$ for $i = 1, \ldots, g$, as are the conventional measures of effect size for two-sample tests: $\hat{d}$, r^2, $\hat{r}^2$, η^2, $\hat{\eta}^2$, ϵ^2, and $\hat{\omega}^2$.

Following Eq. (3.4) on p. 59, the exact expected value of the $M = 210$ δ values is $\mu_\delta = 3.7628$ and, following Eq. (3.3) on p. 59, the observed chance-corrected measure of effect size is

$$\Re_o = 1 - \frac{\delta_o}{\mu_\delta} = 1 - \frac{2.8466}{3.7628} = +0.2435 ,$$

indicating approximately 24 % within-group agreement above that expected by chance.

3.6 Permutation Analogue of One-Way ANOVA

The one-way analysis of variance with $g \geq 3$ treatment groups and univariate response measurements on each object is a popular statistical approach to test for differences among treatment groups. Consider the conventional one-way analysis-of-variance (ANOVA) test statistic,

$$F = \frac{MS_{\text{Between}}}{MS_{\text{Within}}} , \tag{3.23}$$

where

$$MS_{\text{Between}} = \frac{1}{g-1} \sum_{i=1}^{g} n_i (\bar{x}_i - \bar{\bar{x}})^2 ,$$

$$MS_{\text{Within}} = \frac{1}{N-g} \sum_{i=1}^{g} \sum_{j=1}^{n_i} (x_{ij} - \bar{x}_i)^2 ,$$

n_i is the number of objects in the ith of g treatment groups, $N = \sum_{i=1}^{g} n_i$ is the total number of objects in the g treatment groups, x_{ij} is a univariate response measurement score for the jth object in the ith treatment group, $\bar{x}_i$ is the average response measurement score for the ith treatment group, given by

$$\bar{x}_i = \frac{1}{n_i} \sum_{j=1}^{n_i} x_{ij} , \qquad i = 1, \ldots, g ,$$

and $\bar{\bar{x}}$ is the grand mean of the N response measurement scores, given by

$$\bar{\bar{x}} = \frac{1}{N} \sum_{i=1}^{g} \sum_{j=1}^{n_i} x_{ij} \,.$$

Assuming independence, normality, and homogeneity of variance, F is approximately distributed as Snedecor's F under the null hypothesis of no difference among population means with $\nu_1 = g - 1$ and $\nu_2 = N - g$ degrees of freedom. When $r = 1$, $v = 2$, and the treatment-group weights are given by

$$C_i = \frac{n_i - 1}{N - g} \,, \qquad i = 1, \ldots, g \,,$$

it can easily be shown that δ is the permutation analogue of the conventional F-ratio test statistic, as defined in Eq. (3.23). When $v = 2$, employing squared Euclidean distance between response measurement scores, and the treatment-group weights are given by

$$C_i = \frac{n_i - 1}{N - g} \,, \qquad i = 1, \ldots, g \,,$$

the relationships between the MRPP test statistic and the conventional F-ratio are given by

$$\delta = \frac{2(NB - A^2)}{N[N - g + (g - 1)F]} \quad \text{and} \quad F = \frac{2(NB - A^2)}{(g - 1)N\delta} - \frac{N - g}{g - 1} \,,$$

where

$$A = \sum_{i=1}^{N} x_i \,, \qquad B = \sum_{i=1}^{N} x_i^2 \,,$$

and x_i is a univariate response measurement score for the ith of N objects. The permutation analogue of the F test given here is commonly called the Fisher–Pitman permutation test [119, 342]. Note also that

$$\mu_\delta = \frac{2SS_{\text{Total}}}{N - 1} \tag{3.24}$$

and

$$\delta = 2MS_{\text{Within}} \tag{3.25}$$

yield the functional relationship

$$\delta = \frac{1}{N-g}\big[(N-1)\mu_\delta - 2(g-1)MS_{\text{Between}}\big] , \tag{3.26}$$

where

$$SS_{\text{Total}} = (g-1)MS_{\text{Between}} + (N-g)MS_{\text{Within}} = SS_{\text{Between}} + SS_{\text{Within}} .$$

As is readily apparent in Eq. (3.26), test statistic δ depends solely on the differences among (between) the group means since μ_δ is fixed for a given univariate sample and MS_{Between} depends only on the differences among group means. Thus, the permutation statistical inference is completely unaffected by differences in scale among the g treatment groups [33, 194].

3.6.1 Computing Efficiency

It should be noted that it is not necessary to calculate the F-ratio test statistic for each permutation of the observed response measurement scores. As previously, let n_i denote the number of objects in the ith of g treatment groups, $i = 1, \ldots, g$, where

$$N = \sum_{i=1}^{g} n_i$$

is the total number of objects in the g treatment groups. Then the Fisher–Pitman test statistic for g treatment groups, T, is given by

$$T = \sum_{i=1}^{g} n_i \bar{x}_i^2 ,$$

where

$$\bar{x}_i = \frac{1}{n_i}\sum_{j=1}^{n_i} x_{ij}$$

and x_{ij} denotes the univariate response measurement scores of the jth subject in the ith of g treatment groups.

Under the Fisher–Pitman null hypothesis, T and F are equivalent since

$$F = \frac{MS_{\text{Between}}}{MS_{\text{Within}}} = \frac{SS_{\text{Between}}/(g-1)}{SS_{\text{Within}}/(N-g)} = \frac{(T - N\bar{\bar{x}}^2)/(g-1)}{(V-T)/(N-g)} .$$

Note that

$$\bar{\bar{x}} = \frac{1}{N}\sum_{i=1}^{g} n_i \bar{x}_i \,, \qquad V = \sum_{i=1}^{g}\sum_{j=1}^{n_i} x_{ij}^2 \,,$$

N, and g are invariant under permutation. Consequently, T and F are equivalent test statistics for testing the Fisher–Pitman null hypothesis; however, T is computationally more efficient and, more importantly, does not depend on sample estimates of the population variance [38].

3.7 Example Univariate MRPP Analyses with $g = 4$

In this section, three example analyses illustrate the permutation approach to typical one-way analysis-of-variance (ANOVA) problems. The first example is designed to correspond to the conventional F test using a small set of univariate response measurement scores with $v = 2$ and treatment-group weights $C_i = (n_i - 1)/(N - g)$ for $i = 1, \ldots, g$; the second example analyzes the same small set of univariate response measurement scores, but uses $v = 1$ and treatment-group weights $C_i = (n_i - 1)/(N - g)$ for $i = 1, \ldots, g$; and the third example analyzes the same set of univariate response measurement scores using $v = 1$, but adopts a proportional treatment-group weighting function given by $C_i = n_i/N$ for $i = 1, \ldots, g$.

3.7.1 Example 1

Consider the small set of univariate response measurement scores listed in Fig. 3.6, where $r = 1$, $g = 4$, $n_1 = n_2 = 3$, $n_3 = 4$, $n_4 = 5$, and where $N = n_1 + n_2 + n_3 + n_4 = 15$. For this first analysis, let $v = 2$, employing squared Euclidean distance between response measurement scores, and let the treatment-group weights be given by

$$C_i = \frac{n_i - 1}{N - g} \,, \qquad i = 1, \ldots, g \,.$$

Fig. 3.6 Example univariate response measurement scores with $r = 1$, $g = 4$, $n_1 = n_2 = 3$, $n_3 = 4$, $n_4 = 5$, and $N = n_1 + n_2 + n_3 + n_4 = 15$

Treatment			
1	2	3	4
10	11	12	14
11	12	13	15
12	13	14	16
		15	17
			33

Because an exact solution would require generating

$$M = \frac{N!}{\prod_{i=1}^{g} n_i!} = \frac{15!}{3!\;3!\;4!\;5!} = 12{,}612{,}600$$

possible, equally-likely arrangements of the $N = 15$ observed response measurement scores listed in Fig. 3.6, a resampling solution is more practical. In this example analysis, the number of random arrangements of the univariate response measurement scores listed in Fig. 3.6 is set to $L = 1{,}000{,}000$ to ensure an approximate resampling probability value with three decimal places of accuracy. Following Eq. (3.2) on p. 58, the univariate response measurement scores listed in Fig. 3.6 yield $g = 4$ average distance-function values of

$$\xi_1 = \xi_2 = 2.00\;, \quad \xi_3 = 3.3333\;, \quad \text{and} \quad \xi_4 = 125.00\;.$$

Following Eq. (3.1) on p. 58, the observed value of the MRPP test statistic based on $v = 2$ and treatment-group weights

$$C_i = \frac{n_i - 1}{N - g}\;, \qquad i = 1, \ldots, 4\;,$$

is

$$\delta_o = \sum_{i=1}^{g} C_i \xi_i = \frac{1}{15-4}\big[(3-1)(2.00) + (3-1)(2.00) + (4-1)(3.3333) + (5-1)(125.00)\big] = 47.0909\;.$$

If all M possible arrangements of the $N = 15$ observed response measurement scores listed in Fig. 3.6 occur with equal chance, the approximate resampling probability value of $\delta_o = 47.0909$ computed on $L = 1{,}000{,}000$ random arrangements of the observed data with $n_1 = n_2 = 3$, $n_3 = 4$, and $n_4 = 5$ univariate response measurement scores preserved for each arrangement is

$$P(\delta \leq \delta_o | H_0) = \frac{\text{number of } \delta \text{ values} \leq \delta_o}{L} = \frac{53{,}200}{1{,}000{,}000} = 0.0532\;.$$

Following Eq. (3.4) on p. 59, the exact expected value of the $M = 12{,}612{,}600$ δ values is $\mu_\delta = 59.9619$ and, following Eq. (3.3) on p. 59, the observed chance-corrected measure of effect size is

$$\mathfrak{R}_o = 1 - \frac{\delta_o}{\mu_\delta} = 1 - \frac{47.0909}{59.9619} = +0.2147\;,$$

indicating approximately 21 % within-group agreement above that expected by chance.

An Exact Test

Although an exact permutation analysis of the $N = 15$ univariate response measurement scores listed in Fig. 3.6 is impractical, it is not impossible. Following Eq. (3.2) on p. 58, an exact permutation analysis of the univariate response measurement scores listed in Fig. 3.6 yields $g = 4$ average distance-function values of

$$\xi_1 = \xi_2 = 2.00\ , \quad \xi_3 = 3.3333\ , \quad \text{and} \quad \xi_4 = 125.00\ .$$

Following Eq. (3.1) on p. 58, the observed value of the MRPP test statistic based on $v = 2$ and treatment-group weights

$$C_i = \frac{n_i - 1}{N - g}\ , \qquad i = 1, \ldots, 4\ ,$$

is

$$\delta_o = \sum_{i=1}^{g} C_i \xi_i = \frac{1}{15-4}\big[(3-1)(2.00) + (3-1)(2.00) + (4-1)(3.3333) + (5-1)(125.00)\big] = 47.0909\ .$$

Note that the ξ_i values, $i = 1, \ldots, 4$, and the observed δ value, δ_o, are identical for both the resampling and exact tests.

If all arrangements of the $N = 15$ observed response measurement scores listed in Fig. 3.6 occur with equal chance, the exact probability value of $\delta_o = 47.0909$ computed on the $M = 12{,}612{,}600$ possible arrangements of the observed data with $n_1 = n_2 = 3$, $n_3 = 4$, and $n_4 = 5$ univariate response measurement scores preserved for each arrangement is

$$P(\delta \le \delta_o | H_0) = \frac{\text{number of } \delta \text{ values} \le \delta_o}{M} = \frac{673{,}490}{12{,}612{,}600} = 0.0534\ .$$

Carrying the resampling probability value based on $L = 1{,}000{,}000$ and the exact probability value based on $M = 12{,}612{,}600$ to a few extra decimal places allows for a more direct comparison of the resampling and exact permutation approaches. The resampling approximate probability value to six decimal places is $P = 0.053242$ and the corresponding exact probability value is $P = 0.053398$ for a difference of 0.000156, demonstrating the efficiency and accuracy of a resampling approach for permutation methods when L is large.

Following Eq. (3.4) on p. 59, the exact expected value of the $M = 12{,}612{,}600$ δ values is $\mu_\delta = 59.9619$ and, following Eq. (3.3) on p. 59, the observed chance-corrected measure of effect size is

$$\Re_o = 1 - \frac{\delta_o}{\mu_\delta} = 1 - \frac{47.0909}{59.9619} = +0.2147 \; ,$$

indicating approximately 21 % within-group agreement above that expected by chance.

A Conventional Test

For comparison, the univariate response measurement scores listed in Fig. 3.6 yield estimated population means of $\bar{x}_1 = 11.00$, $\bar{x}_2 = 12.00$, $\bar{x}_3 = 13.50$, and $\bar{x}_4 = 19.00$; a grand mean based on all $N = 15$ response measurement scores of $\bar{\bar{x}} = 14.5333$; and estimated population variances of $s_1^2 = 1.00$, $s_2^2 = 1.00$, $s_3^2 = 1.6667$, and $s_4^2 = 62.50$. A conventional F test on the univariate response measurement scores listed in Fig. 3.6 yields $MS_{\text{Between}} = 53.5778$, $MS_{\text{Within}} = 23.5455$, $SS_{\text{Total}} = 419.7333$, and an observed F-ratio value of $F_o = 2.2755$.

Assuming independence, normality, and homogeneity of variance, F is approximately distributed as Snedecor's F under the null hypothesis with $\nu_1 = g - 1 = 4 - 1 = 3$ and $\nu_2 = N - g = 15 - 4 = 11$ degrees of freedom. Under the null hypothesis, the observed value of $F_o = 2.2755$ yields an approximate probability value of $P = 0.1366$.

It is readily apparent that, for the $N = 15$ univariate response measurement scores listed in Fig. 3.6, the assumption of homogeneity of variance has not been met, e.g., $s_1^2 = 1.00$ and $s_4^2 = 62.50$. In 1951 B.L. Welch proposed an adjustment to the conventional F-ratio that compensated for unequal variances [420]. Following Welch, define an adjusted F-ratio as

$$F' = \frac{\dfrac{1}{g-1} \displaystyle\sum_{i=1}^{g} w_i \left(\bar{x}_i - \bar{\bar{x}} \right)^2}{1 + \dfrac{2(g-2)}{g^2 - 1} \displaystyle\sum_{i=1}^{g} \left(\frac{1}{n_i - 1} \right) \left(1 - \frac{w_i}{\displaystyle\sum_{i=1}^{g} w_i} \right)^2} \; , \tag{3.27}$$

where $\bar{x}_i$ is the mean of each of g treatments, $i = 1, \ldots, g$, $\bar{\bar{x}}$ is the grand mean over all treatments, and w_i for $i = 1, \ldots, g$ are weights assigned to each treatment given by

$$w_i = \frac{n_i}{s_i^2} \; , \qquad i = 1, \ldots, g \; ,$$

where n_i is the number of response measurement scores in each of g treatments and s_i^2 is the estimated population variance for each treatment, $i = 1, \ldots, g$.

For the $N = 15$ response measurement scores listed in Fig. 3.6 on p. 92, the $g = 4$ sample means are

$$\bar{x}_1 = \frac{1}{n_1}\sum_{i=1}^{n_1} x_i = \frac{33.00}{3} = 11.00\,, \quad \bar{x}_2 = \frac{1}{n_2}\sum_{i=1}^{n_2} x_i = \frac{36.00}{3} = 12.00\,,$$

$$\bar{x}_3 = \frac{1}{n_3}\sum_{i=1}^{n_3} x_i = \frac{54.00}{4} = 13.50\,, \quad \bar{x}_4 = \frac{1}{n_4}\sum_{i=1}^{n_4} x_i = \frac{95.00}{5} = 19.00\,,$$

and the grand mean is

$$\bar{\bar{x}} = \frac{1}{N}\sum_{i=1}^{g} n_i\bar{x}_i = \frac{1}{15}\Big[(3)(11.00) + (3)(12.00) + (4)(13.50) + (5)(19.00)\Big]$$

$$= \frac{218.00}{15} = 14.5333\,.$$

Also, for the response measurement scores listed in Fig. 3.6, the $g = 4$ weights specified by Welch in Eq. (3.27) are

$$w_1 = \frac{n_1}{s_1^2} = \frac{3}{1.00} = 3.00\,, \qquad w_2 = \frac{n_2}{s_2^2} = \frac{3}{1.00} = 3.00\,,$$

$$w_3 = \frac{n_3}{s_3^2} = \frac{4}{1.6667} = 2.40\,, \qquad w_4 = \frac{n_4}{s_4^2} = \frac{5}{62.50} = 0.08\,,$$

and the sum of the $g = 4$ weights is

$$\sum_{i=1}^{g} w_i = 3.00 + 3.00 + 2.40 + 0.08 = 8.48\,.$$

Then, for the response measurement scores listed in Fig. 3.6, the numerator of Eq. (3.27) is

$$\frac{1}{g-1}\sum_{i=1}^{g} w_i\big(\bar{x}_i - \bar{\bar{x}}\big)^2$$

$$= \frac{1}{4-1}\Big[(3.00)(11.00 - 14.53333)^2 + (3.00)(12.00 - 14.5333)^2$$

$$+ (2.40)(13.50 - 14.53333)^2 + (0.08)(19.00 - 14.5333)^2\Big]$$

$$= \frac{62.8654}{3} = 20.2885\,,$$

and the denominator of Eq. (3.27) is

$$1 + \frac{2(g-2)}{g^2-1} \sum_{i=1}^{g} \left(\frac{1}{n_i - 1} \right) \left(1 - \frac{w_i}{\sum_{i=1}^{g} w_i} \right)^2$$

$$= 1 + \frac{2(4-2)}{4^2-1} \left[\left(\frac{1}{3-1} \right) \left(1 - \frac{3.00}{8.48} \right)^2 + \left(\frac{1}{3-1} \right) \left(1 - \frac{3.00}{8.48} \right)^2 \right.$$

$$\left. + \left(\frac{1}{4-1} \right) \left(1 - \frac{2.40}{8.48} \right)^2 + \left(\frac{1}{5-1} \right) \left(1 - \frac{0.08}{8.48} \right)^2 \right]$$

$$= 1 + (0.2667)(0.2088 + 0.2088 + 0.1714 + 0.2453)$$

$$= 1 + (0.2667)(0.8343) = 1.2225 .$$

Then, following Eq. (3.27), the observed value of Welch's F' is

$$F'_{\text{o}} = \frac{\dfrac{1}{g-1} \displaystyle\sum_{i=1}^{g} w_i \left(\bar{x}_i - \bar{\bar{x}} \right)^2}{1 + \dfrac{2(g-2)}{g^2-1} \displaystyle\sum_{i=1}^{g} \left(\frac{1}{n_i - 1} \right) \left(1 - \frac{w_i}{\sum_{i=1}^{g} w_i} \right)^2}$$

$$= \frac{20.2885}{1.2225} = 16.5963 .$$

Following Welch [420, p. 334], F' is approximately distributed as Snedecor's F with $\nu_1 = g - 1$ and ν_2 degrees of freedom, where ν_2 is given by

$$\nu_2 = \left[\frac{3}{g^2-1} \sum_{i=1}^{g} \left(\frac{1}{n_i - 1} \right) \left(1 - \frac{w_i}{\sum_{i=1}^{g} w_i} \right)^2 \right]^{-1} .$$

For the $N = 15$ response measurement scores listed in Fig. 3.5 on p. 86,

$$\nu_2 = \left[\frac{3}{4^2-1} (0.8343) \right]^{-1} = \left[\frac{2.5029}{15} \right]^{-1} = 5.99 .$$

Under the null hypothesis with $\nu_1 = g - 1 = 4 - 1 = 3$ and $\nu_2 = 5.99$ degrees of freedom, the observed value of $F'_{\text{o}} = 16.5963$ yields an approximate probability value of $P = 0.0026$, which is markedly less than the unadjusted probability value of $P = 0.1366$ based on the observed value of the conventional $F_o = 2.2755$ with $\nu_1 = g - 1 = 3$ and $\nu_2 = N - g = 15 - 4 = 11$ degrees of freedom.

The F-Ratio and MRPP

Note that setting $v = 2$ and $C_i = (n_i - 1)/(N - g)$ for $i = 1, \ldots, g$ yields $\xi_1 = 2s_1^2 = 2(1.00) = 2.00$, $\xi_2 = 2s_2^2 = 2(1.00) = 2.00$, $\xi_3 = 2s_3^2 = 2(1.6667) = 3.3333$, $\xi_4 = 2s_4^2 = 2(62.50) = 125.00$, $\delta = 2MS_{\text{Within}} = 2(23.5455) = 47.0909$, and $\mu_\delta = 2SS_{\text{Total}}/(N - 1) = 2(419.7333)/(15 - 1) = 59.9619$, as shown in Eqs. (3.25) on p. 90 and (3.24) on p. 90.

Given the $N = 15$ univariate response measurement scores listed in Fig. 3.6, the observed values of A and B are

$$A_o = \sum_{i=1}^{N} x_i = 10 + 11 + 12 + \cdots + 33 = 218$$

and

$$B_o = \sum_{i=1}^{N} x_i^2 = 10^2 + 11^2 + 12^2 + \cdots + 33^2 = 3{,}588 \ ,$$

and the relationships between the F-ratio and the MRPP test statistic are

$$F_o = \frac{2(NB_o - A_o^2)}{(g-1)N\delta_o} - \frac{N-g}{g-1} = \frac{2[(15)(3{,}588) - 218^2]}{(4-1)(15)(47.0909)} - \frac{15-4}{4-1} = 2.2755$$

and

$$\delta_o = \frac{2(NB_o - A_o^2)}{N[N - g + (g-1)F_o]} = \frac{2[(15)(3{,}588) - 218^2]}{15[15 - 4 + (4-1)(2.2755)]} = 47.0909 \ .$$

Cohen's Measure of Effect Size

For a one-way analysis of variance, Cohen's $\hat{d}$ is given by

$$\hat{d} = \left\{ \frac{1}{g-1} \left[\frac{\sum_{i=1}^{g} \left(\bar{x}_i - \bar{\bar{x}}\right)^2}{MS_{\text{Within}}} \right] \right\}^{1/2} , \tag{3.28}$$

where $\bar{x}_i$ is the arithmetic mean of the response measurement scores in the ith of g treatment groups and $\bar{\bar{x}}$ denotes the grand (weighted) mean of the g treatment groups.

The observed response measurement scores listed in Fig. 3.5 yield $\bar{x}_1 = 11.00$, $\bar{x}_2 = 12.00$, $\bar{x}_3 = 13.50$, $\bar{x}_4 = 19.00$, $\bar{\bar{x}} = 14.5333$, $MS_{\text{Within}} = 23.5455$, and

$$\sum_{i=1}^{g} \left(\bar{x}_i - \bar{\bar{x}}\right)^2 = (11.00 - 14.5333)^2 + (12.00 - 14.5333)^2$$
$$+ (13.50 - 14.5333)^2 + (19.00 - 14.5333)^2 = 39.9211 \; .$$

Then, following Eq. (3.28), the observed value of $\hat{d}$ is

$$\hat{d} = \left[\frac{1}{4-1} \left(\frac{39.9211}{23.5455} \right) \right]^{1/2} = 0.7518 \; .$$

For comparison, the univariate response measurement scores listed in Fig. 3.6 yield $\eta^2 = 0.3829$, $\hat{\eta}^2 = \hat{r}^2 = \epsilon^2 = \Re = 0.2147$, and $\hat{\omega}^2 = 0.2033$ for a fixed-effects model.

3.7.2 Example 2

For this second analysis of the univariate response measurement scores listed in Fig. 3.6 on p. 92, replicated in Fig. 3.7 for convenience, let the treatment-group weights be given by

$$C_i = \frac{n_i - 1}{N - g} \; , \qquad i = 1, \ldots, g \; ,$$

as in Example 1, but set $v = 1$ instead of $v = 2$, employing ordinary Euclidean distance between response measurement scores, thereby reducing the effects of any extreme values. The $N = 15$ univariate response measurement scores listed in Fig. 3.7 contain one extreme value of $x_{4,5} = 33$, i.e., the fifth response measurement in Treatment 4.

As noted in the discussion of Student's two-sample t test in Sect. 3.3.2, permutation tests based on $v = 1$ are robust to extreme values, while permutation tests

Fig. 3.7 Example univariate response measurement scores with $r = 1$, $g = 4$, $n_1 = n_2 = 3$, $n_3 = 4$, $n_4 = 5$, and $N = n_1 + n_2 + n_3 + n_4 = 15$

Treatment			
1	2	3	4
10	11	12	14
11	12	13	15
12	13	14	16
		15	17
			33

based on $v = 2$ can be severely affected by even a single extreme value [295, pp. 13–15].

Following Eq. (3.2) on p. 58, the univariate response measurement scores listed in Fig. 3.7 yield $g = 4$ average distance-function values of

$$\xi_1 = \xi_2 = 1.3333 , \quad \xi_3 = 1.6667 , \quad \text{and} \quad \xi_4 = 8.00 .$$

Following Eq. (3.1) on p. 58, the observed value of the MRPP test statistic based on $v = 1$ and treatment-group weights

$$C_i = \frac{n_i - 1}{N - g} , \qquad i = 1, \ldots, 4 ,$$

is

$$\delta_o = \sum_{i=1}^{g} C_i \xi_i = \frac{1}{15 - 4} \big[(3 - 1)(1.3333) + (3 - 1)(1.3333) + (4 - 1)(1.6667) + (5 - 1)(8.00) \big] = 3.8485 .$$

Note that the ξ_i values, $i = 1, \ldots, 4$, and the observed δ value, δ_o, are identical for both the resampling and exact tests.

If all M possible arrangements of the $N = 15$ observed response measurement scores listed in Fig. 3.7 occur with equal chance, the approximate resampling probability value of $\delta_o = 3.8485$ computed on $L = 1{,}000{,}000$ random arrangements of the observed data with $n_1 = n_2 = 3$, $n_3 = 4$, and $n_4 = 5$ univariate response measurement scores preserved for each arrangement is

$$P(\delta \leq \delta_o | H_0) = \frac{\text{number of } \delta \text{ values} \leq \delta_o}{L} = \frac{18{,}000}{1{,}000{,}000} = 0.0180 .$$

For comparison, the approximate resampling probability value based on $v = 2$, $L = 1{,}000{,}000$, and $C_i = (n_i - 1)/(N - g)$ for $i = 1, \ldots, 4$ in Example 1 is $P = 0.0532$. No comparison is made with the F-ratio as Fisher's F is undefined for $v = 1$, as are the conventional measures of effect size: $\hat{d}$, r^2, $\hat{r}^2$, η^2, $\hat{\eta}^2$, ϵ^2, and $\hat{\omega}^2$.

Following Eq. (3.3) on p. 59, the exact expected value of the $M = 12{,}612{,}600$ δ values is $\mu_\delta = 4.7238$ and, following Eq. (3.3) on p. 59, the observed chance-corrected measure of effect size is

$$\Re_o = 1 - \frac{\delta_o}{\mu_\delta} = 1 - \frac{3.8485}{4.7238} = +0.1853 ,$$

indicating approximately 19 % within-group agreement above that expected by chance.

An Exact Test

Following Eq. (3.2) on p. 58, an exact permutation analysis of the univariate response measurement scores listed in Fig. 3.7 on p. 99 yields $g = 4$ average distance-function values of

$$\xi_1 = \xi_2 = 1.3333 \,, \quad \xi_3 = 1.6667 \,, \quad \text{and} \quad \xi_4 = 8.00 \,.$$

Following Eq. (3.1) on p. 58, the observed value of the MRPP test statistic based on $v = 1$ and treatment-group weights

$$C_i = \frac{n_i - 1}{N - g} \,, \qquad i = 1, \ldots, 4 \,,$$

is

$$\delta_o = \sum_{i=1}^{g} C_i \xi_i = \frac{1}{15 - 4} \big[(3 - 1)(1.3333) + (3 - 1)(1.3333) + (4 - 1)(1.6667) + (5 - 1)(8.00) \big] = 3.8485 \,.$$

As always, the ξ_i values, $i = 1, \ldots, g$, and the observed δ value, δ_o, are identical for both the resampling and exact tests.

If all arrangements of the $N = 15$ observed response measurement scores listed in Fig. 3.7 occur with equal chance, the exact probability value of $\delta_o = 3.8485$ computed on the $M = 12{,}612{,}600$ possible arrangements of the observed data with $n_1 = n_2 = 3, n_3 = 4$, and $n_4 = 5$ univariate response measurement scores preserved for each arrangement is

$$P(\delta \leq \delta_o | H_0) = \frac{\text{number of } \delta \text{ values} \leq \delta_o}{M} = \frac{225{,}720}{12{,}612{,}600} = 0.0179 \,.$$

Based on $L = 1{,}000{,}000$, the approximate resampling probability value of $P = 0.0180$ compares favorably with the exact probability value of $P = 0.0179$ based on $M = 12{,}612{,}600$. No comparison is made with the F-ratio as Fisher's F test is undefined for $v = 1$, as are the conventional measures of effect size: $\hat{d}$, r^2, $\hat{r}^2$, η^2, $\hat{\eta}^2$, ϵ^2, and $\hat{\omega}^2$.

Following Eq. (3.4) on p. 59, the exact expected value of the $M = 12{,}612{,}600$ δ values is $\mu_\delta = 4.7238$ and, following Eq. (3.3) on p. 59, the observed chance-corrected measure of effect size is

$$\Re_o = 1 - \frac{\delta_o}{\mu_\delta} = 1 - \frac{3.8485}{4.7238} = +0.1853 \,,$$

indicating approximately 19 % within-group agreement above that expected by chance.

Note the effect of the single extreme value ($x_{4,5} = 33$) in Group 4 on the analysis based on $v = 1$, compared with the analysis based on $v = 2$. In the analysis with $v = 2$, the value for ξ_4 was 125.00, but with $v = 1$, ξ_4 was reduced to only $\xi_4 = 8.00$. Also, with $v = 2$ the exact probability value was $P = 0.0534$, but with $v = 1$ the exact probability value was only $P = 0.0180$, a substantial reduction of approximately 66 %.

3.7.3 Example 3

As noted in the discussion of Student's two-sample t test in Sect. 3.3.3, the treatment-group weights given by

$$C_i = \frac{n_i - 1}{N - g} , \qquad i = 1, \ldots, g ,$$

are based on degrees of freedom, are holdovers from the classical F test, and are not appropriate for distribution-free permutation tests. Thus, for this third analysis of the univariate response measurement scores listed in Fig. 3.7, the treatment-group weighting function is set to

$$C_i = \frac{n_i}{N} , \qquad i = 1, \ldots, g ,$$

simply weighting each treatment group proportional to its group size, and v is set to $v = 1$ as in Example 2, employing ordinary Euclidean distance between response measurement scores. Following Eq. (3.2) on p. 58, the univariate response measurement scores listed in Fig. 3.7 yield $g = 4$ average distance-function values of

$$\xi_1 = \xi_2 = 1.3333 , \quad \xi_3 = 1.6667 , \quad \text{and} \quad \xi_4 = 8.00 .$$

Following Eq. (3.1) on p. 58, the observed value of the MRPP test statistic based on $v = 1$ and treatment-group weights

$$C_i = \frac{n_i}{N} \qquad i = 1, \ldots, 4 ,$$

is

$$\delta_o = \sum_{i=1}^{g} C_i \xi_i = \frac{1}{15} \big[(3)(1.3333)(+(3)(1.3333)$$
$$+ (4)(1.6667) + (5)(8.00) \big] = 3.6444 .$$

If all M possible arrangements of the $N = 15$ observed response measurement scores listed in Fig. 3.7 occur with equal chance, the approximate resampling probability value of $\delta_o = 3.6444$ computed on $L = 1{,}000{,}000$ random arrangements of the observed data with $n_1 = n_2 = 3$, $n_3 = 4$, and $n_4 = 5$ univariate response measurement scores preserved for each arrangement is

$$P(\delta \leq \delta_o | H_0) = \frac{\text{number of } \delta \text{ values} \leq \delta_o}{L} = \frac{33{,}000}{1{,}000{,}000} = 0.0033 \, .$$

For comparison, the approximate resampling probability values based on $v = 2$, $L = 1{,}000{,}000$, and $C_i = (n_i - 1)/(N - g)$ for $i = 1, \ldots, 4$ in Example 1 and $v = 1$, $L = 1{,}000{,}000$, and $C_i = (n_i - 1)/(N - g)$ for $i = 1, \ldots, 4$ in Example 2 are $P = 0.0532$ and $P = 0.0179$, respectively. No comparison is made with the conventional F-ratio as Fisher's F test is undefined for both $v = 1$ and $C_i = n_i/N$ for $i = 1, \ldots, g$, as are the conventional measures of effect size: $\hat{d}$, r^2, $\hat{r}^2$, η^2, $\hat{\eta}^2$, ϵ^2, and $\hat{\omega}^2$.

Following Eq. (3.4) on p. 59, the exact expected value of the $M = 12{,}612{,}600$ δ values is $\mu_\delta = 4.7238$ and, following Eq. (3.3) on p. 59, the observed chance-corrected measure of effect size is

$$\Re_o = 1 - \frac{\delta_o}{\mu_\delta} = 1 - \frac{3.6444}{4.7238} = +0.2285 \, ,$$

indicating approximately 23 % within-group agreement above that expected by chance.

An Exact Test

An exact permutation analysis of the univariate response measurement scores listed in Fig. 3.7 with $v = 1$ and proportional treatment-group weights given by

$$C_i = \frac{n_i}{N} \, , \qquad i = 1, \ldots, g \, ,$$

yields $g = 4$ average distance-function values of

$$\xi_1 = \xi_2 = 1.3333 \, , \quad \xi_3 = 1.6667 \, , \quad \text{and} \quad \xi_4 = 8.00 \, .$$

Following Eq. (3.1) on p. 58, the observed value of the MRPP test statistic based on $v = 1$ and treatment-group weights

$$C_i = \frac{n_i}{N} \, , \qquad i = 1, \ldots, 4 \, ,$$

is

$$\delta_o = \sum_{i=1}^{g} C_i \xi_i = \frac{1}{15}\big[(3)(1.3333) + (3)(1.3333)$$
$$+ (4)(1.6667) + (5)(8.00)\big] = 3.6444 \,.$$

If all arrangements of the $N = 15$ observed response measurement scores listed in Fig. 3.7 occur with equal chance, the exact probability value of $\delta_o = 3.6444$ computed on the $M = 12{,}612{,}600$ possible arrangements of the observed data with $n_1 = n_2 = 3$, $n_3 = 4$, and $n_4 = 5$ univariate response measurement scores preserved for each arrangement is

$$P\big(\delta \leq \delta_o | H_0\big) = \frac{\text{number of } \delta \text{ values} \leq \delta_o}{M} = \frac{41{,}172}{12{,}612{,}600} = 0.0033 \,,$$

which is the same, to four decimal places, as the approximate resampling probability value based on $L = 1{,}000{,}000$. For comparison, the exact probability values based on $v = 2$, $M = 12{,}612{,}600$, and $C_i = (n_i - 1)/(N - g)$ for $i = 1, 2$ in Example 1 and $v = 1$, $M = 12{,}612{,}600$, and $C_i = (n_i - 1)/(N - g)$ for $i = 1, 2$ in Example 2 are $P = 0.0532$ and $P = 0.0534$, respectively. No comparison is made with the F-ratio as Fisher's F test is undefined for both $v = 1$ and $C_i = n_i/N$, as are the conventional measures of effect size, $\hat{d}$, r^2, $\hat{r}^2$, η^2, $\hat{\eta}^2$, ϵ^2, and $\hat{\omega}^2$.

Following Eq. (3.4) on p. 59, the exact expected value of the $M = 12{,}612{,}600$ δ values is $\mu_\delta = 4.7238$ and, following Eq. (3.3) on p. 59, the observed chance-corrected measure of effect size is

$$\Re_o = 1 - \frac{\delta_o}{\mu_\delta} = 1 - \frac{3.6444}{4.7238} = +0.2285 \,,$$

indicating approximately 23 % within-group agreement above that expected by chance.

3.8 Permutation Analogue of One-Way MANOVA

It is sometimes desirable to test for differences among $g \geq 3$ independent treatment groups where $r \geq 2$ response measurement scores have been obtained for each object. The conventional approach is one-way multivariate analysis of variance (MANOVA) for which a number of statistical tests have been proposed, including the Bartlett–Nanda–Pillai (BNP) trace test [21, 316, 339], Wilks' likelihood-ratio test [431], Roy's maximum-root test [357, 358], and the Lawley–Hotelling trace test [182, 232, 233]. The Bartlett–Nanda–Pillai trace test is considered to be the most powerful and robust of the four tests [327, 328, 392, p. 269].

To illustrate a multivariate analysis of variance, consider the Bartlett–Nanda–Pillai trace test given by

$$V^{(s)} = \text{trace}\left[\mathbf{H}(\mathbf{E}+\mathbf{H})^{-1}\right]$$

where $\mathbf{E}$ is the error matrix summarizing within-object variability, $\mathbf{H}$ is the hypothesized matrix summarizing between-object variability, and $s = \min(r,\ g-1)$.[20] For a conventional test of significance, the BNP trace statistic, $V^{(s)}$, can be transformed into a conventional F-ratio by

$$F = \frac{2u+s+1}{2t+s+1}\left(\frac{V^{(s)}}{s-V^{(s)}}\right), \tag{3.29}$$

where $s = \min(r,\ g-1)$, $u = 0.50(N-g-r-1)$, $t = 0.50(|r-q|-1)$, and $q = g-1$. Assuming independence, normality, and homogeneity of variance and covariance, F is approximately distributed as Snedecor's F under the null hypothesis with $\nu_1 = s(2t+s+1)$ and $\nu_2 = s(2u+s+1)$ degrees of freedom.

3.9 Example Bivariate MRPP Analyses with $g = 3$

In this section, three example analyses with bivariate response measurement scores illustrate the permutation approach to g-sample problems with multivariate response measurement scores. As with the two-sample example with multivariate response measurement scores illustrated on p. 82, the response measurement scores must be made commensurate prior to analysis (q.v. p. 82). The first example is designed to correspond to the conventional Bartlett–Nanda–Pillai trace test using a small set of bivariate response measurement scores with $v = 2$, Hotelling commensuration, and treatment-group weights given by $C_i = (n_i - 1)/(N-g)$ for $i = 1, \ldots, g$; the second example analyzes the same small set of bivariate response measurement scores, but with $v = 1$, Euclidean commensuration, and treatment-group weights given by $C_i = (n_i - 1)/(N-g)$ for $i = 1, \ldots, g$; and the third example analyzes the same set of bivariate response measurement scores using $v = 1$ and Euclidean commensuration, but adopts proportional treatment-group weighting given by $C_i = n_i/N$ for $i = 1, \ldots, g$.

3.9.1 Example 1

Consider the bivariate response measurement scores listed in Fig. 3.8, where $r = 2$, $g = 3$, $n_1 = 5$, $n_2 = 4$, $n_3 = 3$, and $N = n_1 + n_2 + n_3 = 12$. For this first analysis,

[20] In many textbook presentations, the error matrix, $\mathbf{E}$, is denoted as $\mathbf{W}$ for the within-objects matrix, and the hypothesized matrix, $\mathbf{H}$, is denoted as $\mathbf{B}$ for the between-objects matrix.

Fig. 3.8 Example bivariate response measurement scores with $r = 2$, $g = 3$, $n_1 = 5$, $n_2 = 4$, $n_3 = 3$, and $N = n_1 + n_2 + n_3 = 12$

Treatment		
1	2	3
(5.8, 6.0)	(4.1, 2.9)	(4.2, 7.8)
(6.2, 3.9)	(3.9, 4.1)	(5.1, 5.9)
(3.9, 4.1)	(4.9, 3.9)	(4.8, 7.2)
(5.1, 5.2)	(2.1, 5.1)	
(3.0, 2.8)		

let $v = 2$, employing squared Euclidean distance between response measurement scores, and let the treatment-group weights be given by

$$C_i = \frac{n_i - 1}{N - g}, \qquad i = 1, \ldots, g,$$

to correspond to the BNP trace test. An exact permutation solution is feasible for the response measurement scores listed in Fig. 3.8 since there are only

$$M = \frac{N!}{\prod_{i=1}^{g} n_i!} = \frac{12!}{5!\ 4!\ 3!} = 27{,}720$$

possible, equally-likely arrangements of the $N = 12$ observed scores listed in Fig. 3.8.

Following Eq. (3.2) on p. 58, the bivariate response measurement scores listed in Fig. 3.8 yield $g = 3$ average distance-function values of

$$\xi_1 = 0.3242, \quad \xi_2 = 0.2994, \quad \text{and} \quad \xi_3 = 0.1207.$$

Following Eq. (3.1) on p. 58, the observed value of the MRPP test statistic based on $v = 2$ and treatment-group weights

$$C_i = \frac{n_i - 1}{N - g}, \qquad i = 1, 2, 3,$$

is

$$\delta_o = \sum_{i=1}^{g} C_i \xi_i = \frac{1}{12 - 3}\big[(5 - 1)(0.3242) + (4 - 1)(0.2994)$$
$$+ (3 - 1)(0.1207)\big] = 0.2707.$$

If all arrangements of the $N = 12$ observed bivariate response measurement scores listed in Fig. 3.8 occur with equal chance, the exact probability value of

$\delta_o = 0.2707$ computed on the $M = 27{,}720$ possible arrangements of the observed data with $n_1 = 5$, $n_2 = 4$, and $n_3 = 3$ bivariate response measurement scores preserved for each arrangement is

$$P(\delta \le \delta_o | H_0) = \frac{\text{number of } \delta \text{ values} \le \delta_o}{M} = \frac{967}{27{,}720} = 0.0349 \; .$$

For comparison, a conventional BNP analysis of the bivariate response measurement scores listed in Fig. 3.8 yields

$$\mathbf{E} = \begin{bmatrix} 11.71000 & 1.17000 \\ 1.17000 & 10.42667 \end{bmatrix}, \quad \mathbf{H} = \begin{bmatrix} 2.75250 & 3.19755 \\ 3.19755 & 17.30242 \end{bmatrix},$$

$$\mathbf{E} + \mathbf{H} = \begin{bmatrix} 14.46250 & 4.36755 \\ 4.36755 & 27.72909 \end{bmatrix},$$

$$(\mathbf{E} + \mathbf{H})^{-1} = \begin{bmatrix} 0.07260 & -0.01143 \\ -0.01143 & 0.03786 \end{bmatrix},$$

$$\mathbf{H}(\mathbf{E} + \mathbf{H})^{-1} = \begin{bmatrix} 0.16328 & 0.08960 \\ 0.03476 & 0.61852 \end{bmatrix},$$

and $V^{(2)} = \text{trace}[\mathbf{H}(\mathbf{E} + \mathbf{H})^{-1}] = 0.16328 + 0.61852 = 0.7818$.

Alternatively, $V^{(2)}$ can be defined as

$$V^{(2)} = \sum_{i=1}^{s} \frac{\lambda_i}{1 - \lambda_i} , \tag{3.30}$$

where λ_i for $i = 1, \ldots, s$ are the eigenvalues of the $\mathbf{HE}^{-1}$ matrix given by

$$\mathbf{HE}^{-1} = \begin{bmatrix} 0.20673 & 0.28347 \\ 0.10847 & 1.64727 \end{bmatrix} .$$

The $s = 2$ eigenvalues of $\mathbf{HE}^{-1}$ are $\lambda_1 = 0.18570$ and $\lambda_2 = 1.66831$, and following equation Eq. (3.30),

$$V^{(2)} = \frac{0.18570}{1 + 0.18570} + \frac{1.68831}{1 + 1.68831} = 0.15661 + 0.62523 = 0.7818 \; .$$

Then, $q = g - 1 = 3 - 1 = 2$, $s = \min(r, q) = \min(2, 3 - 1) = 2$, $u = 0.50(N - g - r - 1) = 0.50(12 - 3 - 2 - 1) = 3$, $t = 0.50(|r - q| - 1) = 0.50$

$(|2-2|-1) = -0.50$, and following Eq. (3.29) on p. 105, the observed F-ratio is

$$F_o = \frac{2(3)+2+1}{2(-0.50)+2+1}\left(\frac{0.7818}{2-0.7818}\right) = \frac{9}{2}(0.6414) = 2.8879\,.$$

Assuming independence, normality, and homogeneity of variance and covariance, F is approximately distributed as Snedecor's F with $\nu_1 = s(2t+s+1) = 2[(2)(-0.50)+2+1] = 4$ and $\nu_2 = s(2u+s+1) = 2[(2)(3)+2+1] = 18$ degrees of freedom. Under the null hypothesis, the observed value of $F_o = 2.8879$ yields an approximate probability value of $P = 0.0521$. For comparison, the exact probability value of the observed MRPP test statistic $\delta_o = 0.2707$ based on $v = 2$, $M = 27{,}720$, and $C_i = (n_i-1)/(N-g)$ for $i = 1,2,3$ is $P = 0.0349$.

Following Eq. (3.3) on p. 59, the exact expected value of the $M = 27{,}720$ δ values is $\mu_\delta = 0.3636$ and, following Eq. (3.3) on p. 59, the observed chance-corrected measure of effect size is

$$\Re_o = 1 - \frac{\delta_o}{\mu_\delta} = 1 - \frac{0.2707}{0.3636} = +0.2556\,,$$

indicating approximately 26 % within-group agreement above that expected by chance.

A convenient, although positively biased, measure of effect size for the BNP trace test is given by

$$\eta^2 = \frac{V^{(2)}}{s} = \frac{0.7818}{2} = 0.3909\,,$$

which can be compared with the unbiased chance-corrected measure of effect size, $\Re = +0.2665$. It is perhaps not readily apparent that $V^{(2)}/s = 0.3909$ is simply the squared canonical correlation coefficient, $R^2_{y_1,y_2.x_1,x_2}$, where two sets of dummy-coded variables are required. Figure 3.9 displays the canonical correlation data where variables x_1 and x_2 are the two dummy-coded independent variables and variable y_1 is a dependent variable containing the first of the $r = 2$ response measurement scores for each object, and variable y_2 is a second dependent variable, containing the second of the response measurement scores for each object.

Finally, in this application the MRPP test statistic, δ, is based on the generalized Minkowski distance function given by

$$\Delta(I,J) = \left[\sum_{j=1}^{r}\left(x_{jI}-x_{jJ}\right)^2\right]^{v/2}\,,$$

Fig. 3.9 Example regression data with $N = 12$ cases, independent variables x_1 and x_2, and dependent variables y_1 and y_2

Variable			
x_1	x_2	y_1	y_2
1	0	5.8	6.0
1	0	6.2	3.9
1	0	3.9	4.1
1	0	5.1	5.2
1	0	3.0	2.8
0	1	4.1	2.9
0	1	3.9	4.1
0	1	4.9	3.9
0	1	2.1	5.1
0	0	4.2	7.8
0	0	5.1	5.9
0	0	4.8	7.2

and the functional relationship of the $V^{(2)}$ BNP trace statistic to the MRPP δ test statistic [297, pp. 53–57] is

$$\delta = \frac{2(r - V^{(2)})}{N - g} = \frac{2(2 - 0.7818)}{12 - 3} = \frac{2.4364}{9} = 0.2707 \,.$$

Alternatively,

$$\delta = \frac{2}{N - g} \operatorname{trace}\left[\mathbf{E}(\mathbf{E} + \mathbf{H})^{-1}\right] , \tag{3.31}$$

where

$$\mathbf{E}(\mathbf{E} + \mathbf{H})^{-1} = \begin{bmatrix} 0.83677 & -0.08955 \\ -0.03423 & 0.38138 \end{bmatrix}$$

and the observed value of δ, following Eq. 3.31, is

$$\delta_o = \frac{2}{12 - 3}(0.83677 + 0.38138) = 0.2707 \,.$$

This first example analysis demonstrates that the BNP trace test may be considered a special case of the MRPP test statistic, δ, with $v = 2$. Unlike the conventional multivariate analysis of variance tests such as Roy's maximum-root test [357, 358], Wilks' likelihood-ratio test [431], the Lawley–Hotelling trace test [182, 232, 233], and the Bartlett–Nanda–Pillai trace test [21, 316, 339], the permutation approach illustrated here is not dependent on the assumptions of normality and homogeneity of variance and covariance, making the MRPP δ test statistic and its associated chance-corrected measure of effect size, $\Re$, valuable tools for the analysis of interval-level multivariate data.

3.9.2 Example 2

For a second example analysis of the response measurement scores listed in Fig. 3.8 on p. 106, let the treatment-group weights again be given by

$$C_i = \frac{n_i - 1}{N - g} , \quad i = 1, \ldots, g ,$$

as in Example 1, but set $v = 1$ instead of $v = 2$, employing ordinary Euclidean distance between response measurement scores. Following Eq. (3.2) on p. 58, the bivariate response measurement scores listed in Fig. 3.8 yield $g = 3$ average distance-function values of

$$\xi_1 = 2.3933 , \quad \xi_2 = 1.9326 , \quad \text{and} \quad \xi_3 = 1.4284 .$$

Following Eq. (3.1) on p. 58, the observed value of the MRPP test statistic based on $v = 1$ and treatment-group weights

$$C_i = \frac{n_i - 1}{N - g} , \qquad i = 1, 2, 3 ,$$

is

$$\delta_o = \sum_{i=1}^{g} C_i \xi_i = \frac{1}{12 - 3} \big[(5 - 1)(2.3933) + (4 - 1)(1.9326) + (3 - 1)(1.4284) \big] = 2.0253 .$$

If all arrangements of the $N = 12$ observed bivariate response measurement scores listed in Fig. 3.8 occur with equal chance, the exact probability value of $\delta_o = 2.0253$ computed on the $M = 27{,}720$ possible arrangements of the observed data with $n_1 = 5$, $n_2 = 4$, and $n_3 = 3$ bivariate response measurement scores preserved for each arrangement is

$$P(\delta \leq \delta_o | H_0) = \frac{\text{number of } \delta \text{ values} \leq \delta_o}{M} = \frac{618}{27{,}720} = 0.0223 .$$

For comparison, the exact probability value based on $v = 2$, $M = 27{,}720$, and $C_i = (n_i - 1)/(N - g)$ for $i = 1, 2, 3$ in Example 1 is $P = 0.0349$. No comparison is made with the Bartlett–Nanda–Pillai trace test as the BNP test is undefined for $v = 1$, as is the conventional measure of effect size, η^2.

Following Eq. (3.4) on p. 59, the exact expected value of the $M = 27{,}720$ δ values is $\mu_\delta = 2.5200$ and, following Eq. (3.3) on p. 59, the observed chance-corrected measure of effect size is

$$\mathfrak{R}_o = 1 - \frac{\delta_o}{\mu_\delta} = 1 - \frac{2.0253}{2.5200} = +0.1963 \; ,$$

indicating approximately 20 % within-group agreement above that expected by chance.

3.9.3 Example 3

For a third example analysis of the bivariate response measurement scores listed in Fig. 3.8 on p. 106, let the treatment-group weights be given by

$$C_i = \frac{n_i}{N} \, , \quad i = 1, \ldots, g \; ,$$

weighting each treatment group proportional to its group size, and let $v = 1$, employing ordinary Euclidean distance between response measurement scores, as in Example 2. Following Eq. (3.2) on p. 58, the bivariate response measurement scores listed in Fig. 3.8 yield $g = 3$ average distance-function values of

$$\xi_1 = 2.3933 \; , \quad \xi_2 = 1.9326 \; , \quad \text{and} \quad \xi_3 = 1.4284 \; .$$

Following Eq. (3.1) on p. 58, the observed value of the MRPP test statistic based on $v = 1$ and treatment-group weights

$$C_i = \frac{n_i}{N} \, , \qquad i = 1, 2, 3 \; ,$$

is

$$\delta_o = \sum_{i=1}^{g} C_i \xi_i = \frac{1}{12} \big[(5)(2.3933) + (4)(1.9326) + (3)(1.4284) \big] = 1.9985 \; .$$

If all arrangements of the $N = 12$ observed bivariate response measurement scores listed in Fig. 3.8 occur with equal chance, the exact probability value of $\delta_o = 1.9985$ computed on the $M = 27{,}720$ possible arrangements of the observed data with $n_1 = 5$, $n_2 = 4$, and $n_3 = 3$ bivariate response measurement scores preserved for each arrangement is

$$P(\delta \leq \delta_o | H_0) = \frac{\text{number of } \delta \text{ values} \leq \delta_o}{M} = \frac{518}{27{,}720} = 0.0187 \; .$$

For comparison, the exact probability values based on $v = 2$, $M = 27{,}720$, and $C_i = (n_i - 1)/(N - g)$ for $i = 1, 2, 3$ in Example 1 and $v = 1$, $M = 27{,}720$, and $C_i = (n_i - 1)/(N - g)$ for $i = 1, 2, 3$ in Example 2 are $P = 0.0349$ and $P = 0.0223$, respectively. No comparison is made with the Bartlett–Nanda–Pillai trace test as the BNP test is undefined for both $v = 1$ and $C_i = n_i/N$ for $i = 1, \ldots, g$, as is the conventional measure of effect size, η^2.

Following Eq. (3.4) on p. 59, the exact expected value of the $M = 27{,}720$ δ values is $\mu_\delta = 2.5200$ and, following Eq. (3.3) on p. 59, the observed chance-corrected measure of effect size is

$$\Re_o = 1 - \frac{\delta_o}{\mu_\delta} = 1 - \frac{1.9985}{2.5200} = +0.2070 ,$$

indicating approximately 21 % within-group agreement above that expected by chance.

3.10 Coda

Chapter 3 utilized the Multi-Response Permutation Procedures developed in Chap. 2 to establish relationships between the test statistics of MRPP, δ and $\Re$, and selected conventional tests and measures designed for the analysis of completely randomized data at the interval level of measurement. Considered in this chapter were Student's two-sample t test with interval-level univariate response measurement scores, Hotelling's two-sample T^2 test with interval-level multivariate response measurement scores, one-way fixed-effects analysis of variance (ANOVA) with interval-level univariate response measurement scores, and one-way multivariate analysis of variance (MANOVA) with interval-level multivariate response measurement scores. Also included in Chap. 3 was an introduction of a comprehensive chance-corrected measure of effect size, $\Re$, which is distribution-free, data-dependent, easily interpretable, and applicable to many research designs.

Also discussed in Chap. 3 were several popular measures of effect size, including Cohen's $\hat{d}$, Hedges' g, Pearson's r^2, Kelley's ϵ^2, and Hays' $\hat{\omega}^2$. These five measures were shown to be chance-corrected measures of effect size and, under the population model of statistical inference, biased estimates of the associated population parameters. A permutation-based, unbiased, chance-corrected measure of effect size, $\Re$, was introduced as a universal replacement for five conventional measures of effect size.

Chapter 4

Chapter 4 continues the analysis of interval-level response measurement scores in Chap. 3, applying the test statistics of MRPP, δ and $\Re$, to regression residuals generated by either ordinary least squares (OLS) or least absolute deviation (LAD) regression models. Considered in Chap. 4 are one-way randomized, one-way randomized with a covariate, one-way randomized-block, two-way randomized-block, two-way factorial, Latin square, split-plot, and two-factor nested analysis of variance designs.

Regression Analysis of Interval Data 4

Multi-Response Permutation Procedures (MRPP) were introduced in Chap. 2 and applied to interval-level, completely randomized data in Chap. 3. While multi-response permutation procedures are generally thought of as providing tests of differences among g treatment groups as demonstrated in Chap. 3, they also have applications in ordinary least squares (OLS) linear regression analyses with $v = 2$ and least absolute deviations (LAD) linear regression analyses with $v = 1$. In this fourth chapter of *Permutation Statistical Methods*, MRPP analyses of LAD regression residuals are illustrated with a variety of experimental designs, including one-way completely randomized with and without a covariate, one-way and two-way randomized-block, two-way factorial, Latin square, and two-factor nested analysis-of-variance designs. Also considered are multivariate multiple regression designs.

4.1 LAD Linear Regression

OLS linear regression has long been recognized as a useful tool in many fields of research. The optimal properties of OLS regression are well known when the errors are normally distributed. However, in practice the assumption of multivariate normality is rarely justified. LAD linear regression is an attractive alternative to OLS regression as it is extremely robust to deviations from normality as well as to the presence of extreme values [297, p. 172].

It is widely recognized that estimators of OLS regression parameters can be severely affected by unusual values in either the criterion variable or in one or more of the predictor variables. This is due in large part to the weight given to each data point when minimizing the sum of squared errors. In contrast, LAD regression is much less sensitive to the effects of unusual-value errors due to the fact that the errors are not squared. Moreover, LAD regression has been shown to be superior to OLS regression when errors are generated from heavy-tailed or outlier-

K.J. Berry et al., *Permutation Statistical Methods*,
DOI 10.1007/978-3-319-28770-6_4

producing distributions, such as the Cauchy and double-exponential distributions; see, for example, articles by Blattburg and Sargent [46], Dielman [94, 95], Dielman and Pfaffenberger [96], Dielman and Rose [97], Mathew and Nordström [264], Mielke, Berry, Landsea, and Gray [303], Pfaffenberger and Dinkel [337], Rice and White [346], Rosenberg and Carlson [352], Rousseeuw [355], Taylor [394], and Wilson [432].

As described by Sheynin, the initial known use of regression by Daniel Bernoulli (c. 1734) for astronomical prediction problems involved LAD regression based on ordinary Euclidean distances between the observed and predicted response values [372]. Further developments in LAD regression were due to Roger Joseph (Rogerius Josephus) Boscovich (c. 1755), Pierre-Simon Laplace (c. 1789), and Carl Friedrich Gauss (c. 1809). The American mathematician and astronomer Nathaniel Bowditch (c. 1809) was highly critical of OLS regression because, as he argued, squared regression residuals unduly emphasized questionable observations in comparison with the absolute regression residuals associated with LAD regression [372].

Consider the general multivariate regression model given by

$$\mathbf{y}_i = \boldsymbol{h}\left(\boldsymbol{\beta}, \mathbf{x}_i\right) + \mathbf{e}_i \ ,$$

where $\mathbf{y}_i' = (y_{1i}, \ldots, y_{ri})$ denotes the row vector of r observed response measurements for the ith of N objects, $\mathbf{x}_i' = (x_{1i}, \ldots, x_{si})$ is the row vector of s predictor values for the ith object, $\boldsymbol{\beta}' = (\beta_1, \ldots, \beta_t)$ is the row vector of t parameters, $\boldsymbol{h}' = (h_1, \ldots, h_r)$ is the row vector of r model functions of $\boldsymbol{\beta}$ and $\mathbf{x}_i$ for the ith object, and $\mathbf{e}_i' = (e_{1i}, \ldots, e_{ri})$ denotes the r errors between the response variables and model functions for the ith object, $i = 1, \ldots, N$ objects. The special case of a multivariate linear regression model is given by

$$\mathbf{y}_i = \mathbf{B}\boldsymbol{f}(\mathbf{x}_i) + \mathbf{e}_i \ ,$$

where $\boldsymbol{f}(\mathbf{x}_i)$ denotes a column vector of p distinct functions of s predictors $(\mathbf{x}_i)$ for the ith object, $i = 1, \ldots, N$, and $\mathbf{B}$ is an $r{\times}p$ matrix of parameters in which $(\mathbf{B}_{j1}, \ldots, \mathbf{B}_{jp})$ is the row vector of p parameters associated with the jth response measurement, $j = 1, \ldots, r$.

Let $\mathbf{y}_i$ denote a column vector of r observed response measurement scores and let $\tilde{\mathbf{y}}_i$ denote a column vector of r predicted response values for the ith object, $i = 1, \ldots, N$. Thus, the general and linear predicted multivariate regression models are given by

$$\tilde{\mathbf{y}}_i = \boldsymbol{h}\left(\tilde{\boldsymbol{\beta}}, \mathbf{x}_i\right)$$

and

$$\tilde{\mathbf{y}}_i = \tilde{\mathbf{B}}\boldsymbol{f}\left(\mathbf{x}_i\right) \ ,$$

respectively, where $\tilde{\boldsymbol{\beta}}$ and $\tilde{\mathbf{B}}$ are estimated parameters that are intended to provide good fits between the $\mathbf{y}_i$ and $\tilde{\mathbf{y}}_i$ values relative to a selected goodness-of-fit criterion. The null hypothesis (H_0) underlying each criterion dictates that each of the $N!$ possible, equally-likely pairings of the predicted sequential ordering $(\tilde{\mathbf{y}}_1, \ldots, \tilde{\mathbf{y}}_N)$ with the fixed observed sequential ordering $(\mathbf{y}_1, \ldots, \mathbf{y}_N)$ occurs with equal probability, i.e., $1/N!$.

Let $\Delta(\tilde{\mathbf{y}}_i, \mathbf{y}_i)$ for $i = 1, \ldots, N$ denote the distance function between the predicted and observed response measurement values and consider the generalized Minkowski distance function given by

$$\Delta(\tilde{\mathbf{y}}_i, \mathbf{y}_i) = \left(\sum_{j=1}^{r} \left| \tilde{y}_{ij} - y_{ij} \right|^w \right)^{v/w} ,$$

where $w \geq 1$ and $v > 0$. Since $v = 1$ yields the Minkowski metric [12], the choice of $v = 1$ is preferred since $v > 1$ yields distance functions that do not satisfy the triangle inequality property of a metric. Consequently, the distance function of choice utilizes $v = 1$ and $w = 2$, i.e., an ordinary Euclidean distance function.

Let the average distance function between $(\tilde{\mathbf{y}}_1, \ldots, \tilde{\mathbf{y}}_N)$ and $(\mathbf{y}_1, \ldots, \mathbf{y}_N)$ be given by

$$\delta = \frac{1}{N} \sum_{i=1}^{N} \Delta\left(\tilde{\mathbf{y}}_i, \mathbf{y}_i\right) . \tag{4.1}$$

As noted previously, a distance function with $v > 1$ is not a metric function. If the distance function associated with LAD regression is squared (i.e., $v = 2$), then the estimated parameters that minimize δ yield an OLS regression model.

The criterion for fitting multivariate regression models based on δ is the chance-corrected measure of agreement between the observed and predicted response measurement values given by

$$\Re = 1 - \frac{\delta}{\mu_\delta} , \tag{4.2}$$

where μ_δ is the expected value of δ over the $N!$ possible pairings under the null hypothesis. An efficient computational expression for obtaining μ_δ that involves a sum of N^2 rather than $N!$ terms is given by

$$\mu_\delta = \frac{1}{N^2} \sum_{i=1}^{N} \sum_{j=1}^{N} \Delta\left(\tilde{\mathbf{y}}_i, \mathbf{y}_j\right) . \tag{4.3}$$

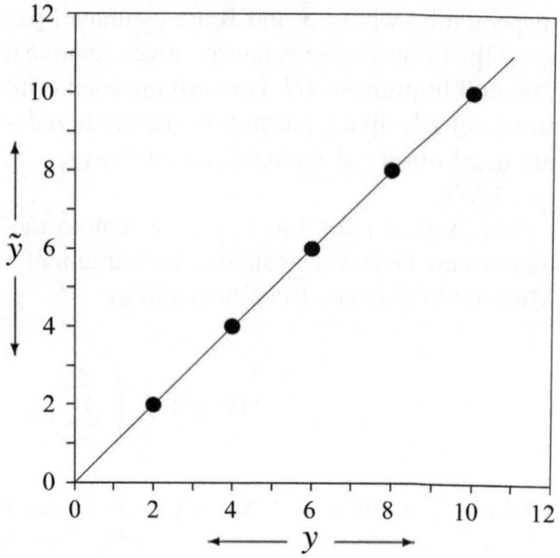

Fig. 4.1 Graphic depicting a regression line with perfect agreement between y and $\tilde{y}$ with intercept equal to 0.00 and slope equal to +1.00

4.1.1 Linear Regression and Agreement

A simple interpretation of $\Re$ can be described for $r = s = 1$ since the same interpretation holds for any r and s. In the case involving perfect agreement, $\tilde{y}_i = y_i$ for $i = 1, \ldots, N$, $\delta = 0.00$, and $\Re = 1.00$. This implies that the functional relationship between $\tilde{y}$ and y can be described by a straight line that passes through the origin with a slope of 45°, as depicted in Fig. 4.1 with $N = 5$ bivariate $(y, \tilde{y})$ values: $(2, 2)$, $(4, 4)$, $(6, 6)$, $(8, 8)$, and $(10, 10)$. For the $N = 5$ data points depicted in Fig. 4.1, the intercept is $\tilde{\beta}_0 = 0.00$, the unstandardized slope is $\tilde{\beta}_1 = +1.00$, the squared Pearson product-moment correlation coefficient is $r^2_{y\tilde{y}} = +1.00$, and the agreement percentage is also 1.00, i.e., all five of the y and $\tilde{y}$ paired values agree.

In this context, the squared Pearson product-moment correlation coefficient, $r^2_{y\tilde{y}}$, has also been used as a measure of agreement. However, $r^2_{y\tilde{y}} = +1.00$ implies a linear relationship between y and $\tilde{y}$, where both the intercept and slope are arbitrary. While perfect agreement is described by $\Re = +1.00$, $r^2_{y\tilde{y}} = +1.00$ describes a linear relationship that may or may not reflect perfect agreement as depicted in Fig. 4.2 with $N = 5$ $(y, \tilde{y})$ values: $(2, 4)$, $(4, 5)$, $(6, 6)$, $(8, 7)$, and $(10, 8)$. For the $N = 5$ bivariate data points depicted in Fig. 4.2, the intercept is $\tilde{\beta}_0 = +3.00$, the unstandardized slope is $\tilde{\beta}_1 = +0.50$, the squared Pearson product-moment correlation coefficient is $r^2_{y\tilde{y}} = +1.00$, and the agreement percentage is 0.20, i.e., only one $(6, 6)$ of the $N = 5$ y and $\tilde{y}$ paired values agree. Comparisons of $\Re$ with other measures of agreement and the advantages of $\Re$ relative to the other agreement measures were detailed in a 1996 article by Watterson [416].

While the agreement measure $\Re$ provides a description of the functional relationship between $(\tilde{\mathbf{y}}_1, \ldots, \tilde{\mathbf{y}}_N)$ and $(\mathbf{y}_1, \ldots, \mathbf{y}_N)$, it does not indicate how extreme an observed value of $\Re$, say $\Re_o$, is relative to the $N!$ possible values of $\Re$ under the null

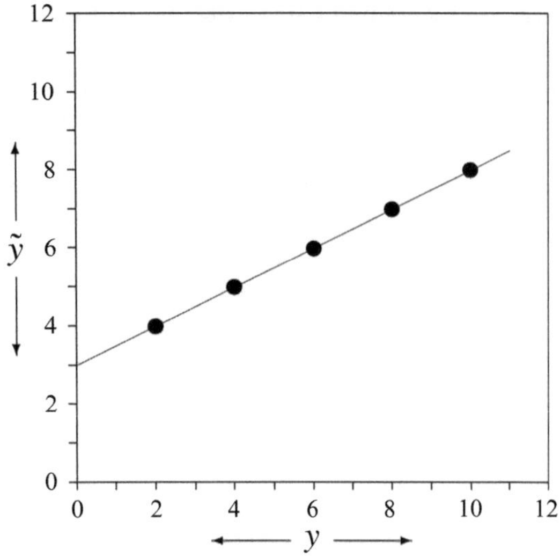

Fig. 4.2 Graphic depicting a regression line with perfect correlation between y and $\tilde{y}$ with intercept equal to +3.00 and slope equal to +0.50

hypothesis. Since μ_δ is invariant under the null hypothesis and the observed value of δ is given by

$$\delta_o = \mu_\delta(1 - \Re_o) ,$$

the exact probability value for $\Re_o$ is given by

$$P(\Re \geq \Re_o|H_0) = P(\delta \leq \delta_o|H_0) = \frac{\text{number of } \delta \text{ values} \leq \delta_o}{M} ,$$

where $M = N!$. Because an exact probability value requires generating $N!$ arrangements of the observed data, calculation of an exact value is prohibitive even for small values of N, e.g., $M = N! = 15! = 1{,}307{,}674{,}368{,}000$.

When M is very large, an approximate probability value for δ may be obtained from a resampling permutation procedure. Let L denote a random sample of all possible arrangements of the observed data, where L is typically a large number, e.g., $L = 1{,}000{,}000$. Then, an approximate resampling probability value is given by

$$P(\Re \geq \Re_o|H_0) = P(\delta \leq \delta_o|H_0) = \frac{\text{number of } \delta \text{ values} \leq \delta_o}{L} .$$

Also, when M is very large and P is exceedingly small, a resampling-approximation permutation procedure based on fitting the first three exact moments of the discrete permutation distribution to a Pearson type III distribution provides approximate probability values, as detailed in Chap. 1, Sect. 1.2.2; see also references [284] and [300].

4.2 Example LAD Regression Analyses

In this section, example analyses illustrate the permutation approach to typical multiple regression problems. The first example analyzes a small set of multivariate response measurement scores using LAD regression and generates a resampling permutation probability value; the second example analyzes the same small set of multivariate response measurement scores using OLS regression and also generates a resampling permutation probability value; the third example analyzes the same set of multivariate response measurement scores using OLS regression, but provides a conventional approximate probability value based on Snedecor's F distribution.

4.2.1 Example Analysis 1

Consider the multiple regression data listed in Fig. 4.3 where $s = 2$ observed response measurement scores have been obtained for each of $N = 12$ objects, $y_1, \ldots, y_N$ denotes the observed response measurement scores for the N objects, and $\mathbf{x}_i' = (x_{1i}, \ldots, x_{2i})$ is the row vector of $s = 2$ predictor variables for the ith of N objects. Because there are $M = 12! = 479{,}001{,}600$ possible, equally-likely arrangements of the $N = 12$ multivariate response measurement scores in Fig. 4.3, an exact permutation approach is impractical and a resampling procedure is mandated.

A LAD regression analysis of the multivariate response measurement scores listed in Fig. 4.3 yields estimated regression coefficients of

$$\tilde{\beta}_0 = +3.8571\,, \quad \tilde{\beta}_1 = +0.4286\,, \quad \text{and} \quad \tilde{\beta}_2 = +0.1429\,.$$ [1]

Fig. 4.3 Example data with $s = 2$ independent variables on $N = 12$ objects

	Variable		
Object	x_1	x_2	y
1	11	22	11
2	11	24	12
3	11	26	13
4	11	26	15
5	12	28	13
6	12	26	11
7	13	22	15
8	13	22	10
9	14	20	16
10	14	22	13
11	15	20	17
12	15	26	14

[1] For the remainder of this chapter, a tilde over a β ($\tilde{\beta}$) indicates an unstandardized LAD regression coefficient, while a caret over a β ($\hat{\beta}$) indicates an unstandardized OLS regression coefficient.

Fig. 4.4 Observed, predicted, and residual LAD regression values for the example data listed in Fig. 4.3

Object	y_i	$\tilde{y}_i$	e_i
1	11	11.7143	−0.7143
2	12	12.0000	0.0000
3	13	12.2857	+0.7143
4	15	12.2857	+2.7143
5	13	13.0000	0.0000
6	11	12.7143	−1.7143
7	15	12.5714	+2.4286
8	10	12.5714	−2.5714
9	16	12.7143	+3.2857
10	13	13.0000	0.0000
11	17	13.1429	+3.8571
12	14	14.0000	0.0000

Figure 4.4 lists the observed y_i values, LAD predicted $\tilde{y}_i$ values, and residual e_i values for $i = 1, \ldots, 12$. Following Eq. (4.1) on p. 117 with $v = 1$, the observed value of the MRPP test statistic calculated on the LAD regression residuals listed in Fig. 4.4 is $\delta_o = 1.50$.

If all M possible arrangements of the $N = 12$ observed LAD regression residuals listed in Fig. 4.4 occur with equal chance, the approximate resampling probability value of $\delta_o = 1.50$ calculated on $L = 1{,}000{,}000$ random arrangements of the observed LAD regression residuals is

$$P(\delta \leq \delta_o | H_0) = \frac{\text{number of } \delta \text{ values} \leq \delta_o}{L} = \frac{191{,}128}{1{,}000{,}000} = 0.0191 \; .$$

Following Eq. (4.3) on p. 117, the exact expected value of the $M = 479{,}001{,}600$ δ values is $\mu_\delta = 1.8294$ and, following Eq. (4.2) on p. 117, the observed chance-corrected measure of effect size for the y_i and $\tilde{y}_i$ values, $i = 1, \ldots, N$, is

$$\mathfrak{R}_o = 1 - \frac{\delta_o}{\mu_\delta} = 1 - \frac{1.50}{1.8294} = +0.1800 \; ,$$

indicating 18 % agreement between the observed and predicted y values above that expected by chance.

4.2.2 Example Analysis 2

For a second example analysis of the multivariate response measurement scores listed in Fig. 4.3 on p. 120, consider an OLS regression analysis based on a resampling permutation procedure. An OLS regression analysis of the multivariate

Fig. 4.5 Observed, predicted, and residual OLS regression values for the example data listed in Fig. 4.3

Object	y_i	$\hat{y}_i$	e_i
1	11	12.3823	−1.3823
2	12	12.2524	−0.2524
3	13	12.1226	+0.8774
4	15	12.1226	+2.8774
5	13	12.6282	+0.3718
6	11	12.7581	−1.7581
7	15	13.6534	+1.3466
8	10	13.6534	−3.6534
9	16	14.4188	+1.5812
10	13	14.2890	−1.2890
11	17	15.0544	+1.9456
12	14	14.6648	−0.6648

response measurement scores listed in Fig. 4.3 yields estimated regression coefficients of

$$\hat{\beta}_0 = +6.8198\ , \quad \hat{\beta}_1 = +0.6356\ , \quad \text{and} \quad \hat{\beta}_2 = -0.0649\ .$$

Figure 4.5 lists the observed y_i values, OLS predicted $\hat{y}_i$ values, and residual e_i values for $i = 1, \ldots, 12$.

Following Eq. (4.1) on p. 117 with $v = 2$, the observed value of the MRPP test statistic computed on the OLS regression residuals listed in Fig. 4.5 is $\delta_o = 3.1502$. If all M possible arrangements of the $N = 12$ observed OLS regression residuals listed in Fig. 4.5 occur with equal chance, the approximate resampling probability value of $\delta_o = 3.1502$ computed on $L = 1{,}000{,}000$ random arrangements of the observed OLS regression residuals is

$$P(\delta \leq \delta_o | H_0) = \frac{\text{number of } \delta_o \text{ values} \leq \delta_o}{L} = \frac{96{,}104}{1{,}000{,}000} = 0.0961\ .$$

For comparison, the approximate resampling probability value based on LAD regression in Example 1 is $P = 0.0191$.

Following Eq. (4.3) on p. 117, the exact expected value of the $M = 479{,}001{,}600$ δ values is $\mu_\delta = 5.2942$ and, following Eq. (4.2) on p. 117, the observed chance-corrected measure of effect size for the y_i and $\hat{y}_i$ values, $i = 1, \ldots, N$, is

$$\Re_o = 1 - \frac{\delta_o}{\mu_\delta} = 1 - \frac{3.1502}{5.2942} = +0.4050\ ,$$

indicating approximately 41 % agreement between the observed and predicted y values above that expected by chance.

4.2.3 Example Analysis 3

Finally, consider a conventional OLS regression analysis of the multivariate response measurement scores listed in Fig. 4.3 on p. 120. An OLS regression analysis yields estimated regression coefficients of

$$\hat{\beta}_0 = +6.8198\,, \quad \hat{\beta}_1 = +0.6356\,, \quad \text{and} \quad \hat{\beta}_2 = -0.0649\,,$$

the regression residuals are listed in Fig. 4.5, and the observed squared multiple correlation coefficient is $R^2_{y.x_1,x_2} = 0.2539$. $R^2_{y.x_1,x_2}$ may be transformed into an F-ratio by

$$F = \frac{(N-s-1)R^2_{y.x_1,x_2}}{s(1-R^2_{y.x_1,x_2})} = \frac{(12-2-1)(0.2539)}{(2)(1-0.2539)} = 1.5313\,.$$

Assuming independence, normality, and homogeneity of variance, F is approximately distributed as Snedecor's F under the null hypothesis with $\nu_1 = s = 2$ and $\nu_2 = N - s - 1 = 12 - 2 - 1 = 9$ degrees of freedom. Under the null hypothesis, the observed value of $F_o = 1.5313$ yields an approximate probability value of $P = 0.2677$.

Note that the asymptotic probability value based on OLS regression in Example 3 is $P = 0.2677$, while a resampling analysis of the same data in Example 2 yielded a probability value, again based on OLS regression, of $P = 0.0961$, a marked difference. Moreover, a LAD regression analysis of the same data in Example 1 yielded an approximate resampling probability value of $P = 0.0191$, once again demonstrating the different results possible with $v = 1$ and $v = 2$, both with and without a permutation analysis.

4.3 LAD Regression and Analysis of Variance Designs

It is well known that experimental designs that would ordinarily be analyzed by some form of analysis of variance can also be analyzed by OLS multiple regression using either dummy- or effect-coding schemes. The same is true of LAD regression. In this section a variety of analysis-of-variance designs are analyzed using MRPP, LAD regression, and either dummy or effect coding of treatment groups; included are one-way randomized, one-way randomized with a covariate, one-way randomized-block, two-way randomized-block, two-way factorial, Latin square, split-plot, and two-factor nested analysis-of-variance designs.

Fig. 4.6 Example data for a one-way randomized design with $g = 3$ treatment groups and univariate response measurement scores on $N = 26$ objects

Treatment		
1	2	3
15	17	6
18	22	9
12	15	12
12	12	11
9	20	11
10	13	8
12	15	13
20	20	30
	21	7

4.3.1 One-Way Randomized Design

Consider a one-way completely randomized experimental design with fixed effects in which $N = 26$ objects have been randomly assigned to one of $g = 3$ treatment groups with $n_1 = 8$ and $n_2 = n_3 = 9$. The design and data are adapted from Stevens [387, p. 70] and are given in Fig. 4.6.

For a one-way randomized experimental design, the appropriate regression model is given by

$$y_i = \sum_{j=1}^{m} x_{ij}\beta_j + e_i \,,$$

where y_i denotes the ith of N responses possibly affected by a treatment; x_{ij} is the jth of m covariates associated with the ith response, where $x_{i1} = 1$ if the model includes an intercept; β_j denotes the jth of m regression parameters; and e_i designates the error associated with the ith of N responses. If the estimates of $\beta_1, \ldots, \beta_m$ that minimize

$$\sum_{i=1}^{N} |e_i|$$

are denoted by $\tilde{\beta}_1, \ldots, \tilde{\beta}_m$, then the N residuals of the LAD regression model are given by $e_i = y_i - \tilde{y}_i$ for $i = 1, \ldots, N$, where the predicted value of y_i is given by

$$\tilde{y}_i = \sum_{j=1}^{m} x_{ij}\tilde{\beta}_j \,, \qquad i = 1, \ldots, N \,.$$

In contrast, OLS regression estimators of $\beta_1, \ldots, \beta_m$ minimize

$$\sum_{i=1}^{N} e_i^2 ,$$

the N residuals of the OLS regression model are given by $e_i = y_i - \hat{y}_i$ for $i = 1, \ldots, N$, and the predicted value of y_i is given by

$$\hat{y}_i = \sum_{j=1}^{m} x_{ij}\hat{\beta}_j , \qquad i = 1, \ldots, N .$$

If the N regression residuals are partitioned into g disjoint treatment groups of sizes $n_1, \ldots, n_g$, where $n_i \geq 2$ for $i = 1, \ldots, g$ and

$$N = \sum_{i=1}^{g} n_i ,$$

then the permutation test depends on test statistic

$$\delta = \sum_{i=1}^{g} C_i \xi_i , \tag{4.4}$$

where

$$C_i = \frac{n_i}{N} , \qquad i = 1, \ldots, g ,$$

is a positive weight for the ith of g treatment groups that minimizes the variability of δ,

$$\sum_{i=1}^{g} C_i = 1 ,$$

and ξ_i is the average pairwise Euclidean difference among the n_i residuals in the ith of g treatment groups defined by

$$\xi_i = \binom{n_i}{2}^{-1} \sum_{j=1}^{N-1} \sum_{k=j+1}^{N} \left[\left(e_j - e_k \right)^2 \right]^{v/2} \Psi_{ji} \, \Psi_{ki} , \tag{4.5}$$

where $v = 1$ for LAD regression and

$$\Psi_{ji} = \begin{cases} 1 & \text{if } e_i \text{ is in the } i\text{th treatment group ,} \\ 0 & \text{otherwise .} \end{cases}$$

The null hypothesis specifies that each of the

$$M = \frac{N!}{\prod_{i=1}^{g} n_i!}$$

allocations of the N residuals to the g treatment groups is equally likely with n_i, $i = 1, \ldots, g$, residuals preserved for each arrangement of the observed data. The exact probability value of an observed value of δ, δ_o, is given by

$$P(\delta \le \delta_o | H_0) = \frac{\text{number of } \delta \text{ values} \le \delta_o}{M} .$$

As previously, when M is large, an approximate probability value of δ may be obtained from a resampling procedure, where

$$P(\delta \le \delta_o | H_0) = \frac{\text{number of } \delta \text{ values} \le \delta_o}{L}$$

and L denotes the number of resampled test statistic values. Typically, L is set to a large number to ensure accuracy, e.g., $L = 1{,}000{,}000$. When M is very large and P is exceedingly small, a resampling-approximation permutation procedure may produce no δ values equal to or less than δ_o, even with $L = 1{,}000{,}000$, yielding an approximate resampling probability value of $P = 0.00$. In such cases, moment-approximation permutation procedures based on fitting the first three exact moments of the discrete permutation distribution to a Pearson type III distribution provide approximate probability values, as detailed in Chap. 1, Sect. 1.2.2 [284, 300].

An index of the effect size for the y_i and $\tilde{y}_i$ values, $i = 1, \ldots, N$, is given by the chance-corrected measure

$$\Re = 1 - \frac{\delta}{\mu_\delta} , \tag{4.6}$$

where μ_δ is the arithmetic average of the δ values calculated on all M equally-likely arrangements of the observed response measurements, i.e.,

$$\mu_\delta = \frac{1}{M} \sum_{i=1}^{M} \delta_i . \tag{4.7}$$

Fig. 4.7 Design matrix and data for a one-way randomized design with $g = 3$ treatment groups and univariate response measurement scores on $N = 26$ objects

Matrix	Score	Matrix	Score	Matrix	Score
1	15	1	17	1	16
1	18	1	22	1	9
1	12	1	15	1	12
1	12	1	12	1	11
1	9	1	20	1	11
1	10	1	14	1	8
1	12	1	15	1	13
1	20	1	20	1	30
		1	21	1	7

A design matrix of dummy codes for an MRPP regression analysis of the $N = 26$ response measurement scores in Fig. 4.6 is given in Fig. 4.7 where the first columns of 1 values provide for an intercept. The second columns contain the $N = 26$ univariate response measurement scores listed according to the original random assignment of the $N = 26$ objects to the $g = 3$ treatment groups with the first $n_1 = 8$ scores, the next $n_2 = 9$ scores, and the last $n_3 = 9$ scores associated with the first, second, and third treatment groups, respectively.

Because the purpose of the analysis is to test for possible differences among the $g = 3$ treatment groups, a reduced regression model is constructed without a variate for treatments. Therefore, for a single-factor experiment the design matrix for the reduced model is composed solely of a code for the intercept. The MRPP regression analysis examines the $N = 26$ regression residuals for possible differences among the $g = 3$ treatment levels; consequently, no dummy codes for treatments are included in Fig. 4.7 as this information is implicit in the ordering of the $g = 3$ treatment groups in the three columns labeled "Score" with $n_1 = 8$ and $n_2 = n_3 = 9$ values.

An exact permutation solution is impractical for the univariate response measurements listed in Fig. 4.7 since there are

$$M = \frac{N!}{\prod_{i=1}^{g} n_i!} = \frac{26!}{8!\ 9!\ 9!} = 75{,}957{,}810{,}500$$

possible, equally-likely arrangements of the $N = 26$ univariate response measurement scores; consequently, a resampling procedure is the default in this case.

LAD Regression Analysis

An MRPP resampling analysis of the LAD regression residuals calculated on the univariate response measurement scores listed in Fig. 4.7 yields an estimated LAD regression coefficient of $\tilde{\beta}_0 = +12.00$. Figure 4.8 lists the observed y_i values, LAD predicted $\tilde{y}_i$ values, and residual e_i values for $i = 1, \ldots, 26$.

Fig. 4.8 Observed, predicted, and residual LAD regression values for the example one-way randomized data listed in Fig. 4.7

Object	y_i	$\tilde{y}_i$	e_i
1	15	12.00	+3.00
2	18	12.00	+6.00
3	12	12.00	0.00
4	12	12.00	0.00
5	9	12.00	−3.00
6	10	12.00	−2.00
7	12	12.00	0.00
8	20	12.00	+8.00
9	17	12.00	+5.00
10	22	12.00	+10.00
11	15	12.00	+3.00
12	12	12.00	0.00
13	20	12.00	+8.00
14	14	12.00	+2.00
15	15	12.00	+3.00
16	20	12.00	+8.00
17	21	12.00	+9.00
18	6	12.00	−6.00
19	9	12.00	−3.00
20	12	12.00	0.00
21	11	12.00	−1.00
22	11	12.00	−1.00
23	8	12.00	−4.00
24	13	12.00	+1.00
25	30	12.00	+18.00
26	7	12.00	−5.00

Following Eq. (4.5) on p. 125 and employing ordinary Euclidean distance between residuals with $v = 1$, the $N = 26$ LAD regression residuals listed in Fig. 4.8 yield $g = 3$ average distance-function values of

$$\xi_1 = 4.50\ , \quad \xi_2 = 4.2222\ , \quad \text{and} \quad \xi_3 = 6.8889\ .$$

Following Eq. (4.4) on p. 125, the observed value of the MRPP test statistic calculated on the LAD regression residuals listed in Fig. 4.8 with $v = 1$ and treatment-group weights

$$C_i = \frac{n_i}{n}\ , \qquad i = 1, 2, 3\ ,$$

is

$$\delta_o = \sum_{i=1}^{g} C_i \xi_i = \frac{1}{26}\big[(8)(4.50) + (9)(4.2222) + (9)(6.8889)\big] = 5.2308\ .$$

If all M possible arrangements of the $N = 26$ observed LAD regression residuals listed in Fig. 4.8 occur with equal chance, the approximate resampling probability value of $\delta_o = 5.2308$ computed on $L = 1{,}000{,}000$ random arrangements of the observed LAD regression residuals with $n_1 = 8$ and $n_2 = n_3 = 9$ residuals preserved for each arrangement is

$$P(\delta \leq \delta_o | H_0) = \frac{\text{number of } \delta \text{ values} \leq \delta_o}{L} = \frac{12{,}062}{1{,}000{,}000} = 0.0121 \; .$$

Following Eq. (4.7) on p. 126, the exact expected value of the M δ values is $\mu_\delta = 6.1262$ and, following Eq. (4.6) on p. 126, the observed chance-corrected measure of effect size for the y_i and $\tilde{y}_i$ values, $i = 1, \ldots, N$, is

$$\Re_o = 1 - \frac{\delta_o}{\mu_\delta} = 1 - \frac{5.2308}{6.1262} = +0.1462 \; ,$$

indicating approximately 15 % agreement between the observed and predicted y values above that expected by chance.

OLS Regression Analysis

For comparison, consider an MRPP resampling analysis of OLS regression residuals calculated on the $N = 26$ univariate response measurement scores listed in Fig. 4.7 on p. 127. The MRPP regression analysis yields an estimated OLS regression coefficient of $\hat{\beta}_0 = +14.2692$. Figure 4.9 lists the observed y_i values, OLS predicted $\hat{y}_i$ values, and residual e_i values for $i = 1, \ldots, 26$.

Following Eq. (4.5) on p. 125 and employing squared Euclidean distance between residuals with $v = 2$, the $N = 26$ OLS regression residuals listed in Fig. 4.9 yield $g = 3$ average distance-function values of

$$\xi_1 = 29.7143 \; , \quad \xi_2 = 25.00 \; , \quad \text{and} \quad \xi_3 = 103.2222 \; .$$

Following Eq. (4.4) on p. 125, the observed value of the MRPP test statistic calculated on the OLS regression residuals listed in Fig. 4.9 with $v = 2$ and treatment-group weights

$$C_i = \frac{n_i - 1}{N - g} \; , \qquad i = 1, 2, 3 \; ,$$

is

$$\delta_o = \sum_{i=1}^{g} C_i \xi_i = \frac{1}{26 - 3} \big[(8 - 1)(29.7143) + (9 - 1)(25.00) + (9 - 1)(103.2222) \big] = 53.6425 \; .$$

Fig. 4.9 Observed, predicted, and residual OLS regression values for the example one-way randomized data listed in Fig. 4.7

Object	y_i	$\hat{y}_i$	e_i
1	15	14.2692	+0.7308
2	18	14.2692	+3.7308
3	12	14.2692	−2.2692
4	12	14.2692	−2.2692
5	9	14.2692	−5.2692
6	10	14.2692	−4.2692
7	12	14.2692	−2.2692
8	20	14.2692	+5.7308
9	17	14.2692	+2.7308
10	22	14.2692	+7.7308
11	15	14.2692	+0.7308
12	12	14.2692	−2.2692
13	20	14.2692	+5.7308
14	14	14.2692	−0.2692
15	15	14.2692	+0.7308
16	20	14.2692	+5.7308
17	21	14.2692	+6.7308
18	6	14.2692	−8.2692
19	9	14.2692	−5.2692
20	12	14.2692	−2.2692
21	11	14.2692	−3.2692
22	11	14.2692	−3.2692
23	8	14.2692	−6.2692
24	13	14.2692	−1.2692
25	30	14.2692	+15.7308
26	7	14.2692	−7.2692

If all M possible arrangements of the $N = 26$ observed OLS regression residuals listed in Fig. 4.9 occur with equal chance, the approximate resampling probability value of $\delta_o = 53.6425$ computed on $L = 1{,}000{,}000$ random arrangements of the observed OLS regression residuals with $n_1 = 8$ and $n_2 = n_3 = 9$ residuals preserved for each arrangement is

$$P(\delta \leq \delta_o|H_0) = \frac{\text{number of } \delta \text{ values} \leq \delta_o}{L} = \frac{91{,}842}{1{,}000{,}000} = 0.0918 \, .$$

For comparison, the approximate resampling probability value based LAD regression, $v = 1$, $L = 1{,}000{,}000$, and $C_i = n_i/N$ for $i = 1, 2, 3$ is $P = 0.0121$.

Following Eq. (4.7) on p. 126, the exact expected value of the $M = 75{,}957{,}810{,}500$ δ values is $\mu_\delta = 60.5692$ and, following Eq. (4.6) on p. 126, the observed chance-corrected measure of effect size for the y_i and $\hat{y}_i$ values, $i = 1, \ldots, N$, is

$$\Re_o = 1 - \frac{\delta_o}{\mu_\delta} = 1 - \frac{53.6425}{60.5692} = +0.1144 \, ,$$

indicating approximately 11 % agreement between the observed and predicted y values above that expected by chance.

Conventional ANOVA Analysis

A conventional fixed-effects one-way analysis of variance calculated on the $N = 26$ univariate response measurement scores listed in Fig. 4.6 on p. 124 yields an observed F-ratio of $F_o = 2.6141$. Assuming independence, normality, and homogeneity of variance, F is approximately distributed as Snedecor's F under the null hypothesis with $\nu_1 = g - 1 = 3 - 1 = 2$ and $\nu_2 = N - g = 26 - 3 = 23$ degrees of freedom. Under the null hypothesis, the observed value of $F_o = 2.6141$ yields an approximate probability value of $P = 0.0948$, which is similar to that produced by the MRPP resampling analysis of the OLS regression residuals.

4.3.2 One-Way Randomized Design with a Covariate

A covariate experimental design permits the testing of differences among the treatment groups after the effect of the covariate has been removed from the analysis. Consider a one-way completely randomized design with a covariate in which $N = 47$ objects are randomly assigned to one of $g = 5$ treatment groups. The experimental data are listed in Table 4.1 and are adapted from a 1984 study by Conti and Musty [78].

A design matrix of dummy codes for analyzing treatments is given in Fig. 4.10, where the first column of 1 values provides for an intercept, the second column contains the covariate (Pre-test) values, and the third column contains the (Post-test) scores listed according to the original random assignment of the $N = 47$ objects to

Table 4.1 Example data for a one-way randomized design with a covariate, consisting of pre-test (Pre) and post-test (Post) response measurement scores on $N = 47$ randomly assigned objects to $g = 5$ treatment groups

Treatment									
1		2		3		4		5	
Pre	Post	Pre	Post	Pre	Post	Pre	Post	Pre	Post
4.34	1.30	1.55	0.93	7.18	5.10	6.94	2.29	4.00	2.93
3.50	0.94	10.56	4.44	8.33	4.16	6.10	4.75	4.10	1.11
4.33	2.25	8.39	4.03	4.05	1.54	4.90	3.48	3.62	2.17
2.76	1.05	3.70	1.92	10.78	6.36	3.69	2.76	3.92	2.00
4.62	0.92	2.40	0.67	6.09	3.96	4.76	1.67	2.90	0.84
5.40	1.90	1.83	1.70	7.78	4.51	4.30	1.51	2.90	0.99
3.95	0.32	2.40	0.77	5.08	3.76	2.32	1.07	1.82	0.44
1.55	0.64	7.67	3.53	2.86	1.92	7.35	2.35	4.94	0.84
1.42	0.69	5.79	3.65	6.30	3.84			5.69	2.84
1.90	0.93	9.58	4.22					5.54	2.93

Fig. 4.10 Design matrix and data, consisting of an intercept and pre- and post-test measurement scores for a one-way randomized design with a covariate

Matrix	Pre	Post	Matrix	Pre	Post
1	4.34	1.30	1	6.94	2.29
1	3.50	0.94	1	6.10	4.75
1	4.33	2.25	1	4.90	3.48
1	2.76	1.05	1	3.69	2.76
1	4.62	0.92	1	4.76	1.67
1	5.40	1.90	1	4.30	1.51
1	3.95	0.32	1	2.32	1.07
1	1.55	0.64	1	7.35	2.35
1	1.42	0.69			
1	1.90	0.93	1	4.00	1.44
			1	4.10	1.11
1	1.55	0.93	1	3.62	2.17
1	10.56	4.44	1	3.92	2.00
1	8.39	4.03	1	2.90	0.84
1	3.70	1.92	1	2.90	0.99
1	2.40	0.67	1	1.82	0.44
1	1.83	1.70	1	4.94	0.84
1	2.40	0.77	1	5.69	2.84
1	7.67	3.53	1	5.54	2.93
1	5.79	3.65			
1	9.58	4.22			
1	7.18	5.10			
1	8.33	4.16			
1	4.05	1.54			
1	10.78	6.36			
1	6.09	3.96			
1	7.78	4.51			
1	5.08	3.76			
1	2.86	1.92			
1	6.30	3.84			

the $g = 5$ treatment groups with the first $n_1 = 10$ scores, the next $n_2 = 10$ scores, the next $n_3 = 9$ scores, the next $n_4 = 8$ scores, and the last $n_5 = 10$ scores associated with the $g = 5$ treatment groups, respectively.

The MRPP regression analysis examines the $N = 47$ regression residuals for possible differences among the $g = 5$ treatment levels; consequently, no dummy codes for treatments are included in Fig. 4.10 as this information is implicit in the ordering of the $g = 5$ treatment groups in the two paired columns labeled "Pre" and "Post."

Because there are

$$M = \frac{N!}{\prod_{i=1}^{g} n_i!} = \frac{47!}{10!\ 10!\ 9!\ 8!\ 10!} = 369{,}908{,}998{,}147{,}203{,}213{,}613{,}129{,}815{,}600$$

possible, equally-likely arrangements of the $N = 47$ univariate response measurement scores listed in Table 4.1, an exact permutation approach is not possible and a resampling analysis is mandated.

LAD Regression Analysis

An MRPP resampling analysis of the LAD regression residuals calculated on the $N = 47$ response measurement scores listed in Fig. 4.10 yields estimated LAD regression coefficients of

$$\tilde{\beta}_0 = -0.1282 \quad \text{and} \quad \tilde{\beta}_1 = +0.4956 \, .$$

Table 4.2 lists the observed y_i values, LAD predicted $\tilde{y}_i$ values, and residual e_i values for $i = 1, \ldots, 47$.

Following Eq. (4.5) on p. 125 and employing ordinary Euclidean distance between residuals $v = 1$, the LAD regression residuals listed in Table 4.2 yield

Table 4.2 Observed, predicted, and residual LAD regression values for the example covariate data listed in Fig. 4.10

Object	y_i	$\tilde{y}_i$	e_i	Object	y_i	$\tilde{y}_i$	e_i
1	1.30	2.0228	−0.7228	25	3.96	2.8901	+1.0699
2	0.94	1.6064	−0.6664	26	4.51	3.7277	+0.7823
3	2.25	2.0178	+0.2322	27	3.76	2.3895	+1.3705
4	1.05	1.2397	−0.1897	28	1.92	1.2893	+0.6307
5	0.92	2.1615	−1.2415	29	3.84	2.9942	+0.8458
6	1.90	2.5481	−0.6481	30	2.29	3.3114	−1.0214
7	0.32	1.8295	−1.5095	31	4.75	2.8950	+1.8550
8	0.64	0.6400	0.0000	32	3.48	2.3003	+1.1797
9	0.69	0.5756	+0.1144	33	2.76	1.7006	+1.0594
10	0.93	0.8135	+0.1165	34	1.67	2.2309	−0.5609
11	0.93	0.6400	+0.2900	35	1.51	2.0029	−0.4929
12	4.44	5.1055	−0.6655	36	1.07	1.0216	+0.0484
13	4.03	4.0300	0.0000	37	2.35	3.5146	−1.1646
14	1.92	1.7056	+0.2144	38	1.44	1.8543	−0.4143
15	0.67	1.0613	−0.3913	39	1.11	1.9038	−0.7938
16	1.70	0.7788	+0.9212	40	2.17	1.6659	+0.5041
17	0.77	1.0613	−0.2913	41	2.00	1.8146	+0.1854
18	3.53	3.6732	−0.1432	42	0.84	1.3091	−0.4691
19	3.65	2.7414	+0.9086	43	0.99	1.3091	−0.3191
20	4.22	4.6198	−0.3998	44	0.44	0.7738	−0.3338
21	5.10	3.4303	+1.6697	45	0.84	2.3201	−1.4801
22	4.16	4.0003	+0.1597	46	2.84	2.6918	+0.1482
23	1.54	1.8790	−0.3390	47	2.93	2.6175	+0.3125
24	6.36	5.2145	+1.1455				

$g = 5$ average distance-function values of

$$\xi_1 = 0.7072\ ,\ \ \xi_2 = 0.6335\ ,\ \ \xi_3 = 0.7213\ ,\ \ \xi_4 = 1.3409\ ,\ \text{and}\ \ \xi_5 = 0.6795\ .$$

Following Eq. (4.4) on p. 125, the observed value of the MRPP test statistic calculated on the LAD regression residuals listed in Table 4.2 with $v = 1$ and treatment-group weights

$$C_i = \frac{n_i}{N}\ , \qquad i = 1, \ldots, 5\ ,$$

is

$$\delta_o = \sum_{i=1}^{g} C_i \xi_i = \frac{1}{47}\big[(10)(0.7072) + (10)(0.6335) + (9)(0.7213)$$
$$+ (8)(1.3409) + (10)(0.6795)\big] = 0.7962\ .$$

If all M possible arrangements of the observed LAD regression residuals listed in Table 4.2 occur with equal chance, the approximate resampling probability value of $\delta_o = 0.7962$ computed on $L =$ 1,000,000 random arrangements of the observed LAD regression residuals with $n_1 = n_2 = n_5 = 10$, $n_3 = 9$, and $n_4 = 8$ residuals preserved for each arrangement is

$$P(\delta \leq \delta_o | H_0) = \frac{\text{number of } \delta \text{ values} \leq \delta_o}{L} = \frac{4{,}095}{1{,}000{,}000} = 0.0041\ .$$

Following Eq. (4.7) on p. 126, the exact expected value of the M δ values is $\mu_\delta =$ 0.9178 and, following Eq. (4.6) on p. 126, the observed chance-corrected measure of effect size for the y_i and $\tilde{y}_i$ values, $i = 1, \ldots, N$, is

$$\Re_o = 1 - \frac{\delta_o}{\mu_\delta} = 1 - \frac{0.7962}{0.9178} = +0.1326\ ,$$

indicating approximately 13 % agreement between the observed and predicted y values above that expected by chance.

OLS Regression Analysis

For comparison, consider an MRPP resampling analysis of the OLS regression residuals calculated on the $N = 47$ univariate response measurement scores listed in Fig. 4.10 on p. 132. The MRPP regression analysis yields estimated OLS regression coefficients of

$$\hat{\beta}_0 = -0.2667 \quad \text{and} \quad \hat{\beta}_1 = +0.5311\ .$$

Table 4.3 Observed, predicted, and residual OLS regression values for the example covariate data listed in Fig. 4.10

Object	y_i	$\hat{y}_i$	e_i	Object	y_i	$\hat{y}_i$	e_i
1	1.30	2.0383	−0.7383	25	3.96	2.9677	+0.9923
2	0.94	1.5922	−0.6522	26	4.51	3.8652	+0.6448
3	2.25	2.0330	+0.2170	27	3.76	2.4313	+1.3287
4	1.05	1.1991	−0.1491	28	1.92	1.2523	+0.6677
5	0.92	2.1870	−1.2670	29	3.84	3.0792	+0.7608
6	1.90	2.6012	−0.7012	30	2.29	3.4191	−1.1291
7	0.32	1.8311	−1.5111	31	4.75	2.9730	+1.7770
8	0.64	0.5565	+0.0835	32	3.48	2.3357	+1.1443
9	0.69	0.4875	+0.2025	33	2.76	1.6931	+1.0669
10	0.93	0.7424	+0.1876	34	1.67	2.2613	−0.5913
11	0.93	0.5565	+0.3735	35	1.51	2.0170	−0.5070
12	4.44	5.3417	−0.9017	36	1.07	0.9655	+0.1045
13	4.03	4.1892	−0.1592	37	2.35	3.6369	−1.2869
14	1.92	1.6984	+0.2216	38	1.44	1.8577	−0.4177
15	0.67	1.0080	−0.3380	39	1.11	1.9108	−0.8008
16	1.70	0.7052	+0.9948	40	2.17	1.6559	+0.5141
17	0.77	1.0080	−0.2380	41	2.00	1.8152	+0.1848
18	3.53	3.8068	−0.2768	42	0.84	1.2735	−0.4335
19	3.65	2.8084	+0.8416	43	0.99	1.2735	−0.2835
20	4.22	4.8212	−0.6012	44	0.44	0.6999	−0.2599
21	5.10	3.5466	+1.5534	45	0.84	2.3569	−1.5169
22	4.16	4.1573	+0.0027	46	2.84	2.7553	+0.0847
23	1.54	1.8843	−0.3443	47	2.93	2.6756	+0.2544
24	6.36	5.4585	+0.9015				

Table 4.3 lists the observed y_i values, OLS predicted $\hat{y}_i$ values, and residual e_i values for $i = 1, \ldots, 47$.

Following Eq. (4.5) on p. 125 and employing squared Euclidean distance between residuals with $v = 2$, the OLS regression residuals listed in Table 4.3 yield $g = 5$ average distance-function values of

$$\xi_1 = 0.8067\,,\quad \xi_2 = 0.7407\,,\quad \xi_3 = 0.7073\,,\quad \xi_4 = 2.6035\,,\quad \text{and}\quad \xi_5 = 0.6906\,.$$

Following Eq. (4.4) on p. 125, the observed value of the MRPP test statistic calculated on the OLS regression residuals listed in Table 4.3 with $v = 2$ and treatment-group weights

$$C_i = \frac{n_i - 1}{N - g}\,,\qquad i = 1, \ldots, 5\,,$$

is

$$\delta_o = \sum_{i=1}^{g} C_i \xi_i = \frac{1}{47-5}[(10-1)(0.8067) + (10-1)(0.7407)$$

$$+ (9-1)(0.7073) + (8-1)(2.6035) + (10-1)(0.6906)] = 1.0482 \, .$$

If all M possible arrangements of the $N = 47$ observed OLS regression residuals listed in Table 4.3 occur with equal chance, the approximate resampling probability value of $\delta_o = 1.0482$ computed on $L = 1{,}000{,}000$ random arrangements of the observed OLS regression residuals with $n_1 = n_2 = n_5 = 10$, $n_3 = 9$, and $n_4 = 8$ residuals preserved for each arrangement is

$$P(\delta \leq \delta_o | H_0) = \frac{\text{number of } \delta \text{ values} \leq \delta_o}{L} = \frac{15{,}301}{1{,}000{,}000} = 0.0153 \, .$$

For comparison, the approximate resampling probability value based on LAD regression, $v = 1$, $L = 1{,}000{,}000$, and $C_i = n_i/N$ for $i = 1, \ldots, 5$ is $P = 0.0041$.

Following Eq. (4.7) on p. 126, the exact expected value of the M δ values is $\mu_\delta = 1.2761$ and, following Eq. (4.6) on p. 126, the observed chance-corrected measure of effect size for the y_i and $\hat{y}_i$ values, $i = 1, \ldots, N$, is

$$\Re_o = 1 - \frac{\delta_o}{\mu_\delta} = 1 - \frac{1.0482}{1.2761} = +0.1785 \, ,$$

indicating approximately 18 % agreement between the observed and predicted y values above that expected by chance.

Conventional ANOVA Analysis

A conventional fixed-effects one-way analysis of covariance calculated on the $N = 47$ univariate response measurement scores listed in Table 4.1 on p. 131 yields an observed F-ratio of $F_o = 4.6978$. Assuming independence, normality, and homogeneity of variance, F is approximately distributed as Snedecor's F under the null hypothesis with $\nu_1 = g - 1 = 5 - 1 = 4$ and $\nu_2 = N - g - 1 = 47 - 5 - 1 = 41$ degrees of freedom. Under the null hypothesis, the observed value of $F_o = 4.6978$ yields an approximate probability value of $P = 0.0033$.

4.3.3 One-Way Randomized-Block Design

One-way randomized-block designs are common in experimental research and have long been valuable statistical tools in such fields as agriculture and genetics. E.J.G. Pitman, for example, developed a permutation approach for one-way randomized-block designs in 1938 [342]. With modern developments in embryo transplants and cloning where subjects can be genetically matched on a large number of important characteristics, randomized-block designs have become very practical and efficient.[2]

Consider a one-way randomized-block design where $b = 6$ objects (blocks) are evaluated over $a = 3$ treatments with $r = 1$ response measurement. The design and data are adapted from a study by Anderson, Sweeney, and Williams [9, p. 471] and are given in Fig. 4.11.

A design matrix of dummy codes for an MRPP regression analysis is given in Fig. 4.12, where the first column of 1 values provides for an intercept, the next five columns contain dummy codes for the $b = 6$ blocks, and the last column contains the univariate response measurement scores listed according to the original random assignment of the $N = 18$ objects to the $a = 3$ treatment levels of Factor A with the first $n_{A_1} = 6$ objects, the next $n_{A_2} = 6$ objects, and the last $n_{A_3} = 6$ objects

Fig. 4.11 Example data for a one-way randomized-block design with $b = 6$ blocks, $a = 3$ treatments, and $r = 1$ response measurement

	Factor A		
Object	A_1	A_2	A_3
1	15	15	18
2	14	14	14
3	10	11	15
4	13	12	17
5	16	13	16
6	13	13	13

Fig. 4.12 Design matrix and data for a one-way randomized-block design with $b = 6$ blocks, $a = 3$ treatments, and $r = 1$ response measurement

Matrix						Score
1	0	0	0	0	0	15
1	1	0	0	0	0	14
1	0	1	0	0	0	10
1	0	0	1	0	0	13
1	0	0	0	1	0	16
1	0	0	0	0	1	13
1	0	0	0	0	0	15
1	1	0	0	0	0	14
1	0	1	0	0	0	11
1	0	0	1	0	0	12
1	0	0	0	1	0	13
1	0	0	0	0	1	13
1	0	0	0	0	0	18
1	1	0	0	0	0	14
1	0	1	0	0	0	15
1	0	0	1	0	0	17
1	0	0	0	1	0	16
1	0	0	0	0	1	13

[2]All the biologically inherited information is not carried in the genes of a cell's nucleus. A small number of genes are carried by intra-cellular bodies, the mitochondria. Thus, the result of cloning is not, strictly speaking, a perfect genetic clone of the donor organism.

associated with treatment levels A_1, A_2, and A_3, respectively. The MRPP regression analysis examines the $N = 18$ regression residuals for possible differences in the $a = 3$ treatment levels; consequently, there are no dummy codes for treatments in Fig. 4.12 as this information is implicit in the ordering of the $a = 3$ treatment levels of Factor A in the last column.

Because there are

$$M = \frac{N!}{\prod_{i=1}^{a} n_{A_i}!} = \frac{18!}{(6!)^3} = 17{,}153{,}136$$

possible, equally-likely arrangements of the $N = 18$ univariate response measurement scores listed in Fig. 4.11, an exact permutation approach is not practical.

LAD Regression Analysis

An MRPP resampling analysis of the LAD regression residuals calculated on the univariate response measurement scores listed in Fig. 4.12 yields estimated LAD regression coefficients of

$$\tilde{\beta}_0 = +15.00\,,\quad \tilde{\beta}_1 = -1.00\,,\quad \tilde{\beta}_2 = -4.00\,,\quad \tilde{\beta}_3 = -2.00\,,$$

$$\tilde{\beta}_4 = +1.00\,,\ \text{and}\quad \tilde{\beta}_5 = -2.00$$

for Factor A. Figure 4.13 lists the observed y_i values, LAD predicted $\tilde{y}_i$ values, and residual e_i values for $i = 1, \ldots, 18$.

Fig. 4.13 Observed, predicted, and residual LAD regression values for the example randomized-block data listed in Fig. 4.12

Object	y_i	$\tilde{y}_i$	e_i
1	15	15.00	0.00
2	14	14.00	0.00
3	10	11.00	−1.00
4	13	13.00	0.00
5	16	16.00	0.00
6	13	13.00	0.00
7	15	15.00	0.00
8	14	14.00	0.00
9	11	11.00	0.00
10	12	13.00	−1.00
11	13	16.00	−3.00
12	13	13.00	0.00
13	18	15.00	+3.00
14	14	14.00	0.00
15	15	11.00	+4.00
16	17	13.00	+4.00
17	16	16.00	0.00
18	13	13.00	0.00

Following Eq. (4.5) on p. 125 and employing ordinary Euclidean distance between residuals with $v = 1$, the $N = 18$ LAD regression residuals listed in Fig. 4.13 yield $a = 3$ average distance-function values of

$$\xi_{A_1} = 0.3333 , \quad \xi_{A_2} = 1.20 , \quad \text{and} \quad \xi_{A_3} = 2.3333 .$$

Following Eq. (4.4) on p. 125, the observed value of the MRPP test statistic calculated on the LAD regression residuals listed in Fig. 4.13 with $v = 1$ and treatment-group weights

$$C_i = \frac{n_{A_i}}{N} , \qquad i = 1, 2, 3 ,$$

is

$$\delta_A = \sum_{i=1}^{a} C_i \xi_i = \frac{6}{18}\big(0.3333 + 1.20 + 2.3333\big) = 1.2889 .$$

If all M possible arrangements of the $N = 18$ observed LAD regression residuals listed in Fig. 4.13 occur with equal chance, the approximate resampling probability value of $\delta_A = 1.2889$ computed on $L = 1{,}000{,}000$ random arrangements of the observed LAD regression residuals with $n_{A_1} = n_{A_2} = n_{A_3} = 6$ residuals preserved for each arrangement is

$$P\big(\delta \leq \delta_A | H_0\big) = \frac{\text{number of } \delta \text{ values} \leq \delta_A}{L} = \frac{56{,}035}{1{,}000{,}000} = 0.0560 .$$

Following Eq. (4.7) on p. 126, the exact expected value of the $M = 17{,}153{,}136$ δ values is $\mu_\delta = 1.6078$ and, following Eq. (4.6) on p. 126, the observed chance-corrected measure of effect size for the y_i and $\tilde{y}_i$ values, $i = 1, \ldots, N$, is

$$\Re_A = 1 - \frac{\delta_A}{\mu_\delta} = 1 - \frac{1.2889}{1.6078} = +0.1984 ,$$

indicating approximately 20 % agreement between the observed and predicted y values above that expected by chance.

An Exact Test

Although an exact permutation analysis of the $N = 18$ LAD regression residuals listed in Fig. 4.13 is impractical, it is not impossible. In fact, exact permutation methods are oftentimes more efficient than resampling permutation methods because the $L = 1{,}000{,}000$ calls to a pseudorandom number generator, necessary for a resampling test, are not required by an exact test.

Following Eq. (4.5) on p. 125, an exact permutation analysis of the $N = 18$ LAD regression residuals listed in Fig. 4.13 yields $a = 3$ average distance-function values of

$$\xi_{A_1} = 0.3333\,, \quad \xi_{A_2} = 1.20\,, \quad \text{and} \quad \xi_{A_3} = 2.3333\,.$$

Following Eq. (4.4) on p. 125, the observed value of the MRPP test statistic based on $v = 1$ and treatment-group weights

$$C_i = \frac{n_{A_i}}{N}\,, \qquad i = 1, 2, 3\,,$$

is

$$\delta_A = \sum_{i=1}^{a} C_i \xi_i = \frac{6}{18}(0.3333 + 1.20 + 2.3333) = 1.2889\,.$$

If all arrangements of the $N = 18$ observed LAD regression residuals listed in Fig. 4.13 occur with equal chance, the exact probability value of $\delta_A = 1.2889$ computed on the $M = 17{,}153{,}136$ possible arrangements of the observed LAD regression residuals with $n_{A_1} = n_{A_2} = n_{A_3} = 6$ residuals preserved for each arrangement is

$$P(\delta \le \delta_A | H_0) = \frac{\text{number of } \delta \text{ values} \le \delta_A}{M} = \frac{961{,}884}{17{,}153{,}136} = 0.0561\,.$$

For comparison, the resampling probability value computed on $L = 1{,}000{,}000$ random arrangements of the observed LAD regression residuals listed in Fig. 4.13 is $P = 0.0560$.

OLS Regression Analysis

For comparison, consider an MRPP resampling analysis of OLS regression residuals calculated on the $N = 18$ univariate response measurement scores listed in Fig. 4.12 on p. 137. The MRPP regression analysis yields estimated OLS regression coefficients of

$$\hat{\beta}_0 = +16.00\,, \quad \hat{\beta}_1 = -2.00\,, \quad \hat{\beta}_2 = -4.00\,, \quad \hat{\beta}_3 = -2.00\,,$$
$$\hat{\beta}_4 = -1.00\,, \quad \text{and} \quad \hat{\beta}_5 = -3.00$$

for Factor A. Figure 4.14 lists the observed y_i values, OLS predicted $\hat{y}_i$ values, and residual e_i values for $i = 1, \ldots, 18$.

Fig. 4.14 Observed, predicted, and residual OLS regression values for the example randomized-block data listed in Fig. 4.12

Object	y_i	$\hat{y}_i$	e_i
1	15	16.00	−1.00
2	14	14.00	0.00
3	10	12.00	−2.00
4	13	14.00	−1.00
5	16	15.00	+1.00
6	13	13.00	0.00
7	15	16.00	−1.00
8	14	14.00	0.00
9	11	12.00	−1.00
10	12	14.00	−2.00
11	13	15.00	−2.00
12	13	13.00	0.00
13	18	16.00	+2.00
14	14	14.00	0.00
15	15	12.00	+3.00
16	17	14.00	+3.00
17	16	15.00	+1.00
18	13	13.00	0.00

Following Eq. (4.5) on p. 125 and employing ordinary Euclidean distance between residuals with $v = 2$, the $N = 18$ OLS regression residuals listed in Fig. 4.14 yield $a = 3$ average distance-function values of

$$\xi_{A_1} = 2.20 \ , \quad \xi_{A_2} = 1.60 \ , \ \text{and} \quad \xi_{A_3} = 3.80 \ .$$

Following Eq. (4.4) on p. 125, the observed value of the MRPP test statistic calculated on the OLS regression residuals listed in Fig. 4.14 with $v = 2$ and treatment-group weights

$$C_i = \frac{n_{A_i} - 1}{N - a} \ , \qquad i = 1, 2, 3 \ ,$$

is

$$\delta_A = \sum_{i=1}^{a} C_i \xi_i = \frac{6 - 1}{18 - 3} \big(2.20 + 1.60 + 3.80\big) = 2.5333 \ .$$

If all M possible arrangements of the $N = 18$ observed OLS regression residuals listed in Fig. 4.14 occur with equal chance, the approximate resampling probability value of $\delta_A = 2.5333$ computed on $L = 1{,}000{,}000$ random arrangements of the observed OLS regression residuals with $n_{A_1} = n_{A_2} = n_{A_3} = 6$ residuals preserved for each arrangement is

$$P(\delta \leq \delta_A | H_0) = \frac{\text{number of } \delta \text{ values} \leq \delta_A}{L} = \frac{4{,}974}{1{,}000{,}000} = 0.0050 \ .$$

For comparison, the approximate resampling probability value based on LAD regression, $v = 1$, $L = 1{,}000{,}000$, and $C_i = n_{A_i}/N$ for $i = 1, 2, 3$ is $P = 0.0560$.

Following Eq. (4.7) on p. 126, the exact expected value of the $M = 17{,}153{,}136$ δ values is $\mu_\delta = 5.5556$ and, following Eq. (4.6) on p. 126, the observed chance-corrected measure of effect size for the y_i and $\hat{y}_i$ values, $i = 1, \ldots, N$, is

$$\Re_A = 1 - \frac{\delta_o}{\mu_\delta} = 1 - \frac{2.5333}{5.5556} = +0.5440 \ ,$$

indicating approximately 54 % agreement between the observed and predicted y values above that expected by chance.

An Exact Test

Although an exact permutation analysis of the $N = 18$ OLS regression residuals listed in Fig. 4.14 is impractical, it is not impossible. Following Eq. (4.5) on p. 125, an exact permutation analysis of the $N = 18$ OLS regression residuals listed in Fig. 4.14 yields $a = 3$ average distance-function values of

$$\xi_{A_1} = 2.20 \ , \quad \xi_{A_2} = 1.60 \ , \text{ and } \quad \xi_{A_3} = 3.80 \ .$$

Following Eq. (4.4) on p. 125, the observed value of the MRPP test statistic based on $v = 2$ and treatment-group weights

$$C_i = \frac{n_{A_i} - 1}{N - a} \ , \qquad i = 1, 2, 3 \ ,$$

is

$$\delta_A = \sum_{i=1}^{a} C_i \xi_i = \frac{6 - 1}{18 - 3}\left(2.20 + 1.60 + 3.80\right) = 2.5333 \ .$$

If all arrangements of the $N = 18$ observed OLS regression residuals listed in Fig. 4.14 occur with equal chance, the exact probability value of $\delta_A = 2.5333$ computed on the $M = 17{,}153{,}136$ possible arrangements of the observed OLS regression residuals with $n_{A_1} = n_{A_2} = n_{A_3} = 6$ residuals preserved for each arrangement is

$$P(\delta \leq \delta_A | H_0) = \frac{\text{number of } \delta \text{ values} \leq \delta_A}{M} = \frac{85{,}188}{17{,}153{,}136} = 0.0050 \ .$$

For comparison, the approximate resampling probability value computed on $L =$ 1,000,000 random arrangements of the observed OLS regression residuals listed in Fig. 4.14 is also $P = 0.0050$.

Conventional ANOVA Analysis

A conventional randomized-block analysis of variance calculated on the $N = 18$ univariate response measurement scores listed in Fig. 4.11 on p. 137 yields an observed F-ratio of $F_A = 5.5263$. Assuming independence and normality, F_A is approximately distributed as Snedecor's F under the null hypothesis with $\nu_1 = a - 1 = 3 - 1 = 2$ and $\nu_2 = (b-1)(a-1) = (6-1)(3-1) = 10$ degrees of freedom. Under the null hypothesis, the observed value of $F_A = 5.5263$ yields an approximate probability value of $P = 0.0242$.

4.3.4 Two-Way Randomized-Block Design

Consider a balanced two-way randomized-block design in which $n = 3$ subjects (S) are tested over $a = 3$ levels of Factor A and the experiment is repeated $b = 3$ times for Factor B. The design and data are adapted from Myers and Well [315, p. 260] and are given in Table 4.4. A complete permutation analysis of a two-way randomized-block design requires three separate analyses comprised of (1) the main effect of Factor A, (2) the main effect of Factor B, and (3) the $A{\times}B$ interaction effect.

Analysis of Factor A

A design matrix of dummy codes for analyzing Factor A is given on the left side of Table 4.5, where the first column of 1 values provides for an intercept and the second and third columns contain dummy codes for Factor B. The last column on the left side of Table 4.5 lists the $N = 9$ response measurement summations over the $b = 3$ levels of Factor B (e.g., $3.10 + 1.90 + 1.60 = 6.60$) and ordered by the $a = 3$ treatment levels of Factor A with the first $n_{A_1} = 3$ summations, the next $n_{A_2} = 3$ summations, and the last $n_{A_3} = 3$ summations associated with treatment levels A_1, A_2, and A_3, respectively. The MRPP regression analysis examines the $N = 9$ regression residuals for possible differences in the $a = 3$ treatment levels of Factor A; consequently, no dummy codes are provided for Factor A as this information is

Table 4.4 Example univariate data for a balanced two-way randomized-block design with $n = 3$ subjects, $a = 3$ levels of Factor A, and $b = 3$ levels of Factor B

	B_1			B_2			B_3		
Subject	A_1	A_2	A_3	A_1	A_2	A_3	A_1	A_2	A_3
S_1	3.10	2.90	2.40	1.90	2.00	1.70	1.60	1.90	1.50
S_2	5.70	6.80	5.30	4.50	5.70	4.40	4.40	5.30	3.90
S_3	9.70	10.90	8.00	7.40	10.50	6.60	6.90	8.90	6.00

Table 4.5 Design matrices and summation data for Factors A and B in a two-way analysis of variance randomized-block design

Factor A				Factor B			
Matrix			Sum over B	Matrix			Sum over A
1	0	0	6.60	1	0	0	8.40
1	1	0	14.60	1	1	0	17.80
1	0	1	24.00	1	0	1	28.60
1	0	0	6.80	1	0	0	5.60
1	1	0	17.80	1	1	0	14.60
1	0	1	30.30	1	0	1	24.50
1	0	0	5.60	1	0	0	5.00
1	1	0	13.60	1	1	0	13.60
1	0	1	20.60	1	0	1	21.80

implicit in the ordering of the $a = 3$ treatment levels of Factor A in the last column on the left side of Table 4.5.

An exact permutation solution is reasonable for the response measurement summations listed on the left side of Table 4.5 since there are only

$$M = \frac{N!}{\prod_{i=1}^{a} n_{A_i}!} = \frac{9!}{(3!)^3} = 1{,}680$$

possible, equally-likely arrangements of the $N = 9$ response measurement summations for Factor A with $n_{A_1} = n_{A_2} = n_{A_3} = 3$ response measurement summations preserved for each arrangement of the observed data.

LAD Regression Analysis

An MRPP analysis of the LAD regression residuals calculated on the $N = 9$ response measurement summations on the left side of Table 4.5 yields estimated LAD regression coefficients of

$$\tilde{\beta}_0 = +6.60 \ , \quad \tilde{\beta}_1 = +8.00 \ , \quad \text{and} \quad \tilde{\beta}_2 = +17.40$$

for Factor A. Figure 4.15 lists the observed y_i values, LAD predicted $\tilde{y}_i$ values, and residual e_i values for $i = 1, \ldots, 9$.

Following Eq. (4.5) on p. 125 and employing squared Euclidean distance between residuals with $v = 1$, the $N = 9$ LAD regression residuals listed in Fig. 4.15 yield $a = 3$ average distance-function values of

$$\xi_{A_1} = 0.00 \ , \quad \xi_{A_2} = 4.0667 \ , \quad \text{and} \quad \xi_{A_3} = 1.60 \ .$$

Fig. 4.15 Observed, predicted, and residual LAD regression values for the summations over Factor B on the left side of Table 4.5

Object	y_i	$\tilde{y}_i$	e_i
1	6.60	6.60	0.00
2	14.60	14.00	0.00
3	24.00	24.00	0.00
4	6.80	6.80	+0.20
5	17.70	17.80	+3.20
6	30.30	30.30	+6.30
7	5.60	5.60	−1.00
8	13.60	13.60	−1.00
9	20.60	20.60	−3.40

Following Eq. (4.4) on p. 125, the observed value of the MRPP test statistic calculated on the LAD regression residuals listed in Fig. 4.15 with $v = 1$ and treatment-group weights

$$C_i = \frac{n_{A_i}}{N}, \qquad i = 1, 2, 3,$$

is

$$\delta_A = \sum_{i=1}^{a} C_i \xi_i = \frac{3}{9}\big(0.00 + 4.0667 + 1.60\big) = 1.8889.$$

If all arrangements of the $N = 9$ observed LAD regression residuals listed in Fig. 4.15 occur with equal chance, the exact probability value of $\delta_A = 1.8889$ computed on the $M = 1{,}680$ possible arrangements of the observed LAD regression residuals with $n_{A_1} = n_{A_2} = n_{A_3} = 3$ residuals preserved for each arrangement is

$$P\big(\delta \leq \delta_A | H_0\big) = \frac{\text{number of } \delta \text{ values} \leq \delta_A}{M} = \frac{6}{1{,}680} = 0.0036.$$

Following Eq. (4.7) on p. 126, the exact expected value of the $M = 1{,}680$ δ values is $\mu_\delta = 2.9889$ and, following Eq. (4.6) on p. 126, the observed chance-corrected measure of effect size for the y_i and $\tilde{y}_i$ values, $i = 1, \ldots, N$, is

$$\Re_A = 1 - \frac{\delta_A}{\mu_\delta} = 1 - \frac{1.8889}{2.9889} = +0.3680,$$

indicating approximately 37 % agreement between the observed and predicted y values above that expected by chance.

OLS Regression Analysis

For comparison, consider an MRPP analysis of OLS regression residuals calculated on the $N = 9$ response measurement summations for Factor A listed on the left

Fig. 4.16 Observed, predicted, and residual OLS regression values for the summations over Factor B on the left side of Table 4.5

Object	y_i	$\hat{y}_i$	e_i
1	6.60	6.3333	+0.2667
2	14.60	15.3333	−0.7333
3	24.00	24.9667	−0.9667
4	6.80	6.3333	+0.4667
5	17.80	15.3333	+2.4667
6	30.30	24.9667	+5.3333
7	5.60	6.3333	−0.7333
8	13.60	15.3333	−1.7333
9	20.60	24.9667	−4.3667

side of Table 4.5. Again, since there are only $M = 1{,}680$ possible arrangements of the response measurement summations, an exact permutation test is selected. The MRPP regression analysis yields estimated OLS regression coefficients of

$$\hat{\beta}_0 = +6.3333\ , \quad \hat{\beta}_1 = +9.00\ , \quad \text{and} \quad \hat{\beta}_2 = +18.6333$$

for Factor A. Figure 4.16 lists the observed y_i values, OLS predicted $\hat{y}_i$ values, and residual e_i values for $i = 1, \ldots, 9$.

Following Eq. (4.5) on p. 125 and employing squared Euclidean distance between residuals with $v = 2$, the $N = 9$ OLS regression residuals listed in Fig. 4.16 yield $a = 3$ average distance-function values of

$$\xi_{A_1} = 0.8585\ , \quad \xi_{A_2} = 11.9674\ , \quad \text{and} \quad \xi_{A_3} = 7.0452\ .$$

Following Eq. (4.4) on p. 125, the observed value of the MRPP test statistic calculated on the OLS regression residuals listed in Fig. 4.16 with $v = 2$ and treatment-group weights

$$C_i = \frac{n_{A_i} - 1}{N - a}\ , \qquad i = 1, 2, 3\ ,$$

is

$$\delta_A = \sum_{i=1}^{a} C_i \xi_i = \frac{3-1}{9-3}(0.8585 + 11.9674 + 7.0452) = 6.6237\ .$$

If all arrangements of the $N = 9$ observed OLS regression residuals listed in Fig. 4.16 occur with equal chance, the exact probability value of $\delta_A = 6.6237$ computed on the $M = 1{,}680$ possible arrangements of the observed OLS regression residuals with $n_{A_1} = n_{A_2} = n_{A_3} = 3$ residuals preserved for each arrangement is

$$P(\delta \leq \delta_A | H_0) = \frac{\text{number of } \delta \text{ values} \leq \delta_A}{M} = \frac{18}{1{,}680} = 0.0107\ .$$

For comparison, the exact probability value based on LAD regression, $v = 1$, $M = 1{,}680$, and $C_i = n_{A_i}/N$ for $i = 1, 2, 3$ is $P = 0.0036$.

Following Eq. (4.7) on p. 126, the exact expected value of the $M = 1{,}680$ δ values is $\mu_\delta = 14.7250$ and, following Eq. (4.6) on p. 126, the observed chance-corrected measure of effect size for the y_i and $\hat{y}_i$ values, $i = 1, \ldots, N$, is

$$\Re_A = 1 - \frac{\delta_A}{\mu_\delta} = 1 - \frac{6.6237}{14.7250} = +0.5512 \,,$$

indicating approximately 55 % agreement between the observed and predicted y values above that expected by chance.

Conventional ANOVA Analysis

A conventional randomized-block analysis of variance calculated on the $N = 27$ univariate response measurement scores for Factor A in Table 4.4 on p. 143 yields an observed F-ratio of $F_A = 3.9282$. Assuming independence and normality, F_A is approximately distributed as Snedecor's F under the null hypothesis with $\nu_1 = a - 1 = 3 - 1 = 2$ and $\nu_2 = (n-1)(a-1) = (3-1)(3-1) = 4$ degrees of freedom. Under the null hypothesis, the observed value of $F_A = 3.9282$ yields an approximate probability value of $P = 0.1138$.

Analysis of Factor *B*

The right side of Table 4.5 on p. 144 contains a design matrix of dummy codes for analyzing Factor B, where the first column of 1 values provides for an intercept and the next two columns contain dummy codes for Factor A. The last column on the right side of Table 4.5 lists the $N = 9$ response measurement summations over the $a = 3$ levels of Factor A (e.g., $3.10 + 2.90 + 2.40 = 8.40$) and ordered by the $b = 3$ treatment levels with the first $n_{B_1} = 3$ summations, the next $n_{B_2} = 3$ summations, and the last $n_{B_3} = 3$ summations associated with treatment levels, B_1, B_2, and B_3, respectively. The MRPP regression analysis examines the $N = 9$ regression residuals for possible differences among the $b = 3$ treatment levels of Factor B; consequently, no dummy codes are provided for Factor B as this information is implicit in the ordering of the $b = 3$ treatment levels of Factor B in the last column on the right side of Table 4.5.

An exact permutation solution is ideal for the response measurement summations on the right side of Table 4.5 since there are only

$$M = \frac{N!}{\displaystyle\prod_{i=1}^{b} n_{B_i}!} = \frac{9!}{(3!)^3} = 1{,}680$$

possible, equally-likely arrangements of the $N = 9$ response measurement summations for Factor B with $n_{B_1} = n_{B_2} = n_{B_3}$ response measurement summations preserved for each arrangement of the observed data.

Fig. 4.17 Observed, predicted, and residual LAD regression values for the summations over Factor *A* on the right side of Table 4.5

Object	y_i	$\tilde{y}_i$	e_i
1	8.40	5.60	+2.80
2	17.80	14.60	+3.20
3	28.60	24.50	+4.10
4	5.60	5.60	0.00
5	14.60	14.60	0.00
6	24.50	24.50	0.00
7	5.00	5.60	−0.60
8	13.60	14.60	−1.00
9	21.80	24.50	−2.70

LAD Regression Analysis

An MRPP analysis of the LAD regression residuals calculated on the $N = 9$ response measurement summations on the right side of Table 4.5 on p. 144 yields estimated LAD regression coefficients of

$$\tilde{\beta}_0 = +5.60\ ,\quad \hat{\beta}_1 = +9.00\ ,\quad \text{and}\quad \tilde{\beta}_2 = +18.90$$

for Factor *B*. Figure 4.17 lists the observed y_i values, LAD predicted $\tilde{y}_i$ values, and residual e_i values for $i = 1, \ldots, 9$.

Following Eq. (4.5) on p. 125 and employing ordinary Euclidean distance between residuals with $v = 1$, the $N = 9$ LAD regression residuals listed in Fig. 4.17 yield $b = 3$ average distance-function values of

$$\xi_{B_1} = 0.8667\ ,\quad \xi_{B_2} = 0.00\ ,\quad \text{and}\quad \xi_{B_3} = 1.40\ .$$

Following Eq. (4.4) on p. 125, the observed value of the MRPP test statistic calculated on the LAD regression residuals listed in Fig. 4.17 with $v = 1$ and treatment-group weights

$$C_i = \frac{n_{B_i}}{N}\ ,\qquad i = 1, 2, 3\ ,$$

is

$$\delta_B = \sum_{i=1}^{b} C_i \xi_i = \frac{3}{9}(0.8667 + 0.00 + 1.40) = 0.7556\ .$$

If all arrangements of the $N = 9$ observed LAD regression residuals listed in Fig. 4.17 occur with equal chance, the exact probability value of $\delta_B = 0.7556$ computed on the $M = 1{,}680$ possible arrangements of the observed LAD regression residuals with $n_{B_1} = n_{B_2} = n_{B_3} = 3$ residuals preserved for each arrangement is

$$P(\delta \leq \delta_B | H_0) = \frac{\text{number of } \delta \text{ values} \leq \delta_B}{M} = \frac{6}{1{,}680} = 0.0036 \, .$$

Following Eq. (4.7) on p. 126, the exact expected value of the $M = 1{,}680$ δ values is $\mu_\delta = 2.5889$ and, following Eq. (4.6) on p. 126, the observed chance-corrected measure of effect size for the y_i and $\tilde{y}_i$ values, $i = 1, \ldots, N$, is

$$\Re_B = 1 - \frac{\delta_B}{\mu_\delta} = 1 - \frac{0.7556}{2.5889} = +0.7082 \, ,$$

indicating approximately 71 % agreement between the observed and predicted y values above that expected by chance.

OLS Regression Analysis

For comparison, consider an MRPP analysis of OLS regression residuals calculated on the $N = 9$ response measurement summations for Factor B listed on the right side of Table 4.5 on p. 144. Again, since there are only $M = 1{,}680$ possible arrangements of the response measurement summations, an exact permutation test is preferred. The MRPP regression analysis yields estimated OLS regression coefficients of

$$\hat{\beta}_0 = +6.3333 \, , \quad \hat{\beta}_1 = +9.00 \, , \quad \text{and} \quad \hat{\beta}_2 = +18.6333$$

for Factor B. Figure 4.18 lists the observed y_i values, OLS predicted $\hat{y}_i$ values, and residual e_i values for $i = 1, \ldots, 9$.

Following Eq. (4.5) on p. 125 and employing squared Euclidean distance between residuals with $v = 2$, the $N = 9$ OLS regression residuals listed in Fig. 4.18 yield $b = 3$ average distance-function values of

$$\xi_{B_1} = 1.3252 \, , \quad \xi_{B_2} = 0.0474 \, , \quad \text{and} \quad \xi_{B_3} = 1.8585 \, .$$

Fig. 4.18 Observed, predicted, and residual OLS regression values for the summations over Factor A on the right side of Table 4.5

Object	y_i	$\hat{y}_i$	e_i
1	8.40	6.3333	+2.0667
2	17.80	15.3333	+2.4667
3	28.60	24.9667	+3.6333
4	5.60	6.3333	−0.7333
5	14.60	15.3333	−0.7333
6	24.50	24.9667	−0.4667
7	5.00	6.3333	−1.3333
8	13.60	15.3333	−1.7333
9	21.80	24.9667	−3.1667

Following Eq. (4.4) on p. 125, the observed value of the MRPP test statistic calculated on the OLS regression residuals listed in Fig. 4.18 with $v = 2$ and treatment-group weights

$$C_i = \frac{n_{B_i} - 1}{N - b} , \qquad i = 1, 2, 3 ,$$

is

$$\delta_B = \sum_{i=1}^{b} C_i \xi_i = \frac{3-1}{9-3}(1.3252 + 0.0474 + 1.8585) = 1.0770 .$$

If all arrangements of the $N = 9$ observed OLS regression residuals listed in Fig. 4.18 occur with equal chance, the exact probability value of $\delta_B = 1.0770$ computed on the $M = 1{,}680$ possible arrangements of the observed OLS regression residuals with $n_{B_1} = n_{B_2} = n_{B_3} = 3$ residuals preserved for each arrangement is

$$P(\delta \leq \delta_B | H_0) = \frac{\text{number of } \delta \text{ values} \leq \delta_B}{M} = \frac{6}{1{,}680} = 0.0036 .$$

For comparison, the exact probability value based on LAD regression, $v = 1$, $M = 1{,}680$, and $C_i = n_{B_i}/N$ for $i = 1, 2, 3$ is also $P = 0.0036$.

Following Eq. (4.7) on p. 126, the exact expected value of the $M = 1{,}680$ δ values is $\mu_\delta = 9.9150$ and, following Eq. (4.6) on p. 126, the observed chance-corrected measure of effect size for the y_i and $\hat{y}_i$ values, $i = 1, \ldots, N$, is

$$\Re_B = 1 - \frac{\delta_B}{\mu_\delta} = 1 - \frac{1.0770}{9.9150} = +0.8914 ,$$

indicating approximately 89 % agreement between the observed and predicted y values above that expected by chance.

Conventional ANOVA Analysis

A conventional randomized-block analysis of variance calculated on the $N = 27$ univariate response measurement scores for Factor B listed in Table 4.4 on p. 143 yields an observed F-ratio of $F_B = 22.5488$. Assuming independence and normality, F_B is approximately distributed as Snedecor's F under the null hypothesis with $\nu_1 = b - 1 = 3 - 1 = 2$ and $\nu_2 = (n-1)(b-1) = (3-1)(3-1) = 4$ degrees of freedom. Under the null hypothesis, the observed value of $F_B = 22.5488$ yields an approximate probability value of $P = 0.0066$, which is similar to the LAD and OLS regression probability value of $P = 0.0036$.

Analysis of the $A \times B$ Interaction

A design matrix of dummy codes for analyzing the interaction of Factors A and B is given in Table 4.6, where the first column of 1 values provides for an intercept and

Table 4.6 Design matrix and univariate response measurement scores for the interaction of Factors *A* and *B* in a two-way randomized-block design with $N = 27$ objects

Matrix															Score
1	0	0	0	0	0	0	0	0	0	0	0	0	0	0	3.10
1	1	0	0	0	0	0	0	0	0	0	0	0	0	0	5.70
1	0	1	0	0	0	0	0	0	0	0	0	0	0	0	9.70
1	0	0	1	0	0	0	0	0	0	0	0	0	0	0	2.90
1	1	0	1	0	0	0	0	0	0	0	0	0	0	0	6.80
1	0	1	1	0	0	0	0	0	1	0	0	0	0	0	10.90
1	0	0	0	1	0	0	0	0	0	0	0	0	0	0	2.40
1	1	0	0	1	0	0	0	1	0	0	0	0	0	0	5.30
1	0	1	0	1	0	0	0	0	1	0	0	0	0	0	8.00
1	0	0	0	0	1	0	0	0	0	0	0	0	0	0	1.90
1	1	0	0	0	1	0	0	0	0	0	1	0	0	0	4.50
1	0	1	0	0	1	0	0	0	0	0	0	0	1	0	7.40
1	0	0	1	0	1	0	0	0	0	0	0	0	0	0	2.00
1	1	0	1	0	1	0	1	0	0	0	1	0	0	0	5.70
1	0	1	1	0	1	0	0	0	1	0	0	0	1	0	10.50
1	0	0	0	1	1	0	0	0	0	0	0	0	0	0	1.70
1	1	0	0	1	1	0	0	1	0	0	1	0	0	0	4.40
1	0	1	0	1	1	0	0	0	0	1	0	0	1	0	6.60
1	0	0	0	0	0	1	0	0	0	0	0	0	0	0	1.60
1	1	0	0	0	0	1	0	0	0	0	0	1	0	0	4.40
1	0	1	0	0	0	1	0	0	0	0	0	0	0	1	6.90
1	0	0	1	0	0	1	0	0	0	0	0	0	0	0	1.90
1	1	0	1	0	0	1	1	0	0	0	0	1	0	0	5.30
1	0	1	1	0	0	1	0	0	1	0	0	0	0	1	8.90
1	0	0	0	1	0	1	0	0	0	0	0	0	0	0	1.50
1	1	0	0	1	0	1	0	1	0	0	0	1	0	0	3.90
1	0	1	0	1	0	1	0	0	0	1	0	0	0	1	6.00

the second and third columns contain dummy codes for subjects (S). The fourth and fifth columns contain dummy codes for Factor A, the sixth and seventh columns contain dummy codes for Factor B, and the next eight columns contain dummy codes for the $S{\times}A$ and $S{\times}B$ interactions. The last column in Table 4.6 lists the response measurement scores ordered by the $ab = (3)(3) = 9$ levels of the $A{\times}B$ interaction.

The MRPP regression analysis examines the $N = 27$ regression residuals for possible differences among the nine treatment levels of the $A{\times}B$ interaction; consequently, no dummy codes are provided for the $A{\times}B$ interaction as this information is implicit in the ordering of the treatment levels of the $A{\times}B$ interaction in the last column of Table 4.6.

Because there are

$$M = \frac{N!}{\prod_{i=1}^{ab} n_{(A\times B)_i}!} = \frac{27!}{(3!)^9} = 1{,}080{,}491{,}954{,}750{,}208{,}000{,}000$$

possible, equally-likely arrangements of the $N = 27$ univariate response measurement scores for the $A\times B$ interaction listed in Table 4.6, an exact permutation solution is not possible.

LAD Regression Analysis

An MRPP resampling analysis of the LAD regression residuals calculated on the $N = 27$ univariate response measurement scores listed in Table 4.6 yields estimated LAD regression coefficients of

$$\tilde{\beta}_0 = +2.70\,, \quad \tilde{\beta}_1 = +3.00\,, \quad \tilde{\beta}_2 = +6.20\,, \quad \tilde{\beta}_3 = +0.20\,,$$
$$\tilde{\beta}_4 = -0.20\,, \quad \tilde{\beta}_5 = -0.80\,, \quad \tilde{\beta}_6 = -1.00\,, \quad \tilde{\beta}_7 = +0.90\,,$$
$$\tilde{\beta}_8 = -0.20\,, \quad \tilde{\beta}_9 = +1.80\,, \quad \tilde{\beta}_{10} = -0.70\,, \quad \tilde{\beta}_{11} = -0.30\,,$$
$$\tilde{\beta}_{12} = -0.40\,, \quad \tilde{\beta}_{13} = -0.60\,, \quad \text{and} \quad \tilde{\beta}_{14} = -1.00$$

for the interaction of Factors A and B. Figure 4.19 lists the observed y_i values, LAD predicted $\tilde{y}_i$ values, and residual e_i values for $i = 1, \ldots, 27$.

Following Eq. (4.5) on p. 125 and employing ordinary Euclidean distance between residuals with $v = 1$, the $N = 27$ LAD regression residuals listed in Fig. 4.19 yield $ab = (3)(3) = 9$ average distance-function values of

$$\xi_{(A\times B)_1} = 0.5333\,, \quad \xi_{(A\times B)_2} = 0.00\,, \quad \xi_{(A\times B)_3} = \xi_{(A\times B)_4} = 0.0667\,,$$
$$\xi_{(A\times B)_5} = 0.7333\,, \quad \xi_{(A\times B)_6} = \xi_{(A\times B)_7} = 0.1333\,, \quad \xi_{(A\times B)_8} = 0.0667\,,$$
$$\text{and} \quad \xi_{(A\times B)_9} = 0.00\,.$$

Following Eq. (4.4) on p. 125, the observed value of the MRPP test statistic calculated on the LAD regression residuals listed in Fig. 4.19 with $v = 1$ and treatment-group weights

$$C_i = \frac{n_{(A\times B)_i}}{N}\,, \qquad i = 1, \ldots, 9\,,$$

is

$$\delta_{A\times B} = \sum_{i=1}^{ab} C_i \xi_i = \frac{3}{9}(0.5333 + 0.00 + \cdots + 0.0667 + 0.00) = 0.1926\,.$$

Fig. 4.19 Observed, predicted, and residual LAD regression values for the univariate response measurement scores listed in Table 4.6

Object	y_i	$\tilde{y}_i$	e_i
1	3.10	2.70	+0.40
2	5.70	5.70	0.00
3	9.70	8.90	+0.80
4	2.90	2.90	0.00
5	6.80	6.80	0.00
6	10.90	10.90	0.00
7	2.40	2.50	−0.10
8	5.30	5.30	0.00
9	8.00	8.00	0.00
10	1.90	1.90	0.00
11	4.50	4.60	−0.10
12	7.40	7.50	−0.10
13	2.00	2.10	−0.10
14	5.70	5.70	0.00
15	10.50	9.50	+1.00
16	1.70	1.70	0.00
17	4.40	4.20	+0.20
18	6.60	6.60	0.00
19	1.60	1.70	−0.10
20	4.40	4.30	+0.10
21	6.90	6.90	0.00
22	1.90	1.90	0.00
23	5.30	5.40	−0.10
24	8.90	8.90	0.00
25	1.50	1.50	0.00
26	3.90	3.90	0.00
27	6.00	6.00	0.00

If all M possible arrangements of the $N = 27$ observed LAD regression residuals listed in Fig. 4.19 occur with equal chance, the approximate resampling probability value of $\delta_{A\times B} = 0.1926$ computed on $L =$ 1,000,000 random arrangements of the observed LAD regression residuals with $n_{(A\times B)_1} = \cdots = n_{(A\times B)_9} = 3$ residuals preserved for each arrangement is

$$P(\delta \leq \delta_{A\times B}|H_0) = \frac{\text{number of } \delta \text{ values} \leq \delta_{A\times B}}{L} = \frac{235{,}542}{1{,}000{,}000} = 0.2355 \; .$$

Following Eq. (4.7) on p. 126, the exact expected value of the M δ values is $\mu_\delta = 0.2063$ and, following Eq. (4.6) on p. 126, the observed chance-corrected measure of effect size for the y_i and $\tilde{y}_i$ values, $i = 1, \ldots, N$, is

$$\Re_{A\times B} = 1 - \frac{\delta_{A\times B}}{\mu_\delta} = 1 - \frac{0.1926}{0.2063} = +0.0663 \; ,$$

indicating approximately 7 % agreement between the observed and predicted y values above that expected by chance.

OLS Regression Analysis

For comparison, consider an MRPP analysis of OLS regression residuals calculated on the $N = 27$ univariate response measurement scores for the $A \times B$ interaction listed in Table 4.6. The MRPP regression analysis yields estimated OLS regression coefficients of

$$\hat{\beta}_0 = +2.8889\,, \quad \hat{\beta}_1 = +2.80\,, \quad \hat{\beta}_2 = +6.3222\,, \quad \hat{\beta}_3 = +0.0667\,,$$

$$\hat{\beta}_4 = -0.3333\,, \quad \hat{\beta}_5 = -0.9333\,, \quad \hat{\beta}_6 = -1.1333\,, \quad \hat{\beta}_7 = +1.00\,,$$

$$\hat{\beta}_8 = 0.00\,, \quad \hat{\beta}_9 = +2.0333\,, \quad \hat{\beta}_{10} = -0.80\,, \quad \hat{\beta}_{11} = -0.1333\,,$$

$$\hat{\beta}_{12} = -0.2667\,, \quad \hat{\beta}_{13} = -0.4333\,, \quad \text{and} \quad \hat{\beta}_{14} = -1.1333$$

for the interaction of Factors A and B. Figure 4.20 lists the observed y_i values, OLS predicted $\hat{y}_i$ values, and residual e_i values for $i = 1, \ldots, 27$.

Fig. 4.20 Observed, predicted, and residual OLS regression values for the univariate response measurement scores listed in Table 4.6

Object	y_i	$\hat{y}_i$	e_i
1	3.10	2.8889	+0.2111
2	5.70	5.6889	+0.0111
3	9.70	9.2111	+0.4889
4	2.90	2.9556	−0.0556
5	6.80	6.7556	+0.0444
6	10.90	11.3111	−0.4111
7	2.40	2.5556	−0.1556
8	5.30	5.3556	−0.0556
9	8.00	8.0778	−0.0778
10	1.90	1.9556	−0.0556
11	4.50	4.6222	−0.1222
12	7.40	7.8444	−0.4444
13	2.00	2.0222	−0.2222
14	5.70	5.6889	+0.0111
15	10.50	9.9444	+0.5556
16	1.70	1.6222	+0.0778
17	4.40	4.2889	+0.1111
18	6.60	6.7111	−0.1111
19	1.60	1.7556	−0.1556
20	4.40	4.2889	+0.1111
21	6.90	6.9444	−0.0444
22	1.90	1.8222	+0.0778
23	5.30	5.3556	−0.0556
24	8.90	9.0444	−0.1444
25	1.50	1.4222	+0.0778
26	3.90	3.9556	−0.0556
27	6.00	5.8111	+0.1889

Following Eq. (4.5) on p. 125 and employing squared Euclidean distance between residuals with $v = 2$, the $N = 27$ OLS regression residuals listed in Fig. 4.20 yield $ab = (3)(3) = 9$ average distance-function values of

$$\xi_{(A\times B)_1} = 0.1151\ , \quad \xi_{(A\times B)_2} = 0.1147\ , \quad \xi_{(A\times B)_3} = 0.0055\ ,$$
$$\xi_{(A\times B)_4} = 0.0865\ , \quad \xi_{(A\times B)_5} = 0.2105\ , \quad \xi_{(A\times B)_6} = 0.0287\ ,$$
$$\xi_{(A\times B)_7} = 0.0359\ , \quad \xi_{(A\times B)_8} = 0.0250\ , \quad \text{and} \quad \xi_{(A\times B)_9} = 0.0300\ .$$

Following Eq. (4.4) on p. 125, the observed value of the MRPP test statistic calculated on the OLS regression residuals listed in Fig. 4.20 with $v = 2$ and treatment-group weights

$$C_i = \frac{n_{(A\times B)_i} - 1}{N - ab}\ , \qquad i = 1, \ldots, 9\ ,$$

is

$$\delta_{A\times B} = \sum_{i=1}^{ab} C_i \xi_i = \frac{3-1}{27-9}\big(0.1151 + 0.1147 + 0.0055 + \cdots + 0.0250 + 0.0300\big) = 0.0724\ .$$

If all M possible arrangements of the $N = 27$ observed OLS regression residuals listed in Fig. 4.20 occur with equal chance, the approximate resampling probability value of $\delta_{A\times B} = 0.0724$ computed on $L = 1{,}000{,}000$ random arrangements of the observed OLS regression residuals with $n_{(A\times B)_1} = \cdots = n_{(A\times B)_9} = 3$ residuals preserved for each arrangement is

$$P\big(\delta \le \delta_{A\times B} | H_0\big) = \frac{\text{number of } \delta \text{ values} \le \delta_{A\times B}}{L} = \frac{141{,}960}{1{,}000{,}000} = 0.1420\ .$$

For comparison, the approximate resampling probability value based on LAD regression, $v = 1$, $L = 1{,}000{,}000$, and $C_i = n_{(A\times B)_i}/N$ for $i = 1, \ldots, 9$ is $P = 0.2355$.

Following Eq. (4.7) on p. 126, the exact expected value of the M δ values is $\mu_\delta = 8.9231$ and, following Eq. (4.6) on p. 126, the observed chance-corrected measure of effect size for the y_i and $\hat{y}_i$ values, $i = 1, \ldots, N$, is

$$\Re_{A\times B} = 1 - \frac{\delta_{A\times B}}{\mu_\delta} = 1 - \frac{0.0724}{8.9231} = +0.1883\ ,$$

indicating approximately 19 % agreement between the observed and predicted y values above that expected by chance.

Conventional ANOVA Analysis

A conventional randomized-block analysis of variance calculated on the $N = 27$ response measurement scores for the $A{\times}B$ interaction listed in Table 4.4 on p. 143 yields an observed F-ratio of $F_{A\times B} = 1.5591$. Assuming independence and normality, $F_{A\times B}$ is approximately distributed as Snedecor's F under the null hypothesis with $\nu_1 = (a-1)(b-1) = (3-1)(3-1) = 4$ and $\nu_2 = (n-1)(a-1)(b-1) = (3-1)(3-1)(3-1) = 8$ degrees of freedom. Under the null hypothesis, the observed value of $F_{A\times B} = 1.5591$ yields an approximate probability value of $P = 0.2744$.

4.3.5 Two-Way Factorial Design

Consider a 2×3 fixed-effects factorial design with $n = 4$ subjects in each treatment combination for a total of $N = 24$ subjects. The univariate response measurement scores for Factors A and B are listed in Fig. 4.21, and the design matrices and data for Factors A and B are given in Table 4.7; the design and data are adapted from Keppel [214, p. 197]. While design matrices of either dummy or effect codes are appropriate for one-way completely randomized and randomized-block designs, the main effects of factorial designs are best analyzed with effect codes when estimation of the effects of each factor is adjusted for all other factors in the model to obtain the unique contribution of each factor [31,37,294].[3] A permutation analysis of factorial designs requires three separate analyses comprising (1) the main effect of Factor A, (2) the main effect of Factor B, and (3) the $A{\times}B$ interaction effect.

Fig. 4.21 Example univariate response measurement scores for Factors A and B in a two-way factorial design

Factor A		Factor B		
A_1	A_2	B_1	B_2	B_3
1	15	1	13	9
4	6	1	5	16
0	10	0	7	18
7	13	7	15	13
13	6	15	6	14
5	18	6	18	7
7	9	10	9	6
15	15	13	15	13
9	14			
16	7			
18	6			
13	13			

[3]This method of estimation is known as Method I as presented in a seminal article by Overall and Spiegel in 1969 [330].

Table 4.7 Design matrices and univariate response measurement scores for the main effects of Factors A and B in a two-way factorial design with $N = 24$ subjects

Factor *A*						Factor *B*				
Matrix					Score	Matrix				Score
1	1	0	1	0	1	1	1	1	0	1
1	1	0	1	0	4	1	1	1	0	4
1	1	0	1	0	0	1	1	1	0	0
1	1	0	1	0	7	1	1	1	0	7
1	0	1	0	1	13	1	−1	−1	0	15
1	0	1	0	1	5	1	−1	−1	0	6
1	0	1	0	1	7	1	−1	−1	0	10
1	0	1	0	1	15	1	−1	−1	0	13
1	−1	−1	−1	−1	9					
1	−1	−1	−1	−1	16	1	1	0	1	13
1	−1	−1	−1	−1	18	1	1	0	1	5
1	−1	−1	−1	−1	13	1	1	0	1	7
						1	1	0	1	15
1	1	0	−1	0	15	1	−1	0	−1	6
1	1	0	−1	0	6	1	−1	0	−1	18
1	1	0	−1	0	10	1	−1	0	−1	9
1	1	0	−1	0	13	1	−1	0	−1	15
1	0	1	0	−1	6					
1	0	1	0	−1	18	1	1	−1	−1	9
1	0	1	0	−1	9	1	1	−1	−1	16
1	0	1	0	−1	15	1	1	−1	−1	18
1	−1	−1	1	1	14	1	1	−1	−1	13
1	−1	−1	1	1	7	1	−1	1	1	14
1	−1	−1	1	1	6	1	−1	1	1	7
1	−1	−1	1	1	13	1	−1	1	1	6
				1	−1	1	1	13		

Analysis of Factor A

A design matrix of effect codes for analyzing Factor A is given on the left side of Table 4.7, where the first column of 1 values provides for an intercept. The second and third columns contain effect codes for Factor B, the fourth and fifth columns contain effect codes for the $A \times B$ interaction, and the last column on the left side of Table 4.7 contains the $N = 24$ univariate response measurement scores listed according to the original random assignment of the subjects to the $a = 2$ levels of Factor A with the first $n_{A_1} = 12$ scores and the last $n_{A_2} = 12$ scores associated with treatment levels A_1 and A_2, respectively. The MRPP regression analysis examines the $N = 24$ regression residuals for possible differences between the $a = 2$ treatment levels of Factor A; consequently, no effect codes are provided for Factor A as this information is implicit in the ordering of the $a = 2$ treatment levels of Factor A in the last column on the left side of Table 4.7.

An exact permutation solution is feasible for the univariate response measurement scores listed on the left side of Table 4.7 since there are only

$$M = \frac{N!}{\prod_{i=1}^{a} n_{A_i}!} = \frac{24!}{(12!)^2} = 2{,}704{,}156$$

possible, equally-likely arrangements of the $N = 24$ response measurement scores for Factor A.

LAD Regression Analysis

An MRPP analysis of the LAD regression residuals calculated on the $N = 24$ univariate response measurement scores on the left side of Table 4.7 yields estimated LAD regression coefficients of

$$\tilde{\beta}_0 = +9.6667\,, \quad \tilde{\beta}_1 = -1.1667\,, \quad \tilde{\beta}_2 = +0.8333\,, \quad \tilde{\beta}_3 = -4.50\,, \quad \text{and}$$

$$\tilde{\beta}_4 = +1.50$$

for Factor A. Figure 4.22 lists the observed y_i values, LAD predicted $\tilde{y}_i$ values, and residual e_i values for $i = 1, \ldots, 24$.

Fig. 4.22 Observed, predicted, and residual LAD regression values for the univariate response measurement scores listed on the left side of Table 4.7

Object	y_i	$\tilde{y}_i$	e_i
1	1	4.00	−3.00
2	4	4.00	0.00
3	0	4.00	−4.00
4	7	4.00	+3.00
5	13	12.00	+1.00
6	5	12.00	−7.00
7	7	12.00	−5.00
8	15	12.00	+3.00
9	9	13.00	−4.00
10	16	13.00	+3.00
11	18	13.00	+5.00
12	13	13.00	0.00
13	15	13.00	+2.00
14	6	13.00	−7.00
15	10	13.00	−3.00
16	13	13.00	0.00
17	6	9.00	−3.00
18	18	9.00	+9.00
19	9	9.00	0.00
20	15	9.00	+6.00
21	14	7.00	+7.00
22	7	7.00	0.00
23	6	7.00	−1.00
24	13	7.00	+6.00

Following Eq. (4.5) on p. 125 and employing ordinary Euclidean distance between residuals with $v = 1$, the $N = 24$ LAD regression residuals listed in Fig. 4.22 yield $a = 2$ average distance-function values of

$$\xi_{A_1} = 4.5455 \quad \text{and} \quad \xi_{A_2} = 5.6061 \; .$$

Following Eq. (4.4) on p. 125, the observed value of the MRPP test statistic calculated on the LAD regression residuals listed in Fig. 4.22 with $v = 1$ and treatment-group weights

$$C_i = \frac{n_{A_i}}{N} \; , \qquad i = 1, 2 \; ,$$

is

$$\delta_A = \sum_{i=1}^{a} C_i \xi_i = \frac{12}{24}\big(4.5455 + 5.6061\big) = 5.0758 \; .$$

If all arrangements of the $N = 24$ observed LAD regression residuals listed in Fig. 4.22 occur with equal chance, the exact probability value of $\delta_A = 5.0758$ computed on the $M = 2{,}704{,}156$ possible arrangements of the observed LAD regression residuals with $n_{A_1} = n_{A_2} = 12$ residuals preserved for each arrangement is

$$P\big(\delta \leq \delta_A | H_0\big) = \frac{\text{number of } \delta \text{ values} \leq \delta_A}{M} = \frac{1{,}039{,}084}{2{,}704{,}156} = 0.3843 \; .$$

Following Eq. (4.7) on p. 126, the exact expected value of the $M = 2{,}704{,}156$ δ values is $\mu_\delta = 5.0725$ and, following Eq. (4.6) on p. 126, the observed chance-corrected measure of effect size for the y_i and $\tilde{y}_i$ values, $i = 1, \ldots, N$, is

$$\Re_A = 1 - \frac{\delta_A}{\mu_\delta} = 1 - \frac{5.0758}{5.0725} = -0.6494 \times 10^{-3} \; ,$$

indicating slightly less than chance agreement between the observed and predicted y values.

OLS Regression Analysis

For comparison, consider an MRPP analysis of OLS regression residuals calculated on the $N = 24$ univariate response measurement scores for Factor A on the left side of Table 4.7. The MRPP regression analysis yields estimated OLS regression coefficients of

$$\hat{\beta}_0 = +10.00 \; , \;\; \hat{\beta}_1 = -3.00 \; , \;\; \hat{\beta}_2 = +1.00 \; , \;\; \hat{\beta}_3 = -3.00 \; , \text{ and } \hat{\beta}_4 = 0.00$$

Fig. 4.23 Observed, predicted, and residual OLS regression values for the univariate response measurement scores listed on the left side of Table 4.7

Object	y_i	$\hat{y}_i$	e_i
1	1	4.00	−3.00
2	4	4.00	0.00
3	0	4.00	−4.00
4	7	4.00	+3.00
5	13	11.00	+2.00
6	5	11.00	−6.00
7	7	11.00	−4.00
8	15	11.00	+4.00
9	9	15.00	−6.00
10	16	15.00	+1.00
11	18	15.00	+3.00
12	13	15.00	−2.00
13	15	10.00	+5.00
14	6	10.00	−4.00
15	10	10.00	0.00
16	13	10.00	+3.00
17	6	11.00	−5.00
18	18	11.00	+7.00
19	9	11.00	−2.00
20	15	11.00	+4.00
21	14	9.00	+5.00
22	7	9.00	−2.00
23	6	9.00	−3.00
24	13	9.00	+4.00

for Factor A. Figure 4.23 lists the observed y_i values, OLS predicted $\hat{y}_i$ values, and residual e_i values for $i = 1, \ldots, 24$.

Following Eq. (4.5) on p. 125 and employing squared Euclidean distance between residuals with $v = 2$, the $N = 24$ OLS regression residuals listed in Fig. 4.23 yield $a = 2$ average distance-function values of

$$\xi_{A_1} = 26.1818 \quad \text{and} \quad \xi_{A_2} = 33.8182 \, .$$

Following Eq. (4.4) on p. 125, the observed value of the MRPP test statistic calculated on the OLS regression residuals listed in Fig. 4.23 with $v = 2$ and treatment-group weights

$$C_i = \frac{n_{A_i} - 1}{N - a} \, , \qquad i = 1, 2 \, ,$$

is

$$\delta_A = \sum_{i=1}^{a} C_i \xi_i = \frac{12 - 1}{24 - 2}(26.1818 + 33.8182) = 30.00 \, .$$

If all arrangements of the $N = 24$ observed OLS regression residuals listed in Fig. 4.23 occur with equal chance, the exact probability value of $\delta_A = 30.00$ computed on the $M = 2{,}704{,}156$ possible arrangements of the observed OLS regression residuals with $n_{A_1} = n_{A_2} = 12$ residuals preserved for each arrangement is

$$P(\delta \leq \delta_A | H_0) = \frac{\text{number of } \delta \text{ values} \leq \delta_A}{M} = \frac{637{,}454}{2{,}704{,}156} = 0.2357 \, .$$

For comparison, the exact probability value based on LAD regression, $v = 1$, $M = 2{,}704{,}156$, and $C_i = n_{A_i}/N$ for $i = 1, 2$ is $P = 0.3843$.

Following Eq. (4.7) on p. 126, the exact expected value of the $M = 2{,}704{,}156$ δ values is $\mu_\delta = 30.7826$ and, following Eq. (4.6) on p. 126, the observed chance-corrected measure of effect size for the y_i and $\hat{y}_i$ values, $i = 1, \ldots, N$, is

$$\Re_A = 1 - \frac{\delta_A}{\mu_\delta} = 1 - \frac{30.00}{30.7826} = +0.0254 \, ,$$

indicating approximately 3 % agreement between the observed and predicted y values above that expected by chance.

Conventional ANOVA Analysis

A conventional fixed-effects factorial analysis of variance calculated on the $N = 24$ Factor A response measurement scores listed in Fig. 4.21 on p. 156 yields an observed F-ratio of $F_A = 1.3091$. Assuming independence, normality, and homogeneity of variance, F_A is approximately distributed as Snedecor's F under the null hypothesis with $\nu_1 = a - 1 = 2 - 1 = 1$ and $\nu_2 = N - ab = 24 - (2)(3) = 18$ degrees of freedom. Under the null hypothesis, the observed value of $F_A = 1.3091$ yields an approximate probability value of $P = 0.2675$, which is similar to the OLS regression probability value of $P = 0.2357$.

Analysis of Factor *B*

The right side of Table 4.7 on p. 157 contains a design matrix of effect codes for analyzing Factor B, where the first column of 1 values provides for an intercept. The second column contains effect codes for Factor A, the third and fourth columns contain effect codes for the $A{\times}B$ interaction, and the last column on the right side of Table 4.7 contains the $N = 24$ univariate response measurement scores listed according to the original random assignment of the subjects to the $b = 3$ levels of Factor B with the first $n_{B_1} = 8$ scores, the next $n_{B_2} = 8$ scores, and the last $n_{B_3} = 8$ scores associated with treatment levels, B_1, B_2, and B_3, respectively. The MRPP regression analysis examines the $N = 24$ regression residuals for possible differences among the $b = 3$ treatment levels of Factor B; consequently, no effect codes are provided for Factor B as this information is implicit in the ordering of the $b = 3$ treatment levels of Factor B in the last column on the right side of Table 4.7.

Because there are

$$M = \frac{N!}{\prod_{i=1}^{b} n_{B_i}!} = \frac{24!}{(8!)^3} = 9{,}465{,}511{,}770$$

possible, equally-likely arrangements of the $N = 24$ response measurement scores for Factor B listed on the right side of Table 4.7, an exact permutation approach is not practical.

LAD Regression Analysis

An MRPP resampling analysis of the $N = 24$ LAD regression residuals calculated on the univariate response measurement scores on the right side of Table 4.7 on p. 157 yields estimated LAD regression coefficients of

$$\tilde{\beta}_0 = +9.50\ , \quad \tilde{\beta}_1 = +0.1667\ , \quad \tilde{\beta}_2 = -3.6667\ , \quad \text{and} \quad \tilde{\beta}_3 = +0.3333$$

for Factor B. Figure 4.24 lists the observed y_i values, LAD predicted $\tilde{y}_i$ values, and residual e_i values for $i = 1, \ldots, 24$.

Fig. 4.24 Observed, predicted, and residual LAD regression values for the univariate response measurement scores listed on the right side of Table 4.7

Object	y_i	$\tilde{y}_i$	e_i
1	1	6.00	−5.00
2	4	6.00	−2.00
3	0	6.00	−6.00
4	7	6.00	+1.00
5	15	13.00	+2.00
6	6	13.00	−7.00
7	10	13.00	−3.00
8	13	13.00	0.00
9	13	10.00	+3.00
10	5	10.00	−5.00
11	7	10.00	−3.00
12	15	10.00	+5.00
13	6	9.00	−3.00
14	18	9.00	+9.00
15	9	9.00	0.00
16	15	9.00	+6.00
17	9	13.00	−4.00
18	16	13.00	+3.00
19	18	13.00	+5.00
20	13	13.00	0.00
21	14	6.00	+8.00
22	7	6.00	+1.00
23	6	6.00	0.00
24	13	6.00	+7.00

Following Eq. (4.5) on p. 125 and employing ordinary Euclidean distance between residuals with $v = 1$, the $N = 24$ LAD regression residuals listed in Fig. 4.24 yield $b = 3$ average distance-function values of

$$\xi_{B_1} = 4.0714 \;, \quad \xi_{B_2} = 6.0714 \;, \quad \text{and} \quad \xi_{B_3} = 4.8571 \;.$$

Following Eq. (4.4) on p. 125, the observed value of the MRPP test statistic calculated on the LAD regression residuals listed in Fig. 4.24 with $v = 1$ and treatment-group weights

$$C_i = \frac{n_{B_i}}{N} \;, \qquad i = 1, 2, 3 \;,$$

is

$$\delta_B = \sum_{i=1}^{b} C_i \xi_i = \frac{12}{24}\big(4.0714 + 6.0714 + 4.8571\big) = 5.00 \;.$$

If all M possible arrangements of the $N = 24$ observed LAD regression residuals listed in Fig. 4.24 occur with equal chance, the approximate resampling probability value of $\delta_B = 5.00$ computed on $L = 1{,}000{,}000$ random arrangements of the observed LAD regression residuals with $n_{B_1} = n_{B_2} = n_{B_3} = 8$ residuals preserved for each arrangement is

$$P\big(\delta \leq \delta_B | H_0\big) = \frac{\text{number of } \delta \text{ values} \leq \delta_B}{L} = \frac{125{,}031}{1{,}000{,}000} = 0.1250 \;.$$

Following Eq. (4.7) on p. 126, the exact expected value of the M δ values is $\mu_\delta = 5.3333$ and, following Eq. (4.6) on p. 126, the observed chance-corrected measure of effect size for the y_i and $\tilde{y}_i$ values, $i = 1, \ldots, N$, is

$$\Re_B = 1 - \frac{\delta_B}{\mu_\delta} = 1 - \frac{5.00}{5.3333} = +0.0625 \;,$$

indicating approximately 6 % agreement between the observed and predicted y values above that expected by chance.

OLS Regression Analysis

For comparison, consider an MRPP analysis of OLS regression residuals calculated on the $N = 24$ univariate response measurement scores for Factor B listed on the right side of Table 4.7 on p. 157. The MRPP regression analysis yields estimated OLS regression coefficients of

$$\hat{\beta}_0 = +10.00 \;, \quad \hat{\beta}_1 = -1.00 \;, \quad \hat{\beta}_2 = -3.00 \;, \quad \text{and} \quad \hat{\beta}_3 = 0.00$$

Fig. 4.25 Observed, predicted, and residual OLS regression values for the univariate response measurement scores listed on the right side of Table 4.7

Object	y_i	$\hat{y}_i$	e_i
1	1	6.00	−5.00
2	4	6.00	−2.00
3	0	6.00	−6.00
4	7	6.00	+1.00
5	15	14.00	+1.00
6	6	14.00	−8.00
7	10	14.00	−4.00
8	13	14.00	−1.00
9	13	9.00	+4.00
10	5	9.00	−4.00
11	7	9.00	−2.00
12	15	9.00	+6.00
13	6	11.00	−5.00
14	18	11.00	+7.00
15	9	11.00	−2.00
16	15	11.00	+4.00
17	9	12.00	−3.00
18	16	12.00	+4.00
19	18	12.00	+6.00
20	13	12.00	+1.00
21	14	8.00	+6.00
22	7	8.00	−1.00
23	6	8.00	−2.00
24	13	8.00	+5.00

for Factor B. Figure 4.25 lists the observed y_i values, OLS predicted $\hat{y}_i$ values, and residual e_i values for $i = 1, \ldots, 24$.

Following Eq. (4.5) on p. 125 and employing squared Euclidean distance between residuals with $v = 2$, the $N = 24$ OLS regression residuals listed in Fig. 4.25 yield $b = 3$ average distance-function values of

$$\xi_{B_1} = 21.7143\,, \quad \xi_{B_2} = 45.1429\,, \quad \text{and} \quad \xi_{B_3} = 27.4286\,.$$

Following Eq. (4.4) on p. 125, the observed value of the MRPP test statistic calculated on the OLS regression residuals listed in Fig. 4.25 with $v = 2$ and treatment-group weights

$$C_i = \frac{n_{B_i} - 1}{N - b}\,, \qquad i = 1, 2, 3\,,$$

is

$$\delta_B = \sum_{i=1}^{b} C_i \xi_i = \frac{8-1}{24-3}(21.7143 + 45.1429 + 27.4286) = 31.4286\,.$$

If all M possible arrangements of the $N = 24$ observed OLS regression residuals listed in Fig. 4.25 occur with equal chance, the approximate resampling probability value of $\delta_B = 31.4286$ computed on $L = 1{,}000{,}000$ random arrangements of the observed OLS regression residuals with $n_{B_1} = n_{B_2} = n_{B_3} = 8$ residuals preserved for each arrangement is

$$P(\delta \leq \delta_B | H_0) = \frac{\text{number of } \delta \text{ values} \leq \delta_B}{L} = \frac{49{,}168}{1{,}000{,}000} = 0.0492 .$$

For comparison, the approximate resampling probability value based on LAD regression, $v = 1$, $L = 1{,}000{,}000$, and $C_i = n_{B_i}/N$ for $i = 1, 2, 3$ is $P = 0.1250$.

Following Eq. (4.7) on p. 126, the exact expected value of the $M = 9{,}465{,}511{,}770$ δ values is $\mu_\delta = 38.4348$ and, following Eq. (4.6) on p. 126, the observed chance-corrected measure of effect size for the y_i and $\hat{y}_i$ values, $i = 1, \ldots, N$, is

$$\Re_B = 1 - \frac{\delta_B}{\mu_\delta} = 1 - \frac{31.4286}{38.4348} = +0.1823 ,$$

indicating approximately 18 % agreement between the observed and predicted y values above that expected by chance.

Conventional ANOVA Analysis

A conventional fixed-effects factorial analysis of variance calculated on the $N = 24$ Factor B response measurement scores listed in Fig. 4.21 on p. 156 yields an observed F-ratio of $F_B = 3.0545$. Assuming independence, normality, and homogeneity of variance, F_B is approximately distributed as Snedecor's F under the null hypothesis with $\nu_1 = b - 1 = 3 - 1 = 2$ and $\nu_2 = N - ab = 24 - (2)(3) = 18$ degrees of freedom. Under the null hypothesis, the observed value of $F_B = 3.0545$ yields an approximate probability value of $P = 0.0721$.

Analysis of the $A{\times}B$ Interaction

A design matrix of effect codes for analyzing the $A{\times}B$ interaction of the data listed in Fig. 4.21 on p. 156 is given in Fig. 4.26, where the first column of 1 values provides for an intercept, the second column contains effect codes for Factor A, the third and fourth columns contain effect codes for Factor B, and the last column lists the $N = 24$ univariate response measurement scores listed according to the original random assignment of the subjects to the $ab = (2)(3) = 6$ levels of the $A{\times}B$ interaction. The MRPP regression analysis examines the $N = 24$ regression residuals for possible differences among the six treatment levels of the $A{\times}B$ interaction; consequently, no effect codes are provided for the $A{\times}B$ interaction as this information is implicit in the ordering of the treatment levels of the $A{\times}B$ interaction in the last column of Fig. 4.26.

Matrix				Score
1	1	1	0	1
1	1	1	0	4
1	1	1	0	0
1	1	1	0	7
1	−1	1	0	15
1	−1	1	0	6
1	−1	1	0	10
1	−1	1	0	13
1	1	0	1	13
1	1	0	1	5
1	1	0	1	7
1	1	0	1	15
1	−1	0	1	6
1	−1	0	1	18
1	−1	0	1	9
1	−1	0	1	15
1	1	−1	−1	9
1	1	−1	−1	16
1	1	−1	−1	18
1	1	−1	−1	13
1	−1	−1	−1	14
1	−1	−1	−1	7
1	−1	−1	−1	6
1	−1	−1	−1	13

Fig. 4.26 Design matrix and univariate response measurement scores for the $A{\times}B$ interaction in a 2×3 factorial design with $N = 24$ subjects

Because there are

$$M = \frac{N!}{\prod_{i=1}^{ab} n_{(A\times B)_i}!} = \frac{24!}{(4!)^6} = 118{,}569{,}536{,}025{,}665{,}614{,}982{,}267{,}535{,}360{,}000$$

possible, equally-likely arrangements of the $N = 24$ univariate response measurement scores for the $A{\times}B$ interaction listed in Fig. 4.26, an exact permutation approach is clearly not possible.

LAD Regression Analysis

An MRPP resampling analysis of the LAD regression residuals calculated on the univariate response measurement scores in Fig. 4.26 yields estimated LAD regression coefficients of

$$\tilde{\beta}_0 = +8.3333\,, \quad \tilde{\beta}_1 = -1.00\,, \quad \tilde{\beta}_2 = -3.3333\,, \text{ and } \quad \tilde{\beta}_3 = -0.3333$$

for the interaction of Factors A and B. Figure 4.27 lists the observed y_i values, LAD predicted $\tilde{y}_i$ values, and residual e_i values for $i = 1, \ldots, 24$.

Fig. 4.27 Observed, predicted, and residual LAD regression values for the univariate response measurement scores listed in Fig. 4.26

Object	y_i	$\tilde{y}_i$	e_i
1	1	4.00	−3.00
2	4	4.00	0.00
3	0	4.00	−4.00
4	7	4.00	+3.00
5	15	6.00	+9.00
6	6	6.00	0.00
7	10	6.00	+4.00
8	13	6.00	+7.00
9	13	7.00	+6.00
10	5	7.00	−2.00
11	7	7.00	0.00
12	15	7.00	+8.00
13	6	9.00	−3.00
14	18	9.00	+9.00
15	9	9.00	0.00
16	15	9.00	+6.00
17	9	11.00	−2.00
18	16	11.00	+5.00
19	18	11.00	+7.00
20	13	11.00	+2.00
21	14	13.00	+1.00
22	7	13.00	−6.00
23	6	13.00	−7.00
24	13	13.00	0.00

Following Eq. (4.5) on p. 125 and employing ordinary Euclidean distance between residuals with $v = 1$, the $N = 24$ LAD regression residuals listed in Fig. 4.27 yield $ab = (2)(3) = 6$ average distance-function values of

$$\xi_{(A\times B)_1} = 4.00\ , \quad \xi_{(A\times B)_2} = 5.00\ , \quad \xi_{(A\times B)_3} = 6.00\ , \quad \xi_{(A\times B)_4} = 7.00\ ,$$
$$\text{and} \quad \xi_{(A\times B)_5} = \xi_{(A\times B)_6} = 5.00\ .$$

Following Eq. (4.4) on p. 125, the observed value of the MRPP test statistic calculated on the LAD regression residuals listed in Fig. 4.27 with $v = 1$ and treatment-group weights

$$C_i = \frac{n_{(A\times B)_i}}{N}\ , \qquad i = 1, \ldots, 6\ ,$$

is

$$\delta_{A\times B} = \sum_{i=1}^{ab} C_i \xi_i = \frac{4}{24}(4.00 + 5.00 + 6.00 + 7.00 + 5.00 + 5.00) = 5.3333\ .$$

If all M possible arrangements of the $N = 24$ observed LAD regression residuals listed in Fig. 4.27 occur with equal chance, the approximate resampling probability value of $\delta_{A\times B} = 5.3333$ computed on $L = 1{,}000{,}000$ random arrangements of the observed LAD regression residuals with $n_{(A\times B)_1} = \cdots = n_{(A\times B)_6} = 4$ residuals preserved for each arrangement is

$$P(\delta \leq \delta_{A\times B}|H_0) = \frac{\text{number of } \delta \text{ values} \leq \delta_{A\times B}}{L} = \frac{347{,}675}{1{,}000{,}000} = 0.3477 \, .$$

Following Eq. (4.7) on p. 126, the exact expected value of the M δ values is $\mu_\delta = 5.50$ and, following Eq. (4.6) on p. 126, the observed chance-corrected measure of effect size for the y_i and $\tilde{y}_i$ values, $i = 1, \ldots, N$, is

$$\Re_{A\times B} = 1 - \frac{\delta_{A\times B}}{\mu_\delta} = 1 - \frac{5.3333}{5.50} = +0.0303 \, ,$$

indicating approximately 3 % agreement between the observed and predicted y values above that expected by chance.

OLS Regression Analysis

For comparison, consider an MRPP analysis of OLS regression residuals calculated on the $N = 24$ univariate response measurement scores of the $A\times B$ interaction listed in Fig. 4.26. The MRPP regression analysis yields estimated OLS regression coefficients of

$$\hat{\beta}_0 = +10.00 \, , \quad \hat{\beta}_1 = -1.00 \, , \quad \hat{\beta}_2 = -3.00 \, , \text{ and } \quad \hat{\beta}_3 = +1.00$$

for the interaction of Factors A and B. Figure 4.28 lists the observed y_i values, OLS predicted $\hat{y}_i$ values, and residual e_i values for $i = 1, \ldots, 24$.

Following Eq. (4.5) on p. 125 and employing squared Euclidean distance between residuals with $v = 2$, the $N = 24$ OLS regression residuals listed in Fig. 4.28 yield $ab = (2)(3) = 6$ average distance-function values of

$$\xi_{(A\times B)_1} = 20.00 \, , \quad \xi_{(A\times B)_2} = 30.6667 \, , \quad \xi_{(A\times B)_3} = 45.3333 \, ,$$
$$\xi_{(A\times B)_4} = 60.00 \, , \quad \xi_{(A\times B)_5} = 30.6667 \, , \quad \text{and} \quad \xi_{(A\times B)_6} = 33.3333 \, .$$

Following Eq. (4.4) on p. 125, the observed value of the MRPP test statistic calculated on the OLS regression residuals listed in Fig. 4.28 with $v = 2$ and treatment-group weights

$$C_i = \frac{n_{(A\times B)_i} - 1}{N - ab} \, , \qquad i = 1, \ldots, 6 \, ,$$

Fig. 4.28 Observed, predicted, and residual OLS regression values for the univariate response measurement scores listed in Fig. 4.26

Object	y_i	$\hat{y}_i$	e_i
1	1	6.00	−5.00
2	4	6.00	−2.00
3	0	6.00	−6.00
4	7	6.00	+1.00
5	15	8.00	+7.00
6	6	8.00	−2.00
7	10	8.00	+2.00
8	13	8.00	+5.00
9	13	10.00	+3.00
10	5	10.00	−5.00
11	7	10.00	−3.00
12	15	10.00	+5.00
13	6	12.00	−6.00
14	18	12.00	+6.00
15	9	12.00	−3.00
16	15	12.00	+3.00
17	9	11.00	−2.00
18	16	11.00	+5.00
19	18	11.00	+7.00
20	13	11.00	+2.00
21	14	13.00	+1.00
22	7	13.00	−6.00
23	6	13.00	−7.00
24	13	13.00	0.00

is

$$\delta_{A\times B} = \sum_{i=1}^{ab} C_i \xi_i = \frac{4-1}{24-6}\big(20.00 + 30.6667 + 45.3333 + 60.00 + 30.6667 + 33.3333\big) = 36.6667 \, .$$

If all M possible arrangements of the observed OLS regression residuals listed in Fig. 4.28 occur with equal chance, the approximate resampling probability value of $\delta_{A\times B} = 36.6666$ computed on $L = 1{,}000{,}000$ random arrangements of the observed OLS regression residuals with $n_{(A\times B)_1} = \cdots = n_{(A\times B)_6} = 4$ residuals preserved for each arrangement is

$$P\big(\delta \le \delta_{A\times B}|H_0\big) = \frac{\text{number of } \delta \text{ values} \le \delta_{A\times B}}{L} = \frac{224{,}204}{1{,}000{,}000} = 0.2242 \, .$$

For comparison, the approximate resampling probability value based on LAD regression, $v = 1$, $L = 1{,}000{,}000$, and $C_i = n_{(A\times B)_i}/N$ for $i = 1, \ldots, 6$ is $P = 0.3477$.

Following Eq. (4.7) on p. 126, the exact expected value of the M δ values is $\mu_\delta = 41.2174$ and, following Eq. (4.6) on p. 126, the observed chance-corrected measure of effect size for the y_i and $\hat{y}_i$ values, $i = 1, \ldots, N$, is

$$\Re_{A\times B} = 1 - \frac{\delta_{A\times B}}{\mu_\delta} = 1 - \frac{36.6667}{41.2174} = +0.1104 \,,$$

indicating approximately 11 % agreement between the observed and predicted y values above that expected by chance.

Conventional ANOVA Analysis

A conventional fixed-effects factorial analysis of variance calculated on the $N = 24$ univariate response measurement scores listed in Fig. 4.21 on p. 156 yields an observed F-ratio of $F_{A\times B} = 3.9273$. Assuming independence, normality, and homogeneity of variance, $F_{A\times B}$ is approximately distributed as Snedecor's F under the null hypothesis with $\nu_1 = (a-1)(b-1) = (2-1)(3-1) = 2$ and $\nu_2 = ab(n-1) = (2)(3)(4-1) = 18$ degrees of freedom. Under the null hypothesis, the observed value of $F_{A\times B} = 3.9273$ yields an approximate probability value of $P = 0.0384$, which differs greatly from the LAD and OLS regression probability values of $P = 0.3477$ and $P = 0.2242$, respectively.

4.3.6 Latin Square Design

A Latin square experimental design assigns treatments to subjects so the treatments occur in a balanced fashion within a square block or field; thus, n treatments appear once in each of n rows and n columns. The Latin square is the design of choice when controlling for two blocking factors. Consider an ordinary balanced Latin square experiment involving repeated measurements in which $n = 4$ subjects (S) are each tested $b = 4$ times on Factor A. The design and data are adapted from Ferguson [115, p. 349] and are given in Table 4.8, where B refers to the ordinal position in which the levels of Factor A are administered. Thus, the first subject, S_1, receives the $b = 4$ treatments in the order A_2, A_4, A_1, A_3, and so on. Due to the balanced nature of Latin square designs, the assumption is that there is no interaction between blocking Factors A and B, or between either blocking factor and the treatments.

Table 4.8 Design and data for a Latin square design with four subjects (S), four treatments (A), and four orders (B)

	Design					Scores			
Subject	B_1	B_2	B_3	B_4	Subject	B_1	B_2	B_3	B_4
S_1	A_2	A_4	A_1	A_3	S_1	10	21	5	14
S_2	A_3	A_1	A_2	A_4	S_2	12	7	11	19
S_3	A_1	A_3	A_4	A_2	S_3	6	16	24	12
S_4	A_4	A_2	A_3	A_1	S_4	22	8	17	9

Fig. 4.29 Design matrix and univariate response measurement scores for treatment (*A*) in a Latin square design

Matrix							Score
1	0	0	0	0	1	0	5
1	1	0	0	1	0	0	7
1	0	1	0	0	0	0	6
1	0	0	1	0	0	1	9
1	0	0	0	0	0	0	10
1	1	0	0	0	1	0	11
1	0	1	0	0	0	1	12
1	0	0	1	1	0	0	8
1	0	0	0	0	0	1	14
1	1	0	0	0	0	0	12
1	0	1	0	1	0	0	16
1	0	0	1	0	1	0	17
1	0	0	0	1	0	0	21
1	1	0	0	0	0	1	19
1	0	1	0	0	1	0	24
1	0	0	1	0	0	0	22

Analysis of Factor *A*

A design matrix of dummy codes for analyzing Factor *A* is given in Fig. 4.29, where the first column of 1 values provides for an intercept, the second through fourth columns contain dummy codes for Subjects, the fifth through seventh columns contain dummy codes for Factor *B*, and the last column lists the univariate response measurement scores ordered by the $a = 4$ levels of Factor *A*, with the first $n_{A_1} = 4$ scores, the next $n_{A_2} = 4$ scores, the next $n_{A_3} = 4$ scores, and the last $n_{A_4} = 4$ scores associated with treatment levels A_1, A_2, A_3, and A_4, respectively. The MRPP regression analysis examines the $N = 16$ regression residuals for possible differences among the $a = 4$ treatment levels of Factor *A*; consequently, no dummy codes are provided for Factor *A* as this information is implicit in the ordering of the $a = 4$ treatment levels of Factor *A* in the last column of Fig. 4.29.

Because there are

$$M = \frac{N!}{\prod_{i=1}^{a} n_{A_i}!} = \frac{16!}{(4!)^4} = 63{,}063{,}000$$

possible, equally-likely arrangements of the $N = 16$ univariate response measurement scores listed in Fig. 4.29, an exact permutation approach is not practical.

LAD Regression Analysis

An MRPP resampling analysis of the $N = 16$ LAD regression residuals calculated on the univariate response measurement scores listed in Fig. 4.29 yields estimated

Fig. 4.30 Observed, predicted, and residual LAD regression values for the univariate response measurement scores listed in Fig. 4.29

Object	y_i	$\tilde{y}_i$	e_i
1	5	22.00	−17.00
2	7	20.00	−13.00
3	6	8.00	−2.00
4	9	9.00	0.00
5	10	10.00	0.00
6	11	24.00	−13.00
7	12	12.00	0.00
8	8	13.00	−5.00
9	14	14.00	0.00
10	12	12.00	0.00
11	16	16.00	0.00
12	17	17.00	0.00
13	21	18.00	+3.00
14	19	16.00	+3.00
15	24	20.00	+4.00
16	22	5.00	+17.00

LAD regression coefficients of

$$\tilde{\beta}_0 = +10.00\ ,\quad \tilde{\beta}_1 = +2.00\ ,\quad \tilde{\beta}_2 = -2.00\ ,\quad \tilde{\beta}_3 = -5.00\ ,$$

$$\tilde{\beta}_4 = +8.00\ ,\quad \tilde{\beta}_5 = +12.00\ ,\ \text{and}\ \ \tilde{\beta}_6 = +4.00$$

for Factor A. Figure 4.30 lists the observed y_i values, LAD predicted $\tilde{y}_i$ values, and residual e_i values for $i = 1, \ldots, 16$.

Following (4.5) on p. 125 and employing ordinary Euclidean distance between residuals with $v = 1$, the $N = 16$ LAD regression residuals listed in Fig. 4.30 yield $a = 4$ average distance-function values of

$$\xi_{A_1} = 10.3333\ ,\quad \xi_{A_2} = 7.3333\ ,\quad \xi_{A_3} = 0.00\ ,\ \text{and}\ \ \xi_{A_4} = 7.1667\ .$$

Following Eq. (4.4) on p. 125, the observed value of the MRPP test statistic calculated on the LAD regression residuals listed in Fig. 4.30 with $v = 1$ and treatment-group weights

$$C_i = \frac{n_{A_i}}{N}\ ,\qquad i = 1, \ldots, 4\ ,$$

is

$$\delta_A = \sum_{i=1}^{a} C_i \xi_i = \frac{4}{16}(10.3333 + 7.3333 + 0.00 + 7.1667) = 6.2083\ .$$

If all M possible arrangements of the $N = 16$ observed LAD regression residuals listed in Fig. 4.30 occur with equal chance, the approximate resampling probability value of $\delta_A = 6.2083$ computed on $L = 1{,}000{,}000$ random arrangements of the observed LAD regression residuals with $n_{A_1} = \cdots = n_{A_4} = 4$ residuals preserved for each arrangement is

$$P(\delta \leq \delta_A | H_0) = \frac{\text{number of } \delta \text{ values} \leq \delta_A}{L} = \frac{27{,}289}{1{,}000{,}000} = 0.0273 \, .$$

Following Eq. (4.7) on p. 126, the exact expected value of the $M = 63{,}063{,}000$ δ values is $\mu_\delta = 8.2750$ and, following Eq. (4.6) on p. 126, the observed chance-corrected measure of effect size for the y_i and $\tilde{y}_i$ values, $i = 1, \ldots, N$, is

$$\Re_A = 1 - \frac{\delta_A}{\mu_\delta} = 1 - \frac{6.2083}{8.2750} = +0.2497 \, ,$$

indicating approximately 25 % agreement between the observed and predicted y values above that expected by chance.

OLS Regression Analysis

For comparison, consider an MRPP resampling analysis of the OLS regression residuals calculated on the $N = 16$ univariate response measurement scores listed in Fig. 4.29 on p. 171. The MRPP regression analysis yields estimated OLS regression coefficients of

$$\hat{\beta}_0 = +11.6875 \, , \quad \hat{\beta}_1 = -0.2500 \, , \quad \hat{\beta}_2 = +2.00 \, , \quad \hat{\beta}_3 = +1.50 \, ,$$
$$\hat{\beta}_4 = +0.50 \, , \quad \hat{\beta}_5 = +1.7500 \, , \quad \text{and} \quad \hat{\beta}_6 = +1.00$$

for Factor A. Figure 4.31 lists the observed y_i values, OLS predicted $\hat{y}_i$ values, and residual e_i values for $i = 1, \ldots, 16$.

Following Eq. (4.5) on p. 125 and employing squared Euclidean distance between residuals with $v = 2$, the $N = 16$ OLS regression residuals listed in Fig. 4.31 yield $a = 4$ average distance-function values of

$$\xi_{A_1} = 6.2083 \, , \quad \xi_{A_2} = 6.4583 \, , \quad \xi_{A_3} = 0.8750 \, , \quad \text{and} \quad \xi_{A_4} = 2.3750 \, .$$

Following Eq. (4.4) on p. 125, the observed value of the MRPP test statistic calculated on the OLS regression residuals listed in Fig. 4.31 with $v = 2$ and treatment-group weights

$$C_i = \frac{n_{A_i} - 1}{N - a} \, , \qquad i = 1, \ldots, 4 \, ,$$

Fig. 4.31 Observed, predicted, and residual LAD regression values for the univariate response measurement scores listed in Fig. 4.29

Object	y_i	$\hat{y}_i$	e_i
1	5	13.4375	−8.4375
2	7	11.9375	−4.9375
3	6	13.6875	−7.6875
4	9	14.1875	−5.1875
5	10	11.6875	−1.6875
6	11	13.1875	−2.1875
7	12	14.6875	−2.6875
8	8	13.6875	−5.6875
9	14	12.6875	+1.3125
10	12	11.4375	+0.5625
11	16	14.1875	+1.8125
12	17	14.9375	+2.0625
13	21	12.1875	+8.8125
14	19	12.4375	+6.5625
15	24	15.4375	+8.5625
16	22	13.1875	+8.8125

is

$$\delta_A = \sum_{i=1}^{a} C_i \xi_i = \frac{4-1}{16-4}(6.2083 + 6.4583 + 0.8750 + 2.3750) = 3.9792 \, .$$

If all M possible arrangements of the $N = 16$ observed OLS regression residuals listed in Fig. 4.31 occur with equal chance, the approximate resampling probability value of $\delta_A = 3.9792$ computed on $L = 1{,}000{,}000$ random arrangements of the observed OLS regression residuals with $n_{A_1} = \cdots = n_{A_4} = 4$ residuals preserved for each arrangement is

$$P(\delta \leq \delta_A | H_0) = \frac{\text{number of } \delta \text{ values} \leq \delta_A}{L} = \frac{1}{1{,}000{,}000} = 0.10 \times 10^{-5} \, .$$

For comparison, the approximate resampling probability value based on LAD regression, $v = 1$, $L = 1{,}000{,}000$, and $C_i = n_{A_i}/N$ for $i = 1, \ldots, 4$ is $P = 0.0273$.

Following Eq. (4.7) on p. 126, the exact expected value of the $M = 63{,}063{,}000$ δ values is $\mu_\delta = 68.0083$ and, following Eq. (4.6) on p. 126, the observed chance-corrected measure of effect size for the y_i and $\hat{y}_i$ values, $i = 1, \ldots, N$, is

$$\Re_A = 1 - \frac{\delta_A}{\mu_\delta} = 1 - \frac{3.9792}{68.0083} = +0.9415 \, ,$$

indicating approximately 95 % agreement between the observed and predicted y values above that expected by chance.

Conventional ANOVA Analysis

A conventional Latin square analysis of variance calculated on the $N = 16$ univariate response measurement scores listed in Table 4.8 on p. 170 yields an observed F-ratio of $F_A = 40.7277$. Assuming independence and normality, F_A is approximately distributed as Snedecor's F under the null hypothesis with $\nu_1 = a - 1 = 4 - 1 = 3$ and $\nu_2 = (a-2)(a-1) = (4-2)(4-1) = 6$ degrees of freedom. Under the null hypothesis, the observed value of $F_A = 40.7277$ yields an approximate probability value of $P = 0.2204 \times 10^{-3}$.

Analysis of Factor *B*

A design matrix of dummy codes for analyzing Factor B is given in Fig. 4.32, where the first column of 1 values provides for an intercept, the second through fourth columns contain dummy codes for Subjects, the fifth through seventh columns contain dummy codes for Factor A, and the last column lists the univariate response measurement scores ordered by the $b = 4$ treatment levels of Factor B, with the first $n_{B_1} = 4$ scores, the next $n_{B_2} = 4$ scores, the next $n_{B_3} = 4$ scores, and the last $n_{B_4} = 4$ associated with treatment levels B_1, B_2, B_3, and B_4, respectively. The MRPP regression analysis examines LAD regression residuals for possible differences among the $b = 4$ treatment levels of Factor B; consequently, no dummy codes are provided for Factor B as this information is implicit in the ordering of the $b = 4$ treatment levels of Factor B in the last column of Fig. 4.32.

Because there are

$$M = \frac{N!}{\prod_{i=1}^{b} n_{B_i}!} = \frac{16!}{(4!)^4} = 63{,}063{,}000$$

Fig. 4.32 Design matrix and univariate response measurement scores for order (B) in a Latin square design

Matrix							Score
1	0	0	0	1	0	0	10
1	1	0	0	0	1	0	12
1	0	1	0	0	0	0	6
1	0	0	1	0	0	1	22
1	0	0	0	0	0	1	21
1	1	0	0	0	0	0	7
1	0	1	0	0	1	0	16
1	0	0	1	1	0	0	8
1	0	0	0	0	0	0	5
1	1	0	0	1	0	0	11
1	0	1	0	0	0	1	24
1	0	0	1	0	1	0	17
1	0	0	0	0	1	0	14
1	1	0	0	0	0	1	19
1	0	1	0	1	0	0	12
1	0	0	1	0	0	0	9

possible, equally-likely arrangements of the $N = 16$ univariate response measurement scores listed in Fig. 4.32, an exact permutation approach is not practical.

LAD Regression Analysis

An MRPP resampling analysis of the $N = 16$ LAD regression residuals calculated on the univariate response measurement scores in Fig. 4.32 yields estimated LAD regression coefficients of

$$\tilde{\beta}_0 = +21.00\,, \quad \tilde{\beta}_1 = -2.00\,, \quad \tilde{\beta}_2 = +2.00\,, \quad \tilde{\beta}_3 = +1.00\,,$$

$$\tilde{\beta}_4 = -13.00\,, \quad \tilde{\beta}_5 = -11.00\,, \text{ and } \quad \tilde{\beta}_6 = -7.00$$

for Factor B. Figure 4.33 lists the observed y_i values, LAD predicted $\tilde{y}_i$ values, and residual e_i values for $i = 1, \ldots, 16$.

Following Eq. (4.5) on p. 125 and employing ordinary Euclidean distance between residuals with $v = 1$, the $N = 16$ LAD regression residuals listed in Fig. 4.33 yield $b = 4$ average distance-function values of

$$\xi_{B_1} = 2.00\,, \quad \xi_{B_2} = 2.00\,, \quad \xi_{B_3} = 3.1667\,, \text{ and } \quad \xi_{B_4} = 0.00\,.$$

Following Eq. (4.4) on p. 125, the observed value of the MRPP test statistic calculated on the LAD regression residuals listed in Fig. 4.33 with $v = 1$ and treatment-group weights

$$C_i = \frac{n_{B_i}}{N}\,, \qquad i = 1, \ldots, 4\,,$$

Fig. 4.33 Observed, predicted, and residual LAD regression values for the univariate response measurement scores listed in Fig. 4.32

Object	y_i	$\tilde{y}_i$	e_i
1	10	10.00	0.00
2	12	12.00	0.00
3	6	10.00	−4.00
4	22	22.00	0.00
5	21	21.00	0.00
6	7	6.00	+1.00
7	16	16.00	0.00
8	8	11.00	−3.00
9	5	8.00	−3.00
10	11	8.00	+3.00
11	24	23.00	+1.00
12	17	15.00	+2.00
13	14	14.00	0.00
14	19	19.00	0.00
15	12	12.00	0.00
16	9	9.00	0.00

is

$$\delta_B = \sum_{i=1}^{b} C_i \xi_i = \frac{4}{16}(2.00 + 2.00 + 3.1667 + 0.00) = 1.7917 .$$

If all M possible arrangements of the $N = 16$ observed LAD regression residuals listed in Fig. 4.33 occur with equal chance, the approximate resampling probability value of $\delta_B = 1.7917$ computed on $L = 1{,}000{,}000$ random arrangements of the observed LAD regression residuals with $n_{B_1} = \cdots = n_{B_4} = 4$ residuals preserved for each arrangement is

$$P(\delta \leq \delta_B | H_0) = \frac{\text{number of } \delta \text{ values} \leq \delta_B}{L} = \frac{495{,}269}{1{,}000{,}000} = 0.4953 .$$

Following Eq. (4.7) on p. 126, the exact expected value of the $M = 63{,}063{,}000$ δ values is $\mu_\delta = 1.8583$ and, following Eq. (4.6) on p. 126, the observed chance-corrected measure of effect size for the y_i and $\tilde{y}_i$ values, $i = 1, \ldots, N$ is

$$\Re_B = 1 - \frac{\delta_B}{\mu_\delta} = 1 - \frac{1.7917}{1.8583} = +0.0359 ,$$

indicating approximately 4 % agreement between the observed and predicted y values above that expected by chance.

OLS Regression Analysis

For comparison, consider an MRPP resampling analysis of the OLS regression residuals calculated on the $N = 16$ univariate response measurement scores listed in Fig. 4.29 on p. 171. The MRPP regression analysis yields estimated OLS regression coefficients of

$$\hat{\beta}_0 = +20.6875 , \quad \hat{\beta}_1 = -0.2500 , \quad \hat{\beta}_2 = +2.00 , \quad \hat{\beta}_3 = +1.50 ,$$

$$\hat{\beta}_4 = -14.7500 , \quad \hat{\beta}_5 = -11.2500 , \text{ and } \hat{\beta}_6 = -6.7500$$

for Factor B. Figure 4.34 lists the observed y_i values, OLS predicted $\hat{y}_i$ values, and residual e_i values for $i = 1, \ldots, 16$.

Following Eq. (4.5) on p. 125 and employing squared Euclidean distance between residuals with $v = 2$, the $N = 16$ OLS regression residuals listed in Fig. 4.34 yield $b = 4$ average distance-function values of

$$\xi_{B_1} = 2.8750 , \quad \xi_{B_2} = 6.7083 , \quad \xi_{B_3} = 3.2083 , \text{ and } \xi_{B_4} = 3.1250 .$$

Fig. 4.34 Observed, predicted, and residual OLS regression values for the univariate response measurement scores listed in Fig. 4.32

Object	y_i	$\hat{y}_i$	e_i
1	10	9.4375	+0.5625
2	12	13.6875	−1.6875
3	6	7.9375	−1.9375
4	22	22.1875	−0.1875
5	21	20.6875	+0.3125
6	7	5.6875	+1.3125
7	16	15.9375	+0.0625
8	8	10.9375	−2.9375
9	5	5.9375	−0.9375
10	11	9.1875	+1.8125
11	24	22.6875	+1.3125
12	17	15.4375	+1.5625
13	14	13.9375	+0.0625
14	19	20.4375	−1.4375
15	12	11.4375	+0.5625
16	9	7.4375	+1.5625

Following Eq. (4.4) on p. 125, the observed value of the MRPP test statistic calculated on the OLS regression residuals listed in Fig. 4.34 with $v = 2$ and treatment-group weights

$$C_i = \frac{n_{B_i} - 1}{N - b} , \qquad i = 1, \ldots, 4 ,$$

is

$$\delta_B = \sum_{i=1}^{b} C_i \xi_i = \frac{4-1}{16-4}\big(2.8750 + 6.7083 + 3.2083 + 3.1250\big) = 3.9792 .$$

If all M possible arrangements of the $N = 16$ observed OLS regression residuals listed in Fig. 4.34 occur with equal chance, the approximate resampling probability value of $\delta_B = 3.9792$ computed on $L = 1{,}000{,}000$ random arrangements of the observed OLS regression residuals with $n_{B_1} = n_{B_2} = n_{B_3} = n_{B_4} = 4$ residuals preserved for each arrangement is

$$P\big(\delta \le \delta_B | H_0\big) = \frac{\text{number of } \delta \text{ values} \le \delta_B}{L} = \frac{378{,}875}{1{,}000{,}000} = 0.3789 .$$

For comparison, the approximate resampling probability value based on LAD regression, $v = 1$, $L = 1{,}000{,}000$, and $C_i = n_{B_i}/N$ for $i = 1, \ldots, 4$ is $P = 0.4953$.

Following Eq. (4.7) on p. 126, the exact expected value of the $M = 63{,}063{,}000$ δ values is $\mu_\delta = 4.0750$ and, following Eq. (4.6) on p. 126, the observed chance-corrected measure of effect size for the y_i and $\hat{y}_i$ values, $i = 1, \ldots, N$, is

$$\Re_B = 1 - \frac{\delta_B}{\mu_\delta} = 1 - \frac{1.7917}{4.0750} = +0.0235 ,$$

indicating only approximately 2 % agreement between the observed and predicted y values above that expected by chance.

Conventional ANOVA Analysis

A conventional Latin square analysis of variance calculated on the $N = 16$ univariate response measurement scores listed in Table 4.8 on p. 170 yields an observed F-ratio of $F_B = 0.5602$. Assuming independence and normality, F_B is approximately distributed as Snedecor's F under the null hypothesis with $\nu_1 = b - 1 = 4 - 1 = 3$ and $\nu_2 = (b-2)(b-1) = (4-2)(4-1) = 6$ degrees of freedom. Under the null hypothesis, the observed value of $F_B = 0.5602$ yields an approximate probability value of $P = 0.6606$. The LAD regression, OLS regression, and F-ratio probability values of $P = 0.4953$, $P = 0.3789$, and $P = 0.6606$, respectively, all indicate that the order in which the treatments were distributed did not matter.

4.3.7 Split-Plot Design

Imagine a testing experiment with two treatment factors, A and B, with a and b treatment levels, respectively, so that there are ab treatment combinations. If each testing session requires h hours of a subject's time and every subject is to be treated under all treatment conditions, each subject will require ab testing sessions and abh hours of testing time. When this is unreasonable, then with S subjects available, assign $n = S/A$ subjects to each level of Factor A and test each subject under all levels of Factor B. The design is a repeated-measures split-plot design in which subjects are randomly assigned to the a treatment levels of Factor A (i.e., plots), and each subject is then tested under all b levels of Factor B (i.e., subplots). The design is also called a mixed factorial design with one between-subjects factor (A) and one within-subjects factor (B), or an $A{\times}(B{\times}S)$ design [214].

Consider a split-plot experiment in which Factor A has $a = 3$ treatment levels, Factor B has $b = 3$ treatment levels, $n = 12$ subjects are randomly assigned to each of the $a = 3$ levels of Factor A, and each subject is tested at all $b = 3$ levels of Factor B. The design and data are adapted from Keppel and Zedeck and are given in Fig. 4.35 [215, p. 303].

Analysis of Factor *A*

A design matrix of effect codes for an MRPP regression analysis of Factor A is given in Fig. 4.36, where the first column of 1 values provides for an intercept and the second column lists the total of response measurement summations over the b

Fig. 4.35 Example univariate response measurements for a split-plot design

Factor A	Subject	B_1	B_2	B_3
		Factor B		
A_1	S_1	53	51	35
	S_2	49	34	18
	S_3	47	44	32
	S_4	42	48	27
A_2	S_5	47	42	16
	S_6	42	33	10
	S_7	39	13	11
	S_8	37	16	6
A_3	S_9	45	35	29
	S_{10}	41	33	21
	S_{11}	38	46	30
	S_{12}	36	40	20

Fig. 4.36 Design matrix and response measurement summations for the main effects of Factor A in a split-plot design

Matrix	Sum over B
1	139
1	101
1	123
1	117
1	105
1	85
1	63
1	59
1	109
1	95
1	114
1	96

levels of Factor B (e.g., $53 + 51 + 35 = 139$). The summations are ordered by the $a = 3$ treatment levels of Factor A with the first $n_{A_1} = 4$ summations, the second $n_{A_2} = 4$ summations, and the last $n_{A_3} = 4$ summations associated with treatment levels A_1, A_2, and A_3, respectively. The MRPP regression analysis examines the $N = 12$ regression residuals for possible differences among the $a = 3$ treatment levels of Factor A; consequently, no effect codes are provided for Factor A as this information is implicit in the ordering of the $a = 3$ treatment levels of Factor A in the second column of Fig. 4.36.

An exact permutation solution is feasible for the response measurement summations listed in Fig. 4.36 since there are only

$$M = \frac{N!}{\prod_{i=1}^{a} n_{A_i}!} = \frac{12!}{(4!)^3} = 34{,}650$$

Fig. 4.37 Observed, predicted, and residual LAD regression values for the response measurement summations listed in Fig. 4.36

Object	y_i	$\tilde{y}_i$	e_i
1	139	101.00	+38.00
2	101	101.00	0.00
3	123	101.00	+22.00
4	117	101.00	+16.00
5	105	101.00	+4.00
6	85	101.00	−16.00
7	63	101.00	−38.00
8	59	101.00	−42.00
9	109	101.00	+8.00
10	95	101.00	−6.00
11	114	101.00	+13.00
12	96	101.00	−5.00

possible, equally-likely arrangements of the $N = 12$ response measurement summations for Factor A.

LAD Regression Analysis

An MRPP analysis of the $N = 12$ LAD regression residuals calculated on the response measurement summations in Fig. 4.36 yields an estimated LAD regression coefficient of $\tilde{\beta}_0 = +101.00$ for Factor A. Figure 4.37 lists the observed y_i values, LAD predicted $\tilde{y}_i$ values, and residual e_i values for $i = 1, \ldots, 12$.

Following Eq. (4.5) on p. 125 and employing ordinary Euclidean distance between residuals with $v = 1$, the $N = 12$ LAD regression residuals listed in Fig. 4.37 yield $a = 3$ average distance-function values of

$$\xi_{A_1} = 20.00\,, \quad \xi_{A_2} = 26.6667\,, \quad \text{and} \quad \xi_{A_3} = 11.6667\,.$$

Following Eq. (4.4) on p. 125, the observed value of the MRPP test statistic calculated on the LAD regression residuals listed in Fig. 4.37 with $v = 1$ and treatment-group weights

$$C_i = \frac{n_{A_i}}{N}\,, \quad i = 1, 2, 3\,,$$

is

$$\delta_A = \sum_{i=1}^{a} C_i \xi_i = \frac{4}{12}\big(20.00 + 26.6667 + 11.6667\big) = 19.4444\,.$$

If all arrangements of the $N = 16$ observed LAD regression residuals listed in Fig. 4.37 occur with equal chance, the exact probability value of $\delta_A = 19.4444$ calculated on the $M = 34{,}650$ possible arrangements of the observed LAD regression residuals with $n_{A_1} = n_{A_2} = n_{A_3} = 4$ residuals preserved for each arrangement is

$$P(\delta \leq \delta_A | H_0) = \frac{\text{number of } \delta \text{ values} \leq \delta_A}{M} = \frac{672}{34{,}650} = 0.0194 \, .$$

Following Eq. (4.7) on p. 126, the exact expected value of the $M = 34{,}650$ δ values is $\mu_\delta = 27.00$ and, following Eq. (4.6) on p. 126, the observed chance-corrected measure of effect size for the y_i and $\tilde{y}_i$ values, $i = 1, \ldots, N$, is

$$\Re_A = 1 - \frac{\delta_A}{\mu_\delta} = 1 - \frac{19.4444}{27.00} = +0.2798 \, ,$$

indicating approximately 28 % agreement between the observed and predicted y values above that expected by chance.

OLS Regression

For comparison, consider an MRPP analysis of OLS regression residuals calculated on the response measurement summations for Factor A in Fig. 4.36. The MRPP regression analysis yields an estimated OLS regression coefficient of $\hat{\beta}_0 = +100.50$ for Factor A. Figure 4.38 lists the observed y_i values, OLS predicted $\hat{y}_i$ values, and residual e_i values for $i = 1, \ldots, 12$.

Following Eq. (4.5) on p. 125 and employing squared Euclidean distance between residuals with $v = 2$, the $N = 12$ OLS regression residuals listed in Fig. 4.38 yield $a = 3$ average distance-function values of

$$\xi_{A_1} = 493.3333 \, , \quad \xi_{A_2} = 909.3333 \, , \quad \text{and} \quad \xi_{A_3} = 179.3333 \, .$$

Following Eq. (4.4) on p. 125, the observed value of the MRPP test statistic calculated on the OLS regression residuals listed in Fig. 4.38 with $v = 2$ and

Fig. 4.38 Observed, predicted, and residual OLS regression values for the response measurement summations listed in Fig. 4.36

Object	y_i	$\hat{y}_i$	e_i
1	139	100.50	+38.50
2	101	100.50	+0.50
3	123	100.50	+22.50
4	117	100.50	+16.50
5	105	100.50	+4.50
6	85	100.50	−15.50
7	63	100.50	−37.50
8	59	100.50	−41.50
9	109	100.50	+8.50
10	95	100.50	−5.50
11	114	100.50	+13.50
12	96	100.50	−4.50

treatment-group weights

$$C_i = \frac{n_{A_i} - 1}{N - a} , \qquad i = 1, 2, 3 ,$$

is

$$\delta_A = \sum_{i=1}^{a} C_i \xi_i = \frac{4-1}{12-3}(493.3333 + 909.3333 + 179.3333) = 527.3333 .$$

If all arrangements of the $N = 12$ observed OLS regression residuals listed in Fig. 4.38 occur with equal chance, the exact probability value of $\delta_A = 527.3333$ computed on the $M = 34{,}650$ possible arrangements of the observed OLS regression residuals with $n_{A_1} = n_{A_2} = n_{A_3} = 4$ residuals preserved for each arrangement is

$$P(\delta \leq \delta_A | H_0) = \frac{\text{number of } \delta \text{ values} \leq \delta_A}{M} = \frac{564}{34{,}650} = 0.0163 .$$

For comparison, the exact probability value based on LAD regression, $v = 1$, $M = 34{,}650$, and $C_i = n_{A_i}/N$ for $i = 1, 2, 3$ is $P = 0.0194$.

Following Eq. (4.7) on p. 126, the exact expected value of the $M = 34{,}650$ δ values is $\mu_\delta = 1{,}082.7273$ and, following Eq. (4.6) on p. 126, the observed chance-corrected measure of effect size for the y_i and $\hat{y}_i$ values, $i = 1, \ldots, N$, is

$$\Re_A = 1 - \frac{\delta_A}{\mu_\delta} = 1 - \frac{527.3333}{1{,}082.7273} = +0.5130 ,$$

indicating approximately 51 % agreement between the observed and predicted y values above that expected by chance.

Conventional ANOVA Analysis

A conventional split-plot analysis of variance calculated on the $N = 12$ univariate response measurement scores listed in Fig. 4.35 on p. 180 yields an observed F-ratio of $F_A = 6.7927$. Assuming independence, normality, and homogeneity of variance, F_A is approximately distributed as Snedecor's F under the null hypothesis with $\nu_1 = a - 1 = 3 - 1 = 2$ and $\nu_2 = a(n-1) = 3(4-1) = 9$ degrees of freedom. Under the null hypothesis, the observed value of $F_A = 6.7927$ yields an approximate probability value of $P = 0.0159$.

Analysis of Factor *B*

A design matrix of effect codes for an MRPP regression analysis of Factor B is given in Table 4.9, where the first column of 1 values provides for an intercept, the next 11 columns contain effect codes for Subjects nested within Factor A, and the next four columns contain effect codes for the $A \times B$ interaction. The last column lists the $N =$

Table 4.9 Design matrix and univariate response measurement scores for the main effects of Factor B

Matrix																Score
1	1	0	0	0	0	0	0	0	0	0	0	1	0	0	0	53
1	0	1	0	0	0	0	0	0	0	0	0	1	0	0	0	49
1	0	0	1	0	0	0	0	0	0	0	0	1	0	0	0	47
1	0	0	0	1	0	0	0	0	0	0	0	1	0	0	0	42
1	0	0	0	0	1	0	0	0	0	0	0	0	1	0	0	47
1	0	0	0	0	0	1	0	0	0	0	0	0	1	0	0	42
1	0	0	0	0	0	0	1	0	0	0	0	0	1	0	0	39
1	0	0	0	0	0	0	0	1	0	0	0	0	1	0	0	37
1	0	0	0	0	0	0	0	0	1	0	0	−1	−1	0	0	45
1	0	0	0	0	0	0	0	0	0	1	0	−1	−1	0	0	41
1	0	0	0	0	0	0	0	0	0	0	1	−1	−1	0	0	38
1	−1	−1	−1	−1	−1	−1	−1	−1	−1	−1	−1	−1	−1	0	0	36
1	1	0	0	0	0	0	0	0	0	0	0	0	0	1	0	51
1	0	1	0	0	0	0	0	0	0	0	0	0	0	1	0	34
1	0	0	1	0	0	0	0	0	0	0	0	0	0	1	0	44
1	0	0	0	1	0	0	0	0	0	0	0	0	0	1	0	48
1	0	0	0	0	1	0	0	0	0	0	0	0	0	0	1	42
1	0	0	0	0	0	1	0	0	0	0	0	0	0	0	1	33
1	0	0	0	0	0	0	1	0	0	0	0	0	0	0	1	13
1	0	0	0	0	0	0	0	1	0	0	0	0	0	0	1	16
1	0	0	0	0	0	0	0	0	1	0	0	0	0	−1	−1	35
1	0	0	0	0	0	0	0	0	0	1	0	0	0	−1	−1	33
1	0	0	0	0	0	0	0	0	0	0	1	0	0	−1	−1	46
1	−1	−1	−1	−1	−1	−1	−1	−1	−1	−1	−1	0	0	−1	−1	40
1	1	0	0	0	0	0	0	0	0	0	0	−1	0	−1	0	35
1	0	1	0	0	0	0	0	0	0	0	0	−1	0	−1	0	18
1	0	0	1	0	0	0	0	0	0	0	0	−1	0	−1	0	32
1	0	0	0	1	0	0	0	0	0	0	0	−1	0	−1	0	27
1	0	0	0	0	1	0	0	0	0	0	0	0	−1	0	−1	16
1	0	0	0	0	0	1	0	0	0	0	0	0	−1	0	−1	10
1	0	0	0	0	0	0	1	0	0	0	0	0	−1	0	−1	11
1	0	0	0	0	0	0	0	1	0	0	0	0	−1	0	−1	6
1	0	0	0	0	0	0	0	0	1	0	0	1	1	1	1	29
1	0	0	0	0	0	0	0	0	0	1	0	1	1	1	1	21
1	0	0	0	0	0	0	0	0	0	0	1	1	1	1	1	30
1	−1	−1	−1	−1	−1	−1	−1	−1	−1	−1	−1	1	1	1	1	20

36 univariate response measurement scores ordered by the $b = 3$ treatment levels of Factor B, with the first $n_{B_1} = 12$ scores, the next $n_{B_2} = 12$ scores, and the last $n_{B_3} = 12$ scores associated with treatment levels B_1, B_2, and B_3, respectively. The

MRPP regression analysis examines the $N = 36$ regression residuals for possible differences among the $b = 3$ treatment levels of Factor B; consequently, no effect codes are provided for Factor B as this information is implicit in the ordering of the $b = 3$ treatment levels of Factor B in the last column of Table 4.9.

Because there are

$$M = \frac{N!}{\prod_{i=1}^{b} n_{B_i}!} = \frac{36!}{(12!)^3} = 3{,}384{,}731{,}762{,}521{,}200$$

possible, equally-likely arrangements of the $N = 36$ univariate response measurement scores listed in Table 4.9, an exact permutation approach is not possible.

LAD Regression Analysis

An MRPP resampling analysis of the LAD regression residuals calculated on the $N = 36$ univariate response measurement scores in Table 4.9 yields estimated LAD regression coefficients of

$$\tilde{\beta}_0 = +35.50\,, \quad \tilde{\beta}_1 = +9.8333\,, \quad \tilde{\beta}_2 = -7.1667\,, \quad \tilde{\beta}_3 = +2.8333\,,$$

$$\tilde{\beta}_4 = +5.8333\,, \quad \tilde{\beta}_5 = +4.8333\,, \quad \tilde{\beta}_6 = -0.1667\,, \quad \tilde{\beta}_7 = -20.1667\,,$$

$$\tilde{\beta}_8 = -17.1667\,, \quad \tilde{\beta}_9 = +2.8333\,, \quad \tilde{\beta}_{10} = +0.8333\,, \quad \tilde{\beta}_{11} = +9.8333\,,$$

$$\tilde{\beta}_{12} = +0.6667\,, \quad \tilde{\beta}_{13} = +6.6667\,, \quad \tilde{\beta}_{14} = +5.6667\,, \quad \text{and}$$

$$\tilde{\beta}_{15} = -2.3333$$

for Factor B. Figure 4.39 lists the observed y_i values, LAD predicted $\tilde{y}_i$ values, and residual e_i values for $i = 1, \ldots, 36$.

Following Eq. (4.5) on p. 125 and employing ordinary Euclidean distance between residuals with $v = 1$, the $N = 36$ LAD regression residuals listed in Fig. 4.39 yield $b = 3$ average distance-function values of

$$\xi_{B_1} = 8.6061\,, \quad \xi_{B_2} = 1.3182\,, \quad \text{and} \quad \xi_{B_3} = 13.5606\,.$$

Following Eq. (4.4) on p. 125, the observed value of the MRPP test statistic calculated on the LAD regression residuals listed in Fig. 4.39 with $v = 1$ and treatment-group weights

$$C_i = \frac{n_{B_i}}{N}\,, \qquad i = 1, 2, 3\,,$$

Fig. 4.39 Observed, predicted, and residual LAD regression values for the univariate response measurement scores listed in Table 4.9

Object	y_i	$\tilde{y}_i$	e_i
1	53	46.00	+7.00
2	49	29.00	+20.00
3	47	39.00	+8.00
4	42	42.00	0.00
5	47	47.00	0.00
6	42	42.00	0.00
7	39	22.00	+17.00
8	37	25.00	+12.00
9	45	31.00	+14.00
10	41	29.00	+12.00
11	38	38.00	0.00
12	36	36.00	0.00
13	51	51.00	0.00
14	34	34.00	0.00
15	44	44.00	0.00
16	48	47.00	+1.00
17	42	38.00	+4.00
18	33	33.00	0.00
19	13	13.00	0.00
20	16	16.00	0.00
21	35	35.00	0.00
22	33	33.00	0.00
23	46	42.00	+4.00
24	40	40.00	0.00
25	35	39.00	−4.00
26	18	22.00	−4.00
27	32	32.00	0.00
28	27	35.00	−8.00
29	16	36.00	−20.00
30	10	31.00	−21.00
31	11	11.00	0.00
32	6	14.00	−8.00
33	29	49.00	−20.00
34	21	47.00	−26.00
35	30	56.00	−26.00
36	20	54.00	−34.00

is

$$\delta_B = \sum_{i=1}^{b} C_i \xi_i = \frac{12}{36}(8.6061 + 1.3182 + 13.5606) = 7.8283 \,.$$

If all M possible arrangements of the $N = 36$ observed LAD regression residuals listed in Fig. 4.39 occur with equal chance, the approximate resampling probability value of $\delta_B = 7.8283$ computed on $L = 1{,}000{,}000$ random arrangements of the observed LAD regression residuals with $n_{B_1} = n_{B_2} = n_{B_3} = 12$ residuals preserved for each arrangement is

$$P(\delta \leq \delta_B | H_0) = \frac{\text{number of } \delta \text{ values} \leq \delta_B}{L} = \frac{0}{1{,}000{,}000} = 0.00 \,,$$

which may be interpreted as a probability of less than one in a million.

When M is very large and the probability of an observed δ is extremely small, as in this case, resampling permutation procedures sometimes result in zero probability, even with $L = 1{,}000{,}000$ random arrangements of the observed regression residuals. A reanalysis of Factor B using $L = 10{,}000{,}000$ random arrangements of the observed data yielded an identical resampling probability value of $P = 0.00$. Moment-approximation permutation procedures, described briefly in Chap. 1, Sect. 1.2.2, can often provide results in these extreme situations. The moment-approximation of a test statistic requires computation of the exact moments of the test statistic, assuming equally-likely arrangements of the observed regression residuals [284, 300]. Usually, the first three exact moments are used: the exact mean, μ_δ, the exact variance, σ_δ^2, and the exact skewness, γ_δ, of δ. The three moments are then used to fit a specified distribution, such as a Pearson type III distribution, that approximates the underlying discrete permutation distribution and provides an approximate probability value. For Factor B, a moment-approximation procedure yields $\delta_B = 7.8283$, $\mu_\delta = 12.5460$, $\sigma_\delta^2 = 0.1675$, $\gamma_\delta = -1.3580$, a standardized test statistic of

$$T_B = \frac{\delta_B - \mu_\delta}{\sigma_\delta} = \frac{7.8283 - 12.5460}{\sqrt{0.1675}} = -11.5272 \,,$$

and a Pearson type III approximate probability value of $P = 0.1495 \times 10^{-6}$.

Following Eq. (4.7) on p. 126, the exact expected value of the M δ values is $\mu_\delta = 12.5460$ and, following Eq. (4.6) on p. 126, the observed chance-corrected measure of effect size for the y_i and $\tilde{y}_i$ values, $i = 1, \ldots, N$, is

$$\Re_B = 1 - \frac{\delta_B}{\mu_\delta} = 1 - \frac{7.8283}{12.5460} = +0.3760 \,,$$

indicating approximately 38 % agreement between the observed and predicted y values above that expected by chance.

OLS Regression Analysis

For comparison, consider an MRPP analysis of OLS regression residuals calculated on the $N = 36$ response measurement summations for Factor B in Table 4.9

on p. 184. The MRPP regression analysis yields estimated OLS regression coefficients of

$$\hat{\beta}_0 = +33.50\,, \quad \hat{\beta}_1 = +12.8333\,, \quad \hat{\beta}_2 = +0.1667\,, \quad \hat{\beta}_3 = +7.50\,,$$

$$\hat{\beta}_4 = +5.50\,, \quad \hat{\beta}_5 = +1.50\,, \quad \hat{\beta}_6 = -5.1667\,, \quad \hat{\beta}_7 = -12.50\,,$$

$$\hat{\beta}_8 = -13.8333\,, \quad \hat{\beta}_9 = +2.8333\,, \quad \hat{\beta}_{10} = -1.8333\,, \quad \hat{\beta}_{11} = +4.50\,,$$

$$\hat{\beta}_{12} = -1.7500\,, \quad \hat{\beta}_{13} = +5.7500\,, \quad \hat{\beta}_{14} = +1.50\,, \quad \text{and}$$

$$\hat{\beta}_{15} = -2.7500$$

for Factor B. Figure 4.40 lists the observed y_i values, OLS predicted $\hat{y}_i$ values, and residual e_i values for $i = 1, \ldots, 36$.

Following Eq. (4.5) on p. 125 and employing squared Euclidean distance between residuals with $v = 2$, the $N = 36$ OLS regression residuals listed in Fig. 4.40 yield $b = 3$ average distance-function values of

$$\xi_{B_1} = 30.2727\,, \quad \xi_{B_2} = 46.4394\,, \quad \text{and} \quad \xi_{B_3} = 16.5606\,.$$

Following Eq. (4.4) on p. 125, the observed value of the MRPP test statistic calculated on the OLS regression residuals listed in Fig. 4.40 with $v = 2$ and treatment-group weights

$$C_i = \frac{n_{B_i} - 1}{N - b}\,, \qquad i = 1, 2, 3\,,$$

is

$$\delta_B = \sum_{i=1}^{b} C_i \xi_i = \frac{12 - 1}{36 - 3}(30.2727 + 46.4394 + 16.5606) = 31.0909\,.$$

If all M possible arrangements of the $N = 36$ observed OLS regression residuals listed in Fig. 4.40 occur with equal chance, the approximate resampling probability value of $\delta_B = 31.0909$ computed on $L = 1{,}000{,}000$ random arrangements of the observed OLS regression residuals with $n_{B_1} = n_{B_2} = n_{B_3} = 12$ residuals preserved for each arrangement is

$$P(\delta \leq \delta_B | H_0) = \frac{\text{number of } \delta \text{ values} \leq \delta_B}{L} = \frac{0}{1{,}000{,}000} = 0.00\,,$$

i.e., a probability of less than one in a million. For comparison, the approximate resampling probability value based on LAD regression, $v = 1$, $L = 1{,}000{,}000$, and $C_i = n_{B_i}/N$ for $i = 1, 2, 3$ is also $P = 0.00$.

As with the analysis of the LAD regression residuals listed in Fig. 4.39 on p. 186, when M is large and the probability of an observed δ is very small, an alternative

Fig. 4.40 Observed, predicted, and residual OLS regression values for the response measurement scores listed in Table 4.9

Object	y_i	$\hat{y}_i$	e_i
1	53	44.5833	+8.4167
2	49	31.9167	+17.0833
3	47	39.2500	+7.7500
4	42	37.2500	+4.7500
5	47	40.7500	+6.2500
6	42	34.0833	+7.9167
7	39	26.7500	+12.2500
8	37	25.4167	+11.5833
9	45	32.3333	+12.6667
10	41	27.6667	+13.3333
11	38	34.0000	+4.0000
12	36	28.0000	+8.0000
13	51	47.8333	+3.1667
14	34	35.1667	−1.1667
15	44	42.5000	+1.5000
16	48	40.5000	+7.5000
17	42	32.2500	+9.7500
18	33	25.5833	+7.4167
19	13	18.2500	−5.2500
20	16	16.9167	−0.9167
21	35	37.5833	−2.5833
22	33	32.9167	+0.0833
23	46	39.2500	+6.7500
24	40	33.2500	+6.7500
25	35	46.5833	−11.5833
26	18	33.9167	−15.9167
27	32	41.2500	−9.2500
28	27	39.2500	−12.2500
29	16	32.0000	−16.0000
30	10	25.3333	−15.3333
31	11	18.0000	−7.0000
32	6	16.6667	−10.6667
33	29	39.0833	−10.0833
34	21	34.4167	−13.4167
35	30	40.7500	−10.7500
36	20	34.7500	−14.7500

moment procedure based on the exact mean, μ_δ, exact variance, σ_δ^2, and exact skewness, γ_δ, of δ can be employed to obtain approximate probability values; see Chap. 1, Sect. 1.2.2. For Factor B, a moment-approximation procedure yields $\delta_B = 31.0909$, $\mu_\delta = 199.2857$, $\sigma_\delta^2 = 134.8578$, $\gamma_\delta = -1.7697$, a standardized test statistic of

$$T_B = \frac{\delta_B - \mu_\delta}{\sigma_\delta} = \frac{31.0909 - 199.2857}{\sqrt{134.8578}} = -14.4835 ,$$

and a Pearson type III approximate probability value of $P = 0.5420\times10^{-7}$.

Following Eq. (4.7) on p. 126, the exact expected value of the M δ values is $\mu_\delta = 199.2857$ and, following Eq. (4.6) on p. 126, the observed chance-corrected measure of effect size for the y_i and $\hat{y}_i$ values, $i = 1, \ldots, N$, is

$$\Re_B = 1 - \frac{\delta_B}{\mu_\delta} = 1 - \frac{31.0909}{199.2857} = +0.8440 ,$$

indicating approximately 84 % agreement between the observed and predicted y values above that expected by chance.

Conventional ANOVA Analysis

A conventional split-plot analysis of variance calculated on the $N = 36$ univariate response measurement scores listed in Fig. 4.35 on p. 180 yields an observed F-ratio of $F_B = 52.1842$. Assuming independence and normality, F_B is approximately distributed as Snedecor's F under the null hypothesis with $\nu_1 = b - 1 = 3 - 1 = 2$ and $\nu_2 = a(n-1)(b-1) = 3(4-1)(3-1) = 18$ degrees of freedom. Under the null hypothesis, the observed value of $F_B = 52.1842$ yields an approximate probability value of $P = 0.3224 \times 10^{-7}$.

Analysis of the $A \times B$ Interaction

A design matrix of effect codes for an MRPP regression analysis of the $A \times B$ interaction is given in Table 4.10, where the first column of 1 values provides for an intercept, the next 11 columns contain effect codes for Subjects nested within Factor A, and the next two columns contain effect codes for Factor B. The last column lists the $N = 36$ univariate response measurement scores ordered by the $ab = (3)(3) = 9$ levels of the $A \times B$ interaction. The MRPP regression analysis examines the $N = 36$ regression residuals for possible differences among the nine treatment levels of the $A \times B$ interaction; consequently, no effect codes are provided for the $A \times B$ interaction as this information is implicit in the ordering of the treatment levels of the $A \times B$ interaction in the last column of Table 4.10.

Because there are

$$M = \frac{N!}{\displaystyle\prod_{i=1}^{ab} n_{(A \times B)_i}!} = \frac{36!}{(4!)^9} = 140{,}810{,}154{,}080{,}474{,}667{,}338{,}550{,}000{,}000$$

possible, equally-likely arrangements of the $N = 36$ univariate response measurement scores listed in Table 4.10, an exact permutation approach is not possible.

Table 4.10 Design matrix and univariate response measurement scores for the interaction effects of Factors *A* and *B*

Matrix														Score
1	1	0	0	0	0	0	0	0	0	0	0	1	0	53
1	0	1	0	0	0	0	0	0	0	0	0	1	0	49
1	0	0	1	0	0	0	0	0	0	0	0	1	0	47
1	0	0	0	1	0	0	0	0	0	0	0	1	0	42
1	0	0	0	0	1	0	0	0	0	0	0	1	0	47
1	0	0	0	0	0	1	0	0	0	0	0	1	0	42
1	0	0	0	0	0	0	1	0	0	0	0	1	0	39
1	0	0	0	0	0	0	0	1	0	0	0	1	0	37
1	0	0	0	0	0	0	0	0	1	0	0	1	0	45
1	0	0	0	0	0	0	0	0	0	1	0	1	0	41
1	0	0	0	0	0	0	0	0	0	0	1	1	0	38
1	−1	−1	−1	−1	−1	−1	−1	−1	−1	−1	−1	1	0	36
1	1	0	0	0	0	0	0	0	0	0	0	0	1	51
1	0	1	0	0	0	0	0	0	0	0	0	0	1	34
1	0	0	1	0	0	0	0	0	0	0	0	0	1	44
1	0	0	0	1	0	0	0	0	0	0	0	0	1	48
1	0	0	0	0	1	0	0	0	0	0	0	0	1	42
1	0	0	0	0	0	1	0	0	0	0	0	0	1	33
1	0	0	0	0	0	0	1	0	0	0	0	0	1	13
1	0	0	0	0	0	0	0	1	0	0	0	0	1	16
1	0	0	0	0	0	0	0	0	1	0	0	0	1	35
1	0	0	0	0	0	0	0	0	0	1	0	0	1	33
1	0	0	0	0	0	0	0	0	0	0	1	0	1	46
1	−1	−1	−1	−1	−1	−1	−1	−1	−1	−1	−1	0	1	40
1	1	0	0	0	0	0	0	0	0	0	0	−1	−1	35
1	0	1	0	0	0	0	0	0	0	0	0	−1	−1	18
1	0	0	1	0	0	0	0	0	0	0	0	−1	−1	32
1	0	0	0	1	0	0	0	0	0	0	0	−1	−1	27
1	0	0	0	0	1	0	0	0	0	0	0	−1	−1	16
1	0	0	0	0	0	1	0	0	0	0	0	−1	−1	10
1	0	0	0	0	0	0	1	0	0	0	0	−1	−1	11
1	0	0	0	0	0	0	0	1	0	0	0	−1	−1	6
1	0	0	0	0	0	0	0	0	1	0	0	−1	−1	29
1	0	0	0	0	0	0	0	0	0	1	0	−1	−1	21
1	0	0	0	0	0	0	0	0	0	0	1	−1	−1	30
1	−1	−1	−1	−1	−1	−1	−1	−1	−1	−1	−1	−1	−1	20

LAD Regression Analysis

An MRPP resampling analysis of the $N = 36$ LAD regression residuals calculated on the univariate response measurement scores in Table 4.10 yields estimated LAD regression coefficients of

$$\tilde{\beta}_0 = +34.00\,, \quad \tilde{\beta}_1 = +12.6667\,, \quad \tilde{\beta}_2 = -3.3333\,, \quad \tilde{\beta}_3 = +6.6667\,,$$
$$\tilde{\beta}_4 = +4.6667\,, \quad \tilde{\beta}_5 = +4.6667\,, \quad \tilde{\beta}_6 = -4.3333\,, \quad \tilde{\beta}_7 = -11.3333\,,$$
$$\tilde{\beta}_8 = -16.3333\,, \quad \tilde{\beta}_9 = +2.6667\,, \quad \tilde{\beta}_{10} = -1.3333\,, \quad \tilde{\beta}_{11} = +7.6667\,,$$
$$\tilde{\beta}_{12} = +8.3333\,, \quad \text{and} \quad \tilde{\beta}_{13} = +3.3333$$

for the interaction of Factors A and B. Figure 4.41 lists the observed y_i values, LAD predicted $\tilde{y}_i$ values, and residual e_i values for $i = 1, \ldots, 36$.

Following Eq. (4.5) on p. 125 and employing ordinary Euclidean distance between residuals with $v = 1$, the $N = 36$ LAD regression residuals listed in Fig. 4.41 yield $ab = (3)(3) = 9$ average distance-function values of

$$\xi_{(A\times B)_1} = 7.50\,, \quad \xi_{(A\times B)_2} = 6.1667\,, \quad \xi_{(A\times B)_3} = 6.6667\,,$$
$$\xi_{(A\times B)_4} = 3.1667\,, \quad \xi_{(A\times B)_5} = 7.3333\,, \quad \xi_{(A\times B)_6} = 5.6667\,,$$
$$\xi_{(A\times B)_7} = 2.00\,, \quad \xi_{(A\times B)_8} = 6.8333\,, \quad \text{and} \quad \xi_{(A\times B)_9} = 2.00\,.$$

Following Eq. (4.4) on p. 125, the observed value of the MRPP test statistic calculated on the LAD regression residuals listed in Fig. 4.41 with $v = 1$ and treatment-group weights

$$C_i = \frac{n_{(A\times B)_i}}{N}\,, \qquad i = 1, \ldots, 9\,,$$

is

$$\delta_{A\times B} = \sum_{i=1}^{ab} C_i \xi_i = \frac{4}{36}(7.50 + 6.1667 + 6.6667$$
$$+ \cdots + 6.8333 + 2.00) = 5.2593\,.$$

If all M possible arrangements of the $N = 36$ observed LAD regression residuals listed in Fig. 4.41 occur with equal chance, the approximate resampling probability value of $\delta_{A\times B} = 5.2593$ computed on $L = 1{,}000{,}000$ random arrangements of the observed LAD regression residuals with $n_{(A\times B)_1} = \cdots = n_{(A\times B)_9} = 4$ residuals preserved for each arrangement is

$$P(\delta \leq \delta_{A\times B}|H_0) = \frac{\text{number of } \delta \text{ values} \leq \delta_{A\times B}}{L} = \frac{140{,}219}{1{,}000{,}000} = 0.1402\,.$$

Following Eq. (4.7) on p. 126, the exact expected value of the M δ values is $\mu_\delta = 5.6825$ and, following Eq. (4.6) on p. 126, the observed chance-corrected measure

Fig. 4.41 Observed, predicted, and residual LAD regression values for the univariate response measurement scores listed in Table 4.10

Object	y_i	$\tilde{y}_i$	e_i
1	53	55.00	−2.00
2	49	39.00	+10.00
3	47	49.00	−2.00
4	42	47.00	−5.00
5	47	47.00	0.00
6	42	38.00	+4.00
7	39	31.00	+8.00
8	37	26.00	+11.00
9	45	45.00	0.00
10	41	41.00	0.00
11	38	50.00	−12.00
12	36	40.00	−4.00
13	51	50.00	+1.00
14	34	34.00	0.00
15	44	44.00	0.00
16	48	42.00	+6.00
17	42	42.00	0.00
18	33	33.00	0.00
19	13	26.00	−13.00
20	16	21.00	−5.00
21	35	40.00	−5.00
22	33	36.00	−3.00
23	46	45.00	+1.00
24	40	35.00	+5.00
25	35	35.00	0.00
26	18	19.00	−1.00
27	32	29.00	+3.00
28	27	27.00	0.00
29	16	27.00	−11.00
30	10	18.00	−8.00
31	11	11.00	0.00
32	6	6.00	0.00
33	29	25.00	+4.00
34	21	21.00	0.00
35	30	30.00	0.00
36	20	20.00	0.00

of effect size for the y_i and $\tilde{y}_i$ values, $i = 1, \ldots, N$, is

$$\Re_{A\times B} = 1 - \frac{\delta_{A\times B}}{\mu_\delta} = 1 - \frac{5.2593}{5.6825} = +0.0745 ,$$

indicating approximately 7 % agreement between the observed and predicted y values above that expected by chance.

OLS Regression Analysis

For comparison, consider an MRPP analysis of OLS regression residuals calculated on the $N = 36$ response measurement scores for the $A{\times}B$ interaction in Table 4.10. The MRPP regression analysis yields estimated OLS regression coefficients of

$$\hat{\beta}_0 = +33.50\,, \quad \hat{\beta}_1 = +12.8333\,, \quad \hat{\beta}_2 = +0.1667\,, \quad \hat{\beta}_3 = +7.50\,,$$

$$\hat{\beta}_4 = +5.50\,, \quad \hat{\beta}_5 = +1.50\,, \quad \hat{\beta}_6 = -5.1667\,, \quad \hat{\beta}_7 = -12.50\,,$$

$$\hat{\beta}_8 = -13.8333\,, \quad \hat{\beta}_9 = +2.8333\,, \quad \hat{\beta}_{10} = -1.8333\,, \quad \hat{\beta}_{11} = +4.50\,,$$

$$\hat{\beta}_{12} = +9.50\,, \quad \text{and} \quad \hat{\beta}_{13} = +2.7500$$

for the interaction of Factors A and B. Figure 4.42 lists the observed y_i values, OLS predicted $\hat{y}_i$ values, and residual e_i values for $i = 1, \ldots, 36$.

Following Eq. (4.5) on p. 125 and employing squared Euclidean distance between residuals with $v = 2$, the $N = 36$ OLS regression residuals listed in Fig. 4.42 yield $ab = (3)(3) = 9$ average distance-function values of

$$\xi_{(A\times B)_1} = 56.2037\,, \quad \xi_{(A\times B)_2} = 16.6481\,, \quad \xi_{(A\times B)_3} = 38.1481\,,$$

$$\xi_{(A\times B)_4} = 26.4259\,, \quad \xi_{(A\times B)_5} = 98.8148\,, \quad \xi_{(A\times B)_6} = 45.0370\,,$$

$$\xi_{(A\times B)_7} = 15.2593\,, \quad \xi_{(A\times B)_8} = 35.7593\,, \quad \text{and} \quad \xi_{(A\times B)_9} = 9.7037\,.$$

Following Eq. (4.4) on p. 125, the observed value of the MRPP test statistic calculated on the OLS regression residuals listed in Fig. 4.42 with $v = 2$ and treatment-group weights

$$C_i = \frac{n_{(A\times B)_i} - 1}{N - ab}\,, \qquad i = 1, \ldots, 9\,,$$

is

$$\delta_{A\times B} = \sum_{i=1}^{ab} C_i \xi_i = \frac{4-1}{36-9}\big(56.2037 + 16.6481 + 38.1481 + \cdots + 35.7593 + 9.7037\big) = 38.00\,.$$

If all M possible arrangements of the $N = 36$ observed OLS regression residuals listed in Fig. 4.42 occur with equal chance, the approximate resampling probability value of $\delta_{A\times B} = 38.00$ calculated on $L = 1{,}000{,}000$ random arrangements of

Fig. 4.42 Observed, predicted, and residual OLS regression values for the univariate response measurement scores listed in Table 4.10

Object	y_i	$\hat{y}_i$	e_i
1	53	55.8333	−2.8333
2	49	43.1667	+5.8333
3	47	50.5000	−3.5000
4	42	48.5000	−6.5000
5	47	44.5000	+2.5000
6	42	37.8333	+4.1667
7	39	30.5000	+8.5000
8	37	29.1667	+7.8333
9	45	45.8333	−0.8333
10	41	41.1667	−0.1667
11	38	47.5000	−9.5000
12	36	41.5000	−5.5000
13	51	49.0833	+1.9167
14	34	36.4167	−2.4167
15	44	43.7500	+0.2500
16	48	41.7500	+6.2500
17	42	37.7500	+4.2500
18	33	31.0833	+1.9167
19	13	23.7500	−10.7500
20	16	22.4167	−6.4167
21	35	39.0833	−4.0833
22	33	34.4167	−1.4167
23	46	40.7500	+5.2500
24	40	34.7500	+5.2500
25	35	34.0833	+0.9167
26	18	21.4167	−3.4167
27	32	28.7500	+3.2500
28	27	26.7500	+0.2500
29	16	22.7500	−6.7500
30	10	16.0833	−6.0833
31	11	8.7500	+2.2500
32	6	7.4167	−1.4167
33	29	24.0833	+4.9167
34	21	19.4167	+1.5833
35	30	25.7500	+4.2500
36	20	19.7500	+0.2500

the observed OLS regression residuals with $n_{(A\times B)_1} = \cdots = n_{(A\times B)_9} = 4$ residuals preserved for each arrangement is

$$P(\delta \le \delta_{A\times B}|H_0) = \frac{\text{number of } \delta \text{ values} \le \delta_{A\times B}}{L} = \frac{72{,}276}{1{,}000{,}000} = 0.0723\ .$$

For comparison, the approximate resampling probability value based on LAD regression, $v = 1$, $L = 1{,}000{,}000$, and $C_i = n_{(A\times B)_i}/N$ for $i = 1, \ldots, 9$ is $P = 0.1402$.

Following Eq. (4.7) on p. 126, the exact expected value of the M δ values is $\mu_\delta = 47.6286$ and, following Eq. (4.6) on p. 126, the observed chance-corrected measure of effect size for the y_i and $\hat{y}_i$ values, $i = 1, \ldots, N$, is

$$\Re_{A\times B} = 1 - \frac{\delta_{A\times B}}{\mu_\delta} = 1 - \frac{38.00}{47.6286} = +0.2022 ,$$

indicating approximately 20 % agreement between the observed and predicted y values above that expected by chance.

Conventional ANOVA Analysis

A conventional split-plot analysis of variance calculated on the $N = 36$ univariate response measurement scores listed in Fig. 4.35 on p. 180 yields an observed F-ratio of $F_{A\times B} = 2.8114$. Assuming independence, normality, and homogeneity of variance, $F_{A\times B}$ is approximately distributed as Snedecor's F under the null hypothesis with $\nu_1 = (a-1)(b-1) = (3-1)(3-1) = 4$ and $\nu_2 = a(n-1)(b-1) = 3(4-1)(3-1) = 18$ degrees of freedom. Under the null hypothesis, the observed value of $F_{A\times B} = 2.8114$ yields an approximate probability value of $P = 0.0565$, which is similar to the probability value of $P = 0.0723$ obtained with the OLS regression analysis.

4.3.8 Nested Design

It is sometimes necessary to compare treatment groups when one independent variable is nested under a second independent variable. Two-factor nested analysis-of-variance designs occur whenever one factor is not completely crossed with the second factor. Consider a nested design to compare $a = 3$ levels of Factor A on scores obtained from $b = 3$ levels of Factor B, with B_1, B_2, and B_3 of Factor B in level A_1; B_4, B_5, and B_6 of Factor B in level A_2, and B_7, B_8, and B_9 of Factor B in level A_3. Thus, Factor B is said to be nested under Factor A. The univariate data for this example are listed in Table 4.11 for a sample of $n = 4$ objects randomly chosen from each of the $ab = (3)(3) = 9$ levels of Factors A and B.

Table 4.11 Example univariate response measurement scores for a nested design with $b = 3$ levels of Factor B nested under $a = 3$ levels of Factor A

A_1			A_2			A_3		
B_1	B_2	B_3	B_4	B_5	B_6	B_7	B_8	B_9
29	30	28	27	33	30	31	27	35
31	32	30	29	35	32	33	29	37
31	32	30	29	35	36	33	29	37
33	34	32	31	37	30	35	31	39

Table 4.12 Design matrix and univariate response measurement scores for an analysis of Factor *A* with Factor *B* nested under Factor *A*

Level A_1				Level A_2				Level A_3			
Matrix			Score	Matrix			Score	Matrix			Score
1	1	0	29	1	0	1	27	1	−1	−1	31
1	1	0	31	1	0	1	29	1	−1	−1	33
1	1	0	31	1	0	1	29	1	−1	−1	33
1	1	0	33	1	0	1	31	1	−1	−1	35
1	1	0	30	1	0	1	33	1	−1	−1	27
1	1	0	32	1	0	1	35	1	−1	−1	29
1	1	0	32	1	0	1	35	1	−1	−1	29
1	1	0	34	1	0	1	37	1	−1	−1	31
1	1	0	28	1	0	1	30	1	−1	−1	35
1	1	0	30	1	0	1	32	1	−1	−1	37
1	1	0	30	1	0	1	32	1	−1	−1	37
1	1	0	32	1	0	1	34	1	−1	−1	39

Analysis of Factor *A*

A design matrix of effect codes for an MRPP regression analysis of Factor *A* is given in Table 4.12, where the first column of 1 values provides for an intercept, the next two columns contain the effect codes for Factor *B*, and the third column contains the univariate response measurement scores listed according to the original random assignment of the $n = 36$ objects to the $a = 3$ levels of Factor *A* with the first $n_{A_1} = 12$ scores, the next $n_{A_2} = 12$ scores, and the last $n_{A_3} = 12$ scores associated with the $a = 3$ levels of Factor *A*, respectively. The MRPP regression analysis examines the $N = 36$ regression residuals for possible differences among the $a = 3$ treatment levels of Factor *A*; consequently, no effect codes are provided for Factor *A* as this information is implicit in the ordering of the $a = 3$ levels of Factor *A* in the rightmost columns of Table 4.12.

Because there are

$$M = \frac{N!}{\prod_{i=1}^{a} n_{A_i}!} = \frac{36!}{(12!)^3} = 3{,}384{,}731{,}762{,}521{,}200$$

possible, equally-likely arrangements of the $N = 36$ univariate response measurement scores listed in Table 4.11, an exact permutation approach is not possible.

LAD Regression Analysis

An MRPP resampling analysis of the $N = 36$ LAD regression residuals calculated on the univariate response measurement scores listed in Table 4.12 yields estimated LAD regression coefficients of

$$\tilde{\beta}_0 = +32.00\,, \quad \tilde{\beta}_1 = -1.00\,, \quad \text{and} \quad \tilde{\beta}_2 = 0.00$$

Fig. 4.43 Observed, predicted, and residual LAD regression values for the nested response measurement scores listed in Table 4.12

Object	y_i	$\tilde{y}_i$	e_i
1	29	31.00	−2.00
2	31	31.00	0.00
3	31	31.00	0.00
4	33	31.00	+2.00
5	30	31.00	−1.00
6	32	31.00	+1.00
7	32	31.00	+1.00
8	34	31.00	+3.00
9	28	31.00	−3.00
10	30	31.00	−1.00
11	30	31.00	−1.00
12	32	31.00	+1.00
13	27	32.00	−5.00
14	29	32.00	−3.00
15	29	32.00	−3.00
16	31	32.00	−1.00
17	33	32.00	+1.00
18	35	32.00	+3.00
19	35	32.00	+3.00
20	37	32.00	+5.00
21	30	32.00	−2.00
22	32	32.00	0.00
23	32	32.00	0.00
24	34	32.00	+2.00
25	31	33.00	−2.00
26	33	33.00	0.00
27	33	33.00	0.00
28	35	33.00	+2.00
29	27	33.00	−6.00
30	29	33.00	−4.00
31	29	33.00	−4.00
32	31	33.00	−2.00
33	35	33.00	+2.00
34	37	33.00	+4.00
35	37	33.00	+4.00
36	39	33.00	+6.00

for Factor *A*. Figure 4.43 lists the observed y_i values, LAD predicted $\tilde{y}_i$ values, and residual e_i values for $i = 1, \ldots, 36$.

Following Eq. (4.5) on p. 125 and employing ordinary Euclidean distance between residuals with $v = 1$, the $N = 36$ LAD regression residuals listed in Fig. 4.43 yield $a = 3$ average distance-function values of

$$\xi_{A_1} = 2.00\,, \quad \xi_{A_2} = 3.5152\,, \quad \text{and} \quad \xi_{A_3} = 4.4242\,.$$

Following Eq. (4.4) on p. 125, the observed value of the MRPP test statistic calculated on the LAD regression residuals listed in Fig. 4.43 with $v = 1$ and treatment-group weights

$$C_i = \frac{n_{A_i}}{N}\ , \qquad i = 1, 2, 3\ ,$$

is

$$\delta_A = \sum_{i=1}^{a} C_i \xi_i = \frac{12}{36}\big(2.00 + 3.5152 + 4.4242\big) = 3.3131\ .$$

If all M possible arrangements of the $N = 36$ observed LAD regression residuals listed in Fig. 4.43 occur with equal chance, the approximate resampling probability value of $\delta_A = 3.3131$ computed on $L = 1{,}000{,}000$ random arrangements of the observed LAD regression residuals with $n_{A_1} = n_{A_2} = n_{A_3} = 12$ residuals preserved for each arrangement is

$$P\big(\delta \le \delta_A | H_0\big) = \frac{\text{number of } \delta \text{ values} \le \delta_A}{L} = \frac{704{,}848}{1{,}000{,}000} = 0.7048\ .$$

Following Eq. (4.7) on p. 126, the exact expected value of the M δ values is $\mu_\delta = 3.2508$ and, following Eq. (4.6) on p. 126, the observed chance-corrected measure of effect size between the y_i and $\tilde{y}_i$ values, $i = 1, \ldots, N$, is

$$\Re_A = 1 - \frac{\delta_A}{\mu_\delta} = 1 - \frac{3.3131}{3.2508} = -0.0192\ ,$$

indicating slightly less than chance agreement between the observed and predicted y values.

OLS Regression Analysis

For comparison, consider an MRPP resampling analysis of OLS regression residuals calculated on the $N = 36$ univariate response measurement scores listed in Table 4.12. The MRPP regression analysis yields estimated OLS regression coefficients of

$$\hat{\beta}_0 = +32.00\ , \quad \hat{\beta}_1 = -1.00\ , \quad \text{and} \quad \hat{\beta}_2 = 0.00$$

for Factor A. Figure 4.44 lists the observed y_i values, OLS predicted $\hat{y}_i$ values, and residual e_i values for $i = 1, \ldots, 36$.[4]

Following Eq. (4.5) on p. 125 and employing squared Euclidean distance between residuals with $v = 2$, the $N = 36$ OLS regression residuals listed in

[4]Note that in the case of Factor A, LAD regression and OLS regression yield the same regression coefficients. Therefore, the observed regression residuals are the same for both analyses.

Fig. 4.44 Observed, predicted, and residual OLS regression values for the nested response measurement scores listed in Table 4.12

Object	y_i	$\hat{y}_i$	e_i
1	29	31.00	−2.00
2	31	31.00	0.00
3	31	31.00	0.00
4	33	31.00	+2.00
5	30	31.00	−1.00
6	32	31.00	+1.00
7	32	31.00	+1.00
8	34	31.00	+3.00
9	28	31.00	−3.00
10	30	31.00	−1.00
11	30	31.00	−1.00
12	32	31.00	+1.00
13	27	32.00	−5.00
14	29	32.00	−3.00
15	29	32.00	−3.00
16	31	32.00	−1.00
17	33	32.00	+1.00
18	35	32.00	+3.00
19	35	32.00	+3.00
20	37	32.00	+5.00
21	30	32.00	−2.00
22	32	32.00	0.00
23	32	32.00	0.00
24	34	32.00	+2.00
25	31	33.00	−2.00
26	33	33.00	0.00
27	33	33.00	0.00
28	35	33.00	+2.00
29	27	33.00	−6.00
30	29	33.00	−4.00
31	29	33.00	−4.00
32	31	33.00	−2.00
33	35	33.00	+2.00
34	37	33.00	+4.00
35	37	33.00	+4.00
36	39	33.00	+6.00

Fig. 4.44 yield $a = 3$ average distance-function values of

$$\xi_{A_1} = 5.8182 \;, \quad \xi_{A_2} = 17.4545 \;, \quad \text{and} \quad \xi_{A_3} = 27.6364 \;.$$

Following Eq. (4.4) on p. 125, the observed value of the MRPP test statistic calculated on the OLS regression residuals listed in Fig. 4.44 with $v = 2$ and treatment-group weights

$$C_i = \frac{n_{A_i} - 1}{N - a} \;, \qquad i = 1, 2, 3 \;,$$

is

$$\delta_A = \sum_{i=1}^{a} C_i \xi_i = \frac{12-1}{36-3}(5.8182 + 17.4545 + 27.6364) = 16.9697.$$

If all M possible arrangements of the $N = 36$ observed OLS regression residuals listed in Fig. 4.44 occur with equal chance, the approximate resampling probability value of $\delta_A = 16.9697$ computed on $L = 1{,}000{,}000$ random arrangements of the observed OLS regression residuals with $n_{A_1} = n_{A_2} = n_{A_3} = 12$ residuals preserved for each arrangement is

$$P(\delta \leq \delta_A | H_0) = \frac{\text{number of } \delta \text{ values} \leq \delta_A}{L} = \frac{1{,}000{,}000}{1{,}000{,}000} = 1.00 \ .$$

A reanalysis of the data based on $L = 10{,}000{,}000$ random arrangements of the $N = 36$ observed regression residuals listed in Fig. 4.44 with $n_{A_1} = n_{A_2} = n_{A_3} = 12$ residuals preserved for each arrangement also yields an approximate resampling probability value of $P = 1.00$.

A probability value of $P = 1.00$ is not very informative. In such cases, an alternative moment procedure based on the exact mean, μ_δ, exact variance, σ_δ^2, and exact skewness, γ_δ, of δ can be employed to obtain approximate probability values; see Chap. 1, Sect. 1.2.2. For Factor A, a moment-approximation procedure yields $\delta_A = 16.9697$, $\mu_\delta = 16.00$, $\sigma_\delta^2 = 0.8472$, $\gamma_\delta = -1.7012$, an observed standardized test statistic of

$$T_B = \frac{\delta_B - \mu_\delta}{\sigma_\delta} = \frac{16.9697 - 16.00}{\sqrt{0.8472}} = +0.0535 \ ,$$

and a Pearson type III approximate probability value of $P = 0.9487$.

For comparison, the approximate resampling probability value based on LAD regression, $v = 1$, $L = 1{,}000{,}000$, and $C_i = n_{A_i}/N$ for $i = 1, \ldots, a$ is $P = 0.7048$.

Following Eq. (4.7) on p. 126, the exact expected value of the M δ values is $\mu_\delta = 16.00$ and, following Eq. (4.6) on p. 126, the observed chance-corrected measure of effect size for the y_i and $\hat{y}_i$ values, $i = 1, \ldots, N$, is

$$\Re_A = 1 - \frac{\delta_A}{\mu_\delta} = 1 - \frac{16.9697}{16.00} = -0.0606 \ ,$$

indicating slightly less than chance agreement between the observed and predicted y values.

Conventional ANOVA Analysis

A conventional fixed-effects nested analysis of variance calculated on the $N = 36$ response measurement scores for Factor A listed in Table 4.11 on p. 196 yields an observed F-ratio of $F_A = 3.6818$. Assuming independence, normality, and homogeneity of variance, F_A is approximately distributed as Snedecor's F under the null

Fig. 4.45 Design matrix and univariate response measurement scores for an analysis of Factor B with Factor B nested under Factor A

Matrix							Score
1	1	0	1	0	0	0	29
1	1	0	1	0	0	0	31
1	1	0	1	0	0	0	31
1	1	0	1	0	0	0	33
1	1	0	0	0	1	0	30
1	1	0	0	0	1	0	32
1	1	0	0	0	1	0	32
1	1	0	0	0	1	0	34
1	1	0	−1	0	−1	0	28
1	1	0	−1	0	−1	0	30
1	1	0	−1	0	−1	0	30
1	1	0	−1	0	−1	0	32
1	0	1	0	1	0	0	27
1	0	1	0	1	0	0	29
1	0	1	0	1	0	0	29
1	0	1	0	1	0	0	31
1	0	1	0	0	0	1	33
1	0	1	0	0	0	1	35
1	0	1	0	0	0	1	35
1	0	1	0	0	0	1	37
1	0	1	0	−1	0	−1	30
1	0	1	0	−1	0	−1	32
1	0	1	0	−1	0	−1	36
1	0	1	0	−1	0	−1	30
1	−1	−1	−1	−1	0	0	31
1	−1	−1	−1	−1	0	0	33
1	−1	−1	−1	−1	0	0	33
1	−1	−1	−1	−1	0	0	35
1	−1	−1	0	0	−1	−1	27
1	−1	−1	0	0	−1	−1	29
1	−1	−1	0	0	−1	−1	29
1	−1	−1	0	0	−1	−1	31
1	−1	−1	1	1	1	1	35
1	−1	−1	1	1	1	1	37
1	−1	−1	1	1	1	1	37
1	−1	−1	1	1	1	1	39

hypothesis with $\nu_1 = a - 1 = 3 - 1 = 2$ and $\nu_2 = ab(n - 1) = (3)(3)(4 - 1) = 27$ degrees of freedom. Under the null hypothesis, the observed value of $F_A = 3.6818$ yields an approximate probability value of $P = 0.0386$.

Analysis of Factor $B|A$

A design matrix of effect codes for an MRPP regression analysis of Factor B, nested under Factor A, is given in Fig. 4.45, where the first column of 1 values provides for an intercept, the next two columns contain effect codes for Factor A, the next four columns contain effect codes for the $A{\times}B$ interaction, and the last column contains the univariate response measurement scores listed according to the $b = 3$ levels of

Factor B with the first $n_{B|A_1} = 12$ scores, the next $n_{B|A_2} = 12$ scores, and the last $n_{B|A_3} = 12$ scores associated with the $b = 3$ levels of Factor B, respectively. The MRPP regression analysis examines the $N = 36$ regression residuals for possible differences among the $b = 3$ treatment levels of Factor B; consequently, no effect codes are provided for Factor B as this information is implicit in the ordering of the $b = 3$ levels of Factor B in the last column of Fig. 4.45.

LAD Regression Analysis

Again, because there are

$$M = \frac{N!}{\prod_{i=1}^{b} n_{B|A_i}!} = \frac{36!}{(12!)^3} = 3{,}384{,}731{,}762{,}521{,}200$$

possible, equally-likely arrangements of the $N = 36$ response measurement scores listed in Fig. 4.45, an exact permutation approach is not possible. An MRPP resampling analysis of the $N = 36$ LAD regression residuals calculated on the univariate response measurement scores in Fig. 4.45 yields estimated LAD regression coefficients of

$$\tilde{\beta}_0 = +32.00\ , \quad \tilde{\beta}_1 = -1.00\ , \quad \tilde{\beta}_2 = 0.6667\ , \quad \tilde{\beta}_3 = +1.00\ ,$$
$$\tilde{\beta}_4 = -1.6667\ , \quad \tilde{\beta}_5 = +1.00\ , \quad \text{and} \quad \tilde{\beta}_6 = +2.3333$$

for Factor $B|A$. Figure 4.46 lists the observed y_i values, LAD predicted $\tilde{y}_i$ values, and residual e_i values for $i = 1, \ldots, 36$.

Following Eq. (4.5) on p. 125 and employing ordinary Euclidean distance between residuals with $v = 1$, the $N = 36$ LAD regression residuals listed in Fig. 4.46 yield $a = 3$ average distance-function values of

$$\xi_{B|A_1} = 2.00\ , \quad \xi_{B|A_2} = 2.00\ , \quad \text{and} \quad \xi_{B|A_3} = 2.00\ .$$

Following Eq. (4.4) on p. 125, the observed value of the MRPP test statistic calculated on the LAD regression residuals listed in Fig. 4.46 with $v = 1$ and treatment-group weights

$$C_i = \frac{n_{B|A_i}}{N}\ , \qquad i = 1, 2, 3\ ,$$

is

$$\delta_{B|A} = \sum_{i=1}^{b} C_i \xi_i = \frac{12}{36}(2.00 + 2.00 + 2.00) = 2.00\ .$$

Fig. 4.46 Observed, predicted, and residual LAD regression values for the nested response measurement scores listed in Table 4.12

Object	y_i	$\tilde{y}_i$	e_i
1	29	32.00	−3.00
2	31	32.00	−1.00
3	31	32.00	−1.00
4	33	32.00	+1.00
5	30	32.00	−2.00
6	32	32.00	0.00
7	32	32.00	0.00
8	34	32.00	+2.00
9	28	29.00	−1.00
10	30	29.00	+1.00
11	30	29.00	+1.00
12	32	29.00	+3.00
13	27	31.00	−4.00
14	29	31.00	−2.00
15	29	31.00	−2.00
16	31	31.00	0.00
17	33	35.00	−2.00
18	35	35.00	0.00
19	35	35.00	0.00
20	37	35.00	+2.00
21	30	32.00	−2.00
22	32	32.00	0.00
23	32	32.00	0.00
24	34	32.00	+2.00
25	31	33.00	−2.00
26	33	33.00	0.00
27	33	33.00	0.00
28	35	33.00	+2.00
29	27	29.00	−2.00
30	29	29.00	0.00
31	29	29.00	0.00
32	31	29.00	+2.00
33	35	35.00	0.00
34	37	35.00	+2.00
35	37	35.00	+2.00
36	39	35.00	+4.00

If all M possible arrangements of the $N = 36$ observed LAD regression residuals listed in Fig. 4.46 occur with equal chance, the approximate resampling probability value of $\delta_{B|A} = 2.00$ computed on $L = 1{,}000{,}000$ random arrangements of the observed LAD regression residuals with $n_{B|A_1} = n_{B|A_2} = n_{B|A_3} = 12$ residuals preserved for each arrangement is

$$P(\delta \leq \delta_{B|A}|H_0) = \frac{\text{number of } \delta \text{ values} \leq \delta_{B|A}}{L} = \frac{361{,}575}{1{,}000{,}000} = 0.3616\ .$$

Following Eq. (4.7) on p. 126, the exact expected value of the M δ values is $\mu_\delta = 2.0127$ and, following Eq. (4.6) on p. 126, the observed chance-corrected measure of effect size for the y_i and $\tilde{y}_i$ values, $i = 1, \ldots, N$, is

$$\Re_{B|A} = 1 - \frac{\delta_{B|A}}{\mu_\delta} = 1 - \frac{2.00}{2.0127} = +0.0063 ,$$

indicating approximately chance agreement between the observed and predicted y values.

OLS Regression Analysis

For comparison, consider an MRPP resampling analysis of OLS regression residuals calculated on the $N = 36$ response measurement scores listed in Table 4.12 on p. 197. The MRPP regression analysis yields estimated OLS regression coefficients of

$$\hat{\beta}_0 = +32.00 , \quad \hat{\beta}_1 = -1.00 , \quad \hat{\beta}_2 = 0.00 , \quad \hat{\beta}_3 = +1.00 ,$$
$$\hat{\beta}_4 = -2.00 , \quad \hat{\beta}_5 = +1.00 , \quad \text{and} \quad \hat{\beta}_6 = +3.00$$

for Factor $B|A$. Figure 4.47 lists the observed y_i values, OLS predicted $\hat{y}_i$ values, and residual e_i values for $i = 1, \ldots, 36$.

Following Eq. (4.5) on p. 125 and employing squared Euclidean distance between residuals with $v = 2$, the $N = 36$ OLS regression residuals listed in Fig. 4.47 yield $a = 3$ average distance-function values of

$$\xi_{B|A_1} = 5.8182 , \quad \xi_{B|A_2} = 5.8182 , \quad \text{and} \quad \xi_{B|A_3} = 5.8182 .$$

Following Eq. (4.4) on p. 125, the observed value of the MRPP test statistic calculated on the OLS regression residuals listed in Fig. 4.47 with $v = 2$ and treatment-group weights

$$C_i = \frac{n_{B|A_i} - 1}{N - b} , \qquad i = 1, 2, 3 ,$$

is

$$\delta_{B|A} = \sum_{i=1}^{b} C_i \xi_i = \frac{12 - 1}{36 - 3} \big(5.8182 + 5.8182 + 5.8182\big) = 5.8182 .$$

If all M possible arrangements of the $N = 36$ observed OLS regression residuals listed in Fig. 4.47 occur with equal chance, the approximate resampling probability value of $\delta_{B|A} = 5.8182$ computed on $L = 1{,}000{,}000$ random arrangements of the observed OLS regression residuals with $n_{B|A_1} = n_{B|A_2} = n_{B|A_3} = 12$ residuals pre-

Fig. 4.47 Observed, predicted, and residual OLS regression values for the nested response measurement scores listed in Table 4.12

Object	y_i	$\hat{y}_i$	e_i
1	29	32.00	−3.00
2	31	32.00	−1.00
3	31	32.00	−1.00
4	33	32.00	+1.00
5	30	32.00	−2.00
6	32	32.00	0.00
7	32	32.00	0.00
8	34	32.00	+2.00
9	28	29.00	−1.00
10	30	29.00	+1.00
11	30	29.00	+1.00
12	32	29.00	+3.00
13	27	30.00	−3.00
14	29	30.00	−1.00
15	29	30.00	−1.00
16	31	30.00	+1.00
17	33	35.00	−2.00
18	35	35.00	0.00
19	35	35.00	0.00
20	37	35.00	+2.00
21	30	31.00	−1.00
22	32	31.00	+1.00
23	32	31.00	+1.00
24	34	31.00	+3.00
25	31	34.00	−3.00
26	33	34.00	−1.00
27	33	34.00	−1.00
28	35	34.00	+1.00
29	27	29.00	−2.00
30	29	29.00	0.00
31	29	29.00	0.00
32	31	29.00	+2.00
33	35	36.00	−1.00
34	37	36.00	+1.00
35	37	36.00	+1.00
36	39	36.00	+3.00

served for each arrangement is

$$P(\delta \leq \delta_{B|A}|H_0) = \frac{\text{number of } \delta \text{ values} \leq \delta_{B|A}}{L} = \frac{7{,}600}{1{,}000{,}000} = 0.0076 \,.$$

For comparison, the approximate resampling probability value based on LAD regression, $v = 1$, $L = 1{,}000{,}000$, and $C_i = n_{B|A_i}/N$ for $i = 1, 2, 3$ is $P = 0.3616$.

Following Eq. (4.7) on p. 126, the exact expected value of the M δ values is $\mu_\delta = 5.4857$ and, following Eq. (4.6) on p. 126, the observed chance-corrected measure

of effect size for the y_i and $\hat{y}_i$ values, $i = 1, \ldots, N$, is

$$\mathfrak{R}_{B|A} = 1 - \frac{\delta_{B|A}}{\mu_\delta} = 1 - \frac{5.8182}{5.4857} = -0.0606 ,$$

indicating slightly less than chance agreement between the observed and predicted y values.

Conventional ANOVA Analysis

A conventional fixed-effects nested analysis of variance calculated on the $N = 36$ univariate response measurement scores for Factor $B|A$ listed in Table 4.11 on p. 196 yields an observed F-ratio of $F_{B|A} = 10.6362$. Assuming independence, normality, and homogeneity of variance, $F_{B|A}$ is approximately distributed as Snedecor's F under the null hypothesis with $\nu_1 = a(b-1) = 3(3-1) = 6$ and $\nu_2 = ab(n-1) = (3)(3)(4-1) = 27$ degrees of freedom. Under the null hypothesis, the observed value of $F_{B|A} = 10.6362$ yields an approximate probability value of $P = 4.5461 \times 10^{-6}$.

4.4 Multivariate Multiple Regression Designs

An extension of LAD multiple regression to include multiple dependent variables, as well as multiple independent variables, i.e., multivariate multiple LAD regression, is developed in this section. The extension was prompted by a multivariate Least Sum (of) Euclidean Distances (LSED) algorithm developed by Kaufman, Taylor, Mielke, and Berry in 2002 [198].

Consider the multivariate multiple regression model given by

$$y_{ik} = \sum_{j=1}^{m} x_{ij}\beta_{jk} + e_{ik}$$

for $i = 1, \ldots, N$ and $k = 1, \ldots, r$, where y_{ik} represents the ith of N measurements for the kth of r response variables, possibly affected by a treatment; x_{ij} is the jth of m covariates associated with the ith response, where $x_{i1} = 1$ if the model includes an intercept; β_{jk} denotes the jth of m regression parameters for the kth of r response variables; and e_{ik} designates the error associated with the ith of N measurements for the k of r response variables.

If estimates of β_{jk} that minimize

$$\sum_{i=1}^{N} \left(\sum_{k=1}^{r} e_{ik}^2 \right)^{1/2}$$

are denoted by $\tilde{\beta}_{jk}$ for $j = 1, \ldots, m$ and $k = 1, \ldots, r$, then the N r-dimensional residuals of the LSED multivariate multiple regression model are given by

$$e_{ik} = y_{ik} - \sum_{j=1}^{m} x_{ij}\tilde{\beta}_{jk}$$

for $i = 1, \ldots, N$ and $k = 1, \ldots, r$.

Let the N r-dimensional residuals, $(e_{i1}, \ldots, e_{ir})$ for $i = 1, \ldots, N$ obtained from an LSED multivariate multiple regression model, be partitioned into g treatment groups of sizes $n_1, \ldots, n_g$, where $n_i \geq 2$ for $i = 1, \ldots, g$ and

$$N = \sum_{i=1}^{g} n_i \, .$$

The MRPP analysis of the multivariate multiple regression residuals depends on statistic

$$\delta = \sum_{i=1}^{g} C_i \xi_i \, , \tag{4.8}$$

where $C_i = n_i/N$ is a positive weight for the ith of g treatment groups and ξ_i is the average pairwise Euclidean distance among the n_i r-dimensional residuals in the ith of g treatment groups defined by

$$\xi_i = \binom{n_i}{2}^{-1} \sum_{k=1}^{N-1} \sum_{l=k+1}^{N} \left[\sum_{j=1}^{r} \left(e_{kj} - e_{lj}\right)^2 \right]^{1/2} \Psi_{ki} \Psi_{li} \, , \tag{4.9}$$

where

$$\Psi_{ki} = \begin{cases} 1 & \text{if } (e_{k1}, \ldots, e_{kr}) \text{ is in the } i\text{th treatment group} \, , \\ 0 & \text{otherwise} \, . \end{cases}$$

The null hypothesis specifies that each of the

$$M = \frac{N!}{\prod_{i=1}^{g} n_i!}$$

possible allocations of the N r-dimensional residuals to the g treatment groups is equally likely. An exact MRPP probability value associated with the observed value of δ, δ_o, is given by

$$P\big(\delta \leq \delta_o | H_0\big) = \frac{\text{number of } \delta \text{ values} \leq \delta_o}{M} \, .$$

As previously, when M is large an approximate probability value may be obtained from a resampling permutation procedure. Let L denote a large random sample drawn from all M possible arrangements of the observed data, then an approximate resampling probability value is given by

$$P(\delta \leq \delta_o | H_0) = \frac{\text{number of } \delta \text{ values} \leq \delta_o}{L} .$$

As with univariate multiple regression models, the criterion for fitting multivariate multiple regression models based on δ is the chance-corrected measure of effect size between the observed and predicted response measurement values given by

$$\Re = 1 - \frac{\delta}{\mu_\delta} , \tag{4.10}$$

where μ_δ is the expected value of δ over the $N!$ possible pairings under the null hypothesis, given by

$$\mu_\delta = \frac{1}{M} \sum_{i=1}^{M} \delta_i . \tag{4.11}$$

4.4.1 Example Analysis

To illustrate a multivariate LSED multiple regression analysis, consider an unbalanced two-way randomized-block experimental design in which $N = 16$ subjects (S) are tested over $a = 3$ levels of Factor A, the experiment is repeated $b = 2$ times for Factor B, and there are $r = 2$ response measurement scores for each subject. The design and data are adapted from Mielke and Berry [297, p. 184] and are given in Fig. 4.48. The design is intentionally kept small to illustrate the multivariate multiple regression procedure.

Analysis of Factor A

A design matrix of dummy codes for an MRPP regression analysis of Factor A is given in Fig. 4.49, where the first column of 1 values provides for an intercept, the next column contains the dummy codes for Factor B, and the third and fourth columns contain the bivariate response measurement scores listed according to the original random assignment of the $N = 16$ subjects to the $a = 3$ levels of Factor A with the first $n_{A_1} = 5$ scores, the next $n_{A_2} = 7$ scores, and the last $n_{A_3} = 4$ scores associated with the $a = 3$ levels of Factor A, respectively. The MRPP regression analysis examines the $N = 16$ regression residuals for possible differences among the $a = 3$ treatment levels of Factor A; consequently, no dummy codes are provided

Fig. 4.48 Example data for a two-way randomized-block design with $a = 3$ blocks and $b = 2$ treatments

	Factor A		
Factor B	A_1	A_2	A_3
B_1	(49, 102)	(63, 84)	(45, 107)
		(60, 89)	(50, 100)
			(42, 111)
			(46, 104)
B_2	(48, 103)	(27, 114)	
	(58, 94)	(66, 83)	
	(51, 100)	(74, 79)	
	(55, 97)	(69, 88)	
		(71, 82)	

Fig. 4.49 Example design matrix and bivariate response measurement scores for multivariate LSED multiple regression analysis of Factor A with $N = 16$

Matrix		Scores	
1	1	49	102
1	0	48	103
1	0	58	94
1	0	51	100
1	0	55	97
1	1	63	84
1	1	60	89
1	0	27	114
1	0	66	83
1	0	74	79
1	0	69	88
1	0	71	82
1	1	45	107
1	1	50	100
1	1	42	111
1	1	46	104

for Factor A as this information is implicit in the ordering of the $a = 3$ levels of Factor A in the last two columns of Fig. 4.49.

Because there are only

$$M = \frac{N!}{\prod_{i=1}^{a} n_{A_i}!} = \frac{16!}{5!\ 7!\ 4!} = 1{,}441{,}440$$

possible, equally-likely arrangements of the $N = 16$ bivariate response measurement scores listed in Fig. 4.49, an exact permutation approach is feasible. An MRPP analysis of the $N = 16$ LAD regression residuals calculated on the bivariate response measurements for Factor A in Fig. 4.49 yields estimated LAD regression coefficients of

$$\tilde{\beta}_{1,1} = +58.00\ ,\ \tilde{\beta}_{2,1} = -9.00\ ,\ \tilde{\beta}_{1,2} = +94.00\ ,\ \text{and}\ \tilde{\beta}_{2,2} = +8.00$$

Fig. 4.50 Observed, predicted, and residual values for a multivariate LSED multiple regression analysis of Factor *A* with $N = 16$

y_{i1}	y_{i2}	$\tilde{y}_{i1}$	$\tilde{y}_{i2}$	e_{i1}	e_{i2}
49	102	49.00	102.00	0.00	0.00
48	103	58.00	94.00	−10.00	+9.00
58	94	58.00	94.00	0.00	0.00
51	100	58.00	94.00	−7.00	+6.00
55	97	58.00	94.00	−3.00	+3.00
63	84	49.00	102.00	+14.00	−18.00
60	89	49.00	102.00	+11.00	−13.00
27	114	58.00	94.00	−31.00	+20.00
66	83	58.00	94.00	+8.00	−11.00
74	79	58.00	94.00	+16.00	−15.00
69	88	58.00	94.00	+11.00	−6.00
71	82	58.00	94.00	+13.00	−12.00
45	107	49.00	102.00	−4.00	+5.00
50	100	49.00	102.00	+1.00	−2.00
42	111	49.00	102.00	−7.00	+9.00
46	104	49.00	102.00	−3.00	+2.00

for Factor *A*. Figure 4.50 lists the observed y_{ik} values, LAD predicted $\tilde{y}_{ik}$ values, and residual e_{ik} values for $i = 1, \ldots, 16$ and $k = 1, 2$.

Following Eq. (4.9) on p. 208 and employing ordinary Euclidean distance between residuals with $v = 1$, the $N = 16$ LAD regression residuals listed in Fig. 4.50 yield $a = 3$ average distance-function values of

$$\xi_{A_1} = 7.2294\,, \quad \xi_{A_2} = 20.0289\,, \quad \text{and} \quad \xi_{A_3} = 7.3475\,.$$

Following Eq. (4.8) on p. 208, the observed value of the MRPP test statistic calculated on the LAD regression residuals listed in Fig. 4.50 with $v = 1$ and treatment-group weights

$$C_i = \frac{n_{A_i}}{N}\,, \qquad i = 1, 2, 3\,,$$

is

$$\delta_A = \sum_{i=1}^{a} C_i \xi_i = \frac{1}{16}\big[(5)(7.2294) + (7)(20.0289) + (4)(7.3475)\big] = 12.8587\,.$$

If all arrangements of the $N = 16$ observed LAD regression residuals listed in Fig. 4.50 occur with equal chance, the exact probability value of $\delta_A = 12.8587$ computed on the $M = 1{,}441{,}440$ possible arrangements of the observed LAD regression residuals with $n_{A_1} = 5$, $n_{A_2} = 7$, and $n_{A_3} = 4$ residuals preserved for each arrangement is

$$P(\delta \le \delta_A | H_0) = \frac{\text{number of } \delta \text{ values} \le \delta_A}{M} = \frac{6{,}676}{1{,}441{,}440} = 0.0046\,.$$

Following Eq. (4.11) on p. 209, the exact expected value of the $M = 1{,}441{,}440$ δ values is $\mu_\delta = 18.1020$ and, following Eq. (4.10) on p. 209, the observed chance-corrected measure of effect size for the y_i and $\tilde{y}_i$ values, $i = 1, \ldots, N$, is

$$\Re_A = 1 - \frac{\delta_A}{\mu_\delta} = 1 - \frac{12.8587}{18.1020} = +0.2897 ,$$

indicating approximately 29 % agreement between the observed and predicted values above that expected by chance.

Analysis of Factor *B*

A design matrix of dummy codes for an MRPP regression analysis of Factor *B* is given in Fig. 4.51, where the first column of 1 values provides for an intercept, the next two columns contain the dummy codes for Factor *A*, and the fourth and fifth columns contain the bivariate response measurement scores listed according to the original random assignment of the $N = 16$ subjects to the $b = 2$ levels of Factor *B* with the first $n_{B_1} = 7$ scores and the last $n_{B_2} = 9$ scores associated with the $b = 2$ levels of Factor *B*, respectively. The MRPP regression analysis examines the $N = 16$ regression residuals for possible differences between the $b = 2$ treatment levels of Factor *B*; consequently, no dummy codes are provided for Factor *B* as this information is implicit in the ordering of the $b = 2$ levels of Factor *B* in the last two columns of Fig. 4.51.

Because there are only

$$M = \frac{N!}{\prod_{i=1}^{b} n_{B_i}!} = \frac{16!}{7!\ 9!} = 11{,}440$$

Fig. 4.51 Example design matrix and bivariate response measurement scores for a multivariate LSED multiple regression analysis of Factor *B* with $N = 16$

Matrix			Scores	
1	1	0	49	102
1	0	1	63	84
1	0	1	60	89
1	0	0	45	107
1	0	0	50	100
1	0	0	42	111
1	0	0	46	104
1	1	0	48	103
1	1	0	58	94
1	1	0	51	100
1	1	0	55	97
1	0	1	27	114
1	0	1	66	83
1	0	1	74	79
1	0	1	69	88
1	0	1	71	82

Fig. 4.52 Observed, predicted, and residual values for a multivariate LSED multiple regression analysis of Factor B with $N = 16$

y_{i1}	y_{i2}	$\tilde{y}_{i1}$	$\tilde{y}_{i2}$	e_{i1}	e_{i2}
49	102	51.00	100.00	−2.00	+2.00
63	84	66.00	84.00	−3.00	0.00
60	89	66.00	84.00	−6.00	+5.00
45	107	46.00	104.00	−1.00	+3.00
50	100	46.00	104.00	+4.00	−4.00
42	111	46.00	104.00	−4.00	+7.00
46	104	46.00	104.00	0.00	0.00
48	103	51.00	100.00	−3.00	+3.00
58	94	51.00	100.00	+7.00	−6.00
51	100	51.00	100.00	0.00	0.00
55	97	51.00	100.00	+4.00	−3.00
27	114	66.00	84.00	−39.00	+30.00
66	83	66.00	84.00	0.00	−1.00
74	79	66.00	84.00	−8.00	−5.00
69	88	66.00	84.00	+3.00	+4.00
71	82	66.00	84.00	+5.00	−2.00

possible, equally-likely arrangements of the $N = 16$ response measurement scores listed in Fig. 4.51, an exact permutation approach is feasible. An MRPP analysis of the $N = 16$ LAD regression residuals calculated on the bivariate response measurements for Factor B in Fig. 4.51 yields estimated LAD regression coefficients of

$$\tilde{\beta}_{1,1} = +46.00 \,, \quad \tilde{\beta}_{2,1} = +5.00 \,, \quad \tilde{\beta}_{3,1} = +20.00 \,, \quad \tilde{\beta}_{1,2} = +104.00 \,,$$

$$\tilde{\beta}_{2,2} = -4.00 \,, \quad \text{and} \quad \tilde{\beta}_{3,2} = -20.00$$

for Factor B. Figure 4.52 lists the observed y_{ik} values, LAD predicted $\tilde{y}_{ik}$ values, and residual e_{ik} values for $i = 1, \ldots, 16$ and $k = 1, 2$.

Following Eq. (4.9) on p. 208 and employing ordinary Euclidean distance between residuals with $v = 1$, the $N = 16$ LAD regression residuals listed in Fig. 4.52 yield $b = 2$ average distance-function values of

$$\xi_{B_1} = 6.0229 \quad \text{and} \quad \xi_{B_2} = 16.7440 \,.$$

Following Eq. (4.4) on p. 208, the observed value of the MRPP test statistic calculated on the LAD regression residuals listed in Fig. 4.52 with $v = 1$ and treatment-group weights

$$C_i = \frac{n_{B_i}}{N} \,, \qquad i = 1, 2 \,,$$

is

$$\delta_B = \sum_{i=1}^{b} C_i \xi_i = \frac{1}{16}\big[(7)(6.0229) + (9)(16.7440)\big] = 12.0535 \,.$$

If all arrangements of the $N = 16$ observed LAD regression residuals listed in Fig. 4.52 occur with equal chance, the exact probability value of $\delta_B = 12.0535$ computed on the $M = 11{,}440$ possible arrangements of the observed LAD regression residuals with $n_{B_1} = 7$ and $n_{B_2} = 9$ residuals preserved for each arrangement is

$$P(\delta \leq \delta_B | H_0) = \frac{\text{number of } \delta \text{ values} \leq \delta_B}{M} = \frac{2{,}090}{11{,}440} = 0.1827 \,.$$

Following Eq. (4.11) on p. 209, the exact expected value of the $M = 11{,}440$ δ values is $\mu_\delta = 12.2923$ and, following Eq. (4.10) on p. 209, the observed chance-corrected measure of effect size for the y_i and $\tilde{y}_i$ values, $i = 1, \ldots, N$, is

$$\Re_B = 1 - \frac{\delta_B}{\mu_\delta} = 1 - \frac{12.0535}{12.2923} = +0.0194 \,,$$

indicating approximately 2 % agreement between the observed and predicted values above that expected by chance.

4.5 Coda

Chapter 4 applied the Multi-Response Permutation Procedures (MRPP) developed in Chap. 2 to interval-level response measurements, utilizing dummy and effect coding of treatment groups to generate regression residuals from LAD regression models, subsequently analyzed with MRPP. Considered in this chapter were one-way randomized, one-way randomized with a covariate, one-way randomized-block, two-way randomized-block, two-way factorial, Latin square, split-plot, and two-factor nested designs. Chapter 4 concluded with example multivariate multiple regression designs.

Comparisons of permutation-based LAD regression with ordinary Euclidean distance between response measurements, permutation-based OLS regression with squared Euclidean distance between response measurements, and conventional OLS regression with squared Euclidean distance between response measurements in Chap. 4, revealed that considerable differences can exist among the three approaches that are not systematic. Oftentimes, one of the three approaches yielded the lowest of the three probability values, while other times the same approach yielded the highest probability value. Sometimes the three approaches yielded the same, or nearly the same, probability value, as was the case with the analysis of Factor B in the two-way randomized-block design example, and other times the three probability values were markedly different, as was the case with the analysis of the $A{\times}B$ interaction in the two-way factorial design example. In general, permutation-based LAD regression, coupled with MRPP and ordinary Euclidean distance between response measurements, is recommended due to the lack of restrictive assumptions and robustness that is possible with extreme values.

Chapter 5

Chapter 5 establishes the relationships between the MRPP test statistics, δ and $\Re$, and selected conventional tests and measures designed for the analysis of completely randomized data at the ordinal level of measurement. Considered in Chap. 5 are the Wilcoxon two-sample rank-sum test, the Kruskal–Wallis multiple-sample rank-sum test, the Mood rank-sum test for dispersion, the Brown–Mood median test, the Mielke power-of-rank functions, the Whitfield two-sample rank-sum test, and the Cureton rank-biserial test.

Randomized Designs: Ordinal Data, I

5

This fifth chapter of *Permutation Statistical Methods* utilizes the Multi-Response Permutation Procedures (MRPP) presented in Chap. 2 to develop the functional relationships between the test statistics of MRPP, δ and $\Re$, and selected conventional tests and measures designed for the analysis of completely randomized data at the ordinal level of measurement. A number of statistical tests and measures are considered in this chapter, including the Wilcoxon two-sample rank-sum test, the Kruskal–Wallis multiple-sample rank-sum test, the Ansari–Bradley rank-sum test for dispersion, the Taha sum-of-squared-ranks test, the Mood rank-sum test for dispersion, the Brown–Mood median test, the Mielke power-of-rank function tests, the Whitfield two-sample rank-sum test, and the Cureton rank-biserial test. Analyses in this chapter are largely limited to univariate rank data. Multivariate extensions for the various tests and measures discussed in Chap. 5 are presented in Chap. 6.

5.1 Introduction

As detailed more completely in Chap. 2, let $\Omega = \{\omega_1, \ldots, \omega_N\}$ denote a finite sample of N objects, let $x_j' = (x_{1j}, \ldots, x_{rj})$ be a transposed vector of r commensurate response measurements for object ω_j, $j = 1, \ldots, N$, and let $S_1, \ldots, S_g$ designate an exhaustive partitioning of the N objects into g disjoint treatment groups. The MRPP test statistic given by

$$\delta = \sum_{i=1}^{g} C_i \xi_i \,, \tag{5.1}$$

K.J. Berry et al., *Permutation Statistical Methods*,
DOI 10.1007/978-3-319-28770-6_5

where $C_i > 0$ is a positive treatment-group weight for group S_i, $i = 1, \ldots, g$,

$$\sum_{i=1}^{g} C_i = 1 ,$$

and

$$\xi_i = \binom{n_i}{2}^{-1} \sum_{j<k} \Delta(j,k)\, \Psi_i(\omega_j)\, \Psi_i(\omega_k) \tag{5.2}$$

is the average distance-function value for all distinct pairs of objects in treatment group S_i for $i = 1, \ldots, g$, $n_i \geq 2$ is the number of objects classified into treatment groups $S_1, \ldots, S_g$,

$$N = \sum_{i=1}^{g} n_i ,$$

$\sum_{j<k}$ is the sum over all j and k such that $1 \leq j < k \leq N$, $\Delta(j,k)$ is the generalized Minkowski distance function,

$$\Delta(j,k) = \left(\sum_{i=1}^{r} \left| x_{ij} - x_{ik} \right|^{p} \right)^{v/p} , \tag{5.3}$$

$p \geq 1$, $v > 0$, and $\Psi_i(\cdot)$ is an indicator function given by

$$\Psi_i(\omega_j) = \begin{cases} 1 & \text{if } \omega_j \in S_i , \\ 0 & \text{otherwise} . \end{cases}$$

The null hypothesis (H_0) states that equal probabilities are assigned to each of the

$$M = \frac{N!}{\displaystyle\prod_{i=1}^{g} n_i!}$$

possible, equally-likely allocations of the N objects to treatment groups $S_1, \ldots, S_g$.

The probability value associated with an observed value of δ, δ_o, is the probability under the null hypothesis (H_0) of observing a value of δ as extreme or more extreme than δ_o. Thus, an exact probability value for δ_o may be expressed as

$$P\big(\delta \leq \delta_o | H_0\big) = \frac{\text{number of } \delta \text{ values} \leq \delta_o}{M} .$$

When M is very large, an approximate probability value for δ may be obtained from a resampling procedure, where

$$P(\delta \leq \delta_o | H_0) = \frac{\text{number of } \delta \text{ values} \leq \delta_o}{L} ,$$

and L denotes the number of randomly sampled test statistic values. Typically, L is set to a large number to ensure accuracy, e.g., $L = 1{,}000{,}000$. Also, when M is very large and P is exceedingly small, a resampling-approximation permutation procedure may produce no δ values equal to or less than δ_o, even with $L = 1{,}000{,}000$, yielding an approximate resampling probability value of $P = 0.00$. In such cases, moment-approximation permutation procedures based on fitting the first three exact moments of the discrete permutation distribution to a Pearson type III distribution provide approximate probability values, as detailed in Chap. 1, Sect. 1.2.2 [284, 300].

A chance-corrected within-group coefficient of agreement is given by

$$\Re = 1 - \frac{\delta}{\mu_\delta} , \tag{5.4}$$

where μ_δ is the arithmetic average of the M δ values calculated on all possible, equally-likely arrangements of the observed response measurements given by

$$\mu_\delta = \frac{1}{M} \sum_{i=1}^{M} \delta_i . \tag{5.5}$$

5.2 Rank-Order Statistics

The conversion of raw scores to rank scores has long been a controversial topic in the statistical literature. In general, two reasons have been advanced for converting raw scores to rank scores: (1) minimize the effect of extreme values and (2) avoid the assumption of normality. Consequently, a plethora of rank tests were proposed during the late 1930s and the 1940s, prior to the advent of high-speed computing and the subsequent development of efficient permutation statistical methods.[1] During this period, many rank tests were proffered by Hotelling and Pabst [183], Friedman [128], Wilcoxon [429], Festinger [116], Kendall [205], Wallis [414], and others. Milton Friedman, in particular, advocated rank tests to avoid the assumption of normality and for parsimony, arguing that the loss of information when converting raw scores to rank scores might be more than compensated for by the greater economy of the rank test [128, p. 675]. W. Allen Wallis echoed Friedman's arguments in a

[1] As Erich Lehmann noted, permutation tests are very tedious to carry out, and as a result they only came into their own after computers made multitudinal calculations possible [235, p. 66].

1939 article wherein he argued for rank tests because of their ease of calculation and the relaxation of the assumption of normality [414].

In general, it is widely recognized that there are two kinds of rank tests: those that analyze pure ranks that have been gathered experimentally, such as when N subjects rank a series of m items, and those that analyze ranks based on transformations of raw scores in order to avoid the assumption of normality or to reduce the effect of outliers. Regarding the latter, many researchers have decried the loss of information in converting raw scores into rank scores.

As early as 1906 Spearman, commenting on the conversion of raw scores to rank scores, stated that "the disadvantage of conversion into rank scores is that thereby a certain amount of the experimental information is lost" [382, p. 92]. In 1940 Kendall and Babington Smith acknowledged that "the use of ranking... destroys what may be valuable information," pointing out that ranking methods suffer from a serious drawback when the data considered are not representable by a linear variable [210, p. 324]. In 1943 Wald and Wolfowitz published a seminal paper on serial correlation in which they noted that observed values could be replaced by their corresponding ranks, but questioned the wisdom in using rank-transformed values instead of the original observations due to the loss of information involved [413, p. 387]. In 1950 F.N. David, in a review of Kendall's *Rank Correlation Methods*, commented:

> It is interesting to note in the univariate case... that while many order statistics have been proposed (all of which are easy to apply and interesting mathematically)... it is rare indeed to find the need to use them in practice. It is customary to twist the observations about and/or to make various assumptions in order that existing techniques may be applied. This, the writer would suggest, is because of the instinctive feeling that tests based on ranks cannot be very discriminating. If, on the other hand, we consider the bivariate case, the order statistics proposed by Spearman and latterly by Kendall are used fairly frequently with little thought of the undoubted loss of information which using them implies [90, p. 190].

In 1952 Kruskal and Wallis observed that a disadvantage of rank methods is the "loss of information about exact magnitudes" [225, p. 601]. In 1954 Bross labeled rank-order statistics as a "mutation" of conventional statistics, observing that rank transformations were first suggested by Spearman in 1904 [381], but were so criticized by mathematical statisticians that no one dared use them for 25 years [58].[2] In 1968 Borgatta concluded that reality is distorted by assigning ranks and performing arithmetic operations on a set of numbers that is not isomorphic with the arithmetic system [47]. In 1973 Feinstein, in an article promoting permutation tests, emphasized the loss of information incurred when converting raw scores to rank scores for the sake of constructing a non-parametric test for analyzing rank scores rather than the observed raw scores [113, p. 911], and in 1975 Arbuckle and Aiken bemoaned the conversion of raw observations to rank scores as a "sacrifice of desirable qualities" [13, p. 381]. In 1993 May and Hunter went so far as to label the practice of

[2] See in this regard, a 2004 article on "Geometric representation of association between categories" in *Psychometrika* by Willem Heiser [171, p. 514].

replacing observations with rank numbers as a "degrading of the original data" [267, p. 404].

In a strongly worded statement in 2000 in reference to converting raw scores to rank scores for the Wilcoxon two-sample rank-sum test, Ludbrook and Dudley argued that "although the [Wilcoxon two-sample rank-sum] test was a brilliant invention by Frank Wilcoxon in the pre-computer era as a way of overcoming the computation difficulties of executing a permutation test for equality of means, it should have little relevance today" [255, p. 87] and in 2008 Ludbrook noted that rank tests are the poor man's substitute for computer-intensive measures, concluding "I see no merit in using this class of test on interval-scale data" [251, p. 673]. Finally, in 2011 Mielke, Berry, and Johnston published an article in *Journal of Applied Statistics* on the robustness of various two-sample statistics. Based on computer simulations, they concluded that permutation methods based on ordinary Euclidean distances between response measurements performed as well or better than methods based on converting raw observations to rank scores [309].

For other similar criticisms of rank transformations, see articles by Friedman in 1937 [128], Feinstein in 1973 [113], Still and White in 1981 [388], Gebhard and Schmitz in 1998 [136], and Lehmann in 2009 [235]. Also, the use of rank transformations can be carried to extremes as noted in the infinite classes of rank tests described by Mielke in 1972 [281].

5.3 Two-Sample Rank-Sum Tests

In the 1940s and 1950s, a variety of two-sample rank-sum tests were developed by a number of different researchers. Among the researchers were chemist Frank Wilcoxon [429], psychologist Leon Festinger [116], mathematicians Henry Mann and Donald Whitney [262], experimental psychologist John Whitfield, geneticists John Haldane and Cedric Smith [164], and statistician Dirk van der Reyden [409]. The tests were essentially variations on a theme and each could easily be transposed into another [41, pp. 132–152].

In 1945 Frank Wilcoxon, at the time a chemist employed by the American Cyanamid Company, published a short article in the first volume of *Biometrics Bulletin* in which he described a new test: the rank-sum test for two independent (unpaired) samples [429].[3] The Wilcoxon test was limited to two samples of equal size, i.e., $n_1 = n_2$. The following year Leon Festinger, a psychologist and statistician at the Massachusetts Institute of Technology at the time, developed a new statistical test to evaluate differences between two independent means by first converting the data to rank scores [116]. The test was equivalent to the Wilcoxon two-sample rank-sum test, but improved upon Wilcoxon's test as Festinger's test

[3]Included in this very brief three-page article was a second new test: the matched-pairs (signed ranks) rank-sum test for two dependent (paired) samples. The Wilcoxon signed-ranks test is discussed in Chap. 10, Sect. 10.2.

could accommodate unequal sample sizes, i.e., $n_1 \neq n_2$. The two-sample rank-sum test developed by Festinger went largely unnoticed by statisticians because it was published in the psychology journal *Psychometrika*, which was not generally read by mathematical statisticians.

In 1947 Henry Mann and Donald Whitney, mathematicians and statisticians at The Ohio State University, published a two-sample rank-sum test that was equivalent to the rank-sum test proposed by Wilcoxon and Festinger,[4] but was easier to calculate, allowed for unequal sample sizes, and permitted larger samples than Wilcoxon's test [262]. That same year John Whitfield, an experimental psychologist at Cambridge University, proposed a measure of rank-order correlation between two variables wherein one variable was composed of ranks and the other variable was dichotomous [424]. Whitfield's proposed rank-sum test was directly related to the Mann–Whitney and Wilcoxon two-sample rank-sum tests, although Whitfield was apparently unaware of the Wilcoxon and Mann–Whitney tests as neither is referenced in Whitfield's 1947 article.

In 1948 John Burdon Sanderson (J.B.S.) Haldane, Professor of Genetics at University College, London, and Cedric Smith, a statistical geneticist at The Francis Galton Laboratory for National Eugenics,[5] proposed a recursively obtained two-sample rank-sum test for birth-order effects, complete with tables [164]. This was an exact permutation test designed to test whether the probability of a child inheriting a certain medical condition, such as phenylketonuria, increased with birth order, and was equivalent to the Wilcoxon two-sample rank-sum test. Like the 1946 Festinger article published in *Psychometrika*, the Haldane–Smith paper went largely unnoticed by statisticians, as it was published in *Annals of Eugenics*.[6]

In 1952 Dirk van der Reyden, an experimental statistician for the Tobacco Research Board of Southern Rhodesia, developed a two-sample rank-sum test that was equivalent to the tests of Wilcoxon, Festinger, Mann and Whitney, Whitfield, and Haldane and Smith [409]. Like the Festinger and Haldane–Smith articles, the van der Reyden article was published in a journal not usually read by statisticians, the *Rhodesia Agricultural Journal*, and went unnoticed for many years. Apparently, the test was independently developed by van der Reyden as the articles by Wilcoxon, Festinger, Mann and Whitney, Whitfield, and Haldane and Smith were not referenced.[7]

[4]Consequently, the test is often referred to as the Wilcoxon–Mann–Whitney (WMW) two-sample rank-sum test.

[5]In 1963 The Francis Galton Laboratory for National Eugenics was renamed The Galton Laboratory of Human Genetics and Biometry.

[6]The *Annals of Eugenics* was renamed the *Annals of Human Genetics* in 1954.

[7]As a matter of fact, there are no references cited in the van der Reyden article whatsoever.

5.4 Example Analyses

Consider g samples with n_i rank scores in each sample, $i = 1, \ldots, g$, and let

$$N = \sum_{i=1}^{g} n_i \,.$$

In this section, three example analyses illustrate a permutation approach to typical two-sample rank-sum problems. The first example is designed to correspond to the conventional Wilcoxon–Mann–Whitney two-sample rank-sum test using a small set of univariate rank scores with $v = 2$ and treatment-group weights $C_i = (n_i - 1)/(N - g)$ for $i = 1, \ldots, g$; the second example analyzes the same small set of univariate rank scores, but uses $v = 1$ and treatment-group weights $C_i = (n_i - 1)/(N - g)$ for $i = 1, \ldots, g$; and the third example analyzes the same set of univariate rank scores using $v = 1$, but adopts a proportional treatment-group weighting function given by $C_i = n_i/N$ for $i = 1, \ldots, g$.[8]

5.4.1 Example 1

Consider a two-sample linear rank test for N univariate rank scores with n_1 and n_2 rank scores in the first and second samples, respectively. The Wilcoxon–Mann–Whitney two-sample rank-sum test statistic is given by

$$W = \sum_{i=1}^{n_1} R_i \,,$$

where R_i denotes the rank function of the ith response measurement and n_1 is, typically, the smaller of the two sample sizes. The identities relating statistic W and the MRPP test statistic δ were first published by Mielke in 1984 [284, p. 818] and are given by

$$\delta = \frac{2}{N(N-2)} \left[NT - S^2 - \frac{(NW - n_1 S)^2}{n_1 n_2} \right] \tag{5.6}$$

and

$$W = \frac{n_1 S}{N} - \left\{ \frac{n_1 n_2}{N^2} \left[NT - S^2 - \frac{N(N-2)\delta}{2} \right] \right\}^{1/2} , \tag{5.7}$$

[8]For detailed descriptions of the Wilcoxon and Mann–Whitney two-sample rank-sum tests, see discussions by Berry, Johnston, and Mielke [41, pp. 134–137, 143–147].

where

$$S = \sum_{i=1}^{N} R_i \quad \text{and} \quad T = \sum_{i=1}^{N} R_i^2 \, .$$

Note that in Eqs. (5.6) and (5.7), N, S, T, n_1, and n_2 are all invariant under permutation.

In the absence of any tied rank scores, it is well known that S and T may simply be expressed as

$$S = \sum_{i=1}^{N} i = \frac{N(N+1)}{2} \quad \text{and} \quad T = \sum_{i=1}^{N} i^2 = \frac{N(N+1)(2N+1)}{6} \, ,$$

as explained in any elementary textbook.[9] Because of the relationship between statistics W and δ, the exact probability value of the realized value of W is given by

$$P\big(W \geq W_{\text{o}}|H_0\big) = P\big(\delta \leq \delta_{\text{o}}|H_0\big) = \frac{\text{number of } \delta \text{ values} \leq \delta_{\text{o}}}{M} \, ,$$

where W_{o} and δ_{o} denote the observed values of W and δ, respectively.

For an example analysis, consider the univariate rank response measurements listed in Table 5.1, where $r = 1$, $g = 2$, $N = n_1 + n_2 = 20$, and there are no tied rank scores. For this application, let $n_1 = 8$ denote the A rank scores and $n_2 = 12$ denote the B rank scores. The data are adapted from Neave and Worthington [317, pp. 111, 113]. For this first analysis, let $v = 2$, employing squared Euclidean distance between the rank scores, and let the treatment-group weights be given by

$$C_i = \frac{n_i - 1}{N - g} \, , \qquad i = 1, \ldots, g \, ,$$

Table 5.1 Example univariate rank-score data for a Wilcoxon–Mann–Whitney two-sample rank-sum test with $n_1 = 8$ A rank scores and $n_2 = 12$ B rank scores

1	2	3	4	5	6	7	8	9	10	11	12	13	14	15	16	17	18	19	20
A	*A*	*A*	*A*	*A*	*A*	*B*	*A*	*B*	*B*	*A*	*B*	*B*	*B*	*B*	*B*	*B*	*B*	*B*	*B*

[9]Technically, $S = N(N+1)/2$ holds for both tied and untied rank scores, but $T = [N(N+1)(2N+1)]/6$ holds only for untied rank scores.

to correspond to the Wilcoxon–Mann–Whitney two-sample rank-sum test [262, 429]. Because there are only

$$M = \frac{N!}{n_1!\; n_2!} = \frac{20!}{8!\; 12!} = 125{,}970$$

possible, equally-likely arrangements of the $N = 20$ univariate rank scores listed in Table 5.1, an exact solution is feasible. Following Eq. (5.2) on p. 218 with $v = 2$, the $N = 20$ univariate rank scores listed in Table 5.1 yield $g = 2$ average distance-function values of

$$\xi_1 = 21.7143 \quad \text{and} \quad \xi_2 = 33.7576 \,.$$

Following Eq. (5.1) on p. 217, the observed value of the MRPP test statistic based on $v = 2$ and treatment-group weights

$$C_i = \frac{n_i - 1}{N - g} \,, \qquad i = 1, 2 \,,$$

is

$$\delta_o = \sum_{i=1}^{g} C_i \xi_i = \frac{1}{20 - 2}\big[(8 - 1)(21.7143) + (12 - 1)(33.7576)\big] = 29.0741 \,.$$

If all arrangements of the $N = 20$ observed rank scores listed in Table 5.1 on p. 224 occur with equal chance, the exact probability value of $\delta_o = 29.0741$ computed on the $M = 125{,}970$ possible arrangements of the observed data with $n_1 = 8$ A univariate rank scores and $n_2 = 12$ B univariate rank scores preserved for each arrangement is

$$P(\delta \le \delta_o | H_0) = \frac{\text{number of } \delta \text{ values} \le \delta_o}{M} = \frac{24}{125{,}970} = 0.1905{\times}10^{-3} \,.$$

For comparison, the conventional Wilcoxon two-sample rank-sum test on the $N = 20$ univariate rank scores listed in Table 5.1 yields an observed Wilcoxon test statistic value of

$$W_o = \sum_{i=1}^{n_1} R_i = 1 + 2 + 3 + 4 + 5 + 6 + 8 + 11 = 40 \,,$$

where Wilcoxon's W is approximately distributed as $N(0, 1)$ under the null hypothesis as $N \to \infty$. For the rank scores listed in Table 5.1, the mean value of W is

$$\mu_W = \frac{n_1(N + 1)}{2} = \frac{8(20 + 1)}{2} = 84 \,,$$

the variance of W is

$$\sigma_W^2 = \frac{n_1 n_2(N+1)}{12} = \frac{(8)(12)(20+1)}{12} = 168 ,$$

the observed standard score, corrected for continuity, is

$$z_o = \frac{W_o - 0.5 - \mu_W}{\sqrt{\sigma_W^2}} = \frac{40 - 0.5 - 84}{\sqrt{168}} = -3.4332 ,$$

and the approximate two-tailed $N(0, 1)$ probability value is $P = 0.5965 \times 10^{-3}$.

The exact probability value of $W_o = 40$ is

$$P(W \geq W_o | H_0) = \frac{\text{number of } W \text{ values} \geq W_o}{M}$$

$$= \frac{24}{125{,}970} = 0.1905 \times 10^{-3} .$$

Following Eq. (5.5) on p. 219, the exact expected value of the $M = 125{,}970$ δ values is $\mu_\delta = 70.00$ and, following Eq. (5.4) on p. 219, the observed chance-corrected measure of effect size is

$$\Re_o = 1 - \frac{\delta_o}{\mu_\delta} = 1 - \frac{29.0741}{70.00} = +0.5847 ,$$

indicating approximately 58 % within-group agreement above that expected by chance.

The relationships between the MRPP test statistic and Wilcoxon's W are confirmed as follows. For the $N = 20$ univariate rank scores listed in Table 5.1 with no tied values, the observed value of S is

$$S_o = \sum_{i=1}^{N} i = \frac{N(N+1)}{2} = \frac{20(20+1)}{2} = 210$$

and the observed value of T is

$$T_o = \sum_{i=1}^{N} i^2 = \frac{N(N+1)(2N+1)}{6} = \frac{20(20+1)[2(20)+1]}{6} = 2{,}870 .$$

Then, following Eq. (5.6) on p. 223, the observed value of δ for the rank scores listed in Table 5.1 is

$$\delta_o = \frac{2}{20(20-2)} \left\{ 20(2{,}870) - (210)^2 - \frac{\left[20(40) - 8(210)\right]^2}{(8)(12)} \right\}$$
$$= \frac{2}{360}\left(13{,}300 - \frac{774{,}400}{96}\right) = 29.0741$$

and, following Eq. (5.7) on p. 223, the observed value of Wilcoxon's W is

$$W_o = \frac{(8)(210)}{20} - \left\{ \frac{(8)(12)}{20^2} \left[(20)(2{,}870) - 210^2 - \frac{20(20-2)(29.0741)}{2} \right] \right\}^{1/2} = 84 - \left[0.24(8{,}066.6667)\right]^{1/2} = 40 \; .$$

5.4.2 Example 2

For this second analysis of the $N = 20$ univariate rank response measurements listed in Table 5.1 on p. 224, let the treatment-group weights be given by

$$C_i = \frac{n_i - 1}{N - g} \; , \qquad i = 1, \ldots, g \; ,$$

as in Example 1, but set $v = 1$ instead of $v = 2$, employing ordinary Euclidean distance between the rank scores. Following Eq. (5.2) on p. 218, the $N = 20$ univariate rank scores listed in Table 5.1 yield $g = 2$ average distance-function values of

$$\xi_1 = 3.9286 \quad \text{and} \quad \xi_2 = 4.9091 \; .$$

Following Eq. (5.1) on p. 217, the observed value of the MRPP test statistic based on $v = 1$ and treatment-group weights

$$C_i = \frac{n_i - 1}{N - g} \; , \qquad i = 1, 2 \; ,$$

is

$$\delta_o = \sum_{i=1}^{g} C_i \xi_i = \frac{1}{20-2}\left[(8-1)(3.9286) + (12-1)(4.9091)\right] = 4.5278 \; .$$

If all arrangements of the $N = 20$ observed rank scores listed in Table 5.1 occur with equal chance, the exact probability value of $\delta_o = 4.5278$ computed on the $M = 125{,}970$ possible arrangements of the observed data with $n_1 = 8$ A univariate rank scores and $n_2 = 12$ B univariate rank scores preserved for each arrangement is

$$P(\delta \leq \delta_o | H_0) = \frac{\text{number of } \delta \text{ values} \leq \delta_o}{M} = \frac{24}{125{,}970} = 0.1905{\times}10^{-3} \ .$$

For comparison, the exact probability value based on $v = 2$, $M = 125{,}970$, and $C_i = (n_i - 1)/(N - g)$ for $i = 1, 2$ in Example 1 is also $P = 0.1905{\times}10^{-3}$. No comparison is made with the conventional Wilcoxon–Mann–Whitney two-sample rank-sum test as the Wilcoxon–Mann–Whitney two-sample test is undefined for $v = 1$.

Following Eq. (5.5) on p. 219, the exact expected value of the $M = 125{,}970$ δ values is $\mu_\delta = 7.00$ and, following Eq. (5.4) on p. 219, the observed chance-corrected measure of effect size is

$$\Re_o = 1 - \frac{\delta_o}{\mu_\delta} = 1 - \frac{4.5278}{7.00} = +0.3532 \ ,$$

indicating approximately 35 % within-group agreement above that expected by chance.

5.4.3 Example 3

The treatment-group weights given by

$$C_i = \frac{n_i - 1}{N - g} \ , \qquad i = 1, \ldots, g \ ,$$

are based on degrees of freedom, are holdovers from classical parametric tests, and are neither necessary nor appropriate for distribution-free permutation methods. Consequently, for this third analysis of the $N = 20$ univariate rank response measurements listed in Table 5.1 on p. 224, let the treatment-group weights be given by

$$C_i = \frac{n_i}{N} \ , \qquad i = 1, \ldots, g \ ,$$

i.e., simply weighting each treatment group proportional to its size, and set $v = 1$, employing ordinary Euclidean distance between the rank scores, as in Example 2. Following Eq. (5.2) on p. 218, the $N = 20$ univariate rank scores listed in Table 5.1 yield $g = 2$ average distance-function values of

$$\xi_1 = 3.9286 \quad \text{and} \quad \xi_2 = 4.9091 \ .$$

Following Eq. (5.1) on p. 217, the observed value of the MRPP test statistic based on $v = 1$ and treatment-group weights

$$C_i = \frac{n_i}{N}\,, \qquad i = 1, 2\,,$$

is

$$\delta_{\text{o}} = \sum_{i=1}^{g} C_i \xi_i = \frac{1}{20}\big[(8)(3.9286) + (12)(4.9091)\big] = 4.5169\,.$$

If all arrangements of the $N = 20$ observed rank scores listed in Table 5.1 occur with equal chance, the exact probability value of $\delta_{\text{o}} = 4.5169$ computed on the $M =$ 125,970 possible arrangements of the observed data with $n_1 = 8$ A univariate rank scores and $n_2 = 12$ B univariate rank scores preserved for each arrangement is

$$P(\delta \leq \delta_{\text{o}}|H_0) = \frac{\text{number of } \delta \text{ values} \leq \delta_{\text{o}}}{M} = \frac{26}{125{,}970} = 0.2064{\times}10^{-3}\,.$$

For comparison, the exact probability values based on $v = 2$, $M =$ 125,970, and $C_i = (n_i - 1)/(N - g)$ for $i = 1, 2$ in Example 1 and $v = 1$, $M =$ 125,970, and $C_i = (n_i - 1)/(N - g)$ for $i = 1, 2$ in Example 2 are both $P = 0.1905{\times}10^{-3}$. No comparison is made with the conventional Wilcoxon–Mann–Whitney two-sample rank-sum test as the Wilcoxon–Mann–Whitney two-sample test is undefined for both $v = 1$ and $C_i = n_i/N$, $i = 1, \ldots, g$.

Following Eq. (5.5) on p. 219, the exact expected value of the $M =$ 125,970 δ values is $\mu_\delta = 7.00$ and, following Eq. (5.4) on p. 219, the observed chance-corrected measure of effect size is

$$\Re_{\text{o}} = 1 - \frac{\delta_{\text{o}}}{\mu_\delta} = 1 - \frac{4.5169}{7.00} = +0.3547\,,$$

indicating approximately 35 % within-group agreement above that expected by chance.

5.5 MRPP and the Kruskal–Wallis Rank-Sum Test

In 1952 William Kruskal and W. Allen Wallis published an exact multi-sample rank-sum test in *Journal of the American Statistical Association* that they denoted as H [225]. Kruskal and Wallis explained that test statistic H stemmed from two statistical methods: rank transformations of the original raw response measurements and permutations of the rank-order statistics.

Consider g random samples of possibly different sizes and denote the size of the ith sample by n_i, $i = 1, \ldots, g$. Let

$$N = \sum_{i=1}^{g} n_i$$

denote the total number of response measurements, assign rank 1 to the smallest of the N measurements, rank 2 to the next smallest measurement, and continue on up to the largest measurement, which is assigned rank N, and let R_i denote the sum of the rank scores in the ith sample, $i = 1, \ldots, g$. When there are no tied rank scores, the Kruskal–Wallis test statistic is given by

$$H = \frac{12}{N(N+1)} \sum_{i=1}^{g} \frac{R_i^2}{n_i} - 3(N+1) \,. \tag{5.8}$$

Kruskal and Wallis showed that when $r = 1$ and $g = 2$, H was equivalent to the Wilcoxon [429], Festinger [116], Mann–Whitney [262], and Haldane–Smith [164] two-sample rank-sum tests [225]. In 1953, in an erratum to their 1952 paper, Kruskal and Wallis documented the equivalence of H with the two-sample rank-sum test developed by van der Reyden in 1952 [409], which had only recently come to their attention.

5.6 Example Analyses

In this section, three example analyses illustrate a permutation approach to typical g-sample rank-sum problems. The first example is designed to correspond to the conventional Kruskal–Wallis g-sample rank-sum test using a small set of univariate rank scores with $v = 2$ and treatment weights $C_i = (n_i - 1)/(N - g)$ for $i = 1, \ldots, g$; the second example analyzes the same small set of univariate rank scores, but uses $v = 1$ and treatment-group weights $C_i = (n_i - 1)/(N - g)$ for $i = 1, \ldots, g$; and the third example analyzes the same set of univariate rank scores using $v = 1$, but adopts a proportional treatment-group weighting given by $C_i = n_i/N$ for $i = 1, \ldots, g$.

5.6.1 Example 1

Consider a g-sample rank-sum test for N rank scores with n_i rank scores in each of the g samples, $i = 1, \ldots, g$. The functional relationships between the Kruskal–Wallis test statistic H and the MRPP test statistic δ, as defined in Eq. (5.1) on p. 217, are given by

$$\delta = \frac{2\left(T - \left\{\frac{S}{6}\left[H + 3(N+1)\right]\right\}\right)}{N - g} \tag{5.9}$$

and

$$H = \frac{6}{S}\left[T - \frac{\delta}{2}(N-g)\right] - 3(N+1)\,, \tag{5.10}$$

where, if no rank scores are tied, S and T may simply be expressed as

$$S = \sum_{i=1}^{N} i = \frac{N(N+1)}{2} \quad \text{and} \quad T = \sum_{i=1}^{N} i^2 = \frac{N(N+1)(2N+1)}{6}\,.$$

Note that in Eqs. (5.9) and (5.10), S, T, N, and g are invariant under permutation, along with the constants 2, 3, and 6.

Because of the relationship between statistics H and δ, the exact probability value of the realized value of H is given by

$$P\big(H \geq H_{\text{o}}|H_0\big) = P\big(\delta \leq \delta_{\text{o}}|H_0\big) = \frac{\text{number of } \delta \text{ values} \leq \delta_{\text{o}}}{M}\,,$$

where H_{o} and δ_{o} denote the observed values of H and δ, respectively.

For an example analysis, consider the univariate rank response measurement scores listed in Fig. 5.1 where $r = 1$, $g = 3$, $n_1 = n_2 = n_3 = 6$, $N = n_1 + n_2 + n_3 = 18$, and there are no tied rank scores. The data are adapted from Kenny [213, p. 317]. For this first analysis, let $v = 2$, employing squared Euclidean distance between the rank scores, and let the treatment-group weights be given by

$$C_i = \frac{n_i - 1}{N - g}\,, \qquad i = 1, \ldots, g\,,$$

to correspond to the Kruskal–Wallis g-sample rank-sum test.

Because there are only

$$M = \frac{N!}{n_1!\; n_2!\; n_3!} = \frac{18!}{(6!)^3} = \frac{6{,}402{,}373{,}705{,}728{,}000}{373{,}248{,}000} = 17{,}153{,}136$$

Fig. 5.1 Ranking of three treatments with $r = 1$, $g = 3$, $n_1 = n_2 = n_3 = 6$, and $N = n_1 + n_2 + n_3 = 18$

	Treatment		
	1	2	3
	4	2	17
	7	3	14
	10	11	12
	15	1	13
	9	8	16
	18	5	6
Total	63	30	78

possible, equally-likely arrangements of the $N = 18$ univariate rank scores listed in Fig. 5.1, an exact solution is feasible. Following Eq. (5.2) on p. 218, the $N = 18$ univariate rank scores listed in Fig. 5.1 yield $g = 3$ average distance-function values of

$$\xi_1 = 53.40\ , \quad \xi_2 = 29.60\ , \quad \text{and} \quad \xi_3 = 30.40\ .$$

Following Eq. (5.1) on p. 217, the observed value of the MRPP test statistic based on $v = 2$ and treatment-group weights

$$C_i = \frac{n_i - 1}{N - g}\ , \qquad i = 1, 2, 3\ ,$$

is

$$\delta_o = \sum_{i=1}^{g} C_i \xi_i = \frac{1}{18-3}\big[(6-1)(53.40) + (6-1)(29.60) + (6-1)(30.40)\big] = 37.80\ .$$

If all arrangements of the $N = 18$ observed rank scores listed in Fig. 5.1 occur with equal chance, the exact probability value of $\delta_o = 37.80$ computed on the $M = 17{,}153{,}136$ possible arrangements of the observed data with $n_1 = n_2 = n_3 = 6$ univariate rank scores preserved for each arrangement is

$$P\big(\delta \leq \delta_o | H_0\big) = \frac{\text{number of } \delta \text{ values} \leq \delta_o}{M} = \frac{376{,}704}{17{,}153{,}136} = 0.0220\ .$$

For comparison, the totals of the rank scores for Treatments 1, 2, and 3 given in Fig. 5.1 are 63, 30, and 78, respectively, and following Eq. (5.8) on p. 230, the observed value of H is

$$H_o = \frac{12}{18(18+1)}\left(\frac{63^2}{6} + \frac{30^2}{6} + \frac{78^2}{6}\right) - 3(18+1) = 7.0526\ ,$$

where H is approximately distributed as chi-squared under the null hypothesis with $g - 1 = 3 - 1 = 2$ degrees of freedom. Under the null hypothesis, the observed value of $H_o = 7.0526$ yields an approximate probability value of $P = 0.0294$. The exact probability value of $H_o = 7.0526$ is

$$P\big(H \geq H_o | H_0\big) = \frac{\text{number of } H \text{ values} \geq H_o}{M} = \frac{376{,}704}{17{,}153{,}136} = 0.0220\ .$$

Note that whereas the Kruskal–Wallis test statistic H, as defined in Eq. (5.8) on p. 230, does not allow for tied rank scores, δ as defined in Eq. (5.1) on p. 217 automatically accommodates tied rank scores.[10]

Following Eq. (5.5) on p. 219, the exact expected value of the $M = 17{,}153{,}136$ δ values is $\mu_\delta = 57.00$ and, following Eq. (5.4) on p. 219, the observed chance-corrected measure of effect size is

$$\Re_o = 1 - \frac{\delta_o}{\mu_\delta} = 1 - \frac{37.80}{57.00} = +0.3368 ,$$

indicating approximately 34 % within-group agreement above that expected by chance.

The relationships between statistics δ and H are confirmed as follows. For the univariate rank scores listed in Fig. 5.1 with no tied values, the observed value of S is

$$S_o = \sum_{i=1}^{N} i = \frac{N(N+1)}{2} = \frac{18(18+1)}{2} = 171 ,$$

and the observed value of T is

$$T_o = \sum_{i=1}^{N} i^2 = \frac{N(N+1)(2N+1)}{6} = \frac{18(18+1)[(2)(18)+1]}{6} = 2{,}109 .$$

Then following Eq. (5.9) on p. 230, the observed value of the MRPP test statistic for the univariate rank scores listed in Fig. 5.1 is

$$\delta_o = \frac{2\left(2{,}109 - \left\{\frac{171}{6}\left[7.0526 + 3(18+1)\right]\right\}\right)}{18-3} = \frac{567}{15} = 37.80$$

and, following Eq. (5.10) on p. 231, the observed value of the Kruskal–Wallis test statistic is

$$H_o = \frac{6}{171}\left[2{,}109 - \frac{37.80}{2}(18-3)\right] - 3(18+1)$$
$$= (0.0351)(1825.50) - 57 = 7.0526 .$$

[10] Many textbooks present rather cumbersome adjustments, permitting H to accommodate tied rank scores.

5.6.2 Example 2

For this second example of the univariate rank response measurements listed in Fig. 5.1, let the treatment-group weights be given by

$$C_i = \frac{n_i - 1}{N - g} , \qquad i = 1, \ldots, g ,$$

as in Example 1, but set $v = 1$ instead of $v = 2$, employing ordinary Euclidean distance between the rank scores. Following Eq. (5.2) on p. 218, the $N = 18$ univariate rank scores listed in Fig. 5.1 yield $g = 3$ average distance-function values of

$$\xi_1 = 6.3333 , \quad \xi_2 = 4.6667 , \quad \text{and} \quad \xi_3 = 4.5333 .$$

Following Eq. (5.1) on p. 217, the observed value of the MRPP test statistic based on $v = 1$ and treatment-group weights

$$C_i = \frac{n_i - 1}{N - g} , \qquad i = 1, 2, 3 ,$$

is

$$\delta_o = \sum_{i=1}^{g} C_i \xi_i = \frac{6 - 1}{18 - 3} \big(6.3333 + 4.6667 + 4.5333\big) = 5.1778 .$$

If all arrangements of the $N = 18$ observed rank scores listed in Fig. 5.1 occur with equal chance, the exact probability value of $\delta_o = 5.1778$ computed on the $M = 17{,}153{,}136$ possible arrangements of the observed data with $n_1 = n_2 = n_3 = 6$ univariate rank scores preserved for each arrangement is

$$P\big(\delta \leq \delta_o | H_0\big) = \frac{\text{number of } \delta \text{ values} \leq \delta_o}{M} = \frac{547{,}662}{17{,}153{,}136} = 0.0319 .$$

For comparison, the exact probability value based on $v = 2$, $M = 17{,}153{,}136$, and $C_i = (n_i - 1)/(N - g)$ for $i = 1, 2, 3$ in Example 1 is $P = 0.0220$. No comparison is made with the conventional Kruskal–Wallis g-sample rank-sum test as the Kruskal–Wallis test is undefined for $v = 1$.

Following Eq. (5.5) on p. 219, the exact expected value of the $M = 17{,}153{,}136$ δ values is $\mu_\delta = 6.3333$ and, following Eq. (5.4) on p. 219, the observed chance-corrected measure of effect size is

$$\Re_o = 1 - \frac{\delta_o}{\mu_\delta} = 1 - \frac{5.1778}{6.3333} = +0.1825 ,$$

indicating approximately 18 % within-group agreement above that expected by chance.

5.6.3 Example 3

For this third example, let the treatment-group weights be given by

$$C_i = \frac{n_i}{N}\,, \qquad i = 1, \ldots, g\,,$$

simply weighting each treatment group proportional to its size, and set $v = 1$, employing ordinary Euclidean distance between the rank scores, as in Example 2. Following Eq. (5.2) on p. 218, the $N = 18$ univariate rank scores listed in Fig. 5.1 on p. 231 yield $g = 3$ average distance-function values of

$$\xi_1 = 6.3333\,, \quad \xi_2 = 4.6667\,, \quad \text{and} \quad \xi_3 = 4.5333\,.$$

Following Eq. (5.1) on p. 217, the observed value of the MRPP test statistic based on $v = 1$ and treatment-group weights

$$C_i = \frac{n_i}{N}\,, \qquad i = 1, 2, 3\,,$$

is

$$\delta_o = \sum_{i=1}^{g} C_i \xi_i = \frac{6}{18}\big(6.3333 + 4.6667 + 4.5333\big) = 5.1778\,.$$

If all arrangements of the $N = 18$ observed rank scores listed in Fig. 5.1 occur with equal chance, the exact probability value of $\delta_o = 5.1778$ computed on the $M =$ 17,153,136 possible arrangements of the observed data with $n_1 = n_2 = n_3 = 6$ univariate rank scores preserved for each arrangement is

$$P\big(\delta \leq \delta_o | H_0\big) = \frac{\text{number of } \delta \text{ values} \leq \delta_o}{M} = \frac{547{,}662}{17{,}153{,}136} = 0.0319\,.$$

For comparison, the exact probability values based on $v = 2$, $M =$ 17,153,136, and $C_i = (n_i - 1)/(N - g)$ for $i = 1, 2, 3$ in Example 1 and $v = 1$, $M =$ 17,153,136, and $C_i = (n_i - 1)/(N - g)$ for $i = 1, 2, 3$ in Example 2 are $P = 0.0220$ and $P = 0.0319$, respectively. No comparison is made with the conventional Kruskal–Wallis g-sample rank-sum test as the Kruskal–Wallis test is undefined for both $v = 1$ and $C_i = n_i/N$, $i = 1, \ldots, g$.

Note that the results in Example 2 with $C_i = (n_i - 1)/(N - g)$ and $v = 1$ are identical to the results of Example 3 with $C_i = n_i/N$ and $v = 1$. When $n_1 = n_2 = \cdots = n_g$, the two weighting functions yield the same result. Thus, for the example data listed in Fig. 5.1 with $g = 3$ and $n_1 = n_2 = n_3 = 6$,

$$C_i = \frac{n_i - 1}{N - g} = \frac{6 - 1}{18 - 3} = 0.3333 \quad \text{and} \quad C_i = \frac{n_i}{N} = \frac{6}{18} = 0.3333$$

for $i = 1, \ldots, g$.

Following Eq. (5.5) on p. 219, the exact expected value of the $M = 17{,}153{,}136$ δ values is $\mu_\delta = 6.3333$ and, following Eq. (5.4) on p. 219, the observed chance-corrected measure of effect size is

$$\Re_o = 1 - \frac{\delta_o}{\mu_\delta} = 1 - \frac{5.1778}{6.3333} = +0.1825 ,$$

indicating approximately 18 % within-group agreement above that expected by chance.

5.7 Three Two-Sample Classes of Rank Tests

In 1972 P.W. Mielke introduced three classes of two-sample tests based on power-of-rank functions [281]. The three classes of tests were described as asymptotically optimum against either scale or location alternatives for specific distributions and designated as A_{Ns}, B_{Ns}, and C_{Ns}, where s denoted the power to which the rank scores were to be raised. Following Mielke, in all three cases let $X_1, \ldots, X_{n_1}$ and $Y_1, \ldots, Y_{n_2}$ denote exchangeable values from $N = n_1 + n_2$ objects that can be arranged in exactly

$$M = \frac{N!}{n_1!\; n_2!}$$

ways [281].

Function A_{Ns} is defined as

$$A_{Ns} = \sum_{i=1}^{N} R_i^s Z_{Ni} , \tag{5.11}$$

where R_i^s is the rank function of the ith response measurement, $s > 0$, and $Z_{Ni} = 1$ or 0 if the ith smallest value in the combined sample of X and Y values is an X or Y, respectively. Mielke showed that the Wilcoxon two-sample rank-sum test [429] and the Taha two-sample sum-of-squared-ranks test [393] were associated with A_{N1} and A_{N2}, respectively [281, p. 850].

Function B_{Ns} is defined as

$$B_{Ns} = \sum_{i=1}^{N} \left| R_i - \frac{N+1}{2} \right|^s Z_{Ni} , \tag{5.12}$$

where R_i is the rank function of the ith response measurement, $s > 0$, and

$$\frac{N+1}{2}$$

is the median value of the consecutive integers, 1, 2, …, N. Mielke demonstrated that the Ansari–Bradley rank-sum test for dispersion [10] and the Mood rank-sum test for dispersion [312] were associated with B_{N1} and B_{N2}, respectively [281, p. 850].

Function C_{Ns} is defined as

$$C_{Ns} = \sum_{i=1}^{N} h(R_i, N, s)\, Z_{Ni} \;, \tag{5.13}$$

where

$$h(R_i, N, s) = \begin{cases} +\left| i - \dfrac{N+1}{2} \right|^s & \text{if } i > \dfrac{N+1}{2} \;, \\ 0 & \text{if } i = \dfrac{N+1}{2} \;, \\ -\left| i - \dfrac{N+1}{2} \right|^s & \text{if } i < \dfrac{N+1}{2} \;, \end{cases} \tag{5.14}$$

R_i is the rank function of the ith response measurement, and $s > -1$. Mielke showed that the Brown–Mood median test [59], the Wilcoxon two-sample rank-sum test [429], and the Mielke two-sample sum-of-squared-ranks test [282] were associated with C_{N0}, C_{N1}, and C_{N2}, respectively [281, p. 852].[11]

5.8 MRPP and Two-Sample Power-of-Rank Functions

Let

$$H = \sum_{i=1}^{N} f(i) Z_{Ni} \;,$$

where $f(i)$ is a score function of the rank-order value of X_i from below, relative to the finite population of N univariate response measurements. If H variously represents A_{N1}, A_{N2}, B_{N1}, B_{N2}, C_{N0}, C_{N1}, and C_{N2}, then the functional relationships between the MRPP test statistic δ and H are given by

$$\delta = \frac{2}{N(N-2)} \left[NT - S^2 - \frac{(NH - n_1 S)^2}{n_1 n_2} \right] \tag{5.15}$$

[11] Two constants have been deleted from the original formulae in Mielke [281] with no loss of generality, as both constants are invariant under permutation.

and

$$H = \frac{n_1 S}{N} - \left\{ \frac{n_1 n_2}{N^2} \left[NT - S^2 - \frac{N(N-2)\delta}{2} \right] \right\}^{1/2} , \tag{5.16}$$

where

$$S = \sum_{i=1}^{N} f(i) \quad \text{and} \quad T = \sum_{i=1}^{N} \left[f(i) \right]^2 .$$

Let n_i denote the number of univariate response measurements in each of g samples, $i = 1, \ldots, g$. Then, following Eq. (5.5) on p. 219, the arithmetic average of the δ values calculated on all

$$M = \frac{N!}{\prod_{i=1}^{g} n_i!}$$

possible arrangements of the observed response measurements is given by

$$\mu_\delta = \frac{1}{M} \sum_{i=1}^{M} \delta_i \tag{5.17}$$

and, following Eq. (5.4) on p. 219, a chance-corrected within-group coefficient of effect size is given by

$$\Re = 1 - \frac{\delta}{\mu_\delta} . \tag{5.18}$$

5.9 Example A_{Ns} Analyses with $s = 1$

Consider $g = 2$ samples with n_1 rank scores in the first sample, n_2 rank scores in the second sample, and $N = n_1 + n_2$. In this section, three example analyses illustrate the A_{Ns} rank function test with $s = 1$. The first example is designed to correspond to the conventional Wilcoxon two-sample rank-sum test using a small set of univariate rank scores with $v = 2$ and treatment-group weights $C_i = (n_i - 1)/(N - g)$ for $i = 1, \ldots, g$; the second example analyzes the same small set of univariate rank scores, but uses $v = 1$ and treatment-group weights $C_i = (n_i - 1)/(N - g)$ for $i = 1, \ldots, g$; and the third example analyzes the same set of univariate rank scores using $v = 1$, but adopts a proportional treatment-group weighting function given by $C_i = n_i/N$ for $i = 1, \ldots, g$.

Rank:	1	2	3	4	5	6	7	8	9
Sample:	A	A	B	A	A	B	B	B	B

Fig. 5.2 Example univariate rank-score data for two-sample rank tests with $n_1 = 4$, $n_2 = 5$, and $N = n_1 + n_2 = 9$

5.9.1 Example 1

Consider the small set of univariate rank scores listed in Fig. 5.2 composed of two samples labeled A and B. For convenience let n_1 denote the smaller of the two sample sizes; in this case, sample A. For this example analysis, $n_1 = 4$, $n_2 = 5$, and $N = n_1 + n_2 = 9$.

The Wilcoxon two-sample rank-sum test statistic is simply the sum of the rank scores in the smaller of the two sample sizes; in this case, n_1 [429]. For the univariate rank scores listed in Fig. 5.2, the observed value of W is

$$W_o = \sum_{i=1}^{n_1} R_i = 1 + 2 + 4 + 5 = 12 \,, \tag{5.19}$$

where R_i denotes a rank score for variable A and $n_1 = 4$ is the smaller of the two sample sizes. For comparison, consider A_{Ns} with $s = 1$, where following Eq. (5.11) on p. 236, the observed value of A_{N1} is

$$\begin{aligned} A_{N1o} &= \sum_{i=1}^{N} R_i^1 Z_{Ni} \\ &= (1^1)(1) + (2^1)(1) + (3^1)(0) + (4^1)(1) + (5^1)(1) \\ &\quad + (6^1)(0) + (7^1)(0) + (8^1)(0) + (9^1)(0) \\ &= 1 + 2 + 0 + 4 + 5 + 0 + 0 + 0 + 0 = 12 \,. \end{aligned} \tag{5.20}$$

It is readily apparent from Eqs. (5.19) and (5.20) that Mielke's A_{N1} and Wilcoxon's W are identical.

Following Eqs. (5.15) on p. 237 and (5.16) on p. 238, the functional relationships between statistics A_{N1} and δ are given by

$$\delta = \frac{2}{N(N-2)} \left[NT - S^2 - \frac{(NA_{N1} - n_1 S)^2}{n_1 n_2} \right] \tag{5.21}$$

and

$$A_{N1} = \frac{n_1 S}{N} - \left\{ \frac{n_1 n_2}{N} \left[NT - S^2 - \frac{N(N-2)\delta}{2} \right] \right\}^{1/2} , \tag{5.22}$$

where

$$S = \sum_{i=1}^{N} R_i\ , \qquad T = \sum_{i=1}^{N} R_i^2\ ,$$

and R_i is the rank function of the ith response measurement, $i = 1, \ldots, N$. Thus, for the univariate rank scores listed in Fig. 5.2, the observed values of S and T are

$$S_o = 1 + 2 + 3 + 4 + 5 + 6 + 7 + 8 + 9 = 45$$

and

$$T_o = 1^2 + 2^2 + 3^2 + 4^2 + 5^2 + 6^2 + 7^2 + 8^2 + 9^2 = 285\ ,$$

respectively. If no rank scores are tied, then S and T may simply be expressed as

$$S = \sum_{i=1}^{N} i = \frac{N(N+1)}{2} \quad \text{and} \quad T = \sum_{i=1}^{N} i^2 = \frac{N(N+1)(2N+1)}{6}\ .$$

Utilizing the univariate rank scores listed in Fig. 5.2, where there are no tied rank scores, the relationships between statistics δ and A_{N1} can be confirmed. Thus, the observed values of S and T are

$$S_o = \sum_{i=1}^{N} i = \frac{9(9+1)}{2} = \frac{90}{2} = 45$$

and

$$T_o = \sum_{i=1}^{N} i^2 = \frac{9(9+1)[2(9)+1]}{6} = \frac{1{,}710}{6} = 285\ ,$$

respectively. Then, following Eq. (5.21) on p. 239, the observed value of the MRPP test statistic for the univariate rank scores listed in Fig. 5.2 is

$$\delta_o = \frac{2}{9(9-2)} \left\{ 9(285) - (45)^2 - \frac{[9(12) - 4(45)]^2}{(4)(5)} \right\}$$

$$= \frac{2}{63} \left(540 - \frac{5{,}184}{20} \right) = 8.9143$$

and, following Eq. (5.22) on p. 239, the observed value of A_{N1} is

$$A_{N1o} = \frac{(4)(45)}{9} - \left\{ \frac{(4)(5)}{9^2} \left[9(285) - 45^2 - \frac{9(9-2)(8.9143)}{2} \right] \right\}^{1/2}$$

$$= 20 - [0.2469(540 - 280.8005)]^{1/2} = 20 - 8 = 12 \; .$$

Because of the relationship between statistics A_{N1} and δ, the exact probability value of the realized value of A_{N1} is given by

$$P(A_{N1} \geq A_{N1o}|H_0) = P(\delta \leq \delta_o|H_0) = \frac{\text{number of } \delta \text{ values} \leq \delta_o}{M} \; ,$$

where A_{N1o} and δ_o denote the observed values of A_{N1} and δ, respectively. In addition, because of the relationships among W, A_{N1}, and δ, the exact probability value of Wilcoxon's W test statistic is given by

$$P(W \geq W_o|H_0) = P(\delta \leq \delta_o|H_0) = \frac{\text{number of } \delta \text{ values} \leq \delta_o}{M} \; ,$$

where W_o and δ_o denote the observed values of W and δ, respectively.

Consider again the univariate rank response measurements listed in Fig. 5.2 where $r = 1$, $g = 2$, $n_1 = 4$, $n_2 = 5$, $N = n_1 + n_2 = 9$, and there are no tied rank scores. In this application, let $v = 2$, employing squared Euclidean distance between the rank scores, and let the treatment-group weights be given by

$$C_i = \frac{n_i - 1}{N - g} \; , \qquad i = 1, \ldots, g \; ,$$

to correspond to the Wilcoxon two-sample rank-sum test. Because there are only

$$M = \frac{N!}{n_1! \, n_2!} = \frac{9!}{4! \; 5!} = 126$$

possible, equally-likely arrangements of the $N = 9$ rank scores, an exact solution is preferred. Following Eq. (5.2) on p. 218, the $N = 9$ univariate rank scores listed in Fig. 5.2 yield $g = 2$ average distance-function values of

$$\xi_1 = 6.6667 \quad \text{and} \quad \xi_2 = 10.60 \; .$$

Following Eq. (5.1) on p. 217, the observed value of the MRPP test statistic based on $v = 2$ and treatment-group weights

$$C_i = \frac{n_i - 1}{N - g} \; , \qquad i = 1, 2 \; ,$$

is

$$\delta_o = \sum_{i=1}^{g} C_i \xi_i = \frac{1}{9-2}\big[(4-1)(6.6667) + (5-1)(10.60)\big] = 8.9143 \; .$$

If all arrangements of the $N = 9$ observed rank scores listed in Fig. 5.2 occur with equal chance, the exact probability value of $\delta_o = 8.9143$ computed on the $M = 126$ possible arrangements of the observed data with $n_1 = 4$ A univariate rank scores and $n_2 = 5$ B univariate rank scores preserved for each arrangement is

$$P\big(\delta \leq \delta_o | H_0\big) = \frac{\text{number of } \delta \text{ values} \leq \delta_o}{M} = \frac{8}{126} = 0.0635 \; .$$

For comparison, the conventional Wilcoxon two-sample rank-sum test computed on the univariate rank scores listed in Fig. 5.2 yields an observed Wilcoxon test statistic value of

$$W_o = \sum_{i=1}^{n_1} R_i = 1 + 2 + 4 + 5 = 12$$

and the exact probability value of $W_o = 12$ is

$$P\big(W \geq W_o | H_0\big) = \frac{\text{number of } W \text{ values} \geq W_o}{M} = \frac{8}{126} = 0.0635 \; .$$

Alternatively, test statistic W is approximately distributed as $N(0, 1)$ under the null hypothesis as $N \to \infty$. For the rank scores listed in Fig. 5.2, the mean value of W is

$$\mu_W = \frac{n_1(N+1)}{2} = \frac{4(9+1)}{2} = 20 \; ,$$

the variance of W is

$$\sigma_W^2 = \frac{n_1 n_2 (N+1)}{12} = \frac{(4)(5)(9+1)}{12} = 16.6667 \; ,$$

the observed standard score is

$$z_o = \frac{W_o - \mu_W}{\sqrt{\sigma_W^2}} = \frac{12 - 20}{\sqrt{16.6667}} = -1.9596 \; ,$$

and the approximate two-tailed $N(0, 1)$ probability value is $P = 0.0500$.

Following Eq. (5.17) on p. 238, the exact expected value of the $M = 126$ δ values is $\mu_\delta = 15.00$ and, following Eq. (5.18) on p. 238, the observed chance-corrected measure of effect size is

$$\Re_o = 1 - \frac{\delta_o}{\mu_\delta} = 1 - \frac{8.9143}{15.00} = +0.4057 \,,$$

indicating approximately 41 % within-group agreement above that expected by chance.

5.9.2 Example 2

As discussed in Chap. 2, permutation statistical tests and measures are distribution-free, data-dependent, and non-parametric; consequently, they require no distributional assumptions and make no estimates of population parameters. Thus, it is not necessary to set $v = 2$ and thereby square the differences between the rank scores. While conventional tests and measures that assume normality must estimate the two parameters of the normal distribution, μ_x and σ_x^2, both of which are based on squared deviations, permutation tests and measures do not assume normality and are not restricted to $v = 2$, whose corresponding distance function is not metric. A distance function based on $v = 1$ is an attractive alternative to $v = 2$ as it is a metric, satisfies the triangle inequality, is robust to extreme values, provides an easy to understand Euclidean distance between the rank scores, and ensures that the data and analysis spaces are congruent. Thus, for this second analysis of the univariate rank scores listed in Fig. 5.2 on p. 239, replicated in Fig. 5.3 for convenience, the distance function is set to $v = 1$.

For this second example of the univariate rank scores, let the treatment-group weights be given by

$$C_i = \frac{n_i - 1}{N - g} \,, \qquad i = 1, \ldots, g \,,$$

as in Example 1, but set $v = 1$ instead of $v = 2$, employing ordinary Euclidean distance between the rank scores.

Rank:	1	2	3	4	5	6	7	8	9
Sample:	*A*	*A*	*B*	*A*	*A*	*B*	*B*	*B*	*B*

Fig. 5.3 Example univariate rank-score data for two-sample rank tests with $n_1 = 4$, $n_2 = 5$, and $N = n_1 + n_2 = 9$

Following Eq. (5.2) on p. 218, the $N = 9$ univariate rank scores listed in Fig. 5.3 yield $g = 2$ average distance-function values of

$$\xi_1 = 2.3333 \quad \text{and} \quad \xi_2 = 2.80 \ .$$

Following Eq. (5.1) on p. 217, the observed value of the MRPP test statistic based on $v = 1$ and treatment-group weights

$$C_i = \frac{n_i - 1}{N - g} \ , \qquad i = 1, 2 \ ,$$

is

$$\delta_o = \sum_{i=1}^{g} C_i \xi_i = \frac{1}{9-2}\big[(4-1)(2.3333) + (5-1)(2.80)\big] = 2.60 \ .$$

If all arrangements of the $N = 9$ observed rank scores listed in Fig. 5.3 occur with equal chance, the exact probability value of $\delta_o = 2.60$ computed on the $M = 126$ possible arrangements of the observed data with $n_1 = 4$ A univariate rank scores and $n_2 = 5$ B univariate rank scores preserved for each arrangement is

$$P\big(\delta \leq \delta_o | H_0\big) = \frac{\text{number of } \delta \text{ values} \leq \delta_o}{M} = \frac{6}{126} = 0.0476 \ .$$

For comparison, the exact probability value based on $v = 2$, $M = 126$, and $C_i = (n_i - 1)/(N - g)$ for $i = 1, 2$ in Example 1 is $P = 0.0635$. No comparison is made with the conventional Wilcoxon two-sample rank-sum test as Wilcoxon's two-sample test is undefined for $v = 1$.

Following Eq. (5.17) on p. 238, the exact expected value of the $M = 126$ δ values is $\mu_\delta = 3.3333$ and, following Eq. (5.18) on p. 238, the observed chance-corrected measure of effect size is

$$\Re_o = 1 - \frac{\delta_o}{\mu_\delta} = 1 - \frac{2.60}{3.3333} = +0.2200 \ ,$$

indicating 22 % within-group agreement above that expected by chance.

5.9.3 Example 3

As discussed in Chap. 3, the treatment-group weighting function given by

$$C_i = \frac{n_i - 1}{N - g} \ , \qquad i = 1, \ldots, g \ ,$$

is, in a permutation context, an unnecessary artifact left over from classical tests and is not appropriate for a distribution-free permutation test, as degrees of freedom are not applicable to permutation methods, except when validating a corresponding statistical test, such as Wilcoxon's two-sample rank-sum test.

For this third example of the univariate rank response measurements listed in Fig. 5.3, let the treatment-group weights be given by

$$C_i = \frac{n_i}{N} , \qquad i = 1, \ldots, g ,$$

simply weighting each treatment group proportional to its size, and set $v = 1$, employing ordinary Euclidean distance between the rank scores, as in Example 2.

Following Eq. (5.2) on p. 218, the $N = 9$ univariate rank scores listed in Fig. 5.3 yield $g = 2$ average distance-function values of

$$\xi_1 = 2.3333 \quad \text{and} \quad \xi_2 = 2.80 .$$

Following Eq. (5.1) on p. 217, the observed value of the MRPP test statistic based on $v = 1$ and treatment-group weights

$$C_i = \frac{n_i}{N} , \qquad i = 1, 2 ,$$

is

$$\delta_o = \sum_{i=1}^{g} C_i \xi_i = \frac{1}{9}\big[(4)(2.3333) + (5)(2.80)\big] = 2.5926 .$$

If all arrangements of the $N = 9$ observed rank scores listed in Fig. 5.3 occur with equal chance, the exact probability value of $\delta_o = 2.5926$ computed on the $M = 126$ possible arrangements of the observed data with $n_1 = 4$ A univariate rank scores and $n_2 = 5$ B univariate rank scores preserved for each arrangement is

$$P(\delta \le \delta_o | H_0) = \frac{\text{number of } \delta \text{ values} \le \delta_o}{M} = \frac{6}{126} = 0.0476 .$$

For comparison, the exact probability values based on $v = 2$, $M = 126$, and $C_i = (n_i - 1)/(N - g)$ for $i = 1, 2$ in Example 1 and $v = 1$, $M = 126$, and $C_i = (n_i - 1)/(N - g)$ for $i = 1, 2$ in Example 2 are $P = 0.0635$ and $P = 0.0476$, respectively. No comparison is made with the conventional Wilcoxon two-sample rank-sum test as Wilcoxon's two-sample test is undefined for both $v = 1$ and $C_i = n_i/N$, $i = 1, \ldots, g$.[12]

[12]Note that the results in Examples 2 and 3 are nearly identical. This occurs whenever n_1 and n_2 are approximately the same and, also, $C_i = (n_i - 1)/(N - g)$ and $C_i = n_i/N$ are equivalent when $n_1 = n_2$.

Following Eq. (5.17) on p. 238, the exact expected value of the $M = 126$ δ values is $\mu_\delta = 3.3333$ and, following Eq. (5.18) on p. 238, the observed chance-corrected measure of effect size is

$$\mathfrak{R}_o = 1 - \frac{\delta_o}{\mu_\delta} = 1 - \frac{2.5926}{3.3333} = +0.2222 ,$$

indicating approximately 22 % within-group agreement above that expected by chance.

5.10 Example $\mathcal{A}_{Ns}$ Analyses with $s = 2$

Consider $g = 2$ samples with n_1 rank scores in the first sample, n_2 rank scores in the second sample, and $N = n_1 + n_2$. In this section, three example analyses illustrate the $\mathcal{A}_{Ns}$ rank function test with $s = 2$. The first example is designed to correspond to the conventional Taha two-sample sum-of-squared-ranks test using a small set of univariate rank scores with $v = 2$ and treatment-group weights $C_i = (n_i - 1)/(N - g)$ for $i = 1, \ldots, g$; the second example analyzes the same small set of univariate rank scores, but uses $v = 1$ and treatment-group weights $C_i = (n_i - 1)/(N - g)$ for $i = 1, \ldots, g$; and the third example analyzes the same set of univariate rank scores using $v = 1$, but adopts a proportional treatment-group weighting function given by $C_i = n_i/N$ for $i = 1, \ldots, g$.

5.10.1 Example 1

In 1964 Mohamed A.H. Taha developed a two-sample test statistic based on squared rank scores, given by

$$L = \sum_{i=1}^{n_1} R_i^2 ,$$

where n_1 denotes the smaller of the two sample sizes and R_i^2 denotes a squared rank score for $i = 1, \ldots, n_1$ [393]. Consider again the univariate rank scores listed in Fig. 5.2 on p. 239, replicated in Fig. 5.4 for convenience, where $n_1 = 4$, $n_2 = 5$, and $N = 9$. Then, the observed value of Taha's rank-sum test statistic is

$$L_o = \sum_{i=1}^{n_1} R_i^2 = 1^2 + 2^2 + 4^2 + 5^2 = 46 ,$$

where $n_1 = 4$ is the smaller of the two sample sizes and R_i^2 denotes a squared rank score for $i = 1, \ldots, n_1$.

Rank:	1	2	3	4	5	6	7	8	9
Sample:	A	A	B	A	A	B	B	B	B

Fig. 5.4 Example univariate rank-score data for two-sample rank tests with $n_1 = 4$, $n_2 = 5$, and $N = n_1 + n_2 = 9$

For comparison, following Eq. (5.11) on p. 236, consider A_{Ns} with $s = 2$, where the observed value of A_{N2} is

$$\begin{aligned} A_{N2\text{o}} &= \sum_{i=1}^{N} R_i^2 Z_{Ni} \\ &= (1^2)(1) + (2^2)(1) + (3^2)(0) + (4^2)(1) \\ &\quad + (5^2)(1) + (6^2)(0) + (7^2)(0) + (8^2)(0) + (9^2)(0) \\ &= 1 + 4 + 0 + 16 + 25 + 0 + 0 + 0 + 0 = 46\,. \end{aligned}$$

Thus, Mielke's A_{N2} and Taha's L are shown to be identical.

Following Eqs. (5.15) on p. 237 and (5.16) on p. 238, the functional relationships between statistics A_{N2} (and L) and δ are given by

$$\delta = \frac{2}{N(N-2)} \left[NT - S^2 - \frac{(NA_{N2} - n_1 S)^2}{n_1 n_2} \right] \tag{5.23}$$

and

$$A_{N2} = \frac{n_1 S}{N} - \left\{ \frac{n_1 n_2}{n^2} \left[NT - S^2 - \frac{N(N-2)\delta}{2} \right] \right\}^{1/2}, \tag{5.24}$$

where

$$S = \sum_{i=1}^{N} R_i^2\,, \qquad T = \sum_{i=1}^{N} \left(R_i^2\right)^2 = \sum_{i=1}^{N} R_i^4\,,$$

and R_i is the rank function of the ith response measurement. Thus, for the univariate rank scores listed in Fig. 5.4, the observed values of S and T are

$$\begin{aligned} S_\text{o} &= 1^2 + 2^2 + 3^2 + 4^2 + 5^2 + 6^2 + 7^2 + 8^2 + 9^2 \\ &= 1 + 4 + 9 + 16 + 25 + 36 + 49 + 64 + 81 = 285 \end{aligned}$$

and

$$\begin{aligned} T_o &= 1^4 + 2^4 + 3^4 + 4^4 + 5^4 + 6^4 + 7^4 + 8^4 + 9^4 \\ &= 1 + 16 + 81 + 256 + 625 + 1{,}296 + 2{,}401 + 4{,}096 + 6{,}561 = 15{,}333 \,. \end{aligned}$$

If no rank scores are tied, then S and T may simply be expressed as

$$S = \sum_{i=1}^{N} i^2 = \frac{N(N+1)(2N+1)}{6} \tag{5.25}$$

and

$$T = \sum_{i=1}^{N} i^4 = \frac{N(N+1)(2N+1)(3N^2+3N-1)}{30} \,. \tag{5.26}$$

Utilizing the univariate rank scores listed in Fig. 5.4, where there are no tied rank scores, the relationships between statistics δ and A_{N2} can be confirmed. Thus, following Eq. (5.25), the observed value of S is

$$S_o = \sum_{i=1}^{N} i^2 = \frac{9(9+1)[2(9)+1]}{6} = \frac{1{,}710}{6} = 285$$

and, following Eq. (5.26), the observed value of T is

$$T_o = \sum_{i=1}^{N} i^4 = \frac{9(9+1)[2(9)+1][3(9^2)+3(9)-1]}{30} = \frac{459{,}990}{30} = 15{,}333 \,.$$

Then, following Eq. (5.23) on p. 247, the observed value of the MRPP test statistic for the univariate rank scores listed in Fig. 5.4 is

$$\begin{aligned} \delta_o &= \frac{2}{9(9-2)} \left\{ 9(15{,}333) - (285)^2 - \frac{[9(46) - 4(285)]^2}{(4)(5)} \right\} \\ &= \frac{2}{63} \left(56{,}772 - \frac{527{,}076}{20} \right) = 965.6571 \end{aligned}$$

and, following Eq. (5.24) on p. 247, the observed value of A_{N2} is

$$A_{N2\text{o}} = \frac{(4)(285)}{9} - \left\{ \frac{(4)(5)}{9^2} \left[(9)(15{,}333) - 285^2 - \frac{9(9-2)(965.6571)}{2} \right] \right\}^{1/2} = 126.6667 - \left[(0.2469)(26{,}303.8013) \right]^{1/2} = 46\ .$$

Because of the relationship between statistics A_{N2} and δ, the exact probability value of the realized value of A_{N2} is given by

$$P\big(A_{N2} \geq A_{N2\text{o}} | H_0\big) = P\big(\delta \leq \delta_\text{o} | H_0\big) = \frac{\text{number of } \delta \text{ values} \leq \delta_\text{o}}{M}\ ,$$

where $A_{N2\text{o}}$ and δ_o denote the observed values of A_{N2} and δ, respectively. In addition, because of the relationships among L, A_{N2}, and δ, the exact probability value of Taha's L test statistic is given by

$$P\big(L \geq L_\text{o} | H_0\big) = P\big(\delta \leq \delta_\text{o} | H_0\big) = \frac{\text{number of } \delta \text{ values} \leq \delta_\text{o}}{M}\ ,$$

where L_o and δ_o denote the observed values of L and δ, respectively.

Consider again the univariate rank response measurements listed in Fig. 5.4 where $r = 1$, $g = 2$, $n_1 = 4$, $n_2 = 5$, $N = n_1 + n_2 = 9$, and there are no tied rank scores. In this application, let $v = 2$, employing squared Euclidean distance between the rank scores, and let the treatment-group weights be given by

$$C_i = \frac{n_i - 1}{N - g}\ , \qquad i = 1, \ldots, g\ ,$$

to correspond to the Taha two-sample sum-of-squared-ranks test [393]. Because there are only

$$M = \frac{N!}{n_1!\ n_2!} = \frac{9!}{4!\ 5!} = 126$$

possible, equally-likely arrangements of the $N = 9$ rank scores, an exact solution is preferred. Following Eq. (5.2) on p. 218, the $N = 9$ univariate rank scores listed in Fig. 5.4 yield $g = 2$ average distance-function values of

$$\xi_1 = 246.00 \quad \text{and} \quad \xi_2 = 1{,}505.40\ .$$

Following Eq. (5.1) on p. 217, the observed value of the MRPP test statistic based on $v = 2$ and treatment-group weights

$$C_i = \frac{n_i - 1}{N - g} , \qquad i = 1, 2 ,$$

is

$$\delta_o = \sum_{i=1}^{g} C_i \xi_i = \frac{1}{9-2}\big[(4-1)(246.00) + (5-1)(1{,}505.40)\big] = 965.6571 .$$

If all arrangements of the $N = 9$ observed rank scores listed in Fig. 5.4 occur with equal chance, the exact probability value of $\delta_o = 965.6571$ computed on the $M = 126$ possible arrangements of the observed data with $n_1 = 4$ A univariate rank scores and $n_2 = 5$ B univariate rank scores preserved for each arrangement is

$$P(\delta \leq \delta_o | H_0) = \frac{\text{number of } \delta \text{ values} \leq \delta_o}{M} = \frac{6}{126} = 0.0476 .$$

For comparison, a conventional Taha two-sample sum-of-squared-ranks test computed on the univariate rank scores listed in Fig. 5.4 yields an observed Taha test statistic value of

$$L_o = \sum_{i=1}^{n_1} R_i^2 = 1^2 + 2^2 + 4^2 + 5^2 = 46$$

and the exact probability value of $L_o = 46$ is

$$P(L \geq L_o | H_0) = \frac{\text{number of } L \text{ values} \geq L_o}{M} = \frac{6}{126} = 0.0476 .$$

Following Eq. (5.17) on p. 238, the exact expected value of the $M = 126$ δ values is $\mu_\delta = 1{,}577.00$ and, following Eq. (5.18) on p. 238, the observed chance-corrected measure of effect size is

$$\Re_o = 1 - \frac{\delta_o}{\mu_\delta} = 1 - \frac{965.6571}{1{,}577.00} = +0.3877 ,$$

indicating approximately 39 % within-group agreement above that expected by chance.

Rank:	1	2	3	4	5	6	7	8	9
Sample:	*A*	*A*	*B*	*A*	*A*	*B*	*B*	*B*	*B*

Fig. 5.5 Example univariate rank-score data for two-sample rank tests with $n_1 = 4$, $n_2 = 5$, and $N = n_1 + n_2 = 9$

5.10.2 Example 2

For this second analysis of the univariate rank response measurements listed in Fig. 5.2 on p. 239, replicated in Fig. 5.5 for convenience, let the treatment-group weights be given by

$$C_i = \frac{n_i - 1}{N - g} , \qquad i = 1, \ldots, g ,$$

as in Example 1, but set $v = 1$ instead of $v = 2$, employing ordinary Euclidean distance between the rank scores.

Following Eq. (5.2) on p. 218, the $N = 9$ univariate rank scores listed in Fig. 5.5 yield $g = 2$ average distance-function values of

$$\xi_1 = 14.00 \quad \text{and} \quad \xi_2 = 34.40 .$$

Following Eq. (5.1) on p. 217, the observed value of the MRPP test statistic based on $v = 1$ and treatment-group weights

$$C_i = \frac{n_i - 1}{N - g} , \qquad i = 1, 2 ,$$

is

$$\delta_o = \sum_{i=1}^{g} C_i \xi_i = \frac{1}{9-2} \big[(4-1)(14.00) + (5-1)(34.40)\big] = 25.6571 .$$

If all arrangements of the $N = 9$ observed rank scores listed in Fig. 5.5 occur with equal chance, the exact probability value of $\delta_o = 25.6571$ computed on the $M = 126$ possible arrangements of the observed data with $n_1 = 4$ A univariate rank scores and $n_2 = 5$ B univariate rank scores preserved for each arrangement is

$$P(\delta \leq \delta_o | H_0) = \frac{\text{number of } \delta \text{ values} \leq \delta_o}{M} = \frac{6}{126} = 0.0476 .$$

For comparison, the exact probability value based on $v = 2$, $M = 126$, and $C_i = (n_i - 1)/(N - g)$ for $i = 1, 2$ in Example 1 is also $P = 0.0476$. No comparison is made with the conventional Taha two-sample squared-ranks test as Taha's two-sample test is undefined for $v = 1$.

Following Eq. (5.17) on p. 238, the exact expected value of the $M = 126$ δ values is $\mu_\delta = 33.3333$ and, following Eq. (5.18) on p. 238, the observed chance-corrected measure of effect size is

$$\Re_o = 1 - \frac{\delta_o}{\mu_\delta} = 1 - \frac{25.6571}{33.3333} = +0.2303 ,$$

indicating approximately 23 % within-group agreement above that expected by chance.

5.10.3 Example 3

The treatment-group weighting function given by

$$C_i = \frac{n_i - 1}{N - g} , \qquad i = 1, \ldots, g ,$$

is essential for classical tests, but is not required for a permutation test, as degrees of freedom need never enter into distribution-free permutation methods. Thus, for this third analysis of the univariate rank scores listed in Fig. 5.5, the treatment-group weighting function is set to

$$C_i = \frac{n_i}{N} , \qquad i = 1, \ldots, g ,$$

simply weighting each treatment group proportional to its size, and setting $v = 1$, employing ordinary Euclidean distance between the rank scores, as in Example 2. Following Eq. (5.2) on p. 218, the $N = 9$ univariate rank scores listed in Fig. 5.5 yield $g = 2$ average distance-function values of

$$\xi_1 = 14.00 \quad \text{and} \quad \xi_2 = 34.40 .$$

Following Eq. (5.1) on p. 217, the observed value of the MRPP test statistic based on $v = 1$ and treatment-group weights

$$C_i = \frac{n_i}{N} , \qquad i = 1, 2 ,$$

is

$$\delta_o = \sum_{i=1}^{g} C_i \xi_i = \frac{1}{9}\big[(4)(14.00) + (5)(34.40)\big] = 25.3333 .$$

If all arrangements of the $N = 9$ observed rank scores listed in Fig. 5.5 occur with equal chance, the exact probability value of $\delta_o = 25.3333$ computed on the $M = 126$ possible arrangements of the observed data with $n_1 = 4$ A univariate rank scores and $n_2 = 5$ B univariate rank scores preserved for each arrangement is

$$P(\delta \leq \delta_o | H_0) = \frac{\text{number of } \delta \text{ values} \leq \delta_o}{M} = \frac{5}{126} = 0.0397 \,.$$

For comparison, the exact probability values based on $v = 2$, $M = 126$, and $C_i = (n_i - 1)/(N - g)$ for $i = 1, 2$ in Example 1 and $v = 1$, $M = 126$, and $C_i = (n_i - 1)/(N - g)$ for $i = 1, 2$ in Example 2 are both $P = 0.0476$. No comparison is made with the conventional Taha two-sample sum-of-squared-ranks test as Taha's two-sample test is undefined for both $v = 1$ and $C_i = n_i/N$, $i = 1, \ldots, g$.

Following Eq. (5.17) on p. 238, the exact expected value of the $M = 126$ δ values is $\mu_\delta = 33.6077$ and, following Eq. (5.18) on p. 238, the observed chance-corrected measure of effect size is

$$\Re_o = 1 - \frac{\delta_o}{\mu_\delta} = 1 - \frac{25.3333}{33.6077} = +0.2462 \,,$$

indicating approximately 25 % within-group agreement above that expected by chance.

5.11 Example B_{Ns} Analyses with $s = 1$

In 1960 Sidney Siegel and John Tukey published a sum-of-ranks test for unpaired samples in *Journal of the American Statistical Association* [376]. The procedure for the Siegel–Tukey test for two samples is:

1. Arrange the observations in the combined data set from smallest to largest.
2. Assign rank 1 to the smallest observation, rank 2 to the largest observation, rank 3 to the next largest observation, rank 4 to the next smallest observation, rank 5 to the next smallest observation, and so on.
3. Apply the Wilcoxon two-sample rank-sum test using the Siegel–Tukey alternating ranking scheme.
4. Evaluate the resulting rank-sum using tables of the Wilcoxon two-sample rank-sum test.

An inherent difficulty with the Siegel–Tukey sum-of-ranks test is that the ranking could just as well start its alternating pattern with the largest observation receiving a rank of 1, instead of the smallest observation. In general, these two ranking procedures yield different values for the Wilcoxon test statistic.

In 1960 Abdur R. Ansari and Ralph A. Bradley, both at the Virginia Agricultural Experiment Station, published a competing rank-sum test for unpaired samples.[13] The Ansari–Bradley two-sample rank-sum test for dispersion overcomes the problem associated with the Siegel–Tukey test by assigning rank 1 to the smallest and largest observations, rank 2 to the next smallest and largest observations, and so on [11]. Specifically:

1. Arrange the observations in the combined data set from smallest to largest.
2. Assign rank 1 to the smallest observation and largest observations, rank 2 to the next smallest and next largest observations, rank 3 to the next smallest and largest observations, and so on.
3. Apply the Wilcoxon two-sample rank-sum test using the Ansari–Bradley alternating ranking scheme.
4. Evaluate the resulting rank-sum using tables of the Ansari–Bradley rank-sum test [376, pp. 1178–1179].

Note that critical values for the Ansari–Bradley rank-sum test can no longer be obtained from the published tables for the Wilcoxon two-sample rank-sum test. Figure 5.6 illustrates the Ansari–Bradley assignment of rank scores applied to the two samples described in Fig. 5.2 on p. 239.

In this section, three example analyses illustrate the B_{Ns} rank function test with $s = 1$. The first example is designed to correspond to the conventional Ansari–Bradley two-sample rank-sum test using a small set of univariate rank scores with $v = 2$ and treatment-group weights $C_i = (n_i - 1)/(N - g)$ for $i = 1, \ldots, g$; the second example analyzes the same small set of univariate rank scores, but uses $v = 1$ and treatment-group weights $C_i = (n_i - 1)/(N - g)$ for $i = 1, \ldots, g$; and the third example analyzes the same set of univariate rank scores using $v = 1$, but adopts a proportional treatment-group weighting function given by $C_i = n_i/N$ for $i = 1, \ldots, g$.

Rank:	1	2	3	4	5	4	3	2	1
Sample:	*A*	*A*	*B*	*A*	*A*	*B*	*B*	*B*	*B*

Fig. 5.6 Example univariate rank-score data for the Ansari–Bradley two-sample rank test with $n_1 = 4$, $n_2 = 5$, and $N = n_1 + n_2 = 9$

[13]Earlier that year, Ralph Bradley accepted a position as Chair of the Department of Statistics at Florida State University in Tallahassee.

5.11.1 Example 1

Consider the univariate rank response measurements listed in Fig. 5.6 with $n_1 = 4$ A rank scores, $n_2 = 5$ B rank scores, and $N = n_1 + n_2 = 9$. For the rank scores listed in Fig. 5.6, the observed Ansari–Bradley test statistic is

$$W_o = \sum_{i=1}^{n_1} R_i = 1 + 2 + 4 + 5 = 12\ ,$$

where R_i denotes a rank score for variable A and $n_1 = 4$ is the smaller of the two sample sizes. For comparison, consider B_{Ns} with $s = 1$, where the median rank is given by

$$\frac{N+1}{2} = \frac{9+1}{2} = 5$$

and, following Eq. (5.12) on p. 236, the observed value of B_{N1} is

$$\begin{aligned} B_{N1o} &= \sum_{i=1}^{N} \left| R_i - \frac{N+1}{2} \right|^1 Z_{Ni} \\ &= |1-5|^1(1) + |2-5|^1(1) + |3-5|^1(0) + |4-5|^1(1) + |5-5|^1(1) \\ &\quad + |6-5|^1(1) + |7-5|^1(0) + |8-5|^1(0) + |9-5|^1(0) \\ &= 4+3+0+1+1+0+0+0+0 = 8\ . \end{aligned}$$

The relationships between Mielke's B_{N1} and the Ansari–Bradley W test statistic are given by

$$B_{N1} = \frac{n_1(N+1)}{2} - W \quad \text{and} \quad W = \frac{n_1(N+1)}{2} - B_{N1}\ .$$

Thus, the observed values of B_{N1} and W are

$$B_{N1o} = \frac{4(9+1)}{2} - 12 = 20 - 12 = 8$$

and

$$W_o = \frac{4(9+1)}{2} - 8 = 20 - 8 = 12\ ,$$

respectively.

Following Eqs. (5.15) on p. 237 and (5.16) on p. 238, the functional relationships between B_{N1} and δ are given by

$$\delta = \frac{2}{N(N-2)}\left[NT - S^2 - \frac{(NB_{N1} - n_1 S)^2}{n_1 n_2}\right] \tag{5.27}$$

and

$$B_{N1} = \frac{n_1 S}{N} - \left\{\frac{n_1 n_2}{N^2}\left[NT - S^2 - \frac{N(N-2)\delta}{2}\right]\right\}^{1/2}, \tag{5.28}$$

where

$$S = \sum_{i=1}^{N} \left|R_i - \frac{N+1}{2}\right|^1 \quad \text{and} \quad T = \sum_{i=1}^{N} \left|R_i - \frac{N+1}{2}\right|^2 .$$

Thus, the median of the $N = 9$ rank scores listed in Fig. 5.6 is

$$\frac{N+1}{2} = \frac{9+1}{2} = 5$$

and the observed values of S and T are

$$\begin{aligned} S_o &= |1-5|^1 + |2-5|^1 + |3-5|^1 + |4-5|^1 + |5-5|^1 + |6-5|^1 \\ &\quad + |7-5|^1 + |8-5|^1 + |9-5|^1 \\ &= 4+3+2+1+0+1+2+3+4 = 20 \end{aligned}$$

and

$$\begin{aligned} T_o &= |1-5|^2 + |2-5|^2 + |3-5|^2 + |4-5|^2 + |5-5|^2 + |6-5|^2 \\ &\quad + |7-5|^2 + |8-5|^2 + |9-5|^2 \\ &= 16+9+4+1+0+1+4+9+16 = 60 , \end{aligned}$$

respectively.

Then, following Eq. (5.27), the observed value of the MRPP test statistic for the univariate rank scores listed in Fig. 5.6 is

$$\delta_o = \frac{2}{9(9-2)}\left\{9(60) - 20^2 - \frac{[9(8) - 4(20)]^2}{(4)(5)}\right\}$$

$$= \frac{2}{63}\left(140 - \frac{64}{20}\right) = 4.3429$$

and, following Eq. (5.28), the observed value of B_{N1} is

$$B_{N1o} = \frac{(4)(20)}{9} - \left\{\frac{(4)(5)}{9^2}\left[(9)(60) - 20^2 - \frac{9(9-2)(4.3429)}{2}\right]\right\}^{1/2}$$

$$= 8.8889 - \left[0.2469(140.00 - 136.80)\right]^{1/2} = 8.00\ .$$

Because of the relationship between B_{N1} and δ, the exact probability value of the realized value of B_{N1} is given by

$$P\left(B_{N1} \geq B_{N1o}|H_0\right) = P\left(\delta \leq \delta_o|H_0\right) = \frac{\text{number of } \delta \text{ values} \leq \delta_o}{M}\ ,$$

where B_{N1o} and δ_o denote the observed values of B_{N1} and δ, respectively. In addition, because of the relationships among W, B_{N1}, and δ, the exact probability value of Ansari–Bradley's W test statistic is given by

$$P\left(W \geq W_o|H_0\right) = P\left(\delta \leq \delta_o|H_0\right) = \frac{\text{number of } \delta \text{ values} \leq \delta_o}{M}\ ,$$

where W_o and δ_o denote the observed values of W and δ, respectively.

Consider again the univariate rank response measurement scores listed in Fig. 5.6 where $r = 1$, $g = 2$, $n_1 = 4$, $n_2 = 5$, $N = n_1 + n_2 = 9$, and there are no tied rank scores. In this application, let $v = 2$, employing squared Euclidean distance between the rank scores, and let the treatment-group weights be given by

$$C_i = \frac{n_i - 1}{N - g}\ , \qquad i = 1, \ldots, g\ ,$$

to correspond to the Ansari–Bradley two-sample rank-sum test. Because there are only

$$M = \frac{N!}{n_1!\ n_2!} = \frac{9!}{4!\ 5!} = 126$$

possible, equally-likely arrangements of the $N = 9$ rank scores, an exact solution is preferred. Following Eq. (5.2) on p. 218, the $N = 9$ univariate rank scores listed in Fig. 5.6 yield $g = 2$ average distance-function values of

$$\xi_1 = 6.6667 \quad \text{and} \quad \xi_2 = 2.60 \; .$$

Following Eq. (5.1) on p. 217, the observed value of the MRPP test statistic based on $v = 2$ and treatment-group weights

$$C_i = \frac{n_i - 1}{N - g} \; , \qquad i = 1, 2 \; ,$$

is

$$\delta_o = \sum_{i=1}^{g} C_i \xi_i = \frac{1}{9-2}\big[(4-1)(6.6667) + (5-1)(2.60)\big] = 4.3429 \; .$$

If all arrangements of the $N = 9$ observed rank scores listed in Fig. 5.6 occur with equal chance, the exact probability value of $\delta_o = 4.3429$ computed on the $M = 126$ possible arrangements of the observed data with $n_1 = 4$ A univariate rank scores and $n_2 = 5$ B univariate rank scores preserved for each arrangement is

$$P\big(\delta \leq \delta_o | H_0\big) = \frac{\text{number of } \delta \text{ values} \leq \delta_o}{M} = \frac{104}{126} = 0.8254 \; .$$

For comparison, a conventional Ansari–Bradley two-sample rank-sum test on the univariate rank scores listed in Fig. 5.6 yields an observed Ansari–Bradley test statistic value of

$$W_o = \sum_{i=1}^{n_1} R_i = 1 + 2 + 4 + 5 = 12$$

and the exact probability value of $W_o = 12$ is

$$P(W \geq W_o | H_0) = \frac{\text{number of } W \text{ values} \geq W_o}{M} = \frac{104}{126} = 0.8254 \; .$$

Alternatively, test statistic W is approximately distributed as $N(0, 1)$ under the null hypothesis as $N \to \infty$. For the rank scores listed in Fig. 5.6, the mean value of W is given by

$$\mu_W = \frac{n_1(N+2)}{4} \; , \qquad \text{if } N \text{ is even} \; ,$$

or

$$\mu_W = \frac{n_1(N+1)^2}{4N} \,, \qquad \text{if } N \text{ is odd} \,,$$

the variance of W is given by

$$\sigma_W^2 = \frac{n_1 n_2(N+2)(N-2)}{48(N-1)} \,, \qquad \text{if } N \text{ is even} \,,$$

or

$$\sigma_W^2 = \frac{n_1 n_2(N+1)(3+N^2)}{48N^2} \,, \qquad \text{if } N \text{ is odd} \,,$$

and the standard score is given by

$$z = \frac{W - \mu_W}{\sqrt{\sigma_W^2}} \,.$$

Since, for this example $N = 9$ is odd, the mean value of W is

$$\mu_W = \frac{n_1(N+1)^2}{4N} = \frac{4(9+1)^2}{(4)(9)} = 11.1111 \,,$$

the variance of W is

$$\sigma_W^2 = \frac{n_1 n_2(N+1)(3+N^2)}{48N^2} = \frac{(4)(5)(9+1)(3+9^2)}{(48)(9^2)} = 4.3210 \,,$$

the observed standard score is

$$z_o = \frac{W_o - \mu_W}{\sqrt{\sigma_W^2}} = \frac{12 - 11.1111}{\sqrt{4.3210}} = +0.4276 \,, \tag{5.29}$$

and the approximate two-tailed $N(0,1)$ probability value is $P = 0.6689$. A correction for continuity applied to Eq. (5.29), as suggested by Ansari and Bradley [10, p. 1181], yields an observed standard score of $z_o = +0.1871$ and an approximate two-tailed $N(0,1)$ probability value of $P = 0.8516$, which is closer to the exact probability value of $P = 0.8254$.

Following Eq. (5.17) on p. 238, the exact expected value of the $M = 126$ δ values is $\mu_\delta = 3.8889$ and, following Eq. (5.18) on p. 238, the observed chance-corrected measure of effect size is

$$\mathfrak{R}_o = 1 - \frac{\delta_o}{\mu_\delta} = 1 - \frac{4.3429}{3.8889} = -0.1167 ,$$

indicating substantially less than chance within-group agreement.

5.11.2 Example 2

For this second analysis of the univariate rank scores listed in Fig. 5.6 on p. 254, replicated in Fig. 5.7 for convenience, let the treatment-group weights be given by

$$C_i = \frac{n_i - 1}{N - g} , \qquad i = 1, \ldots, g ,$$

as in Example 1, but set $v = 1$ instead of $v = 2$, employing ordinary Euclidean distance between the rank scores.

Following Eq. (5.2) on p. 218, the $N = 9$ univariate rank scores listed in Fig. 5.7 yield $g = 2$ average distance-function values of

$$\xi_1 = 2.3333 \quad \text{and} \quad \xi_2 = 1.40 .$$

Following Eq. (5.1) on p. 217, the observed value of the MRPP test statistic based on $v = 1$ and treatment-group weights

$$C_i = \frac{n_i - 1}{N - g} , \qquad i = 1, 2 ,$$

is

$$\delta_o = \sum_{i=1}^{g} C_i \xi_i = \frac{1}{9-2}\big[(4-1)(2.3333) + (5-1)(1.40)\big] = 1.80 .$$

Rank:	1	2	3	4	5	6	7	8	9
Sample:	*A*	*A*	*B*	*A*	*A*	*B*	*B*	*B*	*B*

Fig. 5.7 Example univariate rank-score data for two-sample rank tests with $n_1 = 4$, $n_2 = 5$, and $N = n_1 + n_2 = 9$

If all arrangements of the $N = 9$ observed rank scores listed in Fig. 5.7 occur with equal chance, the exact probability value of $\delta_o = 1.80$ computed on the $M = 126$ possible arrangements of the observed data with $n_1 = 4$ A univariate rank scores and $n_2 = 5$ B univariate rank scores preserved for each arrangement is

$$P(\delta \leq \delta_o|H_0) = \frac{\text{number of } \delta \text{ values} \leq \delta_o}{M} = \frac{98}{126} = 0.7778 \ .$$

For comparison, the exact probability value based on $v = 2$, $M = 126$, and $C_i = (n_i - 1)/(N - g)$ for $i = 1, 2$ in Example 1 is $P = 0.8254$. No comparison is made with the conventional Ansari–Bradley two-sample rank-sum test as the Ansari–Bradley test is undefined for $v = 1$.

Following Eq. (5.17) on p. 238, the exact expected value of the $M = 126$ δ values is $\mu_\delta = 1.6667$ and, following Eq. (5.18) on p. 238, the observed chance-corrected measure of effect size is

$$\mathfrak{R}_o = 1 - \frac{\delta_o}{\mu_\delta} = 1 - \frac{1.80}{1.6667} = -0.0800 \ ,$$

indicating slightly less than chance within-group agreement.

5.11.3 Example 3

The treatment-group weighting function given by

$$C_i = \frac{n_i - 1}{N - g} \ , \qquad i = 1, \ldots, g \ ,$$

is not required for a permutation test. Thus, for this third analysis of the univariate rank scores listed in Fig. 5.7, the treatment-group weighting function is set to

$$C_i = \frac{n_i}{N} \ , \qquad i = 1, \ldots, g \ ,$$

weighting each treatment group proportional to its size, and setting $v = 1$, employing ordinary Euclidean distance between the rank scores, as in Example 2. Following Eq. (5.2) on p. 218, the $N = 9$ univariate rank scores listed in Fig. 5.7 yield $g = 2$ average distance-function values of

$$\xi_1 = 2.3333 \quad \text{and} \quad \xi_2 = 1.40 \ .$$

Following Eq. (5.1) on p. 217, the observed value of the MRPP test statistic based on $v = 1$ and treatment-group weights

$$C_i = \frac{n_i}{N} \ , \qquad i = 1, 2 \ ,$$

is

$$\delta_o = \sum_{i=1}^{g} C_i \xi_i = \frac{1}{9}\big[(4)(2.3333) + (5)(1.40)\big] = 1.8148 \, .$$

If all arrangements of the $N = 9$ observed rank scores listed in Fig. 5.7 occur with equal chance, the exact probability value of $\delta_o = 1.8148$ computed on the $M = 126$ possible arrangements of the observed data with $n_1 = 4$ A univariate rank scores and $n_2 = 5$ B univariate rank scores preserved for each arrangement is

$$P(\delta \leq \delta_o | H_0) = \frac{\text{number of } \delta \text{ values} \leq \delta_o}{M} = \frac{102}{126} = 0.8095 \, .$$

For comparison, the exact probability values based on $v = 2$, $M = 126$, and $C_i = (n_i - 1)/(N - g)$ for $i = 1, 2$ in Example 1 and $v = 1$, $M = 126$, and $C_i = (n_i - 1)/(N - g)$ for $i = 1, 2$ in Example 2 are $P = 0.8254$ and $P = 0.7778$, respectively. No comparison is made with the conventional Ansari–Bradley two-sample rank-sum test as the Ansari–Bradley two-sample test is undefined for both $v = 1$ and $C_i = n_i/N$, $i = 1, \ldots, g$.

Following Eq. (5.17) on p. 238, the exact expected value of the $M = 126$ δ values is $\mu_\delta = 1.6667$ and, following Eq. (5.18) on p. 238, the observed chance-corrected measure of effect size is

$$\Re_o = 1 - \frac{\delta_o}{\mu_\delta} = 1 - \frac{1.8148}{1.6667} = -0.0889 \, ,$$

indicating slightly less than chance within-group agreement.

5.12 Example B_{Ns} Analyses with $s = 2$

In 1954 Alexander Mood published a two-sample rank-sum test for dispersion, given by

$$W = \sum_{i=1}^{n_1} \left(R_i - \frac{N+1}{2} \right)^2 , \tag{5.30}$$

where n_1 is the smaller of the two samples size, N is the total number of rank scores in both samples, and R_i denotes a rank score for $i = 1, \ldots, n_1$ [312]. In this section, three example analyses illustrate the B_{Ns} rank function test with $s = 2$. The first example is designed to correspond to the conventional Mood two-sample rank-sum test using a small set of univariate rank scores with $v = 2$ and treatment-group weights $C_i = (n_i - 1)/(N - g)$ for $i = 1, \ldots, g$; the second example analyzes the same small set of univariate rank scores, but uses $v = 1$ and treatment-group weights $C_i = (n_i - 1)/(N - g)$ for $i = 1, \ldots, g$; and the third example analyzes the same set of univariate rank scores using $v = 1$, but adopts a proportional treatment-group weighting function given by $C_i = n_i/N$ for $i = 1, \ldots, g$.

5.12.1 Example 1

Consider the univariate rank scores listed in Fig. 5.2 on p. 239, replicated in Fig. 5.8 for convenience, where $n_1 = 4$, $n_2 = 5$, and $N = n_1 + n_2 = 9$. The median of the $N = 9$ rank scores listed in Fig. 5.8 is

$$\frac{N+1}{2} = \frac{9+1}{2} = 5$$

and, following Eq. (5.30) on p. 262, the observed value of Mood's W is

$$W_o = \sum_{i=1}^{n_1} \left(R_i - \frac{N+1}{2} \right)^2$$

$$= (1-5)^2 + (2-5)^2 + (4-5)^2 + (5-5)^2$$

$$= 16 + 9 + 1 + 0 = 26 \ ,$$

where $n_1 = 4$ is the smaller of the two sample sizes.

For comparison, following Eq. (5.12) on p. 236, consider B_{Ns} with $s = 2$, where the observed value of B_{N2} is

$$B_{N2o} = \sum_{i=1}^{N} \left| R_i - \frac{N+1}{2} \right|^2 Z_{Ni}$$

$$= |1-5|^2(1) + |2-5|^2(1) + |3-5|^2(0) + |4-5|^2(1) + |5-5|^2(1)$$

$$+ |6-5|^2(0) + |7-5|^2(0) + |8-5|^2(0) + |9-5|^2(0)$$

$$= 16 + 9 + 0 + 1 + 0 + 0 + 0 + 0 + 0 = 26 \ .$$

Thus, Mielke's B_{N2} and Mood's W are shown to be identical.

Rank:	1	2	3	4	5	6	7	8	9
Sample:	A	A	B	A	A	B	B	B	B

Fig. 5.8 Example univariate rank-score data for two-sample rank tests with $n_1 = 4$, $n_2 = 5$, and $N = n_1 + n_2 = 9$

Following Eqs. (5.15) on p. 237 and (5.16) on p. 238, the functional relationships between statistics δ and B_{N2} are given by

$$\delta = \frac{2}{N(N-2)}\left[NT - S^2 - \frac{(NB_{N2} - n_1 S)^2}{n_1 n_2}\right] \tag{5.31}$$

and

$$B_{N2} = \frac{n_1 S}{N} - \left\{\frac{n_1 n_2}{N^2}\left[NT - S^2 - \frac{N(N-2)\delta}{2}\right]\right\}^{1/2}, \tag{5.32}$$

where

$$S = \sum_{i=1}^{N}\left|R_i - \frac{N+1}{2}\right|^2,$$

$$T = \sum_{i=1}^{N}\left[\left|R_i - \frac{N+1}{2}\right|^2\right]^2 = \sum_{i=1}^{N}\left|R_i - \frac{N+1}{2}\right|^4,$$

and R_i is the rank function of the ith response measurement, $i = 1, \ldots, N$. Thus, the median of the $N = 9$ rank scores listed in Fig. 5.8 is

$$\frac{N+1}{2} = \frac{9+1}{2} = 5$$

and the observed values of S and T are

$$\begin{aligned} S_o = {} & |1-5|^2 + |2-5|^2 + |3-5|^2 + |4-5|^2 + |5-5|^2 \\ & + |6-5|^2 + |7-5|^2 + |8-5|^2 + |9-5|^2 \\ & = 16 + 9 + 4 + 1 + 0 + 1 + 4 + 9 + 16 = 60 \end{aligned}$$

and

$$\begin{aligned} T_o = {} & |1-5|^4 + |2-5|^4 + |3-5|^4 + |4-5|^4 + |5-5|^4 \\ & + |6-5|^4 + |7-5|^4 + |8-5|^4 + |9-5|^4 \\ & = 256 + 81 + 16 + 1 + 0 + 1 + 16 + 81 + 256 = 708\,. \end{aligned}$$

Utilizing the univariate rank scores listed in Fig. 5.8, where there are no tied rank scores, the relationship between statistics δ and B_{N2} can be confirmed. Thus, following Eq. (5.31) on p. 264, the observed value of the MRPP test statistic for the univariate rank scores listed in Fig. 5.8 is

$$\delta_o = \frac{2}{9(9-2)} \left\{ 9(708) - (60)^2 - \frac{[9(26) - 4(60)]^2}{(4)(5)} \right\}$$

$$= \frac{2}{63}\left(2{,}772 - \frac{36}{20}\right) = 87.9429$$

and, following Eq. (5.32) on p. 264, the observed value of B_{N2} is

$$B_{N2o} = \frac{(4)(60)}{9} - \left\{ \frac{(4)(5)}{9^2} \left[(9)(708) - 60^2 - \frac{9(9-2)(87.9429)}{2} \right] \right\}^{1/2}$$

$$= 26.6667 - [0.2469(2{,}772.00 - 2770.20)]^{1/2} = 26.00\ .$$

Because of the relationship between statistics B_{N2} and δ, the exact probability value of the realized value of B_{N2} is given by

$$P(B_{N2} \geq B_{N2o}|H_0) = P(\delta \leq \delta_o|H_0) = \frac{\text{number of } \delta \text{ values} \leq \delta_o}{M}\ ,$$

where B_{N2o} and δ_o denote the observed values of B_{N2} and δ, respectively. In addition, because of the relationships among W, B_{N2}, and δ, the exact probability value of Mood's W is given by

$$P(W \geq W_o|H_0) = P(\delta \leq \delta_o|H_0) = \frac{\text{number of } \delta \text{ values} \leq \delta_o}{M}\ ,$$

where W_o and δ_o denote the observed values of W and δ, respectively.

Consider again the univariate rank scores listed in Fig. 5.8 where $r = 1$, $g = 2$, $n_1 = 4$, $n_2 = 5$, $N = n_1 + n_2 = 9$, and there are no tied rank scores. In this application, let $v = 2$, employing squared Euclidean distance between the rank scores, and let the treatment-group weights be given by

$$C_i = \frac{n_i - 1}{N - g}\ , \qquad i = 1, \ldots, g\ ,$$

to correspond to Mood's two-sample rank-sum test. Because there are only

$$M = \frac{N!}{n_1!\, n_2!} = \frac{9!}{4!\ 5!} = 126$$

possible, equally-likely arrangements of the $N = 9$ rank scores, an exact solution is preferred. Following Eq. (5.2) on p. 218, the $N = 9$ univariate rank scores listed in Fig. 5.8 yield $g = 2$ average distance-function values of

$$\xi_1 = 112.6667 \quad \text{and} \quad \xi_2 = 69.40 \; .$$

Following Eq. (5.1) on p. 217, the observed value of the MRPP test statistic based on $v = 2$ and treatment-group weights

$$C_i = \frac{n_i - 1}{N - g} \; , \qquad i = 1, 2 \; ,$$

is

$$\delta_o = \sum_{i=1}^{g} C_i \xi_i = \frac{1}{9-2}\big[(4-1)(112.6667) + (5-1)(69.40)\big] = 87.9429 \; .$$

If all arrangements of the $N = 9$ observed rank scores listed in Fig. 5.8 occur with equal chance, the exact probability value of $\delta_o = 87.9429$ computed on the $M = 126$ possible arrangements of the observed data with $n_1 = 4$ A univariate rank scores and $n_2 = 5$ B univariate rank scores preserved for each arrangement is

$$P(\delta \leq \delta_o | H_0) = \frac{\text{number of } \delta \text{ values} \leq \delta_o}{M} = \frac{122}{126} = 0.9683 \; .$$

For comparison, a conventional Mood two-sample rank-sum test computed on the univariate rank scores listed in Fig. 5.8 yields an observed test statistic value of

$$W_o = \sum_{i=1}^{n_1} \left(R_i - \frac{N+1}{2} \right)^2 = (1-5)^2 + (2-5)^2 + (4-5)^2 + (5-5)^2$$

$$= 16 + 9 + 1 + 0 = 26$$

and the exact probability value of $W_o = 26$ is

$$P(W \geq W_o | H_0) = \frac{\text{number of } W \text{ values} \geq W_o}{M} = \frac{122}{126} = 0.9683 \; .$$

For comparison, Mood's test statistic W is approximately distributed under the null hypothesis as $N(0, 1)$ as $N \to \infty$ with mean given by

$$\mu_W = \frac{n_1(N+1)(N-1)}{12}$$

and variance given by

$$\sigma_W^2 = \frac{n_1 n_2 (N+1)(N+2)(N-2)}{180} ,$$

where n_1 denotes the smaller of the two sample sizes. Thus, for the univariate rank scores listed in Fig. 5.8 on p. 263,

$$\mu_W = \frac{4(9+1)(9-1)}{12} = \frac{320}{12} = 26.6667 ,$$

$$\sigma_W^2 = \frac{(4)(5)(9+1)(9+2)(9-2)}{180} = \frac{15{,}400}{180} = 85.5556 ,$$

and the observed standard score is

$$z_o = \frac{W_o - \mu_W}{\sigma_W} = \frac{26 - 26.6667}{\sqrt{85.5556}} = -0.0721 ,$$

yielding an approximate two-sided probability value of $P = 0.9425$ under the null hypothesis, which is not far removed from the exact probability value of $P = 0.9683$.

Following Eq. (5.17) on p. 238, the exact expected value of the $M = 126$ δ values is $\mu_\delta = 77.00$ and, following Eq. (5.18) on p. 238, the observed chance-corrected measure of effect size is

$$\Re_o = 1 - \frac{\delta_o}{\mu_\delta} = 1 - \frac{87.9429}{77.00} = -0.1421 ,$$

indicating substantially less than chance within-group agreement.

5.12.2 Example 2

For this second analysis of the $N = 9$ univariate rank response measurements listed in Fig. 5.2 on p. 239, replicated in Fig. 5.9 for convenience, let the treatment-group weights be given by

$$C_i = \frac{n_i - 1}{N - g} , \qquad i = 1, \ldots, g ,$$

as in Example 1, but set $v = 1$ instead of $v = 2$, employing ordinary Euclidean distance between the rank scores.

Rank:	1	2	3	4	5	6	7	8	9
Sample:	A	A	B	A	A	B	B	B	B

Fig. 5.9 Example univariate rank-score data for two-sample rank tests with $n_1 = 4$, $n_2 = 5$, and $N = n_1 + n_2 = 9$

Following Eq. (5.2) on p. 218, the $N = 9$ univariate rank scores listed in Fig. 5.9 yield $g = 2$ average distance-function values of

$$\xi_1 = 9.3333 \quad \text{and} \quad \xi_2 = 7.00 \; .$$

Following Eq. (5.1) on p. 217, the observed value of the MRPP test statistic based on $v = 1$ and treatment-group weights

$$C_i = \frac{n_i - 1}{N - g} \; , \qquad i = 1, 2 \; ,$$

is

$$\delta_o = \sum_{i=1}^{g} C_i \xi_i = \frac{1}{9-2}\big[(4-1)(9.3333) + (5-1)(7.00)\big] = 8.00 \; .$$

If all arrangements of the $N = 9$ observed rank scores listed in Fig. 5.9 occur with equal chance, the exact probability value of $\delta_o = 8.00$ computed on the $M = 126$ possible arrangements of the observed data with $n_1 = 4$ A univariate rank scores and $n_2 = 5$ B univariate rank scores preserved for each arrangement is

$$P\big(\delta \leq \delta_o | H_0\big) = \frac{\text{number of } \delta \text{ values} \leq \delta_o}{M} = \frac{98}{126} = 0.7778 \; .$$

For comparison, the exact probability value based on $v = 2$, $M = 126$, and $C_i = (n_i - 1)/(N - g)$ for $i = 1, 2$ in Example 1 is $P = 0.9683$. No comparison is made with the conventional Mood two-sample rank-sum test as Mood's two-sample test is undefined for $v = 1$.

Following Eq. (5.17) on p. 238, the exact expected value of the $M = 126$ δ values is $\mu_\delta = 7.2222$ and, following Eq. (5.18) on p. 238, the observed chance-corrected measure of effect size is

$$\Re_o = 1 - \frac{\delta_o}{\mu_\delta} = 1 - \frac{8.00}{7.2222} = -0.1077 \; ,$$

indicating less than chance within-group agreement.

5.12.3 Example 3

The treatment-group weighting function given by

$$C_i = \frac{n_i - 1}{N - g} , \qquad i = 1, \ldots, g ,$$

is not required for a permutation test. Thus, for this third analysis of the univariate rank scores listed in Fig. 5.9, the treatment-group weighting function is set to

$$C_i = \frac{n_i}{N} , \qquad i = 1, \ldots, g ,$$

simply weighting each treatment group proportional to its size, and setting $v = 1$, employing ordinary Euclidean distance between the rank scores, as in Example 2. Following Eq. (5.2) on p. 218, the $N = 9$ univariate rank scores listed in Fig. 5.9 yield $g = 2$ average distance-function values of

$$\xi_1 = 9.3333 \quad \text{and} \quad \xi_2 = 7.00 .$$

Following Eq. (5.1) on p. 217, the observed value of the MRPP test statistic based on $v = 1$ and treatment-group weights

$$C_i = \frac{n_i}{N} , \qquad i = 1, 2 ,$$

is

$$\delta_o = \sum_{i=1}^{g} C_i \xi_i = \frac{1}{9}\big[(4)(9.3333) + (5)(7.00)\big] = 8.0370 .$$

If all arrangements of the $N = 9$ observed rank scores listed in Fig. 5.9 occur with equal chance, the exact probability value of $\delta_o = 8.0370$ computed on the $M = 126$ possible arrangements of the observed data with $n_1 = 4$ A univariate rank scores and $n_2 = 5$ B univariate rank scores preserved for each arrangement is

$$P\big(\delta \leq \delta_o | H_0\big) = \frac{\text{number of } \delta \text{ values} \leq \delta_o}{M} = \frac{102}{126} = 0.8095 .$$

For comparison, the exact probability values based on $v = 2$, $M = 126$, and $C_i = (n_i - 1)/(N - g)$ for $i = 1, 2$ in Example 1 and $v = 1$, $M = 126$, and $C_i = (n_i - 1)/(N - g)$ for $i = 1, 2$ in Example 2 are $P = 0.9683$ and $P = 0.7778$, respectively. No comparison is made with the conventional Mood two-sample rank-sum test as Mood's two-sample test is undefined for both $v = 1$ and $C_i = n_i/N$, $i = 1, \ldots, g$.

Following Eq. (5.17) on p. 238, the exact expected value of the $M = 126$ δ values is $\mu_\delta = 7.2222$ and, following Eq. (5.18) on p. 238, the observed chance-corrected measure of effect size is

$$\Re_o = 1 - \frac{\delta_o}{\mu_\delta} = 1 - \frac{8.0370}{7.2222} = -0.1128 ,$$

indicating less than chance within-group agreement.

5.13 Example C_{Ns} Analyses with $s = 0$

The Brown–Mood median test provides a test of the null hypothesis that the medians of the populations from which two (or more) samples are drawn are identical [59]. In the case of two independent samples, the data in each sample are assigned to two groups, one consisting of observations with values higher than the median value of the two groups combined, and the other consisting of observations with values equal to or less than the median of the combined samples [77, pp. 218–219].[14]

In this section, three example analyses illustrate the C_{Ns} rank function test with $s = 0$. The first example is designed to correspond to the conventional Brown–Mood median test using a small set of univariate rank scores with $v = 2$ and treatment-group weights $C_i = (n_i - 1)/(N - g)$ for $i = 1, \ldots, g$; the second example analyzes the same small set of univariate rank scores, but uses $v = 1$ and treatment-group weights $C_i = (n_i - 1)/(N - g)$ for $i = 1, \ldots, g$; and the third example analyzes the same set of univariate rank scores using $v = 1$, but adopts a proportional treatment-group weighting function given by $C_i = n_i/N$ for $i = 1, \ldots, g$.

5.13.1 Example 1

Consider the univariate rank scores listed in Fig. 5.2 on p. 239, replicated in Fig. 5.10 for convenience. The combined median is

$$\frac{N+1}{2} = \frac{9+1}{2} = 5$$

[14] An alternative median test that could be considered in this context is the Hodges–Lehmann median test, introduced by Hodges and Lehmann in 1963 [179]. As noted by Newson [318, p. 59], the Hodges–Lehmann median test was later popularized by Conover [77], Campbell and Gardner [62], and Gardner and Altman [133].

Rank:	1	2	3	4	5	6	7	8	9
Sample:	A	A	B	A	A	B	B	B	B

Fig. 5.10 Example univariate rank-score data for two-sample rank tests with $n_1 = 4$, $n_2 = 5$, and $N = n_1 + n_2 = 9$

	Variable	
	A	B
> Median	0	4
≤ Median	4	1

Fig. 5.11 Counts of rank-score values above and below the combined median of 5

and all four A rank scores are equal to or less than the combined median of 5 (ranks 1, 2, 4, and 5) and no A rank scores are greater than the combined median of 5. Likewise, one B rank score is equal to or less than the combined median of 5 (rank 3) and four B rank scores are greater than the combined median of 5 (ranks 6, 7, 8, and 9).

As Brown and Mood explained, the values obtained by counting the positive and negative deviations in each group form a contingency table with all marginal totals fixed and may be evaluated by the ordinary chi-squared criterion when N is large [59, p. 164]. Figure 5.11 displays the deviations obtained from the univariate rank scores listed in Fig. 5.10.

For comparison, following Eq. (5.13) on p. 237, consider C_{Ns} with $s = 0$, where the combined median is given by

$$\frac{N+1}{2} = \frac{9+1}{2} = 5$$

and the observed value of C_{N0} is

$$\begin{aligned} C_{N0o} &= \sum_{i=1}^{N} h(R_i^0, N, 0) Z_{Ni} \\ &= -|1-5|^0(1) - |2-5|^0(1) - |3-5|^0(0) - |4-5|^0(1) - |5-5|^0(1) \\ &\quad + |6-5|^0(0) + |7-5|^0(0) + |8-5|^0(0) + |9-5|^0(0) \\ &= -1 - 1 - 0 - 1 - 0 + 0 + 0 + 0 + 0 = -3\,. \end{aligned}$$

The relationship between C_{N0} and the Brown–Mood median test is complicated by standard presentations in textbooks, e.g., Conover [77, pp. 218–219]. Define D as the number of rank scores greater than the combined median minus the number of rank scores less than or equal to the combined median. For the frequencies listed in

Fig. 5.11, $D = 0 - 4 = -4$ for variable A. This is the standard textbook presentation and it is obviously inconsistent with $C_{NO} = -3$. However, in the original article by Brown and Mood, it was tacitly assumed that no rank score would be exactly equal to the combined median, in this case 5. As seen in Fig. 5.10, $A = 5$ is equal to the combined median and, as noted in Eq. (5.14) on p. 237, $|5 - 5|^0(1)$ is defined as zero.

For comparison, consider variable B in Fig. 5.10 instead of variable A, where no B rank score is equal to the combined median of 5. Then following Eq. (5.13) on p. 237, the observed value of C_{NO} is

$$\begin{aligned} C_{NO_o} &= \sum_{i=1}^{N} h(R_i^0, N, 0) Z_{Ni} \\ &= -|1-5|^0(0) - |2-5|^0(0) - |3-5|^0(1) - |4-5|^0(0) - |5-5|^0(0) \\ &\quad + |6-5|^0(1) + |7-5|^0(1) + |8-5|^0(1) + |9-5|^0(1) \\ &= -0 - 0 - 1 - 0 - 0 + 1 + 1 + 1 + 1 = +3 \,, \end{aligned}$$

which corresponds to $D = 4 - 1 = +3$ for variable B in Fig. 5.11. Thus, C_{NO} is identical to the Brown–Mood median test when no rank is equal to the combined median, as originally described by Brown and Mood [59].

Following Eqs. (5.15) on p. 237 and (5.16) on p. 238, the functional relationships between C_{NO} and δ are given by

$$\delta = \frac{2}{N(N-2)} \left[NT - S^2 - \frac{(NC_{NO} - n_1 S)^2}{n_1 n_2} \right] \tag{5.33}$$

and

$$C_{NO} = \frac{n_1 S}{N} - \left\{ \frac{n_1 n_2}{N^2} \left[NT - S^2 - \frac{N(N-2)\delta}{2} \right] \right\}^{1/2} , \tag{5.34}$$

where

$$S = \sum_{i=1}^{N} h(R_i, N, 0) \,, \qquad T = \sum_{i=1}^{N} \left[h(R_i, N, 0) \right]^2 ,$$

and $h(R_i, N, 0)$ is the rank function of the ith response measurement.

Utilizing the univariate rank scores listed in Fig. 5.10, the relationship between statistics δ and C_{N0} can be confirmed. Thus, for the univariate rank scores listed in Fig. 5.10, the observed values of S and T are

$$\begin{aligned} S_o = &-|1-5|^0 - |2-5|^0 - |3-5|^0 - |4-5|^0 - |5-5|^0 \\ &+ |6-5|^0 + |7-5|^0 + |8-5|^0 + |9-5|^0 \\ &= -1-1-1-1-0+1+1+1+1 = 0 \end{aligned}$$

and

$$\begin{aligned} T_o = &(-|1-5|^0)^2 + (-|2-5|^0)^2 + (-|3-5|^0)^2 + (-|4-5|^0)^2 \\ &+ (-|5-5|^0)^2 + (+|6-5|^0)^2 + (+|7-5|^0)^2 + (+|8-5|^0)^2 \\ &+ (+|9-5|^0)^2 = 1+1+1+1+0+1+1+1+1 = 8\,. \end{aligned}$$

Then, following Eq. (5.33), the observed value of the MRPP test statistic for the univariate rank scores listed in Fig. 5.10 is

$$\begin{aligned} \delta_o &= \frac{2}{9(9-2)} \left\{ 9(8) - (0)^2 - \frac{[9(-3) - 4(0)]^2}{(4)(5)} \right\} \\ &= \frac{2}{63}\left(72 - \frac{729}{20}\right) = 1.1286 \end{aligned}$$

and, following Eq. (5.34), the observed value of C_{N0} is

$$\begin{aligned} C_{N0o} &= \frac{(4)(0)}{9} - \left\{ \frac{(4)(5)}{9^2} \left[9(8) - 0^2 - \frac{9(9-2)(1.1286)}{2} \right] \right\}^{1/2} \\ &= 0 - \left[0.2469(72.00 - 35.5509)\right]^{1/2} = -3.00\,. \end{aligned}$$

Because of the relationship between statistics C_{N0} and δ, the exact probability value of the realized value of C_{N0} is given by

$$P\big(C_{N0} \geq C_{N0o}|H_0\big) = P\big(\delta \leq \delta_o|H_0\big) = \frac{\text{number of } \delta \text{ values} \leq \delta_o}{M}\,,$$

where C_{NOo} and δ_o denote the observed values of C_{NO} and δ, respectively. In addition, because of the relationships among D, C_{NO}, and δ, the exact probability value of Brown–Mood's test statistic D is given by

$$P(D \geq D_o | H_0) = P(\delta \leq \delta_o | H_0) = \frac{\text{number of } \delta \text{ values} \leq \delta_o}{M} ,$$

where D_o and δ_o denote the observed values of D and δ, respectively.

Consider again the univariate rank response measurement scores listed in Fig. 5.10 where $r = 1$, $g = 2$, $n_1 = 4$, $n_2 = 5$, $N = n_1 + n_2 = 9$, and there are no tied rank scores. In this application, let $v = 2$, employing squared Euclidean distance between the rank scores, and let the treatment-group weights be given by

$$C_i = \frac{n_i - 1}{N - g} , \qquad i = 1, \ldots, g ,$$

to correspond to the Brown–Mood median test. Because there are only

$$M = \frac{N!}{n_1!\, n_2!} = \frac{9!}{4!\, 5!} = 126$$

possible, equally-likely arrangements of the $N = 9$ rank scores, an exact solution is preferred. Following Eq. (5.2) on p. 218, the $N = 9$ univariate rank scores listed in Fig. 5.10 yield $g = 2$ average distance-function values of

$$\xi_1 = 0.50 \quad \text{and} \quad \xi_2 = 1.60 .$$

Following Eq. (5.1) on p. 217, the observed value of the MRPP test statistic based on $v = 2$ and treatment-group weights

$$C_i = \frac{n_i - 1}{N - g} , \qquad i = 1, 2 ,$$

is

$$\delta_o = \sum_{i=1}^{g} C_i \xi_i = \frac{1}{9 - 2} \big[(4 - 1)(0.50) + (5 - 1)(1.60) \big] = 1.1286 .$$

If all arrangements of the $N = 9$ observed rank scores listed in Fig. 5.10 occur with equal chance, the exact probability value of $\delta_o = 1.1286$ computed on the $M = 126$ possible arrangements of the observed data with $n_1 = 4$ A univariate rank scores and $n_2 = 5$ B univariate rank scores preserved for each arrangement is

$$P(\delta \leq \delta_o | H_0) = \frac{\text{number of } \delta \text{ values} \leq \delta_o}{M} = \frac{10}{126} = 0.0794 .$$

For comparison, the Brown–Mood median test is conventionally evaluated by the chi-squared test for independence with $g - 1$ degrees of freedom, when N is large. For the frequency data listed in Fig. 5.11, $\chi^2 = 5.76$, and with $g - 1 = 2 - 1 = 1$ degree of freedom, the approximate probability value is $P = 0.0164$. However, in this analysis $N = 9$ rank scores is not considered to be "large." An alternative is an exact chi-squared test based on the sum of the hypergeometric probability values associated with the observed chi-squared value or those chi-squared values that are larger [26]. For the frequency data given in Fig. 5.11 on p. 271, the exact chi-squared probability value is $P = 0.0476$. These probability values ($P = 0.0164$ and $P = 0.0476$) may be compared with the exact probability value of δ_o, which is $P = 0.0794$.

Following Eq. (5.17) on p. 238, the exact expected value of the $M = 126$ δ values is $\mu_\delta = 2.00$ and, following Eq. (5.18) on p. 238, the observed chance-corrected measure of effect size is

$$\Re_o = 1 - \frac{\delta_o}{\mu_\delta} = 1 - \frac{1.1286}{2.00} = +0.4357 ,$$

indicating approximately 44 % within-group agreement above that expected by chance.

5.13.2 Example 2

For this second analysis of the univariate rank scores listed in Fig. 5.2 on p. 239, replicated in Fig. 5.12 for convenience, let the treatment-group weights be given by

$$C_i = \frac{n_i - 1}{N - g} , \qquad i = 1, \ldots, g ,$$

as in Example 1, but set $v = 1$ instead of $v = 2$, employing ordinary Euclidean distance between the rank scores.

Following Eq. (5.2) on p. 218, the $N = 9$ univariate rank scores listed in Fig. 5.12 yield $g = 2$ average distance-function values of

$$\xi_1 = 0.50 \quad \text{and} \quad \xi_2 = 0.80 .$$

Rank:	1	2	3	4	5	6	7	8	9
Sample:	*A*	*A*	*B*	*A*	*A*	*B*	*B*	*B*	*B*

Fig. 5.12 Example univariate rank-score data for two-sample rank tests with $n_1 = 4$, $n_2 = 5$, and $N = n_1 + n_2 = 9$

Following Eq. (5.1) on p. 217, the observed value of the MRPP test statistic based on $v = 1$ and treatment-group weights

$$C_i = \frac{n_i - 1}{N - g}\,, \qquad i = 1, 2\,,$$

is

$$\delta_o = \sum_{i=1}^{g} C_i \xi_i = \frac{1}{9-2}\big[(4-1)(0.50) + (5-1)(0.80)\big] = 0.6714\,.$$

If all arrangements of the $N = 9$ observed rank scores listed in Fig. 5.12 occur with equal chance, the exact probability value of $\delta_o = 0.6714$ computed on the $M = 126$ possible arrangements of the observed data with $n_1 = 4$ A univariate rank scores and $n_2 = 5$ B univariate rank scores preserved for each arrangement is

$$P(\delta \leq \delta_o | H_0) = \frac{\text{number of } \delta \text{ values} \leq \delta_o}{M} = \frac{10}{126} = 0.0794\,.$$

For comparison, the exact probability value based on $v = 2$, $M = 126$, and $C_i = (n_i - 1)/(N - g)$ for $i = 1, 2$ in Example 1 is also $P = 0.0794$. No comparison is made with the conventional Brown–Mood median test as the Brown–Mood test is undefined for $v = 1$.

Following Eq. (5.17) on p. 238, the exact expected value of the $M = 126$ δ values is $\mu_\delta = 1.1111$ and, following Eq. (5.18) on p. 238, the observed chance-corrected measure of effect size is

$$\Re_o = 1 - \frac{\delta_o}{\mu_\delta} = 1 - \frac{0.6714}{1.1111} = +0.3957\,,$$

indicating approximately 40 % within-group agreement above that expected by chance.

5.13.3 Example 3

The treatment-group weighting function given by

$$C_i = \frac{n_i - 1}{N - g}\,, \qquad i = 1, \ldots, g\,,$$

is not required for a permutation test. Thus, for this third analysis of the univariate rank scores listed in Fig. 5.12, the treatment-group weighting function is set to

$$C_i = \frac{n_i}{N}\,, \qquad i = 1, \ldots, g\,,$$

simply weighting each treatment group proportional to its size, and setting $v = 1$, employing ordinary Euclidean distance between the rank scores, as in Example 2. Following Eq. (5.2) on p. 218, the $N = 9$ univariate rank scores listed in Fig. 5.12 yield $g = 2$ average distance-function values of

$$\xi_1 = 0.50 \quad \text{and} \quad \xi_2 = 0.80 \ .$$

Following Eq. (5.1) on p. 217, the observed value of the MRPP test statistic based on $v = 1$ and treatment-group weights

$$C_i = \frac{n_i}{N} \ , \qquad i = 1, 2 \ ,$$

is

$$\delta_o = \sum_{i=1}^{g} C_i \xi_i = \frac{1}{9}\big[(4)(0.50) + (5)(0.80)\big] = 0.6667 \ .$$

If all arrangements of the $N = 9$ observed rank scores listed in Fig. 5.12 occur with equal chance, the exact probability value of $\delta_o = 0.6667$ computed on the $M = 126$ possible arrangements of the observed data with $n_1 = 4$ A univariate rank scores and $n_2 = 5$ B univariate rank scores preserved for each arrangement is

$$P\big(\delta \leq \delta_o | H_0\big) = \frac{\text{number of } \delta \text{ values} \leq \delta_o}{M} = \frac{10}{126} = 0.0794 \ .$$

For comparison, the exact probability values based on $v = 2$, $M = 126$, and $C_i = (n_i - 1)/(N - g)$ for $i = 1, 2$ in Example 1 and $v = 1$, $M = 126$, and $C_i = (n_i - 1)/(N - g)$ for $i = 1, 2$ in Example 2 are both $P = 0.0794$. No comparison is made with the conventional Brown–Mood median test as the Brown–Mood test is undefined for both $v = 1$ and $C_i = n_i/N$, $i = 1, \ldots, g$.

Following Eq. (5.17) on p. 238, the exact expected value of the $M = 126$ δ values is $\mu_\delta = 1.1111$ and, following Eq. (5.18) on p. 238, the observed chance-corrected measure of effect size is

$$\Re_o = 1 - \frac{\delta_o}{\mu_\delta} = 1 - \frac{0.6667}{1.1111} = +0.40 \ ,$$

indicating 40 % within-group agreement above that expected by chance.

5.14 Example C_{Ns} Analyses with $s = 1$

The Wilcoxon–Mann–Whitney two-sample rank-sum test statistic is defined as the sum of the rank scores in the smaller of the two sample sizes. Consider $g = 2$ samples with n_1 rank scores in the first sample, n_2 rank scores in the second sample, and $N = n_1 + n_2$. In this section, three example analyses illustrate the C_{Ns} rank function test with $s = 1$. The first example is designed to correspond to the conventional Wilcoxon–Mann–Whitney two-sample rank-sum test using a small set of univariate rank scores with $v = 2$ and treatment-group weights $C_i = (n_i - 1)/(N - g)$ for $i = 1, \ldots, g$; the second example analyzes the same small set of univariate rank scores, but uses $v = 1$ and treatment-group weights $C_i = (n_i - 1)/(N - g)$ for $i = 1, \ldots, g$; and the third example analyzes the same set of univariate rank scores using $v = 1$, but adopts a proportional treatment-group weighting function given by $C_i = n_i/N$ for $i = 1, \ldots, g$.

5.14.1 Example 1

Consider the $N = 9$ univariate rank response measurements listed in Fig. 5.2 on p. 239, replicated in Fig. 5.13 for convenience. For the univariate rank scores listed in Fig. 5.13, the observed Wilcoxon–Mann–Whitney test statistic is given by

$$W_o = \sum_{i=1}^{n_1} R_i = 1 + 2 + 4 + 5 = 12\,,$$

where $n_1 = 4$ is the smaller of the two sample sizes (variable A) and R_i denotes a rank score for $i = 1, \ldots, n_1$.

For comparison, consider C_{Ns} with $s = 1$, where for variable A the observed value of C_{N1} is

$$\begin{aligned} C_{N1o} &= \sum_{i=1}^{N} h(R_i^1, N, 1)\, Z_{Ni} \\ &= -\left|1-5\right|^1(1) - \left|2-5\right|^1(1) - \left|3-5\right|^1(0) - \left|4-5\right|^1(1) - \left|5-5\right|^1(1) \\ &\quad + \left|6-5\right|^1(0) + \left|7-5\right|^1(0) + \left|8-5\right|^1(0) + \left|9-5\right|^1(0) \\ &= -4 - 3 - 0 - 1 - 0 + 0 + 0 + 0 + 0 = -8\,. \end{aligned}$$

Rank:	1	2	3	4	5	6	7	8	9
Sample:	A	A	B	A	A	B	B	B	B

Fig. 5.13 Example univariate rank-score data for two-sample rank tests with $n_1 = 4$, $n_2 = 5$, and $N = n_1 + n_2 = 9$

The relationships between C_{N1} and the Wilcoxon–Mann–Whitney test statistic are given by

$$C_{N1} = W - \frac{n_1(N+1)}{2} \quad \text{and} \quad W = C_{N1} + \frac{n_1(N+1)}{2} .$$

Thus, the observed values of C_{N1} and W are

$$C_{N1\text{o}} = 12 - \frac{4(9+1)}{2} = 12 - 20 = -8$$

and

$$W_{\text{o}} = -9 + \frac{4(9+1)}{2} = -8 + 20 = 12 .$$

Also, since C_{N1} and A_{N1} are both identical to the Wilcoxon–Mann–Whitney two-sample rank-sum test, then C_{N1} and A_{N1} are necessarily identical to each other. The relationships between statistics C_{N1} and A_{N1} are given by

$$C_{N1} = A_{N1} - \frac{n_1(N+1)}{2} \quad \text{and} \quad A_{N1} = C_{N1} + \frac{n_1(N+1)}{2} ,$$

where n_1 is the smaller of the two sample sizes.

Following Eqs. (5.15) on p. 237 and (5.16) on p. 238, the functional relationships between statistics C_{N1} and δ are given by

$$\delta = \frac{2}{N(N-2)} \left[NT - S^2 - \frac{(NC_{N1} - n_1 S)^2}{n_1 n_2} \right] \tag{5.35}$$

and

$$C_{N1} = \frac{n_1 S}{N} - \left\{ \frac{n_1 n_2}{N^2} \left[NT - S^2 - \frac{N(N-2)\delta}{2} \right] \right\}^{1/2} , \tag{5.36}$$

where

$$S = \sum_{i=1}^{N} h(R_i, N, 1) , \qquad T = \sum_{i=1}^{N} \left[h(R_i, N, 1) \right]^2 ,$$

and $h(R_i, N, 1)$ is the rank function of the ith response measurement.

Utilizing the univariate rank scores listed in Fig. 5.13, the relationships between statistics δ and C_{N1} can be confirmed. Thus, for the univariate rank scores listed in Fig. 5.13, the observed values of S and T are

$$S_o = -|1-5|^1 - |2-5|^1 - |3-5|^1 - |4-5|^1 - |5-5|^1 \\ + |6-5|^1 + |7-5|^1 + |8-5|^1 + |9-5|^1 \\ = -4-3-2-1-0+1+2+3+4 = 0$$

and

$$T_o = (-|1-5|^1)^2 + (-|2-5|^1)^2 + (-|3-5|^1)^2 + (-|4-5|^1)^2 \\ + (-|5-5|^1)^2 + (+|6-5|^1)^2 + (+|7-5|^1)^2 + (+|8-5|^1)^2 \\ + (+|9-5|^1)^2 = 16+9+4+1+0+1+4+9+16 = 60\ .$$

Then, following Eq. (5.35), the observed value of the MRPP test statistic for the univariate rank scores listed in Fig. 5.13 is

$$\delta_o = \frac{2}{9(9-2)}\left\{9(60) - (0)^2 - \frac{\left[9(-8) - 4(0)\right]^2}{(4)(5)}\right\} \\ = \frac{2}{63}\left(540 - \frac{5{,}184}{20}\right) = 8.9143$$

and, following Eq. (5.36), the observed value of C_{N1} is

$$C_{N1o} = \frac{(4)(0)}{9} - \left\{\frac{(4)(5)}{9^2}\left[(9)(60) - 0^2 - \frac{9(9-2)(8.9143)}{2}\right]\right\}^{1/2} \\ = 0 - \left[0.2469(540.00 - 280.8005)\right]^{1/2} = -8.00\ .$$

Because of the relationship between statistics C_{N1} and δ, the exact probability value of the realized value of C_{N1} is given by

$$P\big(C_{N1} \geq C_{N1o}|H_0\big) = P\big(\delta \leq \delta_o|H_0\big) = \frac{\text{number of } \delta \text{ values} \leq \delta_o}{M}\ ,$$

where C_{N1o} and δ_o denote the observed values of C_{N1} and δ, respectively.

Consider again the univariate rank response measurements listed in Fig. 5.13 where $r = 1$, $g = 2$, $n_1 = 4$, $n_2 = 5$, $N = n_1 + n_2 = 9$, and there are no tied rank scores. In this application, let $v = 2$, employing squared Euclidean distance between the rank scores, and let the treatment-group weights be given by

$$C_i = \frac{n_i - 1}{N - g} , \qquad i = 1, \ldots, g ,$$

to correspond to the Wilcoxon–Mann–Whitney two-sample rank-sum test. Because there are only

$$M = \frac{N!}{n_1!\, n_2!} = \frac{9!}{4!\, 5!} = 126$$

possible, equally-likely arrangements of the $N = 9$ rank scores, an exact solution is preferred. Following Eq. (5.2) on p. 218, the $N = 9$ univariate rank scores listed in Fig. 5.13 yield $g = 2$ average distance-function values of

$$\xi_1 = 6.6667 \quad \text{and} \quad \xi_2 = 10.60 .$$

Following Eq. (5.1) on p. 217, the observed value of the MRPP test statistic based on $v = 2$ and treatment-group weights

$$C_i = \frac{n_i - 1}{N - g} , \qquad i = 1, 2 ,$$

is

$$\delta_o = \sum_{i=1}^{g} C_i \xi_i = \frac{1}{9-2}\big[(4-1)(6.6667) + (5-1)(10.60)\big] = 8.9143 .$$

If all arrangements of the $N = 9$ observed rank scores listed in Fig. 5.13 occur with equal chance, the exact probability value of $\delta_o = 8.9143$ computed on the $M = 126$ possible arrangements of the observed data with $n_1 = 4$ A univariate rank scores and $n_2 = 5$ B univariate rank scores preserved for each arrangement is

$$P\big(\delta \le \delta_o | H_0\big) = \frac{\text{number of } \delta \text{ values} \le \delta_o}{M} = \frac{8}{126} = 0.0635 .$$

For comparison, the conventional Wilcoxon–Mann–Whitney two-sample rank-sum test computed on the univariate rank scores listed in Fig. 5.13 yields an observed test statistic value of

$$W_o = \sum_{i=1}^{n_1} R_i = 1 + 2 + 4 + 5 = 12$$

and the exact probability value of $W_o = 12$ is

$$P(W \geq W_o|H_0) = \frac{\text{number of } W \text{ values} \geq W_o}{M} = \frac{8}{126} = 0.0635 .$$

Alternatively, test statistic W is approximately distributed as $N(0, 1)$ under the null hypothesis as $N \to \infty$. For the rank scores listed in Fig. 5.13 on p. 278, the mean value of W is

$$\mu_W = \frac{n_1 N + 1}{2} = \frac{4(9+1)}{2} = 20 ,$$

the variance of W is

$$\sigma_W^2 = \frac{n_1 n_2 (N+1)}{12} = \frac{(4)(5)(9+1)}{12} = 16.6667 ,$$

the observed standard score is

$$z_o = \frac{W_o - \mu_W}{\sqrt{\sigma_W^2}} = \frac{12 - 20}{\sqrt{16.6667}} = -1.9596 ,$$

and the approximate two-tailed $N(0, 1)$ probability value is $P = 0.0500$.

Following Eq. (5.17) on p. 238, the exact expected value of the $M = 126$ δ values is $\mu_\delta = 15.00$ and, following Eq. (5.18) on p. 238, the observed chance-corrected measure of effect size is

$$\Re_o = 1 - \frac{\delta_o}{\mu_\delta} = 1 - \frac{8.9143}{15.00} = +0.4057 ,$$

indicating approximately 41 % within-group agreement above that expected by chance.

5.14.2 Example 2

For this second analysis of the univariate rank scores listed in Fig. 5.2 on p. 239, replicated in Fig. 5.14 for convenience, let the treatment-group weights be given by

$$C_i = \frac{n_i - 1}{N - g} , \qquad i = 1, \ldots, g ,$$

as in Example 1, but set $v = 1$ instead of $v = 2$, employing ordinary Euclidean distance between the rank scores.

Rank:	1	2	3	4	5	6	7	8	9
Sample:	*A*	*A*	*B*	*A*	*A*	*B*	*B*	*B*	*B*

Fig. 5.14 Example univariate rank-score data for two-sample rank tests with $n_1 = 4$, $n_2 = 5$, and $N = n_1 + n_2 = 9$

Following Eq. (5.2) on p. 218, the $N = 9$ univariate rank scores listed in Fig. 5.14 yield $g = 2$ average distance-function values of

$$\xi_1 = 2.3333 \quad \text{and} \quad \xi_2 = 2.80 \, .$$

Following Eq. (5.1) on p. 217, the observed value of the MRPP test statistic based on $v = 1$ and treatment-group weights

$$C_i = \frac{n_i - 1}{N - g} \, , \qquad i = 1, 2 \, ,$$

is

$$\delta_o = \sum_{i=1}^{g} C_i \xi_i = \frac{1}{9-2}\big[(4-1)(2.3333) + (5-1)(2.80)\big] = 2.60 \, .$$

If all arrangements of the $N = 9$ observed rank scores listed in Fig. 5.14 occur with equal chance, the exact probability value of $\delta_o = 2.60$ computed on the $M = 126$ possible arrangements of the observed data with $n_1 = 4$ *A* univariate rank scores and $n_2 = 5$ *B* univariate rank scores preserved for each arrangement is

$$P(\delta \le \delta_o | H_0) = \frac{\text{number of } \delta \text{ values} \le \delta_o}{M} = \frac{6}{126} = 0.0476 \, .$$

For comparison, the exact probability value based on $v = 2$, $M = 126$, and $C_i = (n_i - 1)/(N - g)$ for $i = 1, 2$ in Example 1 is $P = 0.0635$. No comparison is made with the conventional Wilcoxon–Mann–Whitney two-sample rank-sum test as the Wilcoxon–Mann–Whitney test is undefined for $v = 1$.

Following Eq. (5.17) on p. 238, the exact expected value of the $M = 126$ δ values is $\mu_\delta = 3.3333$ and, following Eq. (5.18) on p. 238, the observed chance-corrected measure of effect size is

$$\Re_o = 1 - \frac{\delta_o}{\mu_\delta} = 1 - \frac{2.60}{3.3333} = +0.2200 \, ,$$

indicating 22 % within-group agreement above that expected by chance.

5.14.3 Example 3

The treatment-group weighting function given by

$$C_i = \frac{n_i - 1}{N - g}\,, \qquad i = 1, \ldots, g\,,$$

is not required for a permutation test. Thus, for this third analysis of the univariate rank scores listed in Fig. 5.14, the treatment-group weighting function is set to

$$C_i = \frac{n_i}{N}\,, \qquad i = 1, \ldots, g\,,$$

weighting each treatment group proportional to its size, and setting $v = 1$, employing ordinary Euclidean distance between the rank scores, as in Example 2. Following Eq. (5.2) on p. 218, the $N = 9$ univariate rank scores listed in Fig. 5.14 yield $g = 2$ average distance-function values of

$$\xi_1 = 2.3333 \quad \text{and} \quad \xi_2 = 2.80\,.$$

Following Eq. (5.1) on p. 217, the observed value of the MRPP test statistic based on $v = 1$ and treatment-group weights

$$C_i = \frac{n_i}{N}\,, \qquad i = 1, 2\,,$$

is

$$\delta_o = \sum_{i=1}^{g} C_i \xi_i = \frac{1}{9}\big[(4)(2.3333) + (5)(2.80)\big] = 2.5926\,.$$

If all arrangements of the $N = 9$ observed rank scores listed in Fig. 5.14 occur with equal chance, the exact probability value of $\delta_o = 2.5926$ computed on the $M = 126$ possible arrangements of the observed data with $n_1 = 4$ *A* univariate rank scores and $n_2 = 5$ *B* univariate rank scores preserved for each arrangement is

$$P(\delta \leq \delta_o | H_0) = \frac{\text{number of } \delta \text{ values} \leq \delta_o}{M} = \frac{6}{126} = 0.0476\,.$$

For comparison, the exact probability values based on $v = 2$, $M = 126$, and $C_i = (n_i - 1)/(N - g)$ for $i = 1, 2$ in Example 1 and $v = 1$, $M = 126$, and $C_i = (n_i - 1)/(N - g)$ for $i = 1, 2$ in Example 2 are $P = 0.0635$ and $P = 0.0476$, respectively. No comparison is made with the conventional Wilcoxon–Mann–Whitney two-sample rank-sum test as the Wilcoxon–Mann–Whitney two-sample test is undefined for both $v = 1$ and $C_i = n_i/N$, $i = 1, \ldots, g$.

Following Eq. (5.17) on p. 238, the exact expected value of the $M = 126$ δ values is $\mu_\delta = 3.3333$ and, following Eq. (5.18) on p. 238, the observed chance-corrected measure of effect size is

$$\Re_o = 1 - \frac{\delta_o}{\mu_\delta} = 1 - \frac{2.5926}{3.3333} = +0.2222\,,$$

indicating approximately 22 % within-group agreement above that expected by chance.

5.15 Example C_{Ns} Analyses with $s = 2$

In 1972 P.W. Mielke proposed a new two-sample test based on powers of ranks termed C_{Ns} [281]. In this section, three example analyses illustrate the C_{Ns} rank function test with $s = 2$. The first example uses a small set of univariate rank scores with $v = 2$ and treatment-group weights $C_i = (n_i - 1)/(N - g)$ for $i = 1, \ldots, g$; the second example analyzes the same small set of univariate rank scores, but uses $v = 1$ and treatment-group weights $C_i = (n_i - 1)/(N - g)$ for $i = 1, \ldots, g$; and the third example analyzes the same set of univariate rank scores using $v = 1$, but adopts a proportional treatment-group weighting function given by $C_i = n_i/N$ for $i = 1, \ldots, g$.

5.15.1 Example 1

Utilizing the univariate rank scores listed in Fig. 5.2 on p. 239, replicated in Fig. 5.15 for convenience, consider C_{Ns} with $s = 2$, where for variable A the observed value of C_{N2} is

$$\begin{aligned} C_{N2o} &= \sum_{i=1}^{N} h(R_i, N, 2)\, Z_{Ni} \\ &= -|1-5|^2(1) - |2-5|^2(1) - |3-5|^2(0) - |4-5|^2(1) - |5-5|^2(1) \\ &\quad + |6-5|^2(0) + |7-5|^2(0) + |8-5|^2(0) + |9-5|^2(0) \\ &= -16 - 9 - 0 - 0 - 1 + 0 + 0 + 0 + 0 = -26\,. \end{aligned}$$

Rank:	1	2	3	4	5	6	7	8	9
Sample:	A	A	B	A	A	B	B	B	B

Fig. 5.15 Example univariate rank-score data for two-sample rank tests with $n_1 = 4$, $n_2 = 5$, and $N = n_1 + n_2 = 9$

Following Eqs. (5.15) on p. 237 and (5.16) on p. 238, the functional relationships between statistics C_{N2} and δ are given by

$$\delta = \frac{2}{N(N-2)}\left[NT - S^2 - \frac{(NC_{N2} - n_1S)^2}{n_1n_2}\right] \tag{5.37}$$

and

$$C_{N2} = \frac{n_1S}{N} - \left\{\frac{n_1n_2}{n^2}\left[NT - S^2 - \frac{N(N-2)\delta}{2}\right]\right\}^{1/2}, \tag{5.38}$$

where

$$S = \sum_{i=1}^{N} h(R_i, N, 2)\,, \qquad T = \sum_{i=1}^{N} \left[h(R_i, N, 2)\right]^2,$$

and $h(R_i, N, 2)$ is the rank function of the ith response measurement.

Utilizing the univariate rank scores listed in Fig. 5.15, the relationship between statistics δ and C_{N2} can be confirmed. Thus, for the univariate rank scores listed in Fig. 5.15, the observed values of S and T are

$$\begin{aligned} S_o &= -|1-5|^2 - |2-5|^2 - |3-5|^2 - |4-5|^2 - |5-5|^2 \\ &\quad + |6-5|^2 + |7-5|^2 + |8-5|^2 + |9-5|^2 \\ &= -16 - 9 - 4 - 1 - 0 + 1 + 4 + 9 + 16 = 0 \end{aligned}$$

and

$$\begin{aligned} T_o &= (-|1-5|)^4 + (-|2-5|)^4 + (-|3-5|)^4 + (-|4-5|)^4 \\ &\quad + (-|5-5|)^4 + (+|6-5|)^4 + (+|7-5|)^4 + (+|8-5|)^4 \\ &\quad + (+|9-5|)^4 = 256 + 81 + 16 + 1 + 0 + 1 + 16 + 81 + 256 = 708\,. \end{aligned}$$

Then, following Eq. (5.37), the observed value of the MRPP test statistic for the univariate rank scores listed in Fig. 5.15 is

$$\begin{aligned} \delta_o &= \frac{2}{9(9-2)}\left\{9(708) - (0)^2 - \frac{[9(-26) - 4(0)]^2}{(4)(5)}\right\} \\ &= \frac{2}{63}\left(6{,}372 - \frac{54{,}756}{20}\right) = 115.3714 \end{aligned}$$

and, following Eq. (5.38), the observed value of C_{N2} is

$$C_{N2o} = \frac{(4)(0)}{9} - \left\{ \frac{(4)(5)}{9^2} \left[(9)(708) - 0^2 - \frac{9(9-2)(115.3714)}{2} \right] \right\}^{1/2}$$

$$= 0 - \left[0.2469(6{,}372.00 - 3{,}634.1991) \right]^{1/2} = -26.00\ .$$

Because of the relationship between statistics C_{N2} and δ, the exact probability value of the realized value of C_{N2} is given by

$$P\big(C_{N2} \geq C_{N2o} | H_0\big) = P\big(\delta \leq \delta_o | H_0\big) = \frac{\text{number of } \delta \text{ values} \leq \delta_o}{M}\ ,$$

where C_{N2o} and δ_o denote the observed values of C_{N2} and δ, respectively.

Consider again the univariate rank response measurements listed in Fig. 5.15 where $r = 1$, $g = 2$, $n_1 = 4$, $n_2 = 5$, $N = n_1 + n_2 = 9$, and there are no tied rank scores. In this application, let $v = 2$, employing squared Euclidean distance between the rank scores, and let the treatment-group weights be given by

$$C_i = \frac{n_i - 1}{N - g}\ , \qquad i = 1, \ldots, g\ ,$$

to correspond to the Mielke two-sample sum-of-squared-ranks test. Because there are only

$$M = \frac{N!}{n_1!\ n_2!} = \frac{9!}{4!\ 5!} = 126$$

possible, equally-likely arrangements of the $N = 9$ rank scores, an exact solution is preferred. Following Eq. (5.2) on p. 218, the $N = 9$ univariate rank scores listed in Fig. 5.15 yield $g = 2$ average distance-function values of

$$\xi_1 = 112.6667 \quad \text{and} \quad \xi_2 = 117.40\ .$$

Following Eq. (5.1) on p. 217, the observed value of the MRPP test statistic based on $v = 2$ and treatment-group weights

$$C_i = \frac{n_i - 1}{N - g}\ , \qquad i = 1, 2\ ,$$

is

$$\delta_o = \sum_{i=1}^{g} C_i \xi_i = \frac{1}{9-2} \big[(4-1)(112.6667) + (5-1)(117.40) \big] = 115.3714\ .$$

If all arrangements of the $N = 9$ observed rank scores listed in Fig. 5.15 occur with equal chance, the exact probability value of $\delta_o = 115.3714$ computed on the $M = 126$ possible arrangements of the observed data with $n_1 = 4$ A univariate rank scores and $n_2 = 5$ B univariate rank scores preserved for each arrangement is

$$P(\delta \leq \delta_o|H_0) = \frac{\text{number of } \delta \text{ values} \leq \delta_o}{M} = \frac{8}{126} = 0.0635 \; .$$

Following Eq. (5.17) on p. 238, the exact expected value of the $M = 126$ δ values is $\mu_\delta = 177.00$ and, following Eq. (5.18) on p. 238, the observed chance-corrected measure of effect size is

$$\Re_o = 1 - \frac{\delta_o}{\mu_\delta} = 1 - \frac{115.3714}{177.00} = +0.3482 \; ,$$

indicating approximately 35 % within-group agreement above that expected by chance.

5.15.2 Example 2

For this second analysis of the univariate rank scores listed in Fig. 5.2 on p. 239, replicated in Fig. 5.16 for convenience, let the treatment-group weights be given by

$$C_i = \frac{n_i - 1}{N - g} \; , \qquad i = 1, \ldots, g \; ,$$

as in Example 1, but set $v = 1$ instead of $v = 2$, employing ordinary Euclidean distance between the rank scores.

Following Eq. (5.2) on p. 218, the $N = 9$ univariate rank scores listed in Fig. 5.16 yield $g = 2$ average distance-function values of

$$\xi_1 = 9.3333 \quad \text{and} \quad \xi_2 = 9.60 \; .$$

Following Eq. (5.1) on p. 217, the observed value of the MRPP test statistic based on $v = 1$ and treatment-group weights

$$C_i = \frac{n_i - 1}{N - g} \; , \qquad i = 1, 2 \; ,$$

Rank:	1	2	3	4	5	6	7	8	9
Sample:	*A*	*A*	*B*	*A*	*A*	*B*	*B*	*B*	*B*

Fig. 5.16 Example univariate rank-score data for two-sample rank tests with $n_1 = 4$, $n_2 = 5$, and $N = n_1 + n_2 = 9$

is

$$\delta_o = \sum_{i=1}^{g} C_i \xi_i = \frac{1}{9-2}\big[(4-1)(9.3333) + (5-1)(9.60)\big] = 9.4857 \,.$$

If all arrangements of the $N = 9$ observed rank scores listed in Fig. 5.16 occur with equal chance, the exact probability value of $\delta_o = 9.4857$ computed on the $M = 126$ possible arrangements of the observed data with $n_1 = 4$ A univariate rank scores and $n_2 = 5$ B univariate rank scores preserved for each arrangement is

$$P(\delta \le \delta_o | H_0) = \frac{\text{number of } \delta \text{ values} \le \delta_o}{M} = \frac{8}{126} = 0.0635 \,.$$

For comparison, the exact probability value based on $v = 2$, $M = 126$, and $C_i = (n_i - 1)/(N - g)$ for $i = 1, 2$ in Example 1 is also $P = 0.0635$. No comparison is made with the conventional Mielke two-sample sum-of-squared-ranks test as Mielke's C_{N2} test is undefined for $v = 1$.

Following Eq. (5.17) on p. 238, the exact expected value of the $M = 126$ δ values is $\mu_\delta = 11.1111$ and, following Eq. (5.18) on p. 238, the observed chance-corrected measure of effect size is

$$\Re_o = 1 - \frac{\delta_o}{\mu_\delta} = 1 - \frac{9.4857}{11.1111} = +0.1463 \,,$$

indicating approximately 15 % within-group agreement above that expected by chance.

5.15.3 Example 3

The treatment-group weighting function given by

$$C_i = \frac{n_i - 1}{N - g} \,, \qquad i = 1, \ldots, g \,,$$

is not required for a permutation test. Thus, for this third analysis of the univariate rank scores listed in Fig. 5.16, the treatment-group weighting function is set to

$$C_i = \frac{n_i}{N} \,, \qquad i = 1, \ldots, g \,,$$

simply weighting each treatment group proportional to its size, and setting $v = 1$, employing ordinary Euclidean distance between the rank scores, as in Example 2.

Following Eq. (5.2) on p. 218, the $N = 9$ univariate rank scores listed in Fig. 5.16 yield $g = 2$ average distance-function values of

$$\xi_1 = 9.3333 \quad \text{and} \quad \xi_2 = 9.60\ .$$

Following Eq. (5.1) on p. 217, the observed value of the MRPP test statistic based on $v = 1$ and treatment-group weights

$$C_i = \frac{n_i}{N}\ , \qquad i = 1, 2\ ,$$

is

$$\delta_o = \sum_{i=1}^{g} C_i \xi_i = \frac{1}{9}\big[(4)(9.3333) + (5)(9.60)\big] = 9.4815\ .$$

If all arrangements of the $N = 9$ observed rank scores listed in Fig. 5.16 occur with equal chance, the exact probability value of $\delta_o = 9.4815$ computed on the $M = 126$ possible arrangements of the observed data with $n_1 = 4$ A univariate rank scores and $n_2 = 5$ B univariate rank scores preserved for each arrangement is

$$P\big(\delta \le \delta_o | H_0\big) = \frac{\text{number of } \delta \text{ values} \le \delta_o}{M} = \frac{8}{126} = 0.0635\ .$$

For comparison, the exact probability values based on $v = 2$, $M = 126$, and $C_i = (n_i - 1)/(N - g)$ for $i = 1, 2$ in Example 1 and $v = 1$, $M = 126$, and $C_i = (n_i - 1)/(N - g)$ for $i = 1, 2$ in Example 2 are also $P = 0.0635$. No comparison is made with the conventional Mielke two-sample sum-of-squared-ranks test as Mielke's C_{N2} test is undefined for both $v = 1$ and $C_i = n_i/N$, $i = 1, \ldots, g$.

Following Eq. (5.17) on p. 238, the exact expected value of the $M = 126$ δ values is $\mu_\delta = 11.1111$ and, following Eq. (5.18) on p. 238, the observed chance-corrected measure of effect size is

$$\Re_o = 1 - \frac{\delta_o}{\mu_\delta} = 1 - \frac{9.4815}{11.1111} = +0.1467\ ,$$

indicating approximately 15 % within-group agreement above that expected by chance.

5.16 MRPP and Kendall's S Statistic

In 1947 John Whitfield, an experimental psychologist at Cambridge University,[15] proposed a measure of correlation between two variables in which one variable was composed of N rank scores and the other variable was dichotomous [424]. Whitfield considered the dichotomous variable as composed entirely of two sets of tied rankings. An example will illustrate the Whitfield procedure. Consider the univariate rank scores listed in Fig. 5.17 where the dichotomous variable categories are two samples indicated by the letters A and B and the rank scores are from 1 to 6. Let $n_1 = 4$ denote the number of rank scores in the A category, let $n_2 = 2$ denote the number of rank scores in the B category, and let $N = n_1 + n_2$.[16]

Whitfield designed a procedure to calculate a statistic that he labeled S, following Kendall's notation in a 1945 *Biometrika* article on "The treatment of ties in ranking problems" [206]. Given the $N = 6$ rank scores listed in Fig. 5.17, consider the $n_1 = 4$ rank scores in the category identified by the letter A: 1, 3, 4, and 5. Beginning with rank score 1 with the letter A, there are no rank scores with the letter B to the left of $A = 1$ and two rank scores with the letter B to the right of $A = 1$ (ranks 2 and 6); so Whitfield calculated $0 - 2 = -2$. For rank score 3 with the letter A, there is one rank score to the left of $A = 3$ with the letter B (rank 2) and one rank score to the right of $A = 3$ with the letter B (rank 6); so $1 - 1 = 0$. For rank score 4 with the letter A, there is one rank score to the left of $A = 4$ with the letter B (rank 2) and one rank score to the right of $A = 4$ with the letter B (rank 6); so $1 - 1 = 0$. Finally, for rank score 5 with the letter A, there is one rank score to the left of $A = 5$ with the letter B (rank 2) and one rank score to the right of $A = 5$ with the letter B (rank 6); so $1 - 1 = 0$. The sum of the differences between variables A and B is $S = -2 + 0 + 0 + 0 = -2$. In this manner, Whitfield's approach accommodated unequal sample sizes with $n_1 \neq n_2$ as well as tied rank scores.

Since the number of possible pairs of N consecutive integers is given by

$$\frac{N(N-1)}{2},$$

Rank:	1	2	3	4	5	6
Sample:	*A*	*B*	*A*	*A*	*A*	*B*

Fig. 5.17 Ranking of a dichotomous variable with $n_1 = 4$, $n_2 = 2$, and $N = n_1 + n_2 = 6$

[15]Officially, the University of Cambridge, but universally known simply as Cambridge.

[16]Note that in this example, $n_2 = 2$ is the smaller of the two sample sizes.

Whitfield defined and calculated a measure of rank-order association between variables *A* and *B* as

$$\tau = \frac{2S}{N(N-1)} = \frac{2(-2)}{(6)(6-1)} = \frac{-4}{30} = -0.1333 \,.$$

Whitfield's *S* is identical to Kendall's *S* [206], and is directly related to the two-sample rank-sum *U* statistic of Mann and Whitney [262] and, hence, to the two-sample rank-sum *W* statistic of Wilcoxon [429].[17] This can be demonstrated with a simple comparison. For the univariate rank scores listed in Fig. 5.17, there are $n_1 = 4$ *A* rank scores and $n_2 = 2$ *B* rank scores, so considering the smaller of the two sample sizes (the $n_2 = 2$ *B* rank scores), the first letter *B* (rank 2) precedes three letter *A* rank scores (ranks 3, 4, and 5) and the second letter *B* (rank 6) precedes no letter *A*, so $U = 3 + 0 = 3$. The relationships between Whitfield's *S* and Mann and Whitney's *U* statistics are given by

$$S = 2U - n_1 n_2 \quad \text{and} \quad U = \frac{S + n_1 n_2}{2} \,.$$

Thus, for the rank scores given in Fig. 5.17 the observed values of *S* and *U* are

$$S_o = 2(3) - (4)(2) = -2 \quad \text{and} \quad U_o = \frac{-2 + (4)(2)}{2} = 3 \,,$$

respectively [60, 224]. Also, for the example univariate rank scores listed in Fig. 5.17, the observed Wilcoxon *W* statistic for the smaller of the two sums (the $n_2 = 2$ *B* rank scores) is $W_o = 2 + 6 = 8$ and the relationships between Whitfield's *S* and Wilcoxon's *W* are given by

$$S = n_2(N+1) - 2W \quad \text{and} \quad W = \frac{n_2(N+1) - S}{2} \,.$$

Thus, the observed values of *S* and *W* are

$$S_o = 2(6+1) - (2)(8) = -2 \quad \text{and} \quad W_o = \frac{2(6+1) - (-2)}{2} = 8 \,,$$

respectively [60, 224].

As Whitfield noted, the calculation of *S* was fashioned after a procedure introduced by Maurice Kendall in 1945[18] and Whitfield might have been unaware of the two-sample rank-sum tests previously published by Wilcoxon in 1945 [429],

[17] For clarification, *S* as used by Whitfield and Kendall should not be confused with *S* used as the sum of rank scores earlier in the chapter.

[18] Whitfield lists the date of the Kendall article as 1946, but Kendall's article was actually published in *Biometrika* in 1945.

Table 5.2 Fifteen pairs of observations with concordant/discordant (C/D) pairs and associated rank-pair values

Number	Pair	C/D	Value	Number	Pair	C/D	Value
1	1–2	$-,+$	-1	9	2–6	$+,+$	0
2	1–3	$-,-$	0	10	3–4	$-,-$	0
3	1–4	$-,-$	0	11	3–5	$-,-$	0
4	1–5	$-,-$	0	12	3–6	$-,+$	-1
5	1–6	$-,+$	-1	13	4–5	$-,-$	0
6	2–3	$+,-$	$+1$	14	4–6	$-,+$	-1
7	2–4	$+,-$	$+1$	15	5–6	$-,+$	-1
8	2–5	$+,-$	$+1$				

Festinger in 1946 [116], and Mann and Whitney in 1947 [262], as they are not referenced in the 1947 Whitfield article. Kendall considered the number of concordant (C) and discordant (D) pairs, of which there is a total of $N(N-1)/2$ pairs when there are no ties among the N integers [206]. For the example univariate rank scores listed in Fig. 5.17 there are

$$\frac{N(N-1)}{2} = \frac{6(6-1)}{2} = 15$$

pairs of rank scores. Table 5.2 lists and numbers the 15 rank pairs, the concordant/discordant classification of rank pairs, and the rank-pair values, where concordant pairs ($-,-$ and $+,+$) are given a value of 0, and discordant pairs ($+,-$ and $-,+$) are given values of $+1$ and -1, respectively. The observed sum of the pair values listed in Table 5.2 for the 15 pairs is $S_o = -5 + 3 = -2$.

Today it is well known, although poorly documented, that when one classification is a dichotomy and the other classification is ordered, with or without tied values, the S statistic of Kendall is equivalent to the Mann–Whitney U statistic; see articles on this topic by Lincoln Moses in 1956 [313] and Edmund John Burr in 1960 [60]. Whitfield apparently was the first to uncover the relationship between S, the statistic underlying Kendall's τ rank-order correlation coefficient, and U, the Mann–Whitney two-sample rank-sum statistic for two independent samples.

However, it was Hemelrijk in 1952 [173] and Jonckheere in 1954 [196] who made the relationship between S and U explicit; see also a discussion by Leach in 1979 [234, p. 183]. Because the Jonckheere–Terpstra test, when restricted to two independent samples, is mathematically identical in reverse application to the Wilcoxon and Mann–Whitney tests, see references [196, p. 138] and [343, p. 396], the two-sample rank-sum test is sometimes referred to as the Kendall–Wilcoxon–Mann–Whitney–Jonckheere–Festinger test [313, p. 246].

Because, as Whitfield demonstrated, Kendall's S is related to Wilcoxon's W and Mann and Whitney's U, then Kendall's S is ipso facto related to Mielke's A_{N1} and C_{N1}, when $C_i = (n_i - 1)/(N - g)$ and $v = 2$. The relationships between statistics

Rank:	1	2	3	4	5	6
Sample:	*A*	*B*	*A*	*A*	*A*	*B*

Fig. 5.18 Ranking of a dichotomous variable with $n_1 = 4$, $n_2 = 2$, and $N = n_1 + n_2 = 6$

A_{N1} and S are given by

$$A_{N1} = \frac{n_2(N+1) - S}{2} \quad \text{and} \quad S = n_2(N+1) - 2A_{N1} \ ,$$

and the relationships between statistics C_{N1} and S are given by

$$C_{N1} = -\left(\frac{S}{2}\right) \quad \text{and} \quad S = -2C_{N1} \ .$$

Consider the univariate rank response measurements listed in Fig. 5.17 on p. 291, replicated in Fig. 5.18 for convenience, where $n_1 = 4$, $n_2 = 2$, $N = n_1 + n_2 = 6$, and $S = -2$. For the univariate rank scores listed in Fig. 5.18, and following Eq. (5.11) on p. 236, the observed value of A_{N1} is

$$\begin{aligned} A_{N1o} &= \sum_{i=1}^{N} R_i^1 Z_{Ni} \\ &= (1^1)(0) + (2^1)(1) + (3^1)(0) + (4^1)(0) + (5^1)(0) + (6^1)(1) \\ &= 0 + 2 + 0 + 0 + 0 + 6 = 8 \ . \end{aligned}$$

Alternatively, the observed value of A_{N1o} is given by

$$A_{N1o} = \frac{n_2(N+1) - S_o}{2} = \frac{2(6+1) - (-2)}{2} = \frac{16}{2} = 8 \ ,$$

where the observed value of S is

$$S_o = n_2(N+1) - 2A_{N1} = 2(6+1) - (2)(8) = -2 \ .$$

Also, following Eq. (5.13) on p. 237, the observed value of C_{N1} is

$$\begin{aligned} C_{N1o} &= \sum_{i=1}^{N} h(R_i, N, 1)\, Z_{Ni} \\ &= -|1 - 3.50|^1(0) - |2 - 3.50|^1(1) - |3 - 3.50|^1(0 + |4 - 3.50|^1(0) \\ &\qquad + |5 - 3.50|^1(0) + |6 - 3.50|^1(1) \\ &= -0 - 1.50 - 0 + 0 + 0 + 2.50 = +1 \ . \end{aligned}$$

Alternatively, the observed value of $C_{N1\mathrm{o}}$ is given by

$$C_{N1\mathrm{o}} = -\left(\frac{S_\mathrm{o}}{2}\right) = -\left(\frac{-2}{2}\right) = \frac{2}{2} = +1 \;,$$

and the observed value of S is given by

$$S_\mathrm{o} = -2C_{N1\mathrm{o}} = (-2)(1.00) = -2 \;.$$

5.17 Example Analyses

In this section, three example analyses illustrate Whitfield's S test statistic. The first example is designed to correspond to Whitfield's S statistic using a small set of univariate rank scores with $v = 2$ and treatment-group weights $C_i = (n_i - 1)/(N - g)$ for $i = 1, \ldots, g$; the second example analyzes the same small set of univariate rank scores, but uses $v = 1$ and treatment-group weights $C_i = (n_i - 1)/(N - g)$ for $i = 1, \ldots, g$; and the third example analyzes the same set of univariate rank scores using $v = 1$, but adopts a proportional treatment-group weighting function given by $C_i = n_i/N$ for $i = 1, \ldots, g$.

5.17.1 Example 1

Consider the univariate rank scores listed in Table 5.3 consisting of $n_1 = 9$ rank scores from Sample A and $n_2 = 6$ rank scores from Sample B.[19] Calculating Mann and Whitney's U statistic for the data listed in Table 5.3, the number of A rank scores to the left of (less than) the first B rank score (rank 3) is 2; the number of A rank scores to the left of the second and third B rank scores (ranks 7 and 8) is 5 each; and the number of A rank scores to the left of the last three B rank scores (ranks 13, 14, and 15) is 9 each. Then $U = 2 + 5 + 5 + 9 + 9 + 9 = 39$. To calculate Wilcoxon's W statistic for the rank data listed in Table 5.3, the sum of the rank scores in Sample A is $W = 1 + 2 + 4 + 5 + 6 + 9 + 10 + 11 + 12 = 60$.[20]

Table 5.3 Listing of the $n_1 = 9$ and $n_2 = 6$ univariate rank scores from Samples A and B, respectively

Rank:	1	2	3	4	5	6	7	8	9	10	11	12	13	14	15
Sample:	*A*	*A*	*B*	*A*	*A*	*A*	*B*	*B*	*A*	*A*	*A*	*A*	*B*	*B*	*B*

[19] Note that in this example, $n_2 = 6$ is the smaller of the two sample sizes.

[20] Coincidentally, in this example the sum of the $n_1 = 9$ rank scores in Sample B is also 60.

Fig. 5.19 Contingency table of the frequencies of rank scores in Table 5.3

A:	2	0	3	0	4	0
B:	0	1	0	2	0	3

To calculate Whitfield's S statistic for the data listed in Table 5.3, there are two A rank scores to the left of $B = 3$ (ranks 1 and 2) and seven A rank scores to the right of $B = 3$ (ranks 4, 5, 6, 9, 10 11, and 12), so $2 - 7 = -5$. There are five A rank scores to the left of $B = 7$ and $B = 8$ (ranks 1, 2, 4, 5, and 6) and four A rank scores to the right of $B = 7$ and $B = 8$ (ranks 9, 10, 11, and 12), so $(5 - 4) + (5 - 4) = 2$. There are nine A rank scores to the left of $B = 13$, 14, and 15 (ranks 1, 2, 4, 5, 6, 9, 10, 11, and 12) and zero A rank scores to the right of $B = 13$, 14, and 15, so $(9 - 0) + (9 - 0) + (9 - 0) = 27$. Then $S = -5 + 2 + 27 = 24$. Note that the relationships among Whitfield's S, Mann and Whitney's U, and Wilcoxon's W are given by

$$S = 2U - n_1 n_2 = 2(39) - (9)(6) = 78 - 54 = +24 \,,$$

$$U = \frac{S + n_1 n_2}{2} = \frac{24 + (9)(6)}{2} = \frac{78}{2} = 39 \,,$$

$$S = n_1(N + 1) - 2W = 9(15 + 1) - (2)(60) = 144 - 120 = +24 \,,$$

and

$$W = \frac{n_1(N + 1) - S}{2} = \frac{9(15 + 1) - 24}{2} = \frac{120}{2} = 60 \,.$$

Alternatively, as Whitfield suggested, arrange the two samples into a contingency table with two rows and columns equal to the frequency distribution of the combined samples, as depicted in Fig. 5.19. Here the first row of frequencies in Fig. 5.19 represents the runs in the list of rank scores in Table 5.3 labeled as A, i.e., there are two occurrences of A in ranks 1 and 2; no occurrence of A in rank 3; three occurrences of A in ranks 4, 5, and 6; no occurrence of A in ranks 7 and 8; four occurrences of A in ranks 10, 11, and 12; and no occurrence of A in ranks 13, 14, and 15. The second row of frequencies in Fig. 5.19 represents the runs in the list of rank scores in Table 5.3 labeled as B, i.e., there are no occurrences of B in ranks 1 and 2, one occurrence of B in rank 3, no occurrences of B in ranks 4, 5, and 6, two occurrences of B in ranks 7 and 8, no occurrence of B in ranks 9, 10, 11, and 12, and three occurrences of B in ranks 13, 14, and 15.

Given the $r{\times}c$ contingency table in Fig. 5.19 with $r = 2$ rows and $c = 6$ columns, let x_{ij} indicate a cell frequency for $i = 1, \ldots, r$ and $j = 1, \ldots, c$. Then, as Kendall showed in 1948 [207], the number of concordant pairs is given by

$$C = \sum_{i=1}^{r-1} \sum_{j=1}^{c-1} x_{ij} \left(\sum_{k=i+1}^{r} \sum_{l=j+1}^{c} x_{kl} \right) \qquad (5.39)$$

and the number of discordant pairs is given by

$$D = \sum_{i=1}^{r-1} \sum_{j=1}^{c-1} x_{i,c-j+1} \left(\sum_{k=i+1}^{r} \sum_{l=1}^{c-j} x_{kl} \right) . \tag{5.40}$$

Thus, for the cell frequencies given in Fig. 5.19, C is calculated by proceeding from the upper-left cell with frequency $x_{11} = 2$ downward and to the right, multiplying each cell frequency by the sum of all cell frequencies below and to the right, and summing the products. Thus, following Eq. (5.39) the observed value of C is

$$\begin{aligned} C_o = 2(1 + 0 + 2 + 0 + 3) + 0(0 + 2 + 0 + 3) \\ + 3(2 + 0 + 3) + 0(0 + 3) + (4)(3) \\ = 12 + 0 + 15 + 0 + 12 = 39 , \end{aligned}$$

and D is calculated by proceeding from the upper-right cell with frequency $x_{16} = 0$ downward and to the left, multiplying each cell frequency by the sum of all cell frequencies below and to the left, and summing the products. Thus, following Eq. (5.40) the observed value of D is

$$\begin{aligned} D_o = 0(0 + 1 + 0 + 2 + 0) + 4(0 + 1 + 0 + 2) \\ + 0(0 + 1 + 0) + 3(0 + 1) + (0)(0) \\ = 0 + 12 + 0 + 3 + 0 = 15 . \end{aligned}$$

Then, as defined by Kendall, the observed value of S is $S_o = C_o - D_o = 39 - 15 = +24$.

Now consider the MRPP test statistic δ, as defined in Eq. (5.1) on p. 217, where for this first analysis

$$C_i = \frac{n_i - 1}{N - g} , \qquad i = 1, \ldots, g ,$$

and $v = 2$ in Eq. (5.3) on p. 218, employing squared Euclidean distance between rank scores to correspond to the Kendall–Whitfield S statistic. The functional relationships between statistics δ and S are given by

$$\delta = \frac{N}{2(N-2)} \left[\frac{N^2 - 1}{3} - \frac{S^2}{n_1 n_2} \right] \tag{5.41}$$

and

$$S = \left\{ \frac{n_1 n_2}{3N} \left[N(N^2 - 1) - 6(N - 2)\,\delta \right] \right\}^{1/2} , \tag{5.42}$$

where n_1 and n_2 denote the sizes of the two samples and $N = n_1 + n_2$. Thus,

$$P(S \geq S_o | H_0) = P(\delta \leq \delta_o | H_0) = \frac{\text{number of } \delta \text{ values} \leq \delta_o}{M} ,$$

where S_o and δ_o denote the observed values of S and δ, respectively.

For the univariate rank scores listed in Table 5.3 on p. 295, there are only

$$M = \frac{N!}{n_1!\, n_2!} = \frac{15!}{9!\, 6!} = 5{,}005$$

possible, equally-likely arrangements of the observed rank scores; thus, an exact solution is feasible. Following Eq. (5.2) on p. 218, the $N = 15$ univariate rank scores listed in Table 5.3 yield $g = 2$ average distance-function values of

$$\xi_1 = 32.00 \quad \text{and} \quad \xi_2 = 44.80 .$$

Following Eq. (5.1) on p. 217, the observed value of the MRPP test statistic based on $v = 2$ and treatment-group weights

$$C_i = \frac{n_i - 1}{N - g} , \qquad i = 1, 2 ,$$

is

$$\delta_o = \sum_{i=1}^{g} C_i \xi_i = \frac{1}{15 - 2} \big[(9 - 1)(32.00) + (6 - 1)(44.80) \big] = 36.9231 .$$

If all arrangements of the $N = 15$ observed rank scores listed in Table 5.3 on p. 295 occur with equal chance, the exact probability value of $\delta_o = 36.9231$ computed on the $M = 5{,}005$ possible arrangements of the observed data with $n_1 = 9$ A univariate rank scores and $n_2 = 6$ B univariate rank scores preserved for each arrangement is

$$P(\delta \leq \delta_o | H_0) = \frac{\text{number of } \delta \text{ values} \leq \delta_o}{M} = \frac{906}{5{,}005} = 0.1810 .$$

For comparison, assuming no tied rank scores, the Whitfield test statistic S is approximately distributed under the null hypothesis as $N(0, 1)$ as $N \rightarrow \infty$ with mean $\mu_S = 0$ and variance given by

$$\sigma_S^2 = \frac{1}{18}\left[N(N-1)(2N+5)\right] .$$

Thus, for the univariate rank scores listed in Table 5.3 on p. 295,

$$\sigma_S^2 = \frac{15(15-1)[(2)(15)+5]}{18} = \frac{7{,}350}{18} = 408.3333$$

and the observed standard score is

$$z_o = \frac{S_o - \mu_S}{\sigma_S} = \frac{24-0}{\sqrt{408.3333}} = +1.1877 ,$$

yielding an approximate two-sided $N(0, 1)$ probability value of $P = 0.2350$.

Following Eq. (5.20) on p. 239, the exact expected value of the $M = 5{,}005$ δ values is $\mu_\delta = 40.00$ and, following Eq. (5.18) on p. 238, the observed chance-corrected measure of effect size is

$$\Re_o = 1 - \frac{\delta_o}{\mu_\delta} = 1 - \frac{36.9231}{40.00} = +0.0769 ,$$

indicating approximately 8 % within-group agreement above that expected by chance.

The relationships between statistics δ and S are confirmed as follows. For the univariate rank scores listed in Table 5.3 and, following Eq. (5.41) on p. 297, the observed value of δ is

$$\delta_o = \frac{15}{2(15-2)}\left[\frac{15^2-1}{3} - \frac{24^2}{(9)(6)}\right] = \frac{15}{26}(64) = 36.9231$$

and, following Eq. (5.42) on p. 298, the observed value of S is

$$S_o = \left\{\frac{9(15-9)}{(3)(15)}\left[15(15^2-1) - 6(15-2)(36.9231)\right]\right\}^{1/2}$$

$$= \left[\frac{54}{45}(480)\right]^{1/2} = 24 .$$

5.17.2 Example 2

For this second analysis of the univariate rank scores listed in Table 5.3 on p. 295, replicated in Table 5.4 for convenience, let the treatment-group weights be given by

$$C_i = \frac{n_i - 1}{N - g} \,, \qquad i = 1, \ldots, g \,,$$

as in Example 1, but set $v = 1$ instead of $v = 2$, employing ordinary Euclidean distance between the rank scores.

Following Eq. (5.2) on p. 218, the $N = 15$ univariate rank scores listed in Table 5.4 yield $g = 2$ average distance-function values of

$$\xi_1 = 4.8333 \quad \text{and} \quad \xi_2 = 5.7333 \,.$$

Following Eq. (5.1) on p. 217, the observed value of the MRPP test statistic based on $v = 1$ and treatment-group weights

$$C_i = \frac{n_i - 1}{N - g} \,, \qquad i = 1, 2 \,,$$

is

$$\delta_o = \sum_{i=1}^{g} C_i \xi_i = \frac{1}{15 - 2} \big[(9 - 1)(4.8333) + (6 - 1)(5.7333) \big] = 5.1795 \,.$$

If all arrangements of the $N = 15$ observed rank scores listed in Table 5.4 occur with equal chance, the exact probability value of $\delta_o = 5.1795$ computed on the $M =$ 5,005 possible arrangements of the observed data with $n_1 = 9$ A univariate rank scores and $n_2 = 6$ B univariate rank scores preserved for each arrangement is

$$P\big(\delta \leq \delta_o | H_0\big) = \frac{\text{number of } \delta \text{ values} \leq \delta_o}{M} = \frac{1{,}091}{5{,}005} = 0.2180 \,.$$

For comparison, the exact probability value based on $v = 2$, $M = 5{,}005$, and $C_i = (n_i - 1)/(N - g)$ for $i = 1, 2$ in Example 1 is $P = 0.1810$. No comparison is made with the conventional Whitfield two-sample rank-sum test as Whitfield's two-sample test is undefined for $v = 1$.

Table 5.4 Listing of the $n_1 = 9$ and $n_2 = 6$ rank scores from Samples A and B, respectively

Rank:	1	2	3	4	5	6	7	8	9	10	11	12	13	14	15
Sample:	*A*	*A*	*B*	*A*	*A*	*A*	*B*	*B*	*A*	*A*	*A*	*A*	*B*	*B*	*B*

Following Eq. (5.20) on p. 239, the exact expected value of the $M = 5{,}005$ δ values is $\mu_\delta = 5.3333$ and, following Eq. (5.18) on p. 238, the observed chance-corrected measure of effect size is

$$\Re_o = 1 - \frac{\delta_o}{\mu_\delta} = 1 - \frac{5.1795}{5.3333} = +0.0288 ,$$

indicating approximately 3 % within-group agreement above that expected by chance.

5.17.3 Example 3

The treatment-group weighting function given by

$$C_i = \frac{n_i - 1}{N - g} , \qquad i = 1, \ldots, g ,$$

is not required for a permutation test. Thus, for this third analysis of the univariate rank scores listed in Table 5.4, the treatment-group weighting function is set to

$$C_i = \frac{n_i}{N} , \qquad i = 1, \ldots, g ,$$

weighting each treatment group proportional to its size, and setting $v = 1$, employing ordinary Euclidean distance between the rank scores, as in Example 2. Following Eq. (5.2) on p. 218, the $N = 15$ univariate rank scores listed in Table 5.4 yield $g = 2$ average distance-function values of

$$\xi_1 = 4.8333 \quad \text{and} \quad \xi_2 = 5.7333 .$$

Following Eq. (5.1) on p. 217, the observed value of the MRPP test statistic based on $v = 1$ and treatment-group weights

$$C_i = \frac{n_i}{N} , \qquad i = 1, 2 ,$$

is

$$\delta_o = \sum_{i=1}^{g} C_i \xi_i = \frac{1}{15}\big[(9)(4.8333) + (6)(5.7333)\big] = 5.1933 .$$

If all arrangements of the $N = 15$ observed rank scores listed in Table 5.4 occur with equal chance, the exact probability value of $\delta_o = 5.1933$ computed on the $M = 5{,}005$ possible arrangements of the observed data with $n_1 = 9$ A rank scores and

$n_2 = 6$ B rank scores preserved for each arrangement is

$$P(\delta \leq \delta_o | H_0) = \frac{\text{number of } \delta \text{ values} \leq \delta_o}{M} = \frac{1{,}158}{5{,}005} = 0.2314 .$$

For comparison, the exact probability values based on $v = 2$, $M = 5{,}005$, and $C_i = (n_i - 1)/(N - g)$ for $i = 1, 2$ in Example 1 and $v = 1$, $M = 5{,}005$, and $C_i = (n_1 - 1)/(N - g)$ for $i = 1, 2$ in Example 2 are $P = 0.1810$ and $P = 0.2180$, respectively. No comparison is made with the conventional Whitfield two-sample rank-sum test as Whitfield's two-sample test is undefined for both $v = 1$ and $C_i = n_i/N$, $i = 1, \ldots, g$.

Following Eq. (5.20) on p. 239, the exact expected value of the $M = 5{,}005$ δ values is $\mu_\delta = 5.3333$ and, following Eq. (5.18) on p. 238, the observed chance-corrected measure of effect size is

$$\mathfrak{R}_o = 1 - \frac{\delta_o}{\mu_\delta} = 1 - \frac{5.1933}{5.3333} = +0.0263 ,$$

indicating approximately 3 % within-group agreement above that expected by chance.

5.18 MRPP and Cureton's Rank-Biserial Correlation

Consider two correlated variables, one represented by a ranking and the other by a dichotomy, similar to Whitfield's data (q.v. p. 291). In 1956 psychologist Edward Cureton proposed a new measure of correlation for a ranking and a dichotomous variable that he labeled r_{rb} for rank-biserial correlation [83].[21] The rank-biserial correlation coefficient was introduced by Cureton as a measure of effect size for the Wilcoxon–Mann–Whitney two-sample rank-sum test. Twelve years later, in 1968, Cureton extended r_{rb} to include tied rank scores [84]. In this section, only non-tied rank scores are considered, with no loss of generality. Cureton stated that the new correlation coefficient should norm properly between ± 1 and should be strictly non-parametric, defined solely in terms of inversions and agreements between rank-pairs, without the use of means, variances, covariances, or regression [83, p. 287]. Consequently, as Cureton stated, "clearly r_{rb} is a Kendall-type coefficient" [83, p. 289]. However, Cureton also stated that r_{rb} "is also a Spearman-type coefficient" [83, p. 289].

It is clear that r_{rb} is, indeed, a Kendall-type coefficient as Kendall's S and Cureton's r_{rb} are related, *vide infra*. However, it is less clear that it is a Spearman-type coefficient. Durbin and Stuart [99], along with Daniels [87, 88], investigated this relationship and, although r_{rb} appears to be a Spearman *type* coefficient, the exact relationship with Spearman's ρ seems to have eluded investigators. The difficulty is

[21] Technically, Cureton's r_{rb} is not a measure of correlation [137, p. 629].

that Kendall's τ_a and Spearman's ρ are not equivalent. In general, the relationship between τ and ρ is given by

$$-1 \leq \frac{3(N+2)}{N-2}\tau - \frac{2(N+1)}{N-2}\rho \leq +1$$

[76].

When two variables, x and y, both consist of interval-level response measurements, the correlation between the variables is usually calculated as Pearson's product-moment correlation coefficient given by

$$r_{xy} = \frac{\text{cov}(x, y)}{s_x s_y} ,$$

where

$$s_x = \left[\frac{1}{N-1} \sum_{i=1}^{N} (x_i - \bar{x})^2 \right]^{1/2} , \quad s_y = \left[\frac{1}{N-1} \sum_{i=1}^{N} (y_i - \bar{y})^2 \right]^{1/2} ,$$

$$\text{cov}(x, y) = \frac{1}{N-1} \sum_{i=1}^{N} (x_i - \bar{x})(y_i - \bar{y}) ,$$

$$\bar{x} = \frac{1}{N} \sum_{i=1}^{N} x_i , \quad \text{and} \quad \bar{y} = \frac{1}{N} \sum_{i=1}^{N} y_i .$$

When variables x and y are both coded (0, 1) binary, the correlation between the two variables is usually calculated as Pearson's ϕ given by

$$\phi = \sqrt{\frac{\chi^2}{N}} .$$

It has been well established that ϕ is identical to Pearson's r_{xy} when calculated on (x, y) binary-coded data.

When x and y are both ordinal variables, the correlation between variables x and y is usually calculated as Spearman's rank-order correlation given by

$$\rho_{xy} = 1 - \frac{6 \sum_{i=1}^{N} d^2}{N(N^2 - 1)} ,$$

where $d = x_i - y_i$ for $i = 1, \ldots, N$. It is well known that Spearman's ρ_{xy} is identical to Pearson's r_{xy} when calculated on x, y rank data. An added benefit is that Pearson's r_{xy} automatically corrects for any tied rank scores.

When x is a $(0, 1)$ binary-coded variable and y consists of interval-level response measurements, the correlation between variables x and y is usually calculated as the point-biserial correlation given by

$$r_{pb} = \frac{\bar{y}_1 - \bar{y}_0}{s_y} \sqrt{\frac{n_0 n_1}{N(N-1)}} ,$$

where n_0 and n_1 are the number of y values associated with $x = 0$ and $x = 1$, respectively, $N = n_0 + n_1$, and $\bar{y}_0$ and $\bar{y}_1$ are the sample means of the y values associated with $x = 0$ and $x = 1$, respectively. The point-biserial correlation coefficient is identical to Pearson's r_{xy} calculated on the data where x is a binary-coded variable and y is a continuous, interval-level variable.

On the other hand, Cureton's r_{rb} does not belong to the Pearson r_{xy} family, although a Pearson-type coefficient could easily be defined for a $(0, 1)$ binary-coded variable and a rank-order variable. Cureton's r_{rb} belongs instead to Kendall's tau-like family of measures that is based on $S = C - D$, where C and D denote the number of concordant and discordant pairs of x, y values, respectively.

5.18.1 Example 1

Consider an example data set such as listed in Fig. 5.20 in which $N = 10$ objects are ranked (variable y) and also classified into two groups coded 0 and 1 (variable x). The data are adapted from Glass [139, p. 104]. Cureton defined r_{rb} as

$$r_{rb} = \frac{S}{S_{\max}} ,$$

Fig. 5.20 Example $(0, 1)$ coded data for Cureton's rank-biserial correlation coefficient

	Variable	
Object	x	y
1	0	1
2	1	2
3	0	3
4	0	4
5	0	5
6	0	6
7	1	7
8	0	8
9	1	9
10	1	10

Table 5.5 Paired differences and concordant (C) and discordant (D) values for the univariate rank scores listed in Fig. 5.20

Pair	$x_i - x_j$	$y_i - y_j$	Type	Pair	$x_i - x_j$	$y_i - y_j$	Type
1	$1-0$	$1-2$	C	24	$0-1$	$3-10$	C
2	$0-0$	$1-3$		25	$0-0$	$4-5$	
3	$0-0$	$1-4$		26	$0-0$	$4-6$	
4	$0-0$	$1-5$		27	$0-1$	$4-7$	C
5	$0-0$	$1-6$		28	$0-0$	$4-8$	
6	$0-1$	$1-7$	C	29	$0-1$	$4-9$	C
7	$0-0$	$1-8$		30	$0-1$	$4-10$	C
8	$0-1$	$1-9$	C	31	$0-0$	$5-6$	
9	$0-1$	$1-10$	C	32	$0-1$	$5-7$	C
10	$1-0$	$2-3$	D	33	$0-0$	$5-8$	
11	$1-0$	$2-4$	D	34	$0-1$	$5-9$	C
12	$1-0$	$2-5$	D	35	$0-1$	$5-10$	C
13	$1-0$	$2-6$	D	36	$0-1$	$6-7$	C
14	$1-1$	$2-7$		37	$0-0$	$6-8$	
15	$1-0$	$2-8$	D	38	$0-1$	$6-9$	C
16	$1-1$	$2-9$		39	$0-1$	$6-10$	C
17	$1-1$	$2-10$		40	$1-0$	$7-8$	D
18	$0-0$	$3-4$		41	$1-1$	$7-9$	
19	$0-0$	$3-5$		42	$1-1$	$7-10$	
20	$0-0$	$3-6$		43	$0-1$	$8-9$	C
21	$0-1$	$3-7$	C	44	$0-1$	$8-10$	C
22	$0-0$	$3-8$		45	$1-1$	$9-10$	
23	$0-1$	$3-9$	C				

where $S = C - D$, C is the number of concordant pairs, D is the number of discordant pairs, $S = C - D$ is the S test statistic of Kendall [205] and Whitfield [424], and $S_{\max} = n_0 n_1$, where n_0 is the number of objects coded 0 and n_1 is the number of objects coded 1.

Table 5.5 lists the

$$\binom{N}{2} = \frac{N(N-1)}{2} = \frac{10(10-1)}{2} = 45$$

possible paired comparisons of x_i and x_j with y_i and y_j, where $i < j$ and n_0 and n_1 are the number of objects coded 0 and 1, respectively. Each paired difference is labeled as concordant (C) or discordant (D). Paired differences not labeled as C or D are not relevant in the present context as they are tied by either $x_i = x_j = 0$ or $x_i = x_j = 1$. In Table 5.5 there are 18 concordant and 6 discordant paired differences; thus, for the paired differences listed in Table 5.5, the observed value of S is $S_o = C - D = 18 - 6 = +12$.

0:	1	1	4	0	1	0
1:	0	0	0	1	0	2

Fig. 5.21 Ranking of a dichotomous variable with $n_0 = 6$, $n_1 = 4$, and $N = n_0 + n_1 = 10$

Alternatively, as suggested by Whitfield [424], the rank scores listed in Fig. 5.20 can be rearranged into a contingency table to make calculation of C and D much easier [424]. Consider the data listed in Fig. 5.20 arranged into a 2×6 contingency table, such as given in Fig. 5.21. The top row of frequencies given in Fig. 5.21 represents the runs in the list of rank scores given in Fig. 5.20 coded 0, i.e., there is 1 occurrence of a 0 in rank 1, no occurrence of a 0 in rank 2, 4 occurrences of a 0 in ranks 3, 4, 5, and 6, no occurrence of a 0 in rank 7, 1 occurrence of a 0 in rank 8, and 2 occurrences of a 0 in ranks 9 and 10. The bottom row of frequencies given in Fig. 5.21 represents the runs in the list of rank scores given in Fig. 5.20 coded 1, i.e., there is no occurrence of a 1 in rank 1, one occurrence of a 1 in rank 2, no occurrences of a 1 in ranks 3, 4, 5, and 6, one occurrence of a 1 in rank 7, no occurrence of a 1 in rank 8, and 2 occurrences of a 1 in ranks 9 and 10.

Given the $r \times c$ contingency table presented in Fig. 5.21 with $r = 2$ rows and $c = 6$ columns, let x_{ij} indicate a cell frequency for $i = 1, \ldots, r$ and $j = 1, \ldots, c$. Then, as Kendall showed in 1948 [207], the number of concordant and discordant pairs is given by

$$C = \sum_{i=1}^{r-1} \sum_{j=1}^{c-1} x_{ij} \left(\sum_{k=i+1}^{r} \sum_{l=j+1}^{c} x_{kl} \right)$$

and

$$D = \sum_{i=1}^{r-1} \sum_{j=1}^{c-1} x_{i,c-j+1} \left(\sum_{k=i+1}^{r} \sum_{l=1}^{c-j} x_{kl} \right) ,$$

respectively. Thus, the observed values of C and D are

$$C_o = 1(1 + 1 + 2) + 4(1 + 2) + 1(2) = 18 ,$$

$$D_o = 1(1 + 1) + 4(1) = 6 ,$$

and the observed value of S is $S_o = C_o - D_o = 18 - 6 = +12$. It can easily be shown that $S_{\max}$ is given by $n_0 n_1$, where n_0 is the number of objects coded 0 and n_1 is the number of objects coded 1. Then, Cureton's rank-biserial coefficient is given by

$$r_{rb} = \frac{S_o}{S_{\max}} = \frac{S_o}{n_0 n_1} = \frac{+12}{(6)(4)} = +0.50 .$$

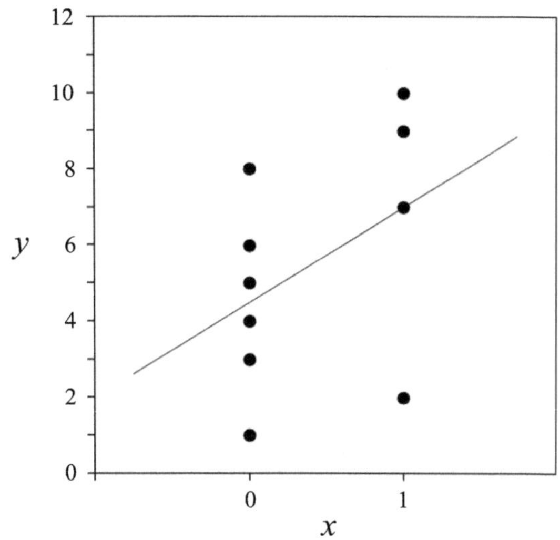

Fig. 5.22 Graphic depicting the regression line for the data listed in Fig. 5.20 with intercept equal to $\bar{y}_0 = 4.50$ and slope equal to $\bar{y}_1 - \bar{y}_0 = 7.00 - 4.50 = 2.50$

In 1966 Glass derived a simplified formula for r_{rb}, assuming no tied rank scores [137]. Glass's formula is given by

$$r_{rb} = \frac{2}{N}(\bar{y}_1 - \bar{y}_0) \ ,$$

where $\bar{y}_0$ and $\bar{y}_1$ are the arithmetic averages of the y values coded 0 and 1, respectively. In this case, $\bar{y}_0 = 4.50$ and $\bar{y}_1 = 7.00$. Note that under (0, 1) binary coding, $\bar{y}_0$ and $\bar{y}_1 - \bar{y}_0$ are the intercept (a_{yx}) and slope (b_{yx}), respectively, of a regression line passing through the two points ($x = 0, \bar{y}_0 = 4.40$) and ($x = 1, \bar{y}_1 = 7.00$), as illustrated in Fig. 5.22.

Glass provided two alternate calculating formulæ given by

$$r_{rb} = \frac{2}{n_0}\left(\bar{y}_1 - \frac{N+1}{2}\right) \quad \text{or} \quad r_{rb} = \frac{2}{n_1}\left(\frac{N+1}{2} - \bar{y}_0\right) .$$

Thus, for the data listed in Fig. 5.20 on p. 304 where

$$\bar{y}_0 = \frac{1}{n_0}\sum_{i=1}^{n_0} = \frac{1}{6}(1+3+4+5+6+8) = \frac{1}{6}(27) = 4.50$$

$$\bar{y}_1 = \frac{1}{n_1}\sum_{i=1}^{n_1} = \frac{1}{4}(2+7+9+10) = \frac{1}{4}(28) = 7.00 \ ,$$

$$r_{rb} = \frac{2}{n_0}\left(\bar{y}_1 - \frac{N+1}{2}\right) = \frac{2}{6}\left(7.00 - \frac{10+1}{2}\right) = +0.50$$

Fig. 5.23 Example data from Fig. 5.20 for the MRPP analyses

y_0	y_1
1	2
3	7
4	9
5	10
6	
8	

and

$$r_{rb} = \frac{2}{n_1}\left(\frac{N+1}{2} - \bar{y}_0\right) = \frac{2}{4}\left(\frac{10+1}{2} - 4.50\right) = +0.50\,.$$

Now consider the MRPP test statistic δ, as defined in Eq. (5.1) on p. 217, where for this first example analysis,

$$C_i = \frac{n_i - 1}{N - g}\,, \qquad i = 1, \ldots, g\,,$$

and $v = 2$ in Eq. (5.3) on p. 218, employing squared Euclidean distance between the rank scores to correspond to Cureton's r_{rb} test statistic. Figure 5.23 rearranges the data listed in Fig. 5.20 into $g = 2$ groups. The functional relationships between statistics δ and r_{rb} are given by

$$\delta = \frac{N}{2(N-2)}\left(\frac{N^2-1}{3} - n_0 n_1 r_{rb}^2\right) \tag{5.43}$$

and

$$r_{rb} = \left[\frac{N(N^2-1) - 6\delta(N-2)}{3Nn_0n_1}\right]^{1/2}\,, \tag{5.44}$$

where n_0 and n_1 denote the sizes of the two groups and $N = n_0 + n_1$. Since n_0, n_1, and N are all constants under permutation,

$$P\big(r_{rb} \geq r_o | H_0\big) = P\big(\delta \leq \delta_o | H_0\big) = \frac{\text{number of } \delta \text{ values} \leq \delta_o}{M}\,,$$

where r_o and δ_o denote the observed value of r_{rb} and δ, respectively.

For the univariate rank scores listed in Fig. 5.23, there are only

$$M = \frac{N!}{n_0!\; n_1!} = \frac{10!}{6!\; 4!} = 210$$

possible, equally-likely arrangements of the observed rank scores; thus, an exact solution is feasible. Following Eq. (5.2) on p. 218, the $N = 10$ univariate rank scores listed in Fig. 5.23 yield $g = 2$ average distance-function values of

$$\xi_0 = 11.80 \quad \text{and} \quad \xi_1 = 25.3333 \ .$$

Following Eq. (5.1) on p. 217, the observed value of the MRPP test statistic based on $v = 2$ and treatment-group weights

$$C_i = \frac{n_i - 1}{N - g} \ , \qquad i = 1, 2 \ ,$$

is

$$\delta = \sum_{i=1}^{g} C_i \xi_i = \frac{1}{10-2}(6-1)(11.80) + (4-1)(25.3333)] = 16.8750 \ .$$

If all arrangements of the $N = 10$ observed rank scores listed in Fig. 5.23 occur with equal chance, the exact probability value of $\delta = 16.8750$ computed on the $M = 210$ possible arrangements of the observed data with $n_0 = 6$ univariate rank scores and $n_1 = 4$ univariate rank scores preserved for each arrangement is

$$P(\delta \le \delta_o | H_0) = \frac{\text{number of } \delta \text{ values} \le \delta_o}{M} = \frac{54}{210} = 0.2571 \ .$$

Following Eq. (5.5) on p. 219, the exact expected value of the $M = 210$ δ values is $\mu_\delta = 18.3333$ and, following Eq. (5.4) on p. 219, the observed chance-corrected measure of effect size is

$$\Re_o = 1 - \frac{\delta_o}{\mu_\delta} = 1 - \frac{16.8750}{18.3333} = +0.0795 \ ,$$

indicating approximately 8 % within-group agreement above that expected by chance.

The relationships between δ and Cureton's r_{rb} are confirmed as follows. For the $N = 10$ univariate rank scores listed in Fig. 5.20 on p. 304 and, following Eq. (5.43) on p. 308, the observed value of δ is

$$\delta_o = \frac{10}{2(10-2)} \left[\frac{10^2 - 1}{3} - (6)(4)(0.50)^2 \right] = \frac{5}{8}(33 - 6) = 16.8750$$

and, following Eq. (5.44) on p. 308, the observed value of r_{rb} is

$$r_{rb} = \left[\frac{10(10^2 - 1) - 6(16.8750)(10 - 2)}{3(10)(6)(4)} \right]^{1/2} = \left(\frac{990 - 810}{720} \right)^{1/2} = +0.50 \ .$$

Since δ is related to r_{rb} and, as shown previously, δ is related to Wilcoxon's W, Mann and Whitney's U, and Kendall's τ_a, then r_{rb} is ipso facto related to W, U, and τ_a. For the univariate rank scores listed in Fig. 5.23, Wilcoxon's W is simply the smaller of the sums of the rank scores of the two samples, i.e.,

$$W_o = \sum_{i=1}^{n_0} = 1 + 3 + 4 + 5 + 6 + 8 = 27 \; .$$

The relationships between Wilcoxon's W and Cureton's r_{rb} are given by

$$W = \frac{n_0(N+1) - n_0 n_1 r_{rb}}{2} \quad \text{and} \quad r_{rb} = \frac{n_0(N+1) - 2W}{n_0 n_1} \; ,$$

where n_0 is the number of objects in the group with the smaller of the two sums; in this case, 27. Thus, the observed value of Wilcoxon's W is

$$W_o = \frac{6(10+1) - (6)(4)(0.50)}{2} = \frac{54}{2} = 27$$

and the observed value of Cureton's r_{rb} is

$$r_{rb} = \frac{6(10+1) - (2)(27)}{(6)(4)} = \frac{+12}{24} = +0.50 \; .$$

For the univariate rank scores listed in Fig. 5.23, Mann and Whitney's U is the sum of the number of values in one group, preceded by the number of values in the other group. Thus, for the univariate rank scores listed in Fig. 5.23, the value of 1 in Group 0 is less than values 2, 7, 9, and 10 in Group 1, yielding $U = 4$. Then, the value of 3 in Group 0 is less than values 7, 9, and 10 in Group 1, yielding $U = 3 + 4 = 7$. Next, the value of 4 in Group 0 is less than values 7, 9, and 10 in Group 1, yielding $U = 3 + 3 + 4 = 10$. Next, the value of 5 in Group 0 is less than values 7, 9, and 10 in Group 1, yielding $U = 3 + 3 + 3 + 4 = 13$. Next, the value of 6 in Group 0 is less than values 7, 9, and 10 in Group 1, yielding $U = 3 + 3 + 3 + 3 + 4 = 16$. Finally, the value of 8 in Group 0 is less than values 9 and 10 in Group 1, yielding $U = 3 + 3 + 3 + 3 + 4 + 2 = 18$. Alternatively,

$$U = n_0 n_1 + \frac{n_0(n_0+1)}{2} - W = (6)(4) + \frac{6(6+1)}{2} - 27 = 18 \; .$$

The relationships between Mann and Whitney's U and Cureton's r_{rb} are given by

$$U = \frac{n_0 n_1 (1 + r_{rb})}{2} \quad \text{and} \quad r_{rb} = \frac{2U}{n_0 n_1} - 1 \; .$$

Thus, the observed value of Mann and Whitney's U is

$$U = \frac{(6)(4)(1 + 0.50)}{2} = \frac{36}{2} = 18$$

and the observed value of Cureton's r_{rb} is

$$r_{rb} = \frac{(2)(18)}{(6)(4)} - 1 = 1.50 - 1 = +0.50\ .$$

For the univariate rank scores listed in Fig. 5.23, Kendall's τ_a is

$$\tau_a = \frac{2S}{N(N-1)} = \frac{(2)(12)}{10(10-1)} = \frac{24}{90} = 0.2667\ .$$

The relationships between Kendall's τ_a and Cureton's r_{rb} are given by

$$\tau_a = \frac{2n_0 n_1 r_{rb}}{N(N-1)} \quad \text{and} \quad r_{rb} = \frac{\tau_a N(N-1)}{2n_0 n_1}\ .$$

Thus, the observed value of Kendall's τ_a is

$$\tau_a = \frac{(2)(6)(4)(0.50)}{10(10-1)} = \frac{24}{90} = 0.2667$$

and the observed value of Cureton's r_{rb} is

$$r_{rb} = \frac{(0.2667)(10)(10-1)}{(2)(6)(4)} = \frac{24}{48} = +0.50\ .$$

5.18.2 Example 2

For this second analysis of the univariate rank scores listed in Fig. 5.23 on p. 308, replicated in Fig. 5.24 for convenience, let the treatment-group weights be given by

Fig. 5.24 Example data from Fig. 5.20 for the MRPP analyses

y_0	y_1
1	2
3	7
4	9
5	10
6	
8	

$$C_i = \frac{n_i - 1}{N - g}\,, \qquad i = 1, \ldots, g\,,$$

as in Example 1, but set $v = 1$ instead of $v = 2$, employing ordinary Euclidean distance between the rank scores.

Following Eq. (5.2) on p. 218, the $N = 10$ univariate rank response measurements listed in Fig. 5.24 yield $g = 2$ average distance-function values of

$$\xi_0 = 3.00 \quad \text{and} \quad \xi_1 = 4.3333\,.$$

Following Eq. (5.1) on p. 217, the observed value of the MRPP test statistic based on $v = 1$ and treatment-group weights

$$C_i = \frac{n_i - 1}{N - g}\,, \qquad i = 1, 2\,,$$

is

$$\delta_o = \sum_{i=1}^{g} C_i \xi_i = \frac{1}{10 - 2}\big[(6 - 1)(3.00) + (4 - 1)(4.3333)\big] = 3.50\,.$$

If all arrangements of the $N = 10$ observed rank scores listed in Fig. 5.24 occur with equal chance, the exact probability value of $\delta_o = 3.50$ computed on the $M = 210$ possible arrangements of the observed data with $n_0 = 6$ univariate rank scores and $n_1 = 4$ univariate rank scores preserved for each arrangement is

$$P\big(\delta \leq \delta_o | H_0\big) = \frac{\text{number of } \delta \text{ values} \leq \delta_o}{M} = \frac{52}{210} = 0.2476\,.$$

For comparison, the exact probability value based on $v = 2$, $M = 210$, and $C_i = (n_i - 1)/(N - g)$ for $i = 1, 2$ in Example 1 is $P = 0.2571$. No comparison is made with Cureton's rank-biserial test as Cureton's test is undefined for $v = 1$.

Following Eq. (5.5) on p. 219, the exact expected value of the $M = 210$ δ values is $\mu_\delta = 3.6667$ and, following Eq. (5.4) on p. 219, the observed chance-corrected measure of effect size is

$$\Re_o = 1 - \frac{\delta_o}{\mu_\delta} = 1 - \frac{3.50}{3.6667} = +0.0455\,,$$

indicating approximately 5 % within-group agreement above that expected by chance.

5.18.3 Example 3

The treatment-group weighting function given by

$$C_i = \frac{n_i - 1}{N - g} , \qquad i = 1, \ldots, g ,$$

is not required for a permutation test. Thus, for this third analysis of the univariate rank scores listed in Fig. 5.24, the treatment-group weighting function is set to

$$C_i = \frac{n_i}{N} , \qquad i = 1, \ldots, g ,$$

simply weighting each treatment group proportional to its size, and setting $v = 1$, employing ordinary Euclidean distance between the rank scores, as in Example 2. Following Eq. (5.2) on p. 218, the $N = 10$ univariate rank scores listed in Fig. 5.24 yield $g = 2$ average distance-function values of

$$\xi_0 = 3.00 \quad \text{and} \quad \xi_1 = 4.3333 .$$

Following Eq. (5.1) on p. 217, the observed value of the MRPP test statistic based on $v = 1$ and treatment-group weights

$$C_i = \frac{n_i}{N} , \qquad i = 1, 2 ,$$

is

$$\delta_o = \sum_{i=1}^{g} C_i \xi_i = \frac{1}{10}(6)(3.00) + (4)(4.3333)\big] = 3.5333 .$$

If all arrangements of the $N = 10$ observed rank scores listed in Fig. 5.24 occur with equal chance, the exact probability value of $\delta_o = 3.5333$ computed on the $M = 210$ possible arrangements of the observed data with $n_0 = 6$ univariate rank scores and $n_1 = 4$ univariate rank scores preserved for each arrangement is

$$P\big(\delta \le \delta_o | H_0\big) = \frac{\text{number of } \delta \text{ values} \le \delta_o}{M} = \frac{53}{210} = 0.2524 .$$

For comparison, the exact probability values based on $v = 2$, $M = 210$, and $C_i = (n_i - 1)/(N - g)$ for $i = 1, 2$ in Example 1 and $v = 1$, $M = 210$, and $C_i = (n_i - 1)/(N - g)$ for $i = 1, 2$ in Example 2 are $P = 0.2571$ and $P = 0.2476$, respectively. No comparison is made with Cureton's rank-biserial test as Cureton's test is undefined for both $v = 1$ and $C_i = n_i/N$, $i = 1, \ldots, g$.

Following Eq. (5.5) on p. 219, the exact expected value of the $M = 210$ δ values is $\mu_\delta = 3.6667$ and, following Eq. (5.4) on p. 219, the observed chance-corrected measure of effect size is

$$\Re_o = 1 - \frac{\delta_o}{\mu_\delta} = 1 - \frac{3.5333}{3.6667} = +0.0364 \, ,$$

indicating approximately 4 % within-group agreement above that expected by chance.

5.19 Coda

Chapter 5 applied the Multi-Response Permutation Procedures (MRPP) developed in Chap. 2 to ordinal-level response measurement data. Chapter 5 also established the relationships between the MRPP test statistics, δ and $\Re$, and selected conventional tests and measures designed for the analysis of completely randomized data at the ordinal level of measurement. Considered in this chapter were the Wilcoxon two-sample rank-sum test, the Kruskal–Wallis multiple-sample rank-sum test, the Ansari–Bradley rank-sum test for dispersion, the Taha sum-of-squared-ranks test, the Mood rank-sum test for dispersion, the Brown–Mood median test, the Mielke power-of-rank function tests, the Whitfield two-sample rank-sum test, and the Cureton rank-biserial test.

Comparisons of the MRPP test statistic based on ordinary Euclidean distance and squared Euclidean distance between the rank scores with the conventional statistics listed above revealed small differences among the observed probability values, which were not always consistent. As demonstrated in this chapter, ordinary Euclidean distance between the rank scores with $v = 1$ and squared Euclidean distance between the rank scores with $v = 2$ often yield similar results due to the elimination of extreme values by transforming the raw data to rank scores. However, substantial differences in probability values can still be obtained, as noted by Mielke, Berry, and Johnston in 2011 [309]. While conventional statistics, under the population model, require restrictive assumptions and are based on squared Euclidean distance between the rank scores, permutation methods based on ordinary Euclidean distance between the rank scores yield accurate probability values, are free of any distributional assumptions, and are completely data-dependent.

Chapter 6
Chapter 6 continues the ordinal-level analyses presented in Chap. 5, but applies the Multi-Response Permutation Procedures developed in Chap. 2 to multivariate ordinal-level response measurements. Tests and measures presented in Chap. 6 include MRPP multivariate extensions of the Wilcoxon two-sample rank-sum test, the Kruskal–Wallis multiple-sample rank-sum test, the Ansari–Bradley rank-sum test for dispersion, the Taha sum-of-squared-ranks test, the Mielke power-of-rank function tests, the Whitfield two-sample rank-sum test, and the Cureton rank-biserial test.

Randomized Designs: Ordinal Data, II

6

Chapter 5 of *Permutation Statistical Methods* utilized the multi-response permutation methods developed in Chap. 2 to establish relationships between the test statistics of MRPP, δ and $\Re$, and selected tests and measures designed for the analysis of completely randomized response measurements at the ordinal level of measurement. Considered in Chap. 5 were permutation versions of the Wilcoxon two-sample rank-sum test, the Kruskal–Wallis multi-sample rank-sum test, the Ansari–Bradley rank-sum test for dispersion, the Taha sum-of-squared-ranks test, the Mood rank-sum test for dispersion, the Brown–Mood median test, the Mielke A_{Ns}, B_{Ns}, and C_{Ns} power-of-rank function tests, the Whitfield two-sample rank-sum test, and the Cureton rank-biserial test. While Chap. 5 considered only univariate response measurements, i.e., $r = 1$, this sixth chapter of *Permutation Statistical Methods* extends the tests and measures considered in Chap. 5 to multivariate rank data, i.e., $r \geq 2$. Thus, for example, in this chapter the Wilcoxon two-sample rank-sum test and the Kruskal–Wallis multi-sample rank-sum tests are generalized to accommodate multivariate ordinal response measurements.

6.1 Introduction

Multivariate analysis can often be an improvement over univariate analysis, especially when the univariate response measurements are comprised of an average or index of several univariate variables, since averages or indices usually result in some significant loss of information. The loss of information is compounded when one or more of the univariate variables included in an index have previously been converted from raw response measurements to rank statistics—the subject of this chapter. Consider, for example, three ranked variables comprising socio-economic status: (1) income measured in grouped intervals, such as \$0 to \$4,999, \$5,000 to \$9,999, \$10,000 to \$19,999, and so on; (2) education, measured as elementary school, some high school, high school, some college, college and so on; and (3) occupation, mea-

K.J. Berry et al., *Permutation Statistical Methods*,
DOI 10.1007/978-3-319-28770-6_6

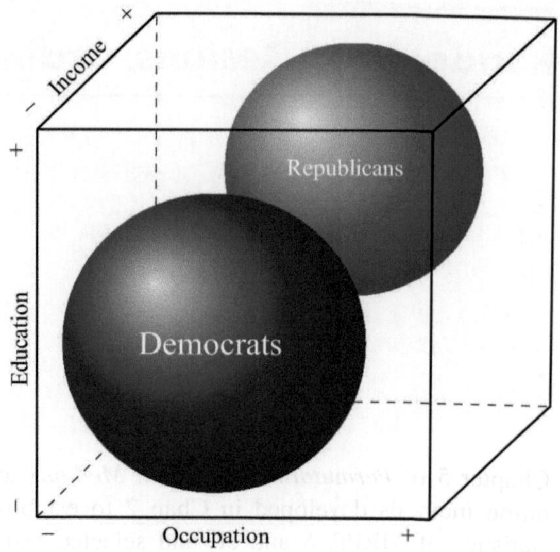

Fig. 6.1 Graphic depicting Democrats and Republicans in a three-dimensional space structured by Income, Occupation, and Education

sured as service, working, blue-collar, white-collar, professional, and so on. The usual practice of comparing two groups on socio-economic status, say Republicans and Democrats, is to combine the three variables into one measure, typically a weighted average, usually termed socioeconomic status (SES), and test for the average difference between the two groups under the null hypothesis of no difference. The alternative, advocated here, is to utilize a multivariate approach and examine the difference between Democrats and Republicans in a 3-dimensional space defined by three variables: Income, Occupation, and Education. Figure 6.1 graphically depicts a 3-dimensional space structured by Income, Occupation, and Education and containing simulated Democrats and Republicans, where for this illustration, Democrats have a wider range of Income, Occupation, and Education than Republicans, but generally average lower than Republicans on all three dimensions.

To illustrate the problems with aggregating discrete variables into an average or index, consider two subjects, A and B, and their percentile scores on three variables: Income, Occupation, and Education. Subject A scores in the 80th percentile on Income, the 50th percentile on Occupation, and the 20th percentile on Education. In contrast, subject B scores in the 20th percentile on Income, the 40th percentile on Occupation, and the 90th percentile on Education. Define SES as the simple unweighted average of the percentile ranks on Income, Occupation, and Education. Then, subject A scores SES $= (80 + 50 + 20)/3 = 50$ and subject B scores SES $= (90 + 40 + 20)/3 = 50$. On the average, subjects A and B are identical on SES, but in terms of Income and Education they are very different, although similar on Occupation. Figure 6.2 graphically illustrates the differences between subjects A and B, where subject A is represented by black nodes (●) connected by a dotted (···) line, and subject B is represented by gray nodes (●) connected by a dashed (– – –) line.

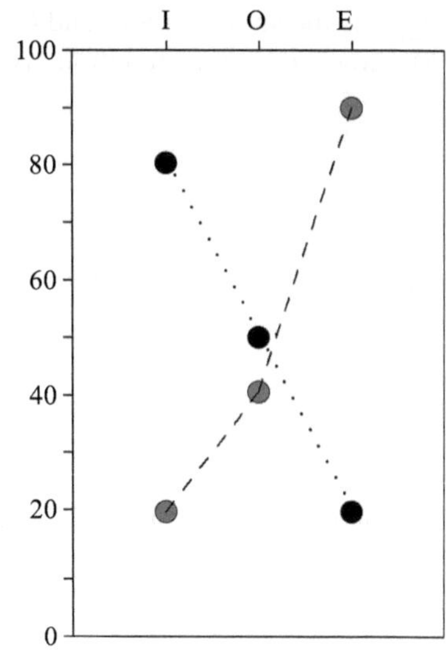

Fig. 6.2 Graphic of two subjects scoring on percentile ranks of Income (I), Occupation (O), and Education (E)

As developed in Chap. 2, let $\Omega = \{\omega_1, \ldots, \omega_N\}$ denote a finite sample of N objects, let $x_j' = (x_{1j}, \ldots, x_{rj})$ be a transposed vector of r commensurate response measurements for object ω_j, $j = 1, \ldots, N$, and let $S_1, \ldots, S_g$ designate an exhaustive partitioning of the N objects into g disjoint treatment groups. The MRPP test statistic given by

$$\delta = \sum_{i=1}^{g} C_i \xi_i \;, \tag{6.1}$$

where $C_i > 0$ is a positive treatment-group weight for $S_1, \ldots, S_g$,

$$\sum_{i=1}^{g} C_i = 1 \;,$$

$$\xi_i = \binom{n_i}{2}^{-1} \sum_{j<k} \Delta(j,k) \Psi_i(\omega_j) \Psi_i(\omega_k) \tag{6.2}$$

is the average distance-function value for all distinct pairs of objects in treatment group S_i for $i = 1, \ldots, g$,

$$N = \sum_{i=1}^{g} n_i \;,$$

$\sum_{j<k}$ is the sum over all j and k such that $1 \leq j < k \leq N$, $\Delta(j,k)$ is the generalized Minkowski distance function given by

$$\Delta(j,k) = \left(\sum_{i=1}^{r} \left| x_{ij} - x_{ik} \right|^{p} \right)^{v/p} , \tag{6.3}$$

where $p \geq 1$, $v > 0$, and $\Psi_i(\cdot)$ is an indicator function given by

$$\Psi_i(\omega_j) = \begin{cases} 1 & \text{if } \omega_j \in S_i \ , \\ 0 & \text{otherwise} \ . \end{cases}$$

The null hypothesis (H_0) states that equal probabilities are assigned to each of the

$$M = \frac{N!}{\prod_{i=1}^{g} n_i!}$$

possible, equally-likely allocations of the N objects to the g treatment groups. The probability value associated with an observed value of δ, δ_o, is the probability under the null hypothesis (H_0) of observing a value of δ as extreme or more extreme than δ_o. Thus, as detailed in Chap. 2, an exact probability value for δ_o may be expressed as

$$P(\delta \leq \delta_o | H_0) = \frac{\text{number of } \delta \text{ values} \leq \delta_o}{M} \ .$$

When M is very large, an approximate probability value for δ may be obtained from a resampling procedure, where

$$P(\delta \leq \delta_o | H_0) = \frac{\text{number of } \delta \text{ values} \leq \delta_o}{L} \ ,$$

and L denotes the number of randomly sampled test statistic values. Typically, L is set to a large number to ensure accuracy, e.g., $L = 1{,}000{,}000$. Also, when M is very large and P is exceedingly small, a resampling-approximation permutation procedure may produce no δ values equal to or less than δ_o, even with $L = 1{,}000{,}000$, yielding an approximate resampling probability value of $P = 0.00$. In such cases, moment-approximation permutation procedures based on fitting the first three exact moments of the discrete permutation distribution to a Pearson type III distribution provide approximate probability values, as detailed in Chap. 1, Sect. 1.2.2 [284, 300].

Finally, a chance-corrected within-group coefficient of effect size is given by

$$\Re = 1 - \frac{\delta}{\mu_\delta} \,, \tag{6.4}$$

where μ_δ is the arithmetic average of the M δ values calculated on all possible arrangements of the observed response measurements given by

$$\mu_\delta = \frac{1}{M} \sum_{i=1}^{M} \delta_i \,. \tag{6.5}$$

6.2 MRPP with $r = 2$ and $g = 2$

The analysis of multivariate rank data can most simply be illustrated with bivariate response measurement scores, with no loss of generality, and easily extended to $r > 2$. Figure 6.3 portrays the possible locations of two treatment groups in a 2-dimensional space consisting of a possible six rank scores on variables x_1 and x_2. The two circles, gray (●) and white (○), represent $g = 2$ collections of a multitude of bivariate ($r = 2$) response measurements. The problem is to determine if the two sets of response measurements differ statistically in location, given a substantial overlap.

To demonstrate the computation of MRPP with bivariate rank scores, consider a finite sample of $N = 7$ objects and let S_1 and S_2 denote an exhaustive partitioning of the N objects into $g = 2$ disjoint treatment groups. For simplicity, let S_1 consist of

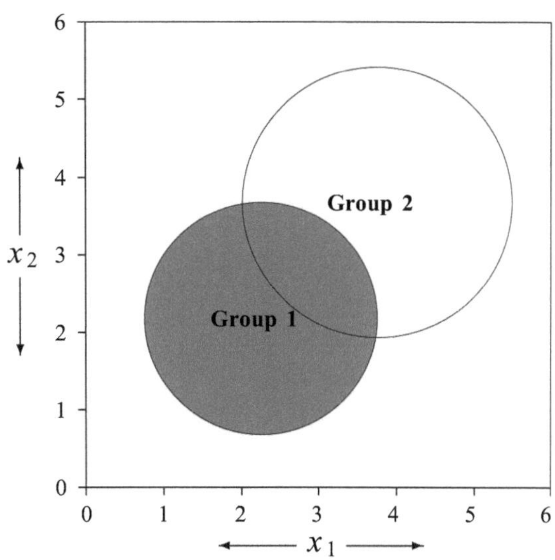

Fig. 6.3 Graphic depicting $g = 2$ treatment groups with simulated bivariate rank measurement scores

$n_1 = 3$ objects for $r = 2$ commensurate rank response measurements (x_{1i} and x_{2i}) on each object for $i = 1, 2, 3$, with $x'_1 = (4, 5)$, $x'_2 = (3, 4)$, and $x'_3 = (4, 3)$, and let S_2 consist of $n_2 = 4$ objects with $r = 2$ commensurate rank response measurements (x_{1i} and x_{2i}) on each object for $i = 1, \ldots, 4$, with $x'_4 = (2, 3)$, $x'_5 = (2, 2)$, $x'_6 = (3, 2)$, and $x'_7 = (3, 1)$.

Example bivariate rank data for the $N = 7$ objects are adapted from Mielke et al. [301, p. 121]. The bivariate rank scores are listed in Fig. 6.4 and are graphically displayed in Fig. 6.5, where the responses of the $n_1 = 3$ objects in treatment group S_1 are plotted as black circles (●) and labeled ω_1, ω_2, and ω_3, and the responses of the $n_2 = 4$ objects in treatment group S_2 are plotted as gray circles (●) and labeled ω_4, ω_5, ω_6, and ω_7. Although a visual impression of Fig. 6.5 suggests that the $g = 2$ treatment groups are separated in the two-dimensional space, a more rigorous characterization of the separation is needed before a quantitative evaluation

Fig. 6.4 Example data set with $r = 2$, $g = 2$, $n_1 = 3$, $n_2 = 4$, and $N = n_1 + n_2 = 7$

		Values	
Group	Object	x_1	x_2
S_1	ω_1	4	5
	ω_2	3	4
	ω_3	4	3
S_2	ω_4	2	3
	ω_5	2	2
	ω_6	3	2
	ω_7	3	1

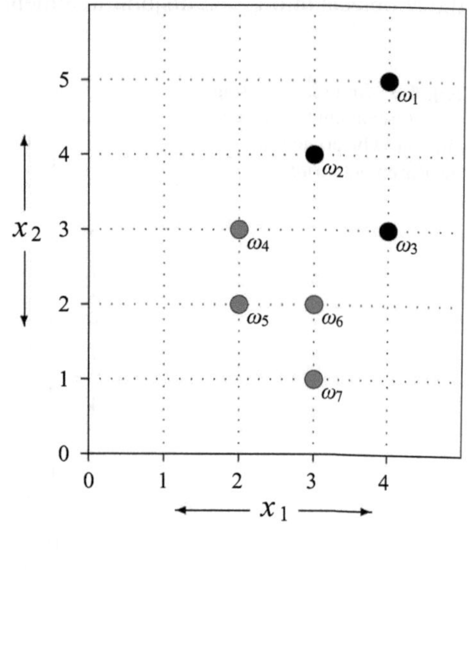

Fig. 6.5 Two-dimensional scatter diagram showing the bivariate rank measurement scores of the $g = 2$ treatment groups listed in Fig. 6.4

can be made. A classical approach would involve Hotelling's two-sample T^2 test, which has the disadvantage of requiring the assumptions that the response measurements of the two groups are distributed as multivariate normal with equal variances and covariances [181]. Since these conditions are never met in practice, it is desirable to consider methods that do not require such assumptions; in this case, exact permutation statistical methods.

For the MRPP analysis of the $N = 7$ bivariate rank scores listed in Fig. 6.4, let $g = 2$, $r = 2$, $p = 2$, $v = 1$, and $C_i = n_i/N$ for $i = 1, \ldots, g$. For this example analysis, $C_i = n_i/N$ is selected as it simply weights each treatment group as proportional to its size and $v = 1$ is chosen as it provides ordinary Euclidean (Pythagorean) distance between the response measurements of two selected objects in an r-dimensional space.

Thus, for the bivariate rank scores listed in Fig. 6.4,

$$C_1 = \frac{n_1}{N} = \frac{3}{7} \quad \text{and} \quad C_2 = \frac{n_2}{N} = \frac{4}{7} \,.$$

Following Eq. (6.3) on p. 318 for treatment group S_1 with $n_1 = 3$ objects, $p = 2$, and $v = 1$, the generalized Minkowski distance function yields

$$\Delta(1,2) = \left(|4-3|^2 + |5-4|^2\right)^{1/2} = 1.4142 \,,$$

$$\Delta(1,3) = \left(|4-4|^2 + |5-3|^2\right)^{1/2} = 2.0000 \,,$$

and

$$\Delta(2,3) = \left(|3-4|^2 + |4-3|^2\right)^{1/2} = 1.4142 \,.$$

For treatment group S_2 with $n_2 = 4$ objects, $p = 2$, and $v = 1$, the generalized Minkowski distance function yields

$$\Delta(4,5) = \left(|2-2|^2 + |3-2|^2\right)^{1/2} = 1.0000 \,,$$

$$\Delta(4,6) = \left(|2-3|^2 + |3-2|^2\right)^{1/2} = 1.4142 \,,$$

$$\Delta(4,7) = \left(|2-3|^2 + |3-1|^2\right)^{1/2} = 2.2361 \,,$$

$$\Delta(5,6) = \left(|2-3|^2 + |2-2|^2\right)^{1/2} = 1.0000 \,,$$

$$\Delta(5,7) = \left(|2-3|^2 + |2-1|^2\right)^{1/2} = 1.4142 \,,$$

and

$$\Delta(6,7) = \left(|3-3|^2 + |2-1|^2\right)^{1/2} = 1.0000\,.$$

Then, following Eq. (6.2) on p. 317, the $N = 7$ bivariate rank scores listed in Fig. 6.4 yield $g = 2$ average-distance function values of

$$\xi_1 = \binom{n_1}{2}^{-1} \left[\Delta(1,2) + \Delta(1,3) + \Delta(2,3)\right]$$

$$= \binom{3}{2}^{-1} (1.4142 + 2.0000 + 1.4142) = 1.6095$$

and

$$\xi_2 = \binom{n_2}{2}^{-1} \left[\Delta(4,5) + \Delta(4,6) + \Delta(4,7) + \Delta(5,6) + \Delta(5,7) + \Delta(6,7)\right]$$

$$= \binom{4}{2}^{-1} (1.0000 + 1.4142 + 2.2361 + 1.0000 + 1.4142 + 1.0000)$$

$$= 1.3441\,.$$

Following Eq. (6.1) on p. 317, the observed MRPP test statistic based on $v = 1$ and $C_i = n_i/N$ for $i = 1, \ldots, g$ is

$$\delta_o = C_1\xi_1 + C_2\xi_2 = \left(\frac{3}{7}\right)(1.6095) + \left(\frac{4}{7}\right)(1.3441) = 1.4578\,.$$

The $N = 7$ objects can be partitioned into $g = 2$ treatment groups, S_1 and S_2, with $n_1 = 3$ and $n_2 = 4$ bivariate rank scores preserved for each arrangement in

$$M = \frac{N!}{n_1!\; n_2!} = \frac{7!}{3!\; 4!} = 35$$

possible, equally-likely ways. The ξ_1, ξ_2, and δ values for each of the $M = 35$ arrangements are listed in Table 6.1 and are ordered from lowest to highest by the δ values.

The observed MRPP test statistic, $\delta_o = 1.4578$, obtained from the realized arrangement of the $N = 7$ bivariate response measurement scores in groups S_1 and S_2,

$$\{(4,5)(3,4)(4,3)\} \quad \{(2,3)(2,2)(3,2)(3,1)\}\,,$$

Table 6.1 Permutations of the observed bivariate response measurement scores listed in Fig. 6.4 for treatment groups S_1 and S_2 with values for ξ_1, ξ_2, and δ based on $v = 1$, ordered by values of δ from lowest to highest

Order	Group S_1	Group S_2	ξ_1	ξ_2	δ
1	{(4, 5)(3, 4)(4, 3)}	{(2, 3)(2, 2)(3, 2)(3, 1)}	1.6095	1.3341	1.4578
2	{(2, 2)(3, 2)(3, 1)}	{(4, 5)(3, 4)(4, 3)(2, 3)}	1.1381	1.8452	1.5421
3	{(4, 5)(3, 4)(2, 3)}	{(4, 3)(2, 2)(3, 2)(3, 1)}	1.8856	1.5501	1.6939
4	{(2, 3)(2, 2)(3, 1)}	{(4, 5)(3, 4)(4, 3)(3, 2)}	1.5501	1.9008	1.7505
5	{(2, 3)(2, 2)(3, 2)}	{(4, 5)(3, 4)(4, 3)(3, 1)}	1.1381	2.3646	1.8389
6	{(4, 3)(3, 2)(3, 1)}	{(4, 5)(3, 4)(2, 3)(2, 2)}	1.5501	2.0831	1.8547
7	{(2, 3)(3, 2)(3, 1)}	{(4, 5)(3, 4)(4, 3)(2, 2)}	1.5501	2.1510	1.8935
8	{(4, 5)(4, 3)(2, 3)}	{(3, 4)(2, 2)(3, 2)(3, 1)}	2.2761	1.7750	1.9898
9	{(3, 4)(2, 3)(2, 2)}	{(4, 5)(4, 3)(3, 2)(3, 1)}	1.5501	2.3226	1.9915
10	{(4, 5)(3, 4)(3, 2)}	{(4, 3)(2, 3)(2, 2)(3, 1)}	2.1922	1.8537	1.9988
11	{(4, 3)(2, 2)(3, 1)}	{(4, 5)(3, 4)(2, 3)(3, 2)}	1.9621	2.0389	2.0060
12	{(4, 5)(4, 3)(3, 2)}	{(3, 4)(2, 3)(2, 2)(3, 1)}	2.1922	1.8834	2.0157
13	{(4, 5)(3, 4)(2, 2)}	{(4, 3)(2, 3)(3, 2)(3, 1)}	2.4186	1.7168	2.0176
14	{(3, 4)(4, 3)(2, 3)}	{(4, 5)(2, 2)(3, 2)(3, 1)}	1.6095	2.3842	2.0522
15	{(4, 5)(4, 3)(3, 1)}	{(3, 4)(2, 3)(2, 2)(3, 2)}	2.7864	1.5107	2.0575
16	{(4, 5)(3, 4)(3, 1)}	{(4, 3)(2, 3)(2, 2)(3, 2)}	2.8458	1.5107	2.0829
17	{(4, 3)(2, 2)(3, 2)}	{(4, 5)(3, 4)(2, 3)(3, 1)}	1.5501	2.5027	2.0944
18	{(4, 5)(2, 3)(2, 2)}	{(3, 4)(4, 3)(3, 2)(3, 1)}	2.4780	1.8441	2.1158
19	{(3, 4)(4, 3)(3, 2)}	{(4, 5)(2, 3)(2, 2)(3, 1)}	1.6095	2.5346	2.1381
20	{(4, 3)(2, 3)(2, 2)}	{(4, 5)(3, 4)(3, 2)(3, 1)}	1.7454	2.4499	2.1480
21	{(3, 4)(3, 2)(3, 1)}	{(4, 5)(4, 3)(2, 3)(2, 2)}	2.0000	2.2783	2.1591
22	{(4, 5)(3, 2)(3, 1)}	{(3, 4)(4, 3)(2, 2)(2, 3)}	2.7618	1.7168	2.1646
23	{(3, 4)(2, 2)(3, 1)}	{(4, 5)(4, 3)(2, 3)(3, 2)}	2.2168	2.1365	2.1709
24	{(4, 5)(4, 3)(2, 2)}	{(3, 4)(2, 3)(3, 2)(3, 1)}	2.6139	1.8441	2.1740
25	{(3, 4)(2, 3)(3, 2)}	{(4, 5)(4, 3)(2, 2)(3, 1)}	1.6095	2.6025	2.1769
26	{(3, 4)(4, 3)(3, 1)}	{(4, 5)(2, 3)(2, 2)(3, 2)}	2.2168	2.1684	2.1891
27	{(4, 3)(2, 3)(3, 2)}	{(4, 5)(3, 4)(2, 2)(3, 1)}	1.6095	2.6322	2.1939
28	{(4, 3)(2, 3)(3, 1)}	{(4, 5)(3, 4)(2, 2)(3, 2)}	2.1574	2.2364	2.2025
29	{(3, 4)(2, 2)(3, 2)}	{(4, 5)(4, 3)(2, 3)(3, 1)}	1.7454	2.5706	2.2169
30	{(4, 5)(2, 2)(3, 1)}	{(3, 4)(4, 3)(2, 3)(3, 2)}	3.0476	1.6095	2.2258
31	{(3, 4)(2, 3)(3, 1)}	{(4, 5)(4, 3)(2, 2)(3, 2)}	2.2168	2.2364	2.2280
32	{(3, 4)(4, 3)(2, 2)}	{(4, 5)(2, 3)(3, 2)(3, 1)}	1.9621	2.4607	2.2470
33	{(4, 5)(2, 3)(3, 2)}	{(3, 4)(4, 3)(2, 2)(3, 1)}	2.4683	2.0894	2.2518
34	{(4, 5)(2, 2)(3, 2)}	{(3, 4)(4, 3)(2, 3)(3, 1)}	2.5893	2.0501	2.2812
35	{(4, 5)(2, 3)(3, 1)}	{(3, 4)(4, 3)(2, 2)(3, 2)}	3.0625	1.7168	2.2935

(Order 1 in Table 6.1) is unusual since each of the remaining 34 δ values listed in Table 6.1 exceeds the observed value of $\delta_o = 1.4578$ and none is less than the observed value. If all arrangements of the observed rank scores occur with equal chance, the exact probability value of $\delta_o = 1.4578$ computed on the $M = 35$ possible arrangements of the observed data with $n_1 = 3$ and $n_2 = 4$ bivariate rank scores preserved for each arrangement is

$$P(\delta \leq \delta_o | H_0) = \frac{\text{number of } \delta \text{ values} \leq \delta_o}{M} = \frac{1}{35} = 0.0286 .$$

Following Eq. (6.5) on p. 319, the exact expected value of the $M = 35$ δ values is $\mu_\delta = 2.0547$ and, following Eq. (6.4) on p. 319, the observed chance-corrected measure of effect size is

$$\Re_o = 1 - \frac{\delta_o}{\mu_\delta} = 1 - \frac{1.4578}{2.0547} = +0.2905 ,$$

indicating approximately 29 % within-group agreement above that expected by chance.

6.3 MRPP for the WMW Rank-Sum Test with $r = 2$

In this section the Wilcoxon–Mann–Whitney (WMW) two-sample rank-sum test [262, 429], discussed in Chap. 5, Sects. 5.3 and 5.4, is generalized to $r \geq 2$ response measurement scores. Three example analyses illustrate a permutation approach to two-sample rank-sum problems with multivariate response measurements. The first example utilizes MRPP to extend the conventional Wilcoxon–Mann–Whitney two-sample rank-sum test to a multivariate two-sample rank-sum test using a small set of bivariate rank scores with $v = 2$ and treatment-group weights $C_i = (n_i - 1)/(N - g)$ for $i = 1, \ldots, g$; the second example analyzes the same small set of bivariate rank scores, but uses $v = 1$ and treatment-group weights $C_i = (n_i - 1)/(N - g)$ for $i = 1, \ldots, g$; and the third example analyzes the same set of bivariate rank scores using $v = 1$, but adopts a proportional treatment-group weighting function given by $C_i = n_i/N$ for $i = 1, \ldots, g$.

6.3.1 Example 1

Consider a two-sample rank test for N objects with n_1 and n_2 objects in the first and second samples, respectively, and $r = 2$ ordinal response measurements obtained on each object. Example bivariate rank scores are listed in Fig. 6.6 where $g = 2$,

Fig. 6.6 Bivariate rank response measurement scores for two treatment groups with $r = 2$, $g = 2$, $n_1 = 15$, $n_2 = 10$, and $N = n_1 + n_2 = 25$

Group 1	Group 2
(2, 2)	(5, 3)
(2, 3)	(5, 4)
(3, 2)	(5, 7)
(3, 3)	(6, 6)
(3, 4)	(7, 5)
(4, 3)	(7, 8)
(4, 4)	(8, 6)
(4, 5)	(8, 7)
(5, 6)	(8, 9)
(6, 4)	(9, 8)
(6, 5)	
(7, 6)	
(7, 7)	
(8, 8)	
(9, 9)	

$r = 2$, $n_1 = 15$, $n_2 = 10$, and $N = n_1 + n_2 = 25$. Figure 6.7 graphically displays the bivariate rank scores listed in Fig. 6.6 with white circles (○) representing the $n_1 = 15$ objects in Group 1 and gray circles (●) representing the $n_2 = 10$ objects in Group 2. For the first analysis of the $N = 25$ bivariate rank response measurements listed in Fig. 6.6, let $v = 2$, employing squared Euclidean distance between the bivariate rank scores, and let the treatment-group weights be given by

$$C_i = \frac{n_i - 1}{N - g} , \qquad i = 1, \ldots, g ,$$

to correspond to a multivariate Wilcoxon–Mann–Whitney two-sample rank-sum test [262, 429].

Because there are only

$$M = \frac{N!}{n_1!\; n_2!} = \frac{25!}{15!\; 10!} = 3{,}268{,}760$$

possible, equally-likely arrangements of the $N = 25$ bivariate rank scores listed in Fig. 6.6, an exact solution is possible. Following Eq. (6.2) on p. 317, the $N = 25$ bivariate rank scores listed in Fig. 6.6 yield $g = 2$ average distance-function values of

$$\xi_1 = 18.6667 \quad \text{and} \quad \xi_2 = 11.4889 .$$

Following Eq. (6.1) on p. 317, the MRPP test statistic based on $v = 2$ and treatment-group weights

$$C_i = \frac{n_i - 1}{N - g} , \qquad i = 1, 2 ,$$

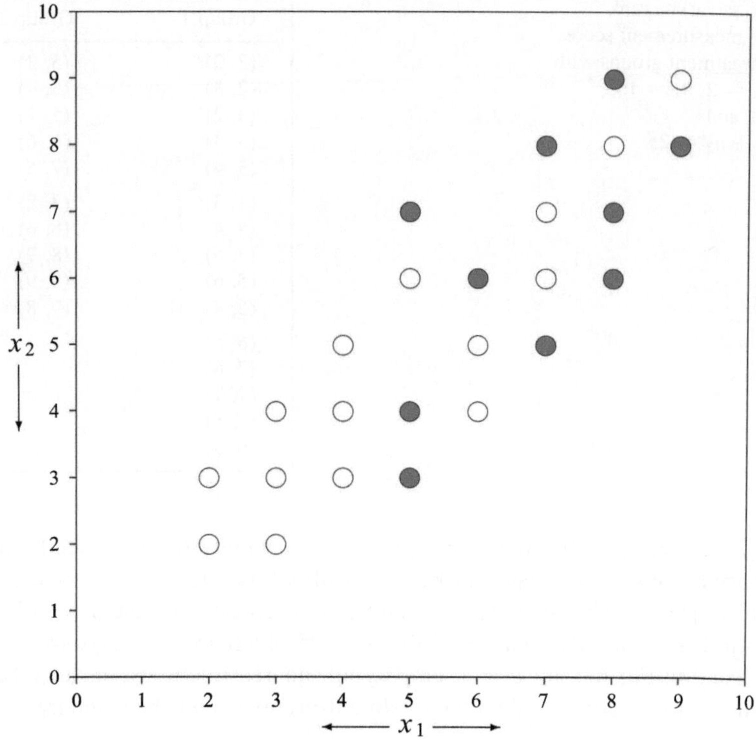

Fig. 6.7 Graphic depicting the bivariate rank measurement scores listed in Fig. 6.6 for $g = 2$ treatment groups with $n_1 = 15$ objects in Group 1, indicated by *white circles*, and $n_2 = 10$ objects in Group 2, indicated by *gray circles*

is

$$\delta_o = \sum_{i=1}^{g} C_i \xi_i = \frac{1}{25 - 2} \big[(15 - 1)(18.6667) + (10 - 1)(11.4889)\big] = 15.8580 \, .$$

If all arrangements of the $N = 25$ observed bivariate rank scores listed in Fig. 6.6 occur with equal chance, the exact probability value of $\delta_o = 15.8580$ computed on the $M = 3{,}268{,}760$ possible arrangements of the observed data with $n_1 = 15$ and $n_2 = 10$ bivariate rank scores preserved for each arrangement is

$$P(\delta \leq \delta_o | H_0) = \frac{\text{number of } \delta \text{ values} \leq \delta_o}{M} = \frac{116{,}998}{3{,}268{,}760} = 0.0358 \, .$$

No comparison is made with the conventional Wilcoxon–Mann–Whitney two-sample rank-sum test as the Wilcoxon–Mann–Whitney test is undefined for $r > 1$.

Following Eq. (6.5) on p. 319, the exact expected value of the $M = 3{,}268{,}760$ δ values is $\mu_\delta = 18.2933$ and, following Eq. (6.4) on p. 319, the observed chance-corrected measure of effect size is

$$\Re_o = 1 - \frac{\delta_o}{\mu_\delta} = 1 - \frac{15.8580}{18.2933} = +0.1331 \; ,$$

indicating approximately 13 % within-group agreement above that expected by chance.

6.3.2 Example 2

For the second analysis of the bivariate rank scores listed in Fig. 6.6, let the treatment-group weights be given by

$$C_i = \frac{n_i - 1}{N - g} \; , \qquad i = 1, \ldots, g \; ,$$

as in Example 1, but set $v = 1$ instead of $v = 2$, employing ordinary Euclidean distance between the bivariate rank scores. Following Eq. (6.2) on p. 317, the $N = 25$ bivariate rank scores listed in Fig. 6.6 yield $g = 2$ average distance-function values of

$$\xi_1 = 3.6963 \quad \text{and} \quad \xi_2 = 3.0663 \; .$$

Following Eq. (6.1) on p. 317, the observed value of the MRPP test statistic based on $v = 1$ and treatment-group weights

$$C_i = \frac{n_i - 1}{N - g} \; , \qquad i = 1, 2 \; ,$$

is

$$\delta_o = \sum_{i=1}^{g} C_i \xi_i = \frac{1}{25 - 2}\big[(15 - 1)(3.6963) + (10 - 1)(3.0663)\big] = 3.4498 \; .$$

If all arrangements of the $N = 25$ observed bivariate rank scores listed in Fig. 6.6 occur with equal chance, the exact probability value of $\delta_o = 3.4498$ computed on the $M = 3{,}268{,}760$ possible arrangements of the observed data with $n_1 = 15$ and $n_2 = 10$ bivariate rank scores preserved for each arrangement is

$$P(\delta \le \delta_o | H_0) = \frac{\text{number of } \delta \text{ values} \le \delta_o}{M} = \frac{130{,}370}{3{,}268{,}760} = 0.0399 \; .$$

For comparison, the exact probability value based on $v = 2$, $M = 3{,}268{,}760$, and $C_i = (n_i - 1)/(N - g)$ for $i = 1, 2$ in Example 1 is $P = 0.0358$. No comparison is made with the conventional Wilcoxon–Mann–Whitney two-sample rank-sum test as the Wilcoxon–Mann–Whitney two-sample test is undefined for both $v = 1$ and $r > 1$.

Following Eq. (6.5) on p. 319, the exact expected value of the $M = 3{,}268{,}760$ δ values is $\mu_\delta = 3.7103$ and, following Eq. (6.4) on p. 319, the observed chance-corrected measure of effect size is

$$\Re_o = 1 - \frac{\delta_o}{\mu_\delta} = 1 - \frac{3.4498}{3.7103} = +0.0702 ,$$

indicating approximately 7 % within-group agreement above that expected by chance.

6.3.3 Example 3

For the third analysis of the bivariate rank scores listed in Fig. 6.6 on p. 325, let the treatment-group weights be given by

$$C_i = \frac{n_i}{N} , \qquad i = 1, \ldots, g ,$$

simply weighting each treatment group proportional to its size, and let $v = 1$, employing ordinary Euclidean distance between the bivariate rank scores, as in Example 2. Following Eq. (6.2) on p. 317, the $N = 25$ bivariate rank scores listed in Fig. 6.6 yield $g = 2$ average distance-function values of

$$\xi_1 = 3.6963 \quad \text{and} \quad \xi_2 = 3.0663 ,$$

and following Eq. (6.1) on p. 317, the observed value of the MRPP test statistic based on $v = 1$ and treatment-group weights

$$C_i = \frac{n_i}{N} , \qquad i = 1, 2 ,$$

is

$$\delta_o = \sum_{i=1}^{g} C_i \xi_i = \frac{1}{25}\big[(15)(3.6963) + (10)(3.0663)\big] = 3.4443 .$$

If all arrangements of the $N = 25$ observed bivariate rank scores listed in Fig. 6.6 occur with equal chance, the exact probability value of $\delta_o = 3.4443$ computed on the $M = 3{,}268{,}760$ possible arrangements of the observed data with $n_1 = 15$ and

$n_2 = 10$ bivariate rank scores preserved for each arrangement is

$$P(\delta \leq \delta_o|H_0) = \frac{\text{number of } \delta \text{ values} \leq \delta_o}{M} = \frac{126{,}392}{3{,}268{,}760} = 0.0387 .$$

For comparison, the exact probability values based on $v = 2$, $M = 3{,}268{,}760$, and $C_i = (n_i - 1)/(N - g)$ for $i = 1, 2$ in Example 1 and $v = 1$, $M = 3{,}268{,}760$, and $C_i = (n_i - 1)/(N - g)$ for $i = 1, 2$ in Example 2 are $P = 0.0358$ and $P = 0.0399$, respectively. No comparison is made with the conventional Wilcoxon–Mann–Whitney two-sample rank-sum test as the Wilcoxon–Mann–Whitney test is undefined for $v = 1$, $r > 1$, and $C_i = n_i/N$, $i = 1, \ldots, g$.

Following Eq. (6.5) on p. 319, the exact expected value of the $M = 3{,}268{,}760$ δ values is $\mu_\delta = 3.7103$ and, following Eq. (6.4) on p. 319, the observed chance-corrected measure of effect size is

$$\Re_o = 1 - \frac{\delta_o}{\mu_\delta} = 1 - \frac{3.4443}{3.7103} = +0.0717 ,$$

indicating approximately 7 % within-group agreement above that expected by chance.

6.4 MRPP for the KW Rank-Sum Test with $r = 2$

In this section the Kruskal–Wallis (KW) multi-sample rank-sum test [225], discussed in Chap. 5, Sects. 5.5 and 5.6, is generalized to $r \geq 2$ response measurements. Three example analyses illustrate a permutation approach to multi-sample rank-sum problems with multivariate response measurements. The first example utilizes MRPP to extend the conventional Kruskal–Wallis multi-sample rank-sum test to a multivariate g-sample rank-sum test using a small set of bivariate rank scores with $v = 2$ and treatment-group weights $C_i = (n_i - 1)/(N - g)$ for $i = 1, \ldots, g$; the second example analyzes the same small set of bivariate rank scores, but uses $v = 1$ and treatment-group weights $C_i = (n_i - 1)/(N - g)$ for $i = 1, \ldots, g$; and the third example analyzes the same set of bivariate rank scores using $v = 1$, but adopts a proportional treatment-group weighting function given by $C_i = n_i/N$ for $i = 1, \ldots, g$.

6.4.1 Example 1

Consider a three-sample rank test for N objects with n_1, n_2, and n_3 objects in the first, second, and third samples, respectively, and $r = 2$ ordinal response measurements obtained on each object. Example bivariate rank scores are listed in Fig. 6.8 where $g = 3$, $r = 2$, $n_1 = n_2 = n_3 = 10$, and $N = n_1 + n_2 + n_3 = 30$. Figure 6.9 graphically displays the bivariate rank scores listed in Fig. 6.8 with white circles

Group 1	Group 2	Group 3
(2, 2)	(5, 3)	(3, 6)
(2, 3)	(5, 4)	(4, 7)
(3, 3)	(5, 7)	(4, 8)
(3, 4)	(6, 6)	(4, 9)
(4, 5)	(7, 5)	(5, 5)
(5, 6)	(7, 8)	(5, 8)
(6, 5)	(8, 6)	(6, 4)
(7, 7)	(8, 7)	(6, 7)
(8, 8)	(8, 9)	(6, 8)
(9, 9)	(9, 8)	(7, 6)

Fig. 6.8 Bivariate rank response measurement scores for three treatment groups with $r = 2$, $g = 3$, $n_1 = n_2 = n_3 = 10$, and $N = n_1 + n_2 + n_3 = 30$

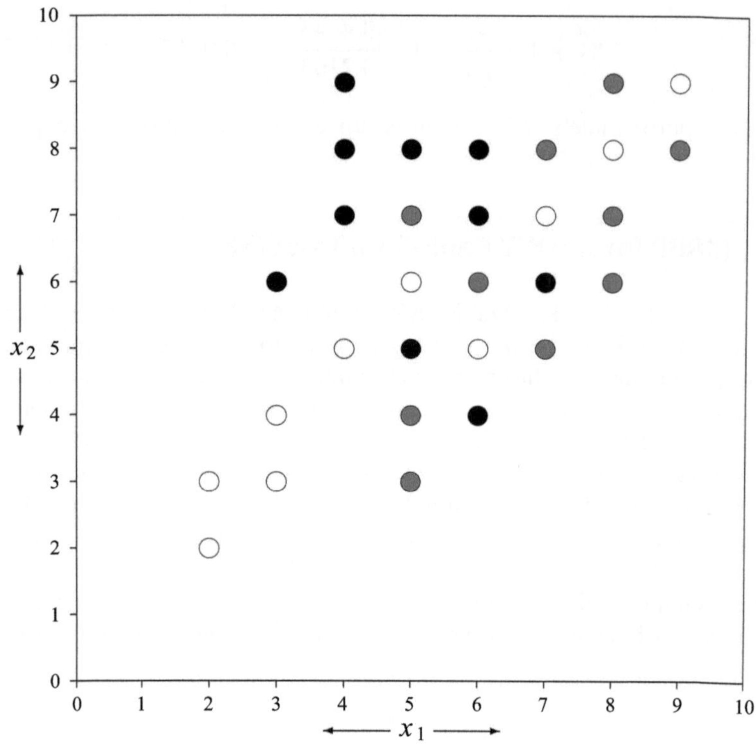

Fig. 6.9 Graphic depicting the bivariate rank measurement scores listed in Fig. 6.8 for $g = 3$ treatment groups with $n_1 = 10$ objects in Group 1, indicated by *white circles*, $n_2 = 10$ objects in Group 2, indicated by *gray circles*, and $n_3 = 10$ objects in Group 3, indicated by *black circles*

(○) representing the $n_1 = 10$ objects in Group 1, gray circles (●) representing the $n_2 = 10$ objects in Group 2, and black circles (●) representing the $n_3 = 10$ objects in Group 3.

For the first analysis of the $N = 30$ bivariate rank scores listed in Fig. 6.8, let $v = 2$, employing squared Euclidean distance between the bivariate rank scores, and let the treatment-group weights be given by

$$C_i = \frac{n_i - 1}{N - g} , \qquad i = 1, \ldots, g ,$$

to correspond to a multivariate Kruskal–Wallis g-sample rank-sum test [225].

Because there are

$$M = \frac{N!}{\prod_{i=1}^{g} n_i!} = \frac{30!}{(10!)^3} = 5{,}550{,}996{,}791{,}340$$

possible, equally-likely arrangements of the $N = 30$ bivariate rank scores listed in Fig. 6.8, an exact solution is not feasible. Following Eq. (6.2) on p. 317, the $N = 30$ bivariate rank scores listed in Fig. 6.8 yield $g = 3$ average distance-function values of

$$\xi_1 = 23.2222 , \quad \xi_2 = 11.4889 , \quad \text{and} \quad \xi_3 = 7.9111 .$$

Following Eq. (6.1) on p. 317, the observed value of the MRPP test statistic based on $v = 2$ and treatment-group weights

$$C_i = \frac{n_i - 1}{N - g} , \qquad i = 1, 2, 3 ,$$

is

$$\delta_o = \sum_{i=1}^{g} C_i \xi_i = \frac{10 - 1}{30 - 3} \big(23.2222 + 11.4889 + 7.9111\big) = 14.2074 .$$

If all M possible arrangements of the $N = 30$ observed bivariate rank scores listed in Fig. 6.8 occur with equal chance, the approximate resampling probability value of $\delta_o = 14.2074$ computed on $L = 1{,}000{,}000$ random arrangements of the observed data with $n_1 = n_2 = n_3 = 10$ bivariate rank scores preserved for each arrangement is

$$P\big(\delta \le \delta_o | H_0\big) = \frac{\text{number of } \delta \text{ values} \le \delta_o}{L} = \frac{69{,}672}{1{,}000{,}000} = 0.0697 .$$

No comparison is made with the conventional Kruskal–Wallis g-sample rank-sum test as the Kruskal–Wallis test is undefined for $r > 1$.

Following Eq. (6.5) on p. 319, the exact expected value of the M δ values is $\mu_\delta = 15.7287$ and, following Eq. (6.4) on p. 319, the observed chance-corrected measure

of effect size is

$$\Re_o = 1 - \frac{\delta_o}{\mu_\delta} = 1 - \frac{14.2074}{15.7287} = +0.0967 ,$$

indicating approximately 10 % within-group agreement above that expected by chance.

6.4.2 Example 2

For the second analysis of the bivariate rank scores listed in Fig. 6.8, let the treatment-group weights be given by

$$C_i = \frac{n_i - 1}{N - g} , \qquad i = 1, \ldots, g ,$$

as in Example 1, but set $v = 1$ instead of $v = 2$, employing ordinary Euclidean distance between the bivariate rank scores. Following Eq. (6.2) on p. 317, the $N = 30$ bivariate rank scores listed in Fig. 6.8 yield $g = 3$ average distance-function values of

$$\xi_1 = 4.1457 , \quad \xi_2 = 3.0663 , \quad \text{and} \quad \xi_3 = 2.5980 .$$

Following Eq. (6.1) on p. 317, the observed value of the MRPP test statistic based on $v = 1$ and treatment-group weights

$$C_i = \frac{n_i - 1}{N - g} , \qquad i = 1, 2, 3 ,$$

is

$$\delta_o = \sum_{i=1}^{g} C_i \xi_i = \frac{10 - 1}{30 - 3}(4.1457 + 3.0663 + 2.5980) = 3.2700 .$$

If all M possible arrangements of the $N = 30$ observed bivariate rank scores listed in Fig. 6.8 occur with equal chance, the approximate resampling probability value of $\delta_o = 3.2700$ computed on $L = 1{,}000{,}000$ random arrangements of the observed data with $n_1 = n_2 = n_3 = 10$ bivariate rank scores preserved for each arrangement is

$$P(\delta \leq \delta_o | H_0) = \frac{\text{number of } \delta \text{ values} \leq \delta_o}{L} = \frac{30{,}092}{1{,}000{,}000} = 0.0301 .$$

For comparison, the approximate resampling probability value based on $v = 2$, $L = 1{,}000{,}000$, and $C_i = (n_i - 1)/(N - g)$ for $i = 1, 2, 3$ in Example 1 is $P = 0.0697$. No comparison is made with the conventional Kruskal–Wallis g-sample rank-sum test as the Kruskal–Wallis test is undefined for both $v = 1$ and $r > 1$.

Following Eq. (6.5) on p. 319, the exact expected value of the M δ values is $\mu_\delta = 3.5203$ and, following Eq. (6.4) on p. 319, the observed chance-corrected measure of effect size is

$$\Re_o = 1 - \frac{\delta_o}{\mu_\delta} = 1 - \frac{3.2700}{3.5203} = +0.0711 \;,$$

indicating approximately 7 % within-group agreement above that expected by chance.

6.4.3 Example 3

For the third analysis of the bivariate rank scores listed in Fig. 6.8 on p. 330, let the treatment-group weights be given by

$$C_i = \frac{n_i}{N} \;, \qquad i = 1, \ldots, g \;,$$

simply weighting each treatment group proportional to its size, and let $v = 1$, employing ordinary Euclidean distance between the bivariate rank scores, as in Example 2. Following Eq. (6.2) on p. 317, the $N = 30$ bivariate rank scores listed in Fig. 6.8 yield $g = 3$ average distance-function values of

$$\xi_1 = 4.1457 \;, \quad \xi_2 = 3.0663 \;, \quad \text{and} \quad \xi_3 = 2.5980 \;.$$

Following Eq. (6.1) on p. 317, the observed value of the MRPP test statistic based on $v = 1$ and treatment-group weights

$$C_i = \frac{n_i}{N} \;, \qquad i = 1, 2, 3 \;,$$

is

$$\delta_o = \sum_{i=1}^{g} C_i \xi_i = \frac{10}{30}\big(4.1457 + 3.0663 + 2.5980\big) = 3.2700 \;.$$

If all M possible arrangements of the $N = 30$ observed bivariate rank scores listed in Fig. 6.8 occur with equal chance, the approximate resampling probability value of $\delta_o = 3.2700$ computed on $L = 1{,}000{,}000$ random arrangements of the observed data with $n_1 = n_2 = n_3 = 10$ bivariate rank scores preserved for each arrangement is

$$P(\delta \leq \delta_o | H_0) = \frac{\text{number of } \delta \text{ values} \leq \delta_o}{L} = \frac{30{,}092}{1{,}000{,}000} = 0.0301 \; .$$

For comparison, the approximate resampling probability values based on $v = 2$, $L = 1{,}000{,}000$, and $C_i = (n_i - 1)/(N - g)$ for $i = 1, 2, 3$ in Example 1 and $v = 1$, $L = 1{,}000{,}000$, and $C_i = (n_i - 1)/(N - g)$ for $i = 1, 2, 3$ in Example 2 are $P = 0.0697$ and $P = 0.0301$, respectively. Since $n_1 = n_2 = n_3$, the probability values based on $v = 1$ and $C_i = (n_i - 1)/(N - g)$ and $v = 1$ and $C_i = n_i/N$ are the same. No comparison is made with the conventional Kruskal–Wallis g-sample rank-sum test as the Kruskal–Wallis test is undefined for $v = 1$, $r > 1$, and $C_i = n_i/N$, $i = 1, \ldots, g$.

Following Eq. (6.5) on p. 319, the exact expected value of the M δ values is $\mu_\delta = 3.5203$ and, following Eq. (6.4) on p. 319, the observed chance-corrected measure of effect size is

$$\Re_o = 1 - \frac{\delta_o}{\mu_\delta} = 1 - \frac{3.2700}{3.5203} = +0.0711 \; ,$$

indicating approximately 7 % within-group agreement above that expected by chance.

6.5 MRPP for the A_{Ns} Function with $s = 1$

In this section the A_{Ns} power-of-rank function test with $s = 1$ [281], discussed in Chap. 5, Sects. 5.8 and 5.9, is generalized to $r \geq 2$ response measurements. Because A_{N1} is associated with the Wilcoxon two-sample rank-sum test [429] when $r = 1$, as described in Chap. 5, Sects. 5.7 and 5.9, this A_{N1} analysis can be considered as a multivariate generalization of the Wilcoxon test, i.e., $r \geq 2$.

Three example analyses illustrate a permutation approach to two-sample power-of-rank problems with multivariate response measurements. The first example utilizes MRPP to extend the conventional A_{Ns} function with $s = 1$ to a multivariate power-of-rank test using a small set of bivariate rank scores with $v = 2$ and treatment-group weights $C_i = (n_i - 1)/(N - g)$ for $i = 1, \ldots, g$; the second example analyzes the same small set of bivariate rank scores, but uses $v = 1$ and treatment-group weights $C_i = (n_i - 1)/(N - g)$ for $i = 1, \ldots, g$; and the third example analyzes the same set of bivariate rank scores using $v = 1$, but adopts a proportional treatment-group weighting function given by $C_i = n_i/N$ for $i = 1, \ldots, g$.

6.5.1 Example 1

Consider a two-sample rank test for N objects with n_1 and n_2 objects in the first and second samples, respectively, and $r = 2$ ordinal response measurements obtained on each object. Example bivariate rank scores are listed in Fig. 6.10 where $g = 2$, $r = 2$, $n_1 = n_2 = 9$, and $N = n_1 + n_2 = 18$.

Figure 6.11 graphically displays the bivariate rank scores listed in Fig. 6.10 with white circles (○) representing the $n_1 = 9$ objects in Group 1 and gray circles (●) representing the $n_2 = 9$ objects in Group 2.

Group 1	Group 2
(1, 1)	(1, 3)
(1, 2)	(2, 1)
(2, 3)	(4, 3)
(3, 1)	(5, 3)
(3, 2)	(5, 5)
(3, 3)	(5, 6)
(3, 4)	(5, 7)
(5, 4)	(6, 4)
(6, 5)	(7, 8)

Fig. 6.10 Bivariate rank response measurement scores for two treatment groups with $r = 2$, $g = 2$, $n_1 = n_2 = 9$, and $N = n_1 + n_2 = 18$

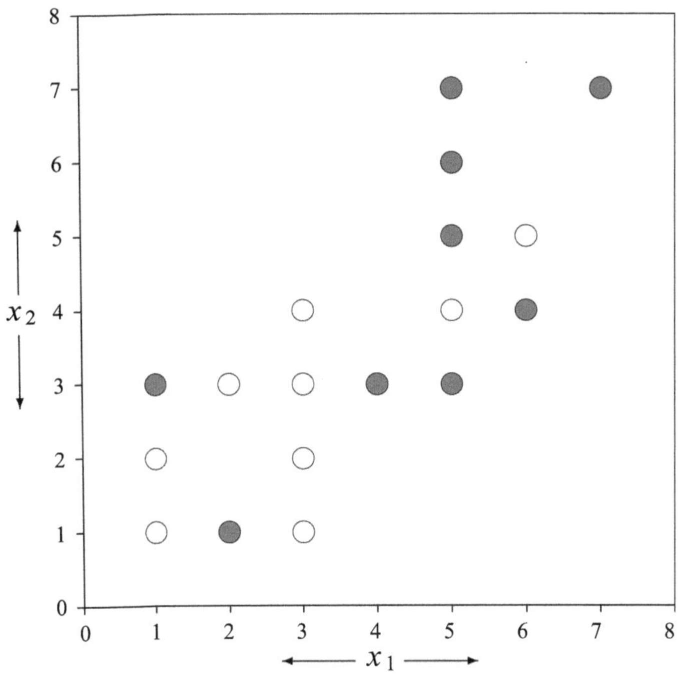

Fig. 6.11 Graphic depicting the bivariate rank measurement scores listed in Fig. 6.10 for $g = 2$ treatment groups with $n_1 = 9$ objects in Group 1, indicated by *white circles*, and $n_2 = 9$ objects in Group 2, indicated by *gray circles*

For the first analysis of the bivariate rank response measurements listed in Fig. 6.10, let $v = 2$, employing squared Euclidean distance between the bivariate rank scores, and let the treatment-group weights be given by

$$C_i = \frac{n_i - 1}{N - g} , \qquad i = 1, \ldots, g ,$$

to correspond to a multivariate Wilcoxon two-sample rank-sum test.

Because there are only

$$M = \frac{N!}{\prod_{i=1}^{g} n_i!} = \frac{18!}{(9!)^2} = 48{,}620$$

possible, equally-likely arrangements of the $N = 18$ bivariate rank scores listed in Fig. 6.10, an exact solution is feasible. Following Eq. (6.2) on p. 317, the $N = 18$ bivariate rank scores listed in Fig. 6.10 yield $g = 2$ average distance-function values of

$$\xi_1 = 9.3889 \quad \text{and} \quad \xi_2 = 23.9444 .$$

Following Eq. (6.1) on p. 317, the observed value of the MRPP test statistic based on $v = 2$ and treatment-group weights

$$C_i = \frac{n_i - 1}{N - g} , \qquad i = 1, 2 ,$$

is

$$\delta_o = \sum_{i=1}^{g} C_i \xi_i = \frac{9 - 1}{18 - 2}(9.3889 + 23.9444) = 16.6667 .$$

If all arrangements of the $N = 18$ observed bivariate rank scores listed in Fig. 6.10 occur with equal chance, the exact probability value of $\delta_o = 16.6667$ computed on the $M = 48{,}620$ possible arrangements of the observed data with $n_1 = n_2 = 9$ bivariate rank scores preserved for each arrangement is

$$P(\delta \leq \delta_o | H_0) = \frac{\text{number of } \delta \text{ values} \leq \delta_o}{M} = \frac{2{,}536}{48{,}620} = 0.0522 .$$

No comparison is made with the conventional Wilcoxon two-sample rank-sum test as Wilcoxon's test is undefined for $r > 1$.

Following Eq. (6.5) on p. 319, the exact expected value of the $M = 48{,}620$ δ values is $\mu_\delta = 19.7712$ and, following Eq. (6.4) on p. 319, the observed chance-corrected measure of effect size is

$$\mathfrak{R}_o = 1 - \frac{\delta_o}{\mu_\delta} = 1 - \frac{16.6667}{19.7712} = +0.1570 \, ,$$

indicating approximately 16 % within-group agreement above that expected by chance.

6.5.2 Example 2

For the second analysis of the bivariate rank scores listed in Fig. 6.10, let the treatment-group weights be given by

$$C_i = \frac{n_i - 1}{N - g} \, , \qquad i = 1, \ldots, g \, ,$$

as in Example 1, but set $v = 1$ instead of $v = 2$, employing ordinary Euclidean distance between the bivariate rank scores. Following Eq. (6.2) on p. 317, the $N = 18$ bivariate rank scores listed in Fig. 6.10 yield $g = 2$ average distance-function values of

$$\xi_1 = 2.7256 \quad \text{and} \quad \xi_2 = 4.3440 \, .$$

Following Eq. (6.1) on p. 317, the observed value of the MRPP test statistic based on $v = 1$ and treatment-group weights

$$C_i = \frac{n_i - 1}{N - g} \, , \qquad i = 1, 2 \, ,$$

is

$$\delta_o = \sum_{i=1}^{g} C_i \xi_i = \frac{9 - 1}{18 - 2}(2.7256 + 4.3440) = 3.5348 \, .$$

If all arrangements of the $N = 18$ observed bivariate rank scores listed in Fig. 6.10 occur with equal chance, the exact probability value of $\delta_o = 3.5348$ computed on the $M = 48{,}620$ possible arrangements of the observed data with $n_1 = n_2 = 9$ bivariate rank scores preserved for each arrangement is

$$P(\delta \leq \delta_o | H_0) = \frac{\text{number of } \delta \text{ values} \leq \delta_o}{M} = \frac{3{,}138}{48{,}620} = 0.0645 \, .$$

For comparison, the exact probability value based on $v = 2$, $M = 48{,}620$, and $C_i = (n_i - 1)/(N - g)$ for $i = 1, 2$ in Example 1 is $P = 0.0522$. No comparison is made with the conventional Wilcoxon two-sample rank-sum test as Wilcoxon's test is undefined for both $r > 1$ and $v = 1$.

Following Eq. (6.5) on p. 319, the exact expected value of the $M = 48{,}620$ δ values is $\mu_\delta = 3.8128$ and, following Eq. (6.4) on p. 319, the observed chance-corrected measure of effect size is

$$\Re_o = 1 - \frac{\delta_o}{\mu_\delta} = 1 - \frac{3.5348}{3.8128} = +0.0729 ,$$

indicating approximately 7 % within-group agreement above that expected by chance.

6.5.3 Example 3

For the third analysis of the bivariate rank scores listed in Fig. 6.10, let the treatment-group weights be given by

$$C_i = \frac{n_i}{N} , \qquad i = 1, \ldots, g ,$$

simply weighting each treatment group proportional to its size, and let $v = 1$, employing ordinary Euclidean distance between the bivariate rank scores, as in Example 2. Following Eq. (6.2) on p. 317, the $N = 18$ bivariate rank scores listed in Fig. 6.10 yield $g = 2$ average distance-function values of

$$\xi_1 = 2.7256 \quad \text{and} \quad \xi_2 = 4.3440 .$$

Following Eq. (6.1) on p. 317, the observed value of the MRPP test statistic based on $v = 1$ and treatment-group weights

$$C_i = \frac{n_i}{N} , \qquad i = 1, 2 ,$$

is

$$\delta_o = \sum_{i=1}^{g} C_i \xi_i = \frac{9}{18}\big(2.7256 + 4.3440\big) = 3.5348 .$$

If all arrangements of the $N = 18$ observed bivariate rank scores listed in Fig. 6.10 occur with equal chance, the exact probability value of $\delta_o = 3.5348$ computed on the $M = 48{,}620$ possible arrangements of the observed data with $n_1 = n_2 = 9$ bivariate rank scores preserved for each arrangement is

$$P\big(\delta \leq \delta_o | H_0\big) = \frac{\text{number of } \delta \text{ values} \leq \delta_o}{M} = \frac{3{,}138}{48{,}620} = 0.0645 .$$

For comparison, the exact probability values based on $v = 2$, $M = 48{,}620$, and $C_i = (n_i - 1)/(N - g)$ for $i = 1, 2$ in Example 1 and $v = 1$, $M = 48{,}620$, and $C_i = (n_i - 1)/(N - g)$ for $i = 1, 2$ in Example 2 are $P = 0.0522$ and $P = 0.0645$, respectively.[1] No comparison is made with the conventional Wilcoxon two-sample rank-sum test as Wilcoxon's test is undefined for $r > 1$, $v = 1$, and $C_i = n_i/N$, $i = 1, \ldots, g$.

Following Eq. (6.5) on p. 319, the exact expected value of the $M = 48{,}620$ δ values is $\mu_\delta = 3.8128$ and, following Eq. (6.4) on p. 319, the observed chance-corrected measure of effect size is

$$\mathfrak{R}_o = 1 - \frac{\delta_o}{\mu_\delta} = 1 - \frac{3.5348}{3.8128} = +0.0729 ,$$

indicating approximately 7 % within-group agreement above that expected by chance.

6.6 MRPP for the A_{Ns} Function with $s = 2$

In this section the A_{Ns} power-of-rank function test with $s = 2$ [281], discussed in Chap. 5, Sect. 5.10, is generalized to $r \geq 2$ response measurements. Because A_{N2} is associated with the Taha sum-of-squared-ranks test [393] when $r = 1$, as described in Chap. 5, Sects. 5.7 and 5.10, this A_{N2} analysis can be considered as a multivariate generalization of the Taha test, i.e., $r \geq 2$.

Three example analyses illustrate a permutation approach to two-sample power-of-rank problems with multivariate response measurements. The first example utilizes MRPP to extend the conventional A_{Ns} function with $s = 2$ to a multivariate power-of-rank test using a small set of bivariate rank scores with $v = 2$ and treatment-group weights $C_i = (n_i - 1)/(N - g)$ for $i = 1, \ldots, g$; the second example analyzes the same small set of bivariate rank scores, but uses $v = 1$ and treatment-group weights $C_i = (n_i - 1)/(N - g)$ for $i = 1, \ldots, g$; and the third example analyzes the same set of bivariate rank scores using $v = 1$, but adopts a proportional treatment-group weighting function given by $C_i = n_i/N$ for $i = 1, \ldots, g$.

6.6.1 Example 1

Consider a two-sample rank test for N objects with n_1 and n_2 objects in the first and second samples, respectively, and $r = 2$ ordinal response measurements obtained on each object. For the $N = 18$ bivariate rank scores listed in Fig. 6.10 on p. 335, replicated in Fig. 6.12 for convenience, $g = 2$, $r = 2$, $n_1 = n_2 = 9$, and $N = n_1 + n_2 = 18$.

[1] Note that when $n_1 = n_2$, as in this case with $n_1 = n_2 = 9$, $C_i = (n_i - 1)/(N - g)$ and $C_i = n_i/N$ for $i = 1, \ldots, g$ yield identical results.

Fig. 6.12 Bivariate rank response measurement scores for two treatment groups with $r = 2$, $g = 2$, $n_1 = n_2 = 9$, and $N = n_1 + n_2 = 18$

Group 1	Group 2
(1, 1)	(1, 3)
(1, 2)	(2, 1)
(2, 3)	(4, 3)
(3, 1)	(5, 3)
(3, 2)	(5, 5)
(3, 3)	(5, 6)
(3, 4)	(5, 7)
(5, 4)	(6, 4)
(6, 5)	(7, 8)

For the first analysis of the bivariate rank response measurements listed in Fig. 6.12, let $v = 2$, employing squared Euclidean distance between the bivariate rank scores, and let the treatment-group weights be given by

$$C_i = \frac{n_i - 1}{N - g}, \qquad i = 1, \ldots, g,$$

to correspond to a multivariate Taha sum-of-squared-ranks test [393].

Because there are only

$$M = \frac{N!}{\prod_{i=1}^{g} n_i!} = \frac{18!}{(9!)^2} = 48{,}620$$

possible, equally-likely arrangements of the $N = 18$ bivariate rank scores listed in Fig. 6.12, an exact solution is practical. Following Eq. (6.2) on p. 317, the $N = 18$ bivariate rank scores listed in Fig. 6.12 yield $g = 2$ average distance-function values of

$$\xi_1 = 358.0556 \quad \text{and} \quad \xi_2 = 1{,}348.1111 .$$

Following Eq. (6.1) on p. 317, the observed value of the MRPP test statistic based on $v = 2$ and treatment-group weights

$$C_i = \frac{n_i - 1}{N - g}, \qquad i = 1, 2,$$

is

$$\delta_o = \sum_{i=1}^{g} C_i \xi_i = \frac{9 - 1}{18 - 2}\big(358.0556 + 1{,}348.1111\big) = 853.0833 .$$

If all arrangements of the $N = 18$ observed bivariate rank scores listed in Fig. 6.12 occur with equal chance, the exact probability value of $\delta_o = 853.0833$ computed on the $M = 48{,}620$ possible arrangements of the observed data with $n_1 = n_2 = 9$ bivariate rank scores preserved for each arrangement is

$$P(\delta \leq \delta_o | H_0) = \frac{\text{number of } \delta \text{ values} \leq \delta_o}{M} = \frac{2{,}318}{48{,}620} = 0.0477 \; .$$

No comparison is made with the conventional Taha sum-of-squared-ranks test as Taha's test is undefined for $r > 1$.

Following Eq. (6.5) on p. 319, the exact expected value of the $M = 48{,}620$ δ values is $\mu_\delta = 1{,}001.9739$ and, following Eq. (6.4) on p. 319, the observed chance-corrected measure of effect size is

$$\Re_o = 1 - \frac{\delta_o}{\mu_\delta} = 1 - \frac{853.0833}{1{,}001.7712} = +0.1486 \; ,$$

indicating approximately 15 % within-group agreement above that expected by chance.

6.6.2 Example 2

For the second analysis of the $N = 18$ bivariate rank scores listed in Fig. 6.12, let the treatment-group weights be given by

$$C_i = \frac{n_i - 1}{N - g} \; , \qquad i = 1, \ldots, g \; ,$$

as in Example 1, but set $v = 1$ instead of $v = 2$, employing ordinary Euclidean distance between the bivariate rank scores. Following Eq. (6.2) on p. 317, the $N = 18$ bivariate rank scores listed in Fig. 6.12 yield $g = 2$ average distance-function values of

$$\xi_1 = 15.3217 \quad \text{and} \quad \xi_2 = 32.2592 \; .$$

Following Eq. (6.1) on p. 317, the observed value of the MRPP test statistic based on $v = 1$ and treatment-group weights

$$C_i = \frac{n_i - 1}{N - g} \; , \qquad i = 1, 2 \; ,$$

is

$$\delta_o = \sum_{i=1}^{g} C_i \xi_i = \frac{9 - 1}{18 - 2}(15.3217 + 32.2592) = 23.7905 \; .$$

If all arrangements of the $N = 18$ observed bivariate rank scores listed in Fig. 6.12 occur with equal chance, the exact probability value of $\delta_o = 23.7905$ computed on the $M = 48{,}620$ possible arrangements of the observed data with $n_1 = n_2 = 9$ bivariate rank scores preserved for each arrangement is

$$P(\delta \leq \delta_o | H_0) = \frac{\text{number of } \delta \text{ values} \leq \delta_o}{M} = \frac{2{,}328}{48{,}620} = 0.0479 \, .$$

For comparison, the exact probability value based on $v = 2$, $M = 48{,}620$, and $C_i = (n_i - 1)/(N - g)$ for $i = 1, 2$ in Example 1 is $P = 0.0477$. No comparison is made with the conventional Taha sum-of-squared-ranks test as Taha's test is undefined for both $r > 1$ and $v = 1$.

Following Eq. (6.5) on p. 319, the exact expected value of the $M = 48{,}620$ δ values is $\mu_\delta = 25.8923$ and, following Eq. (6.4) on p. 319, the observed chance-corrected measure of effect size is

$$\Re_o = 1 - \frac{\delta_o}{\mu_\delta} = 1 - \frac{23.7905}{25.8923} = +0.0812 \, ,$$

indicating approximately 8 % within-group agreement above that expected by chance.

6.6.3 Example 3

For the third analysis of the bivariate rank scores listed in Fig. 6.12, let the treatment-group weights be given by

$$C_i = \frac{n_i}{N} \, , \qquad i = 1, \ldots, g \, ,$$

simply weighting each treatment group proportional to its size, and let $v = 1$, employing ordinary Euclidean distance between the bivariate rank scores, as in Example 2. Because

$$C_i = \frac{n_i}{N} \quad \text{and} \quad C_i = \frac{n_i - 1}{N - g} \, , \qquad i = 1, \ldots, g \, ,$$

are equivalent when $n_1 = n_2$, as in this case, the results are the same as Example 2, with $\xi_1 = 15.3217$, $\xi_2 = 32.2592$, an observed MRPP test statistic of $\delta_o = 23.7905$, an exact probability value of $P = 0.0479$, and an observed chance-corrected effect size of $\Re_o = +0.0812$.

6.7 MRPP for the B_{Ns} Function with $s = 1$

In this section the B_{Ns} power-of-rank function test with $s = 1$ [281], discussed in Chap. 5, Sect. 5.11, is generalized to $r \geq 2$ response measurements. Because B_{N1} is associated with the Ansari–Bradley rank-sum test [10] when $r = 1$, as described in Chap. 5, Sects. 5.7 and 5.11, this B_{N1} analysis can be considered as a multivariate generalization of the Ansari–Bradley rank-sum test, i.e., $r \geq 2$.

Three example analyses illustrate a permutation approach to two-sample power-of-rank problems with multivariate response measurements. The first example utilizes MRPP to extend the conventional B_{Ns} function with $s = 1$ to a multivariate power-of-rank test using a small set of bivariate rank scores with $v = 2$ and treatment-group weights $C_i = (n_i - 1)/(N - g)$ for $i = 1, \ldots, g$; the second example analyzes the same small set of bivariate rank scores, but uses $v = 1$ and treatment-group weights $C_i = (n_i - 1)/(N - g)$ for $i = 1, \ldots, g$; and the third example analyzes the same set of bivariate rank scores using $v = 1$, but adopts a proportional treatment-group weighting function given by $C_i = n_i/N$ for $i = 1, \ldots, g$.

6.7.1 Example 1

Consider a two-sample rank test for N objects with n_1 and n_2 objects in the first and second samples, respectively, and $r = 2$ ordinal response measurements obtained on each object. For the bivariate rank scores listed in Fig. 6.10 on p. 335, replicated in Fig. 6.13 for convenience, $g = 2$, $r = 2$, $n_1 = n_2 = 9$, and $N = n_1 + n_2 = 18$.

For the first analysis of the bivariate rank response measurements listed in Fig. 6.13, let $v = 2$, employing squared Euclidean distance between the bivariate rank scores, and let the treatment-group weights be given by

$$C_i = \frac{n_i - 1}{N - g}\,, \qquad i = 1, \ldots, g\,,$$

to correspond to a multivariate Ansari–Bradley rank-sum test [10].

Fig. 6.13 Bivariate rank response measurement scores for two treatment groups with $r = 2$, $g = 2$, $n_1 = n_2 = 9$, and $N = n_1 + n_2 = 18$

Group 1	Group 2
(1, 1)	(1, 3)
(1, 2)	(2, 1)
(2, 3)	(4, 3)
(3, 1)	(5, 3)
(3, 2)	(5, 5)
(3, 3)	(5, 6)
(3, 4)	(5, 7)
(5, 4)	(6, 4)
(6, 5)	(7, 8)

Because there are only

$$M = \frac{N!}{\prod_{i=1}^{g} n_i!} = \frac{18!}{(9!)^2} = 48{,}620$$

possible, equally-likely arrangements of the $N = 18$ bivariate rank scores listed in Fig. 6.13, an exact solution is practical. Following Eq. (6.2) on p. 317, the $N = 18$ bivariate rank scores listed in Fig. 6.13 yield $g = 2$ average distance-function values of

$$\xi_1 = 6.8889 \quad \text{and} \quad \xi_2 = 11.0278 \ .$$

Following Eq. (6.1) on p. 317, the observed value of the MRPP test statistic based on $v = 2$ and treatment-group weights

$$C_i = \frac{n_i - 1}{N - g} \ , \qquad i = 1, 2 \ ,$$

is

$$\delta_o = \sum_{i=1}^{g} C_i \xi_i = \frac{9-1}{18-2}\left(6.8889 + 11.0278\right) = 8.9583 \ .$$

If all arrangements of the $N = 18$ observed bivariate rank scores listed in Fig. 6.13 occur with equal chance, the exact probability value of $\delta_o = 8.9583$ computed on the $M = 48{,}620$ possible arrangements of the observed data with $n_1 = n_2 = 9$ bivariate rank scores preserved for each arrangement is

$$P\left(\delta \le \delta_o | H_0\right) = \frac{\text{number of } \delta \text{ values} \le \delta_o}{M} = \frac{1{,}602}{48{,}620} = 0.0329 \ .$$

No comparison is made with the conventional Ansari–Bradley rank-sum test as the Ansari–Bradley test is undefined for $r > 1$.

Following Eq. (6.5) on p. 319, the exact expected value of the $M = 48{,}620$ δ values is $\mu_\delta = 11.0866$ and, following Eq. (6.4) on p. 319, the observed chance-corrected measure of effect size is

$$\Re_o = 1 - \frac{\delta_o}{\mu_\delta} = 1 - \frac{8.9583}{11.0866} = +0.1920 \ ,$$

indicating approximately 19 % within-group agreement above that expected by chance.

6.7.2 Example 2

For the second analysis of the $N = 18$ bivariate rank scores listed in Fig. 6.13, let the treatment-group weights be given by

$$C_i = \frac{n_i - 1}{N - g}\ , \qquad i = 1, \ldots, g\ ,$$

as in Example 1, but set $v = 1$ instead of $v = 2$, employing ordinary Euclidean distance between the bivariate rank scores. Following Eq. (6.2) on p. 317, the $N = 18$ bivariate rank scores listed in Fig. 6.13 yield $g = 2$ average distance-function values of

$$\xi_1 = 2.2331 \quad \text{and} \quad \xi_2 = 2.8262\ .$$

Following Eq. (6.1) on p. 317, the observed value of the MRPP test statistic based on $v = 1$ and treatment-group weights

$$C_i = \frac{n_i - 1}{N - g}\ , \qquad i = 1, 2\ ,$$

is

$$\delta_o = \sum_{i=1}^{g} C_i \xi_i = \frac{9 - 1}{18 - 2}(2.2331 + 2.8262) = 2.5297\ .$$

If all arrangements of the $N = 18$ observed bivariate rank scores listed in Fig. 6.13 occur with equal chance, the exact probability value of $\delta_o = 2.5297$ computed on the $M = 48{,}620$ possible arrangements of the observed data with $n_1 = n_2 = 9$ bivariate rank scores preserved for each arrangement is

$$P(\delta \leq \delta_o | H_0) = \frac{\text{number of } \delta \text{ values} \leq \delta_o}{M} = \frac{1{,}634}{48{,}620} = 0.0336\ .$$

For comparison, the exact probability value based on $v = 2$, $M = 48{,}620$, and $C_i = (n_i - 1)/(N - g)$ for $i = 1, 2$ in Example 1 is $P = 0.0329$. No comparison is made with the conventional Ansari–Bradley rank-sum test as the Ansari–Bradley test is undefined for both $r > 1$ and $v = 1$.

Following Eq. (6.5) on p. 319, the exact expected value of the $M = 48{,}620$ δ values is $\mu_\delta = 2.8931$ and, following Eq. (6.4) on p. 319, the observed chance-corrected measure of effect size is

$$\Re_o = 1 - \frac{\delta_o}{\mu_\delta} = 1 - \frac{2.5297}{2.8931} = +0.1256\ ,$$

indicating approximately 13 % within-group agreement above that expected by chance.

6.7.3 Example 3

For the third analysis of the bivariate rank scores listed in Fig. 6.13, let the treatment-group weights be given by

$$C_i = \frac{n_i}{N} , \qquad i = 1, \ldots, g ,$$

simply weighting each treatment group proportional to its size, and let $v = 1$, employing ordinary Euclidean distance between the bivariate rank scores, as in Example 2. Because

$$C_i = \frac{n_i}{N} \quad \text{and} \quad C_i = \frac{n_i - 1}{N - g} , \qquad i = 1, \ldots, g ,$$

are equivalent when $n_1 = n_2$, as in this case, the results are the same as Example 2, with $\xi_1 = 2.2331$, $\xi_2 = 2.8262$, an observed MRPP test statistic of $\delta_o = 2.5297$, an exact probability value of $P = 0.0336$, and an observed chance-corrected effect size of $\mathfrak{R}_o = +0.1256$.

6.8 MRPP for the B_{Ns} Function with $s = 2$

In this section the B_{Ns} power-of-rank function test with $s = 2$ [281], discussed in Chap. 5, Sect. 5.12, is generalized to $r \geq 2$ response measurements. Because B_{N2} is associated with the Mood two-sample rank test [312] when $r = 1$, as described in Chap. 5, Sects. 5.7 and 5.12, this B_{N2} analysis can be considered as a multivariate generalization of the Mood test, i.e., $r \geq 2$.

Three example analyses illustrate a permutation approach to two-sample power-of-rank problems with multivariate response measurements. The first example utilizes MRPP to extend the conventional B_{Ns} function with $s = 2$ to a multivariate power-of-rank test using a small set of bivariate rank scores with $v = 2$ and treatment-group weights $C_i = (n_i - 1)/(N - g)$ for $i = 1, \ldots, g$; the second example analyzes the same small set of bivariate rank scores, but uses $v = 1$ and treatment-group weights $C_i = (n_i - 1)/(N - g)$ for $i = 1, \ldots, g$; and the third example analyzes the same set of bivariate rank scores using $v = 1$, but adopts a proportional treatment-group weighting function given by $C_i = n_i/N$ for $i = 1, \ldots, g$.

6.8.1 Example 1

Consider a two-sample rank test for N objects with n_1 and n_2 objects in the first and second samples, respectively, and $r = 2$ ordinal response measurements obtained on each object. The bivariate rank scores listed in Fig. 6.10 on p. 335 are replicated

Fig. 6.14 Bivariate rank response measurement scores for two treatment groups with $r = 2$, $g = 2$, $n_1 = n_2 = 9$, and $N = n_1 + n_2 = 18$

Group 1	Group 2
(1, 1)	(1, 3)
(1, 2)	(2, 1)
(2, 3)	(4, 3)
(3, 1)	(5, 3)
(3, 2)	(5, 5)
(3, 3)	(5, 6)
(3, 4)	(5, 7)
(5, 4)	(6, 4)
(6, 5)	(7, 8)

in Fig. 6.14 for convenience. For the bivariate data listed in Fig. 6.14, $g = 2$, $r = 2$, $n_1 = n_2 = 9$, and $N = n_1 + n_2 = 18$.

For the first analysis of the $N = 18$ bivariate rank response measurements listed in Fig. 6.14, let $v = 2$, employing squared Euclidean distance between the bivariate rank scores, and let the treatment-group weights be given by

$$C_i = \frac{n_i - 1}{N - g} , \qquad i = 1, \ldots, g ,$$

to correspond to a multivariate Mood two-sample rank test [312].

Because there are only

$$M = \frac{N!}{\prod_{i=1}^{g} n_i!} = \frac{18!}{(9!)^2} = 48{,}620$$

possible, equally-likely arrangements of the $N = 18$ bivariate rank scores listed in Fig. 6.14, an exact solution is practical. Following Eq. (6.2) on p. 317, the $N = 18$ bivariate rank scores listed in Fig. 6.14 yield $g = 2$ average distance-function values of

$$\xi_1 = 21.2778 \quad \text{and} \quad \xi_2 = 172.7778 .$$

Following Eq. (6.1) on p. 317, the observed value of the MRPP test statistic based on $v = 2$ and treatment-group weights

$$C_i = \frac{n_i - 1}{N - g} , \qquad i = 1, 2 ,$$

is

$$\delta_o = \sum_{i=1}^{g} C_i \xi_i = \frac{9 - 1}{18 - 2} (21.2778 + 172.7778) = 97.0278 .$$

If all arrangements of the $N = 18$ observed bivariate rank scores listed in Fig. 6.14 occur with equal chance, the exact probability value of $\delta_o = 97.0278$ computed on the $M = 48{,}620$ possible arrangements of the observed data with $n_1 = n_2 = 9$ bivariate rank scores preserved for each arrangement is

$$P(\delta \leq \delta_o | H_0) = \frac{\text{number of } \delta \text{ values} \leq \delta_o}{M} = \frac{2{,}776}{48{,}620} = 0.0571 .$$

No comparison is made with the conventional Mood two-sample rank test as Mood's test is undefined for $r > 1$.

Following Eq. (6.5) on p. 319, the exact expected value of the $M = 48{,}620$ δ values is $\mu_\delta = 109.2614$ and, following Eq. (6.4) on p. 319, the observed chance-corrected measure of effect size is

$$\Re_o = 1 - \frac{\delta_o}{\mu_\delta} = 1 - \frac{97.0278}{109.2614} = +0.0481 ,$$

indicating approximately 5 % within-group agreement above that expected by chance.

6.8.2 Example 2

For the second analysis of the bivariate rank scores listed in Fig. 6.14, let the treatment-group weights be given by

$$C_i = \frac{n_i - 1}{N - g} , \qquad i = 1, \ldots, g ,$$

as in Example 1, but set $v = 1$ instead of $v = 2$, employing ordinary Euclidean distance between the bivariate rank scores. Following Eq. (6.2) on p. 317, the $N = 18$ bivariate rank scores listed in Fig. 6.14 yield $g = 2$ average distance-function values of

$$\xi_1 = 4.0910 \quad \text{and} \quad \xi_2 = 10.2850 .$$

Following Eq. (6.1) on p. 317, the observed value of the MRPP test statistic based on $v = 1$ and treatment-group weights

$$C_i = \frac{n_i - 1}{N - g} , \qquad i = 1, 2 ,$$

is

$$\delta_o = \sum_{i=1}^{g} C_i \xi_i = \frac{9-1}{18-2}\big(4.0910 + 10.2850\big) = 7.1880 .$$

If all arrangements of the $N = 18$ observed bivariate rank scores listed in Fig. 6.14 occur with equal chance, the exact probability value of $\delta_o = 7.1880$ computed on the $M = 48{,}620$ possible arrangements of the observed data with $n_1 = n_2 = 9$ bivariate rank scores preserved for each arrangement is

$$P(\delta \leq \delta_o | H_0) = \frac{\text{number of } \delta \text{ values} \leq \delta_o}{M} = \frac{4{,}490}{48{,}620} = 0.0923 \ .$$

For comparison, the exact probability value based on $v = 2$, $M = 48{,}620$, and $C_i = (n_i - 1)/(N - g)$ for $i = 1, 2$ in Example 1 is $P = 0.0571$. No comparison is made with the conventional Mood two-sample rank test as Mood's test is undefined for both $r > 1$ and $v = 1$.

Following Eq. (6.5) on p. 319, the exact expected value of the $M = 48{,}620$ δ values is $\mu_\delta = 7.7590$ and, following Eq. (6.4) on p. 319, the observed chance-corrected measure of effect size is

$$\Re_o = 1 - \frac{\delta_o}{\mu_\delta} = 1 - \frac{7.1880}{7.7590} = +0.0491 \ ,$$

indicating approximately 5 % within-group agreement above that expected by chance.

6.8.3 Example 3

For the third analysis of the bivariate rank scores listed in Fig. 6.14, let the treatment-group weights be given by

$$C_i = \frac{n_i}{N} \ , \qquad i = 1, \ldots, g \ ,$$

simply weighting each treatment group proportional to its size, and let $v = 1$, employing ordinary Euclidean distance between the bivariate rank scores, as in Example 2. Because

$$C_i = \frac{n_i}{N} \quad \text{and} \quad C_i = \frac{n_i - 1}{N - g} \ , \qquad i = 1, \ldots, g \ ,$$

are equivalent when $n_1 = n_2$, as in this case, the results are the same as Example 2, with $\xi_1 = 4.0910$, $\xi_2 = 10.2850$, an observed MRPP test statistic of $\delta_o = 7.1880$, an exact probability value of $P = 0.0923$, and an observed chance-corrected effect size of $\Re_o = +0.0491$.

6.9 MRPP for the C_{Ns} Function with $s = 0$

In this section the C_{Ns} power-of-rank function test with $s = 0$ [281], discussed in Chap. 5, Sect. 5.13, is generalized to $r \geq 2$ response measurements. Because C_{N0} is associated with the Brown–Mood median test [59] when $r = 1$, as described in Chap. 5, Sects. 5.7 and 5.13, this C_{N0} analysis can be considered as a multivariate generalization of the Brown–Mood test, i.e., $r \geq 2$.

Three example analyses illustrate a permutation approach to two-sample power-of-rank problems with multivariate response measurements. The first example utilizes MRPP to extend the conventional C_{Ns} function with $s = 0$ to a multivariate power-of-rank test using a small set of bivariate rank scores with $v = 2$ and treatment-group weights $C_i = (n_i - 1)/(N - g)$ for $i = 1, \ldots, g$; the second example analyzes the same small set of bivariate rank scores, but uses $v = 1$ and treatment-group weights $C_i = (n_i - 1)/(N - g)$ for $i = 1, \ldots, g$; and the third example analyzes the same set of bivariate rank scores using $v = 1$, but adopts a proportional treatment-group weighting function given by $C_i = n_i/N$ for $i = 1, \ldots, g$.

6.9.1 Example 1

Consider a two-sample rank test for N objects with n_1 and n_2 objects in the first and second samples, respectively, and $r = 2$ ordinal response measurements obtained on each object. The bivariate rank scores listed in Fig. 6.10 on p. 335 are replicated in Fig. 6.15 for convenience. For the bivariate rank scores listed in Fig. 6.15, $g = 2$, $r = 2$, $n_1 = n_2 = 9$, and $N = n_1 + n_2 = 18$.

For the first analysis of the bivariate rank response measurements listed in Fig. 6.15, let $v = 2$, employing squared Euclidean distance between the bivariate rank scores, and let the treatment-group weights be given by

$$C_i = \frac{n_i - 1}{N - g}\,, \qquad i = 1, \ldots, g\,,$$

to correspond to a multivariate Brown–Mood median test [59].

Fig. 6.15 Bivariate rank response measurements for two treatment groups with $r = 2$, $g = 2$, $n_1 = n_2 = 9$, and $N = n_1 + n_2 = 18$

Group 1	Group 2
(1, 1)	(1, 3)
(1, 2)	(2, 1)
(2, 3)	(4, 3)
(3, 1)	(5, 3)
(3, 2)	(5, 5)
(3, 3)	(5, 6)
(3, 4)	(5, 7)
(5, 4)	(6, 4)
(6, 5)	(7, 8)

Because there are only

$$M = \frac{N!}{\prod_{i=1}^{g} n_i!} = \frac{18!}{(9!)^2} = 48{,}620$$

possible, equally-likely arrangements of the $N = 18$ bivariate rank scores listed in Fig. 6.15, an exact solution is practical. Following Eq. (6.2) on p. 317, the $N = 18$ bivariate rank scores listed in Fig. 6.15 yield $g = 2$ average distance-function values of

$$\xi_1 = 2.1111 \quad \text{and} \quad \xi_2 = 1.7778 \; .$$

Following Eq. (6.1) on p. 317, the observed value of the MRPP test statistic based on $v = 2$ and treatment-group weights

$$C_i = \frac{n_i - 1}{N - g} \,, \qquad i = 1, 2 \,,$$

is

$$\delta_o = \sum_{i=1}^{g} C_i \xi_i = \frac{9-1}{18-2}\big(2.1111 + 1.7778\big) = 1.9444 \; .$$

If all arrangements of the $N = 18$ observed bivariate rank scores listed in Fig. 6.15 occur with equal chance, the exact probability value of $\delta_o = 1.9444$ computed on the $M = 48{,}620$ possible arrangements of the observed data with $n_1 = n_2 = 9$ bivariate rank scores preserved for each arrangement is

$$P\big(\delta \leq \delta_o | H_0\big) = \frac{\text{number of } \delta \text{ values} \leq \delta_o}{M} = \frac{116}{48{,}620} = 0.0024 \; .$$

No comparison is made with the conventional Brown–Mood median test as the Brown–Mood test is undefined for $r > 1$.

Following Eq. (6.5) on p. 319, the exact expected value of the $M = 48{,}620$ δ values is $\mu_\delta = 3.4248$ and, following Eq. (6.4) on p. 319, the observed chance-corrected measure of effect size is

$$\Re_o = 1 - \frac{\delta_o}{\mu_\delta} = 1 - \frac{1.9444}{3.4248} = +0.4325 \,,$$

indicating approximately 43 % within-group agreement above that expected by chance.

6.9.2 Example 2

For the second analysis of the bivariate rank scores listed in Fig. 6.15, let the treatment-group weights be given by

$$C_i = \frac{n_i - 1}{N - g} , \qquad i = 1, \ldots, g ,$$

as in Example 1, but set $v = 1$ instead of $v = 2$, employing ordinary Euclidean distance between the bivariate rank scores. Following Eq. (6.2) on p. 317, the $N = 18$ bivariate rank scores listed in Fig. 6.15 yield $g = 2$ average distance-function values of

$$\xi_1 = 1.1636 \quad \text{and} \quad \xi_2 = 0.8515 .$$

Following Eq. (6.1) on p. 317, the observed value of the MRPP test statistic based on $v = 1$ and treatment-group weights

$$C_i = \frac{n_i - 1}{N - g} , \qquad i = 1, 2 ,$$

is

$$\delta_o = \sum_{i=1}^{g} C_i \xi_i = \frac{9-1}{18-2}\big(1.1636 + 0.8515\big) = 1.0075 .$$

If all arrangements of the $N = 18$ observed bivariate rank scores listed in Fig. 6.15 occur with equal chance, the exact probability value of $\delta_o = 1.0075$ computed on the $M = 48{,}620$ possible arrangements of the observed data with $n_1 = n_2 = 9$ bivariate rank scores preserved for each arrangement is

$$P\big(\delta \leq \delta_o | H_0\big) = \frac{\text{number of } \delta \text{ values} \leq \delta_o}{M} = \frac{116}{48{,}620} = 0.0024 .$$

For comparison, the exact probability value based on $v = 2$, $M = 48{,}620$, and $C_i = (n_i - 1)/(N - g)$ for $i = 1, 2$ in Example 1 is also $P = 0.0024$. No comparison is made with the conventional Brown–Mood median test as the Brown–Mood test is undefined for both $r > 1$ and $v = 1$.

Following Eq. (6.5) on p. 319, the exact expected value of the $M = 48{,}620$ δ values is $\mu_\delta = 1.5436$ and, following Eq. (6.4) on p. 319, the observed chance-corrected measure of effect size is

$$\Re_o = 1 - \frac{\delta_o}{\mu_\delta} = 1 - \frac{1.0075}{1.5436} = +0.3473 ,$$

indicating approximately 35 % within-group agreement above that expected by chance.

6.9.3 Example 3

For the third analysis of the bivariate rank scores listed in Fig. 6.15, let the treatment-group weights be given by

$$C_i = \frac{n_i}{N}, \qquad i = 1, \ldots, g,$$

simply weighting each treatment group proportional to its size, and let $v = 1$, employing ordinary Euclidean distance between the bivariate rank scores, as in Example 2. Because

$$C_i = \frac{n_i}{N} \quad \text{and} \quad C_i = \frac{n_i - 1}{N - g}, \qquad i = 1, \ldots, g,$$

are equivalent when $n_1 = n_2$, as in this case, the results are the same as Example 2, with $\xi_1 = 1.1636$, $\xi_2 = 0.8515$, an observed MRPP test statistic of $\delta_o = 1.0075$, an exact probability value of $P = 0.0024$, and an observed chance-corrected effect size of $\Re_o = +0.3473$.

6.10 MRPP for the C_{Ns} Function with $s = 1$

In this section the C_{Ns} power-of-rank function test with $s = 1$, discussed in Chap. 5, Sect. 5.14, is generalized to $r \geq 2$ response measurements. Because C_{N1} is associated with the Wilcoxon–Mann–Whitney two-sample rank-sum test [262, 429] when $r = 1$, as described in Chap. 5, Sects. 5.7 and 5.14, this C_{N1} analysis can be considered as a multivariate generalization of the Wilcoxon–Mann–Whitney test, i.e., $r \geq 2$.

Three example analyses illustrate a permutation approach to two-sample power-of-rank problems with multivariate response measurements. The first example utilizes MRPP to extend the conventional C_{Ns} function with $s = 1$ to a multivariate power-of-rank test using a small set of bivariate rank scores with $v = 2$ and treatment-group weights $C_i = (n_i - 1)/(N - g)$ for $i = 1, \ldots, g$; the second example analyzes the same small set of bivariate rank scores, but uses $v = 1$ and treatment-group weights $C_i = (n_i - 1)/(N - g)$ for $i = 1, \ldots, g$; and the third example analyzes the same set of bivariate rank scores using $v = 1$, but adopts a proportional treatment-group weighting function given by $C_i = n_i/N$ for $i = 1, \ldots, g$.

Fig. 6.16 Bivariate rank response measurement scores for two treatment groups with $r = 2$, $g = 2$, $n_1 = n_2 = 9$, and $N = n_1 + n_2 = 18$

Group 1	Group 2
(1, 1)	(1, 3)
(1, 2)	(2, 1)
(2, 3)	(4, 3)
(3, 1)	(5, 3)
(3, 2)	(5, 5)
(3, 3)	(5, 6)
(3, 4)	(5, 7)
(5, 4)	(6, 4)
(6, 5)	(7, 8)

6.10.1 Example 1

Consider a two-sample rank test for N objects with n_1 and n_2 objects in the first and second samples, respectively, and $r = 2$ ordinal response measurements obtained on each object. The bivariate rank scores listed in Fig. 6.10 on p. 335 are replicated in Fig. 6.16 for convenience. For the bivariate rank scores listed in Fig. 6.16, $g = 2$, $r = 2$, $n_1 = n_2 = 9$, and $N = n_1 + n_2 = 18$.

For the first analysis of the bivariate rank scores listed in Fig. 6.16, let $v = 2$, employing squared Euclidean distance between the bivariate rank scores, and let the treatment-group weights be given by

$$C_i = \frac{n_i - 1}{N - g} , \qquad i = 1, \ldots, g ,$$

to correspond to a multivariate Wilcoxon–Mann–Whitney two-sample test [262, 429].

Because there are only

$$M = \frac{N!}{\prod_{i=1}^{g} n_i!} = \frac{18!}{(9!)^2} = 48{,}620$$

possible, equally-likely arrangements of the $N = 18$ bivariate rank scores listed in Fig. 6.16, an exact solution is practical. Following Eq. (6.2) on p. 317, the $N = 18$ bivariate rank scores listed in Fig. 6.16 yield $g = 2$ average distance-function values of

$$\xi_1 = 9.3889 \quad \text{and} \quad \xi_2 = 23.9444 .$$

Following Eq. (6.1) on p. 317, the observed value of the MRPP test statistic based on $v = 2$ and treatment-group weights

$$C_i = \frac{n_i - 1}{N - g} , \qquad i = 1, 2 ,$$

is

$$\delta_o = \sum_{i=1}^{g} C_i \xi_i = \frac{9-1}{18-2}(9.3889 + 23.9444) = 16.6667 .$$

If all arrangements of the $N = 18$ observed bivariate rank scores listed in Fig. 6.16 occur with equal chance, the exact probability value of $\delta_o = 16.6667$ computed on the $M = 48{,}620$ possible arrangements of the observed data with $n_1 = n_2 = 9$ bivariate rank scores preserved for each arrangement is

$$P(\delta \le \delta_o | H_0) = \frac{\text{number of } \delta \text{ values} \le \delta_o}{M} = \frac{2{,}536}{48{,}620} = 0.0522 .$$

No comparison is made with the conventional Wilcoxon–Mann–Whitney two-sample rank-sum test as the Wilcoxon–Mann–Whitney test is undefined for $r > 1$.

Following Eq. (6.5) on p. 319, the exact expected value of the $M = 48{,}620$ δ values is $\mu_\delta = 19.7712$ and, following Eq. (6.4) on p. 319, the observed chance-corrected measure of effect size is

$$\Re_o = 1 - \frac{\delta_o}{\mu_\delta} = 1 - \frac{16.6667}{19.7712} = +0.1570 ,$$

indicating approximately 16 % within-group agreement above that expected by chance.

6.10.2 Example 2

For the second analysis of the bivariate rank scores listed in Fig. 6.16, let the treatment-group weights be given by

$$C_i = \frac{n_i - 1}{N - g} , \qquad i = 1, \ldots, g ,$$

as in Example 1, but set $v = 1$ instead of $v = 2$, employing ordinary Euclidean distance between the bivariate rank scores. Following Eq. (6.2) on p. 317, the $N = 18$ bivariate rank scores listed in Fig. 6.16 yield $g = 2$ average distance-function values of

$$\xi_1 = 2.7256 \quad \text{and} \quad \xi_2 = 4.3440 .$$

Following Eq. (6.1) on p. 317, the observed value of the MRPP test statistic based on $v = 1$ and treatment-group weights

$$C_i = \frac{n_i - 1}{N - g} , \qquad i = 1, 2 ,$$

is

$$\delta_o = \sum_{i=1}^{g} C_i \xi_i = \frac{9-1}{18-2}(2.7256 + 4.3440) = 3.5348 \; .$$

If all arrangements of the $N = 18$ observed bivariate rank scores listed in Fig. 6.16 occur with equal chance, the exact probability value of $\delta_o = 3.5348$ computed on the $M = 48{,}620$ possible arrangements of the observed data with $n_1 = n_2 = 9$ bivariate rank scores preserved for each arrangement is

$$P(\delta \leq \delta_o | H_0) = \frac{\text{number of } \delta \text{ values} \leq \delta_o}{M} = \frac{3{,}138}{48{,}620} = 0.0645 \; .$$

For comparison, the exact probability value based on $v = 2$, $M = 48{,}620$, and $C_i = (n_i - 1)/(N - g)$ for $i = 1, 2$ in Example 1 is $P = 0.0522$. No comparison is made with the conventional Wilcoxon–Mann–Whitney two-sample rank-sum test as the Wilcoxon–Mann–Whitney test is undefined for both $r > 1$ and $v = 1$.

Following Eq. (6.5) on p. 319, the exact expected value of the $M = 48{,}620$ δ values is $\mu_\delta = 3.8128$ and, following Eq. (6.4) on p. 319, the observed chance-corrected measure of effect size is

$$\Re_o = 1 - \frac{\delta_o}{\mu_\delta} = 1 - \frac{3.5348}{3.8128} = +0.0729 \; ,$$

indicating approximately 7 % within-group agreement above that expected by chance.

6.10.3 Example 3

For the third analysis of the bivariate rank scores listed in Fig. 6.16, let the treatment-group weights be given by

$$C_i = \frac{n_i}{N} \; , \qquad i = 1, \ldots, g \; ,$$

simply weighting each treatment group proportional to its size, and let $v = 1$, employing ordinary Euclidean distance between the bivariate rank scores, as in Example 2. Because

$$C_i = \frac{n_i}{N} \quad \text{and} \quad C_i = \frac{n_i - 1}{N - g} \; , \qquad i = 1, \ldots, g \; ,$$

are equivalent when $n_1 = n_2$, as in this case, the results are the same as Example 2, with $\xi_1 = 2.7256$, $\xi_2 = 4.3440$, an observed MRPP test statistic of $\delta_o = 3.5348$, an exact probability value of $P = 0.0645$, and an observed chance-corrected effect size of $\Re_o = +0.0729$.

6.11 MRPP for the C_{Ns} Function with $s = 2$

In this section the C_{Ns} power-of-rank function test with $s = 2$ [281], discussed in Chap. 5, Sect. 5.15, is generalized to $r \geq 2$ response measurements. Finally, because C_{N2} is associated with the Mielke two-sample sum-of-squared-ranks test [282] when $r = 1$, as described in Chap. 5, Sects. 5.7 and 5.15, this C_{N2} analysis can be considered as a multivariate generalization of the Mielke test, i.e., $r \geq 2$.

Three example analyses illustrate a permutation approach to two-sample power-of-rank problems with multivariate response measurements. The first example utilizes MRPP to extend the conventional C_{Ns} function with $s = 2$ to a multivariate power-of-rank test using a small set of bivariate rank scores with $v = 2$ and treatment-group weights $C_i = (n_i - 1)/(N - g)$ for $i = 1, \ldots, g$; the second example analyzes the same small set of bivariate rank scores, but uses $v = 1$ and treatment-group weights $C_i = (n_i - 1)/(N - g)$ for $i = 1, \ldots, g$; and the third example analyzes the same set of bivariate rank scores using $v = 1$, but adopts a proportional treatment-group weighting function given by $C_i = n_i/N$ for $i = 1, \ldots, g$.

6.11.1 Example 1

Consider a two-sample rank test for N objects with n_1 and n_2 objects in the first and second samples, respectively, and $r = 2$ ordinal response measurements obtained on each object. The bivariate rank scores listed in Fig. 6.10 on p. 335 are replicated in Fig. 6.17 for convenience. For the bivariate rank scores listed in Fig. 6.17, $g = 2$, $r = 2$, $n_1 = n_2 = 9$, and $N = n_1 + n_2 = 18$.

For the first analysis of the bivariate rank response measurements listed in Fig. 6.17, let $v = 2$, employing squared Euclidean distance between the bivariate rank scores, and let the treatment-group weights be given by

$$C_i = \frac{n_i - 1}{N - g}, \qquad i = 1, \ldots, g,$$

to correspond to a multivariate Mielke two-sample sum-of-squared-ranks test [282].

Fig. 6.17 Bivariate rank response measurement scores for two treatment groups with $r = 2$, $g = 2$, $n_1 = n_2 = 9$, and $N = n_1 + n_2 = 18$

Group 1	Group 2
(1, 1)	(1, 3)
(1, 2)	(2, 1)
(2, 3)	(4, 3)
(3, 1)	(5, 3)
(3, 2)	(5, 5)
(3, 3)	(5, 6)
(3, 4)	(5, 7)
(5, 4)	(6, 4)
(6, 5)	(7, 8)

Because there are only

$$M = \frac{N!}{\prod_{i=1}^{g} n_i!} = \frac{18!}{(9!)^2} = 48{,}620$$

possible, equally-likely arrangements of the $N = 18$ bivariate rank scores listed in Fig. 6.17, an exact solution is practical. Following Eq. (6.2) on p. 317, the $N = 18$ bivariate rank scores listed in Fig. 6.17 yield $g = 2$ average distance-function values of

$$\xi_1 = 36.1528 \quad \text{and} \quad \xi_2 = 198.9861 \, .$$

Following Eq. (6.1) on p. 317, the observed value of the MRPP test statistic based on $v = 2$ and treatment-group weights

$$C_i = \frac{n_i - 1}{N - g} \, , \qquad i = 1, 2 \, ,$$

is

$$\delta_o = \sum_{i=1}^{g} C_i \xi_i = \frac{9-1}{18-2}(36.1528 + 198.9861) = 117.5694 \, .$$

If all arrangements of the $N = 18$ observed bivariate rank scores listed in Fig. 6.17 occur with equal chance, the exact probability value of $\delta_o = 117.5694$ computed on the $M = 48{,}620$ possible arrangements of the observed data with $n_1 = n_2 = 9$ bivariate rank scores preserved for each arrangement is

$$P(\delta \leq \delta_o | H_0) = \frac{\text{number of } \delta \text{ values} \leq \delta_o}{M} = \frac{412}{48{,}620} = 0.0085 \, .$$

No comparison is made with the conventional Mielke two-sample sum-of-squared-ranks test as Mielke's test is undefined for $r > 1$.

Following Eq. (6.5) on p. 319, the exact expected value of the $M = 48{,}620$ δ values is $\mu_\delta = 148.1912$ and, following Eq. (6.4) on p. 319, the observed chance-corrected measure of effect size is

$$\Re_o = 1 - \frac{\delta_o}{\mu_\delta} = 1 - \frac{117.5694}{148.1912} = +0.2066 \, ,$$

indicating approximately 21 % within-group agreement above that expected by chance.

6.11.2 Example 2

For the second analysis of the bivariate rank scores listed in Fig. 6.17, let the treatment-group weights be given by

$$C_i = \frac{n_i - 1}{N - g}\,, \qquad i = 1, \ldots, g\,,$$

as in Example 1, but set $v = 1$ instead of $v = 2$, employing ordinary Euclidean distance between the bivariate rank scores. Following Eq. (6.2) on p. 317, the $N = 18$ bivariate rank scores listed in Fig. 6.17 yield $g = 2$ average distance-function values of

$$\xi_1 = 5.2424 \quad \text{and} \quad \xi_2 = 11.3789\,.$$

Following Eq. (6.1) on p. 317, the observed value of the MRPP test statistic based on $v = 1$ and treatment-group weights

$$C_i = \frac{n_i - 1}{N - g}\,, \qquad i = 1, 2\,,$$

is

$$\delta_o = \sum_{i=1}^{g} C_i \xi_i = \frac{9-1}{18-2}\big(5.2424 + 11.3789\big) = 8.3107\,.$$

If all arrangements of the $N = 18$ observed bivariate rank scores listed in Fig. 6.17 occur with equal chance, the exact probability value of $\delta_o = 8.3107$ computed on the $M = 48{,}620$ possible arrangements of the observed data with $n_1 = n_2 = 9$ bivariate rank scores preserved for each arrangement is

$$P\big(\delta \le \delta_o | H_0\big) = \frac{\text{number of } \delta \text{ values} \le \delta_o}{M} = \frac{444}{48{,}620} = 0.0091\,.$$

For comparison, the exact probability value based on $v = 2$, $M = 48{,}620$, $C_i = (n_i - 1)/(N - g)$ for $i = 1, 2$ in Example 1 is $P = 0.0085$. No comparison is made with the conventional Mielke two-sample sum-of-squared-ranks test as Mielke's test is undefined for both $r > 1$ and $v = 1$.

Following Eq. (6.5) on p. 319, the exact expected value of the $M = 48{,}620$ δ values is $\mu_\delta = 9.3039$ and, following Eq. (6.4) on p. 319, the observed chance-corrected measure of effect size is

$$\Re_o = 1 - \frac{\delta_o}{\mu_\delta} = 1 - \frac{8.3107}{9.3039} = +0.1068\,,$$

indicating approximately 11 % within-group agreement above that expected by chance.

6.11.3 Example 3

For the third analysis of the bivariate rank scores listed in Fig. 6.17, let the treatment-group weights be given by

$$C_i = \frac{n_i}{N}\,, \qquad i = 1, \ldots, g\,,$$

simply weighting each treatment group proportional to its size, and let $v = 1$, employing ordinary Euclidean distance between the bivariate rank scores, as in Example 2. Because

$$C_i = \frac{n_i}{N} \quad \text{and} \quad C_i = \frac{n_i - 1}{N - g}\,, \qquad i = 1, \ldots, g\,,$$

are equivalent when $n_1 = n_2$, as in this case, the results are the same as Example 2, with $\xi_1 = 5.2424$, $\xi_2 = 11.3789$, an observed MRPP test statistic of $\delta_o = 8.3107$, an exact probability value of $P = 0.0091$, and an observed chance-corrected effect size of $\Re_o = +0.1068$.

6.12 MRPP for Cureton's Rank-Biserial Statistic

Cureton's rank-biserial test [83], discussed in Chap. 5, Sect. 5.18, was designed for $g = 2$ groups and $r = 1$ response measurement. In this section r_{rb} is generalized to $r \geq 2$ response measurements. Three example analyses illustrate a permutation approach to two-sample rank-biserial tests with multivariate response measurements. The first example utilizes MRPP to extend the conventional r_{rb} statistic to a multivariate rank-biserial test using a small set of bivariate rank scores with $v = 2$ and treatment-group weights $C_i = (n_i - 1)/(N - g)$ for $i = 1, \ldots, g$; the second example analyzes the same small set of bivariate rank scores, but uses $v = 1$ and treatment-group weights $C_i = (n_i - 1)/(N - g)$ for $i = 1, \ldots, g$; and the third example analyzes the same set of bivariate rank scores using $v = 1$, but adopts a proportional treatment-group weighting function given by $C_i = n_i/N$ for $i = 1, \ldots, g$.

6.12.1 Example 1

Consider a two-sample rank-biserial test for N objects with n_0 and n_1 objects in the first and second samples, respectively, and $r = 2$ ordinal response measurements

Fig. 6.18 Example data set with $r = 2$, $g = 2$, $n_0 = 7$, $n_1 = 5$, and $N = n_0 + n_1 = 12$

	Variable	
Object	x	y
1	0	(1, 3)
2	0	(2, 1)
3	0	(3, 2)
4	1	(4, 6)
5	0	(5, 4)
6	0	(6, 5)
7	0	(7, 7)
8	1	(8, 8)
9	1	(9, 10)
10	1	(10, 12)
11	0	(11, 9)
12	1	(12, 11)

obtained on each object. Let variable y contain the bivariate rank response measurements and let variable x indicate the group to which each response measurement score belongs, indicated by either 0 or 1. Example bivariate rank scores are listed in Fig. 6.18, where $g = 2$, $r = 2$, $n_0 = 7$, $n_1 = 5$, and $N = n_0 + n_1 = 12$.

For the first analysis of the bivariate rank response measurements listed in Fig. 6.18, let $v = 2$, employing squared Euclidean distance between the bivariate rank scores, and let the treatment group weights be given by

$$C_i = \frac{n_i - 1}{N - g} , \qquad i = 1, \ldots, g ,$$

to correspond to a multivariate Cureton rank-biserial test [83].

Because there are only

$$M = \frac{N!}{\prod_{i=1}^{g} n_i!} = \frac{12!}{7!\ 5!} = 792$$

possible, equally-likely arrangements of the $N = 12$ bivariate rank scores listed in Fig. 6.18, an exact solution is feasible. Following Eq. (6.2) on p. 317, the $N = 12$ bivariate rank scores listed in Fig. 6.18 yield $g = 2$ average distance-function values of

$$\xi_0 = 39.2381 \quad \text{and} \quad \xi_1 = 29.20 .$$

Following Eq. (6.1) on p. 317, the observed value of the MRPP test statistic based on $v = 2$ and treatment-group weights

$$C_i = \frac{n_i - 1}{N - g} , \qquad i = 1, 2 ,$$

is

$$\delta_o = \sum_{i=1}^{g} C_i \xi_i = \frac{1}{12-2}\big[(7-1)(39.2381) + (5-1)(29.20)\big] = 35.2229\,.$$

If all arrangements of the $N = 12$ observed bivariate rank scores listed in Fig. 6.18 occur with equal chance, the exact probability value of $\delta_o = 35.2229$ computed on the $M = 792$ possible arrangements of the observed data with $n_0 = 7$ and $n_1 = 5$ bivariate rank scores preserved for each arrangement is

$$P(\delta \leq \delta_o | H_0) = \frac{\text{number of } \delta \text{ values} \leq \delta_o}{M} = \frac{25}{792} = 0.0316\,.$$

No comparison is made with the Cureton rank-biserial test as Cureton's test is undefined for $r > 1$.

Following Eq. (6.5) on p. 319, the exact expected value of the $M = 792$ δ values is $\mu_\delta = 52.00$ and, following Eq. (6.4) on p. 319, the observed chance-corrected measure of effect size is

$$\Re_o = 1 - \frac{\delta_o}{\mu_\delta} = 1 - \frac{35.2229}{52.00} = +0.3226\,,$$

indicating approximately 32 % within-group agreement above that expected by chance.

6.12.2 Example 2

For the second analysis of the bivariate rank scores listed in Fig. 6.18, let the treatment-group weights be given by

$$C_i = \frac{n_i - 1}{N - g}\,, \qquad i = 1, \ldots, g\,,$$

as in Example 1, but set $v = 1$ instead of $v = 2$, employing ordinary Euclidean distance between the bivariate rank scores. Following Eq. (6.2) on p. 317, the $N = 12$ bivariate rank scores listed in Fig. 6.18 yield $g = 2$ average distance-function values of

$$\xi_0 = 5.4316 \quad \text{and} \quad \xi_1 = 4.8137\,.$$

Following Eq. (6.1) on p. 317, the observed value of the MRPP test statistic based on $v = 1$ and treatment-group weights

$$C_i = \frac{n_i - 1}{N - g}\,, \qquad i = 1, 2\,,$$

is

$$\delta_o = \sum_{i=1}^{g} C_i \xi_i = \frac{1}{12-2}\big[(7-1)(5.4316) + (5-1)(4.8137)\big] = 5.1845 \, .$$

If all arrangements of the $N = 12$ observed bivariate rank scores listed in Fig. 6.18 occur with equal chance, the exact probability value of $\delta_o = 5.1845$ computed on the $M = 792$ possible arrangements of the observed data with $n_0 = 7$ and $n_1 = 5$ bivariate rank scores preserved for each arrangement is

$$P(\delta \leq \delta_o | H_0) = \frac{\text{number of } \delta \text{ values} \leq \delta_o}{M} = \frac{29}{792} = 0.0366 \, .$$

For comparison, the exact probability value based on $v = 2$, $M = 792$, and $C_i = (n_i - 1)/(N - g)$ for $i = 1, 2$ in Example 1 is $P = 0.0316$. No comparison is made with the Cureton rank-biserial test as Cureton's test is undefined for both $r > 1$ and $v = 1$.

Following Eq. (6.5) on p. 319, the exact expected value of the $M = 792$ δ values is $\mu_\delta = 6.2644$ and, following Eq. (6.4) on p. 319, the observed chance-corrected measure of effect size is

$$\Re_o = 1 - \frac{\delta_o}{\mu_\delta} = 1 - \frac{5.1845}{6.2644} = +0.1724 \, ,$$

indicating approximately 17 % within-group agreement above that expected by chance.

6.12.3 Example 3

For the third analysis of the bivariate rank scores listed in Fig. 6.18, let the treatment-group weights be given by

$$C_i = \frac{n_i}{N} \, , \qquad i = 1, \ldots, g \, ,$$

simply weighting each treatment group proportional to its size, and let $v = 1$, employing ordinary Euclidean distance between the bivariate rank scores, as in Example 2. Following Eq. (6.2) on p. 317, the $N = 12$ bivariate rank scores listed in Fig. 6.18 yield $g = 2$ average distance-function values of

$$\xi_0 = 5.4316 \quad \text{and} \quad \xi_1 = 4.8137 \, .$$

Following Eq. (6.1) on p. 317, the observed value of the MRPP test statistic based on $v = 1$ and treatment-group weights

$$C_i = \frac{n_i}{N} \, , \qquad i = 1, 2 \, ,$$

is

$$\delta_o = \sum_{i=1}^{g} C_i \xi_i = \frac{1}{12}\big[(7)(5.4316) + (5)(4.8137)\big] = 5.1742 \,.$$

If all arrangements of the $N = 12$ observed bivariate rank scores listed in Fig. 6.18 occur with equal chance, the exact probability value of $\delta_o = 5.1742$ computed on the $M = 792$ possible arrangements of the observed data with $n_O = 7$ and $n_1 = 5$ bivariate rank scores preserved for each arrangement is

$$P(\delta \leq \delta_o | H_0) = \frac{\text{number of } \delta \text{ values} \leq \delta_o}{M} = \frac{26}{792} = 0.0328 \,.$$

For comparison, the exact probability value based on $v = 2$, $M = 792$, and $C_i = (n_i - 1)/(N - g)$ for $i = 1, 2$ in Example 1 and $v = 1$, $M = 792$, and $C_i = (n_i - 1)/(N - g)$ for $i = 1, 2$ in Example 2 are $P = 0.0316$ and $P = 0.0366$, respectively. No comparison is made with the Cureton rank-biserial test as Cureton's test is undefined for $r > 1$, $v = 1$, and $C_i = n_i/N$, $i = 1, \ldots, g$.

Following Eq. (6.5) on p. 319, the exact expected value of the $M = 792$ δ values is $\mu_\delta = 6.2644$ and, following Eq. (6.4) on p. 319, the observed chance-corrected measure of effect size is

$$\Re_o = 1 - \frac{\delta_o}{\mu_\delta} = 1 - \frac{5.1742}{6.2644} = +0.1740 \,,$$

indicating approximately 17 % within-group agreement above that expected by chance.

6.13 Coda

Chapter 6 utilized the Multi-Response Permutation Procedures (MRPP) developed in Chap. 2 to establish relationships between the test statistics of MRPP, δ and $\Re$, and multivariate generalizations of selected conventional tests and measures designed for the analysis of completely randomized data at the ordinal level of measurement. Considered in this chapter were multivariate extensions of the Wilcoxon two-sample rank-sum test, the Kruskal–Wallis multi-sample rank sum test, the Ansari–Bradley rank sum test for dispersion, the Taha sum-of-squared-ranks test, the Mood rank-sum test for dispersion, the Brown–Mood median test, the Mielke A_{Ns}, B_{Ns}, and C_{Ns} power-of-rank function tests, the Whitfield two-sample rank-sum test, and the Cureton rank-biserial test.

Because MRPP is inherently multivariate, it is convenient to extend a variety of classical statistical tests designed for univariate data. Comparisons of the MRPP test statistic δ based on ordinary Euclidean distance with $v = 1$ and squared Euclidean distance with $v = 2$ with the conventional statistics listed above revealed only small differences among the observed probability values due to the transformation of raw scores to rank scores. While conventional statistics, under the population model, require restrictive assumptions and are based on squared Euclidean distance between rank scores, permutation methods based on ordinary Euclidean distance between rank scores yield exact probability values, are free of any distributional assumptions, and are completely data-dependent. It should be noted that although the permutation-based and conventional probability values in Chap. 6 were often very similar due to the fact that the raw data had been transformed to rank scores and extreme value eliminated, permutation tests mitigate the need for such transformations and the attendant loss of information and are therefore preferred over conventional rank tests.

Chapter 7

Chapter 7 establishes the relationships between the MRPP test statistics, δ and $\Re$, and selected conventional tests and measures designed for the analysis of completely randomized data at the nominal level of measurement. Considered in Chap. 7 are Goodman and Kruskal's t_a and t_b asymmetric measures of categorical association, Light and Margolin's categorical analysis of variance, tests to analyze multiple binary choices, and various multivariate measures of association for a nominal-level independent variable and nominal-, ordinal-, and interval-level dependent variables.

Randomized Designs: Nominal Data

7

This seventh chapter of *Permutation Statistical Methods* utilizes the Multi-Response Permutation Procedures (MRPP) presented in Chap. 2 to develop the functional relationships between the test statistics of MRPP, δ and $\Re$, and selected conventional tests and measures designed for the analysis of completely randomized data at the nominal (categorical) level of measurement. Nominal-level variables, such as gender, political affiliation, and marital status, are notoriously difficult to analyze. As Heiser noted in 2004, "Categories can be counted, rated, or ranked, but they cannot be measured" [171, p. 514]. In a 2014 article, de Mast, Akkerhuis, and Erdmann detailed the complexity of evaluating categorical measurements, noting (1) the underlying empirical reality is usually very complex, which is difficult to capture with a simple mathematical structure, especially when the measurement structure is binary; (2) categorical statistics typically evaluate measurement systems in terms of concepts not clearly related to a notion of measurement error; and (3) interpretations of categorical statistics typically depend on rather strict assumptions about conditional independence and the representativeness of samples, assumptions that de Mast, Akkerhuis, and Erdmann argued are almost always violated in practice [91].

Because of the limitations of categorical data analysis, only a small variety of tests are described to illustrate the application of the MRPP test statistics, δ and $\Re$, to nominal-level data. The tests described in this chapter include Goodman and Kruskal's t_a and t_b asymmetric measures of nominal association, Light and Margolin's categorical analysis of variance, Berry and Mielke's permutation test to analyze multiple binary choices, and various multivariate measures of association for a nominal-level independent variable and nominal-, ordinal-, and interval-level dependent variables.

K.J. Berry et al., *Permutation Statistical Methods*,
DOI 10.1007/978-3-319-28770-6_7

7.1 Introduction

As detailed in Chap. 2, let $\Omega = \{\omega_1, \ldots, \omega_N\}$ denote a finite sample of N objects, let $x'_j = (x_{1j}, \ldots, x_{rj})$ be a transposed vector of r commensurate response measurements for object ω_j, $j = 1, \ldots, N$, and let $S_1, \ldots, S_g$ designate an exhaustive partitioning of the N objects into g disjoint treatment groups. The MRPP test statistic given by

$$\delta = \sum_{i=1}^{g} C_i \xi_i \,, \tag{7.1}$$

where $C_i > 0$ is a positive treatment-group weight for $S_1, \ldots, S_g$,

$$\sum_{i=1}^{g} C_i = 1 \,,$$

and

$$\xi_i = \binom{n_i}{2}^{-1} \sum_{j<k} \Delta(j,k)\, \Psi_i(\omega_j)\, \Psi_i(\omega_k) \tag{7.2}$$

is the average distance-function value for all distinct pairs of objects in treatment groups $S_1, \ldots, S_g$, $n_i \geq 2$ is the number of objects classified into treatment group S_i, $i = 1, \ldots, g$,

$$N = \sum_{i=1}^{g} n_i \,,$$

$\sum_{j<k}$ is the sum over all j and k such that $1 \leq j < k \leq N$, $\Delta(j,k)$ is the generalized Minkowski distance function,

$$\Delta(j,k) = \left(\sum_{i=1}^{r} \left| x_{ij} - x_{ik} \right|^p \right)^{v/p} \,, \tag{7.3}$$

where $p \geq 1$, $v > 0$, and $\Psi_i(\cdot)$ is an indicator function given by

$$\Psi_i(\omega_j) = \begin{cases} 1 & \text{if } \omega_j \in S_i \,, \\ 0 & \text{otherwise} \,. \end{cases}$$

The null hypothesis (H_0) states that equal probabilities are assigned to each of the

$$M = \frac{N!}{\prod_{i=1}^{g} n_i!}$$

possible, equally-likely allocations of the N objects to treatment groups $S_1, \ldots, S_g$.

The probability value associated with an observed value of δ, δ_o, is the probability under the null hypothesis (H_0) of observing a value of δ as extreme or more extreme than δ_o. Thus, an exact probability value for δ_o may be expressed as

$$P(\delta \leq \delta_o | H_0) = \frac{\text{number of } \delta \text{ values} \leq \delta_o}{M} .$$

When M is very large, an approximate probability value for δ may be obtained from a resampling procedure, where

$$P(\delta \leq \delta_o | H_0) = \frac{\text{number of } \delta \text{ values} \leq \delta_o}{L}$$

and L denotes the number of randomly-sampled test statistic values. Typically, L is set to a large number to ensure accuracy, e.g., $L = 1{,}000{,}000$. Also, when M is very large and P is exceedingly small, a resampling-approximation permutation procedure may produce no δ values equal to or less than δ_o, even with $L = 1{,}000{,}000$, yielding an approximate resampling probability value of $P = 0.00$. In such cases, moment-approximation permutation procedures based on fitting the first three exact moments of the discrete permutation distribution to a Pearson type III distribution provide approximate probability values, as detailed in Chap. 1, Sect. 1.2.2 [284, 300].

A chance-corrected within-group coefficient of agreement is given by

$$\Re = 1 - \frac{\delta}{\mu_\delta} , \tag{7.4}$$

where μ_δ is the arithmetic average of the M δ values calculated on all possible arrangements of the observed response measurement scores given by

$$\mu_\delta = \frac{1}{M} \sum_{i=1}^{M} \delta_i . \tag{7.5}$$

Permutation analogues of three selected tests are examined in this chapter: (1) Goodman and Kruskal's asymmetric measures of association for nominal-level response measurements [151], which is identical to Light and Margolin's analysis of variance for nominal-level dependent variables [243], (2) Berry and Mielke's test to analyze multiple binary choices [34], and (3) Berry and Mielke's multivariate measures of association. The tests are illustrated with examples analyzed with $C_i = (n_i - 1)/(N - g)$ and $C_i = n_i/N$ for $i = 1, \ldots, g$. When the observed data are coded (0, 1) binary, $v = 2$ and $v = 1$ yield the same result.

7.2 Goodman and Kruskal's t_a and t_b Statistics

A common problem that many researchers confront is the analysis of a cross-classification table where both variables are categorical [242, p. 534]. The usual measures of association based on chi-squared, such as Pearson's ϕ^2 [334], Tschuprov's (Čuprov's) T^2 [402], and Cramér's V^2 [81], have proven to be less than satisfactory due to difficulties in interpretation; see, for example, discussions by Agresti and Finley [3, p. 284], Berry, Martin, and Olson [35, 36], Berry, Johnston, and Mielke [39, 40], Costner [80], Ferguson [115, p. 422], Guilford [161, p. 342], and Wickens [425, p. 226].

In 1954 Leo Goodman and William Kruskal proposed several new measures of association. Among the measures was an asymmetric proportional-reduction-in-error (PRE) prediction measure, t_a, for the analysis of a random sample of two categorical variables [151]. Consider two cross-classified unordered polytomies, A and B, with variable A the dependent variable and variable B the independent variable. Figure 7.1 provides notation for the cross-classification, where a_j for $j = 1, \ldots, g$ denotes the g categories for dependent variable A, b_i for $i = 1, \ldots, r$ denotes the r categories for independent variable B, and N is the total number of observations.

The Goodman and Kruskal t_a statistic is a measure of the relative reduction in prediction error where two types of errors are defined. The first type is the error in prediction based solely on knowledge of the distribution of the dependent variable, termed error of the first kind (E_1), consisting of the expected number of errors when predicting the g dependent variable categories ($a_1, \ldots, a_g$) from the observed distribution of the marginals of the dependent variable ($n_{.1}, \ldots, n_{.g}$). The second type is the error in prediction based on knowledge of the distributions of both the inde-

Fig. 7.1 Notation for the cross-classification of two categorical variables, A_j for $j = 1, \ldots, g$ and B_i for $i = 1, \ldots, r$

	A				
B	a_1	a_2	$\cdots$	a_g	Total
b_1	n_{11}	n_{12}	$\cdots$	n_{1g}	$n_{1.}$
b_2	n_{21}	n_{22}	$\cdots$	n_{2g}	$n_{2.}$
$\vdots$	$\vdots$	$\vdots$	$\ddots$	$\vdots$	$\vdots$
b_r	n_{r1}	n_{r2}	$\cdots$	n_{rg}	$n_{r.}$
Total	$n_{.1}$	$n_{.2}$	$\cdots$	$n_{.g}$	N

pendent and dependent variables, termed error of the second kind (E_2), consisting of the expected number or errors when predicting the g dependent variable categories $(a_1, \ldots, a_g)$ from knowledge of the r independent variable categories $(b_1, \ldots, b_r)$.

To illustrate the two error types, consider predicting category a_1 only from knowledge of its marginal distribution, $n_{.1}, \ldots, n_{.g}$. Clearly, $n_{.1}$ out of the N total cases are in category a_1, but exactly which $n_{.1}$ of the N cases is unknown. The probability of incorrectly identifying one of the N cases in category a_1 by chance alone is given by

$$\frac{N - n_{.1}}{N} .$$

Since there are $n_{.1}$ such classifications required, the number of expected incorrect classifications is

$$\frac{n_{.1}(N - n_{.1})}{N}$$

and, for all g categories of variable A, the number of expected errors of the first kind is given by

$$E_1 = \sum_{j=1}^{g} \frac{n_{.j}(N - n_{.j})}{N} .$$

Likewise, to predict $n_{11}, \ldots, n_{1g}$ from the independent category b_1, the probability of incorrectly classifying one of the $n_{1.}$ cases in cell n_{11} by chance alone is

$$\frac{n_{1.} - n_{11}}{n_{1.}} .$$

Since there are n_{11} such classifications required, the number of incorrect classifications is

$$\frac{n_{11}(n_{1.} - n_{11})}{n_{1.}}$$

and, for all gr cells, the number of expected errors of the second kind is given by

$$E_2 = \sum_{j=1}^{g} \sum_{i=1}^{r} \frac{n_{ij}(n_{i.} - n_{ij})}{n_{i.}} .$$

Goodman and Kruskal's t_a statistic is then defined as

$$t_a = \frac{E_1 - E_2}{E_1} .$$

An efficient computation form for Goodman and Kruskal's t_a is given by

$$t_a = \frac{N \sum_{i=1}^{r} \sum_{j=1}^{g} \frac{n_{ij}^2}{n_{.j}} - \sum_{i=1}^{r} n_{i.}^2}{N^2 - \sum_{i=1}^{r} n_{i.}^2} . \tag{7.6}$$

A computed value of t_a indicates the proportional reduction in prediction error given knowledge of the distribution of independent variable B over and above knowledge of only the distribution of dependent variable A. As defined, t_a is a point estimator of Goodman and Kruskal's population parameter τ_a for the population from which the sample of N cases was obtained. If variable B is considered the dependent variable and variable A the independent variable, then Goodman and Kruskal's test statistic t_b and associated population parameter τ_b are analogously defined.

While parameter τ_a norms properly from 0 to 1, possesses a clear and meaningful proportional-reduction-in-error interpretation [80], and is characterized by high intuitive and factorial validity [188], statistic t_a poses difficulties whenever the null hypothesis posits that $\tau_a = 0$ [263]. The problem is that the distribution of t_a is not asymptotically normal when $\tau_a = 0$. Consequently, the applicability of Goodman and Kruskal's t_a to typical tests of null hypotheses is severely circumscribed.

Although t_a was developed by Goodman and Kruskal in 1954, it was not until 1963 that the asymptotic normality for t_a was established and an asymptotic variance was given for t_a, but only for $0 < \tau_a < 1$ [152]. Unfortunately, the asymptotic variance for t_a given in 1963 was later found to be incorrect, and it was not until 1972 that the correct asymptotic variance for t_a was obtained, but again, only for $0 < \tau_a < 1$.

In 1971, Richard Light and Barry Margolin developed R^2, an analysis-of-variance technique for categorical response variables, called CATANOVA for CATegorical ANalysis Of VAriance [243]. They apparently were unaware that R^2 was identical to Goodman and Kruskal's t_a and that they had asymptotically solved the longstanding problem of testing H_0: $\tau_a = 0$. The identity between R^2 and t_a was first recognized by Särndal in 1974 [362] and later discussed by Margolin and Light [263], where they showed that $t_a(N-1)(r-1)$ was asymptotically distributed as chi-squared with $(r-1)(g-1)$ degrees of freedom under H_0: $\tau_a = 0$ as $N \to \infty$.

7.2.1 Goodman and Kruskal's t_a and δ

Consider two cross-classified unordered polytomies, A and B, with A the dependent variable where $r \geq 2$ and $g \geq 2$. If each of the N cases is represented by a binary column vector h of dimension r, with a single row entry set to 1 to indicate the classification of the case and the remaining $r - 1$ row entries set to 0, then an $r \times N$ matrix may be defined by $\mathbf{H} = \{h_1, h_2, \ldots, h_N\}$.

If $\Delta(I, J) = 1 - h_I' h_J$, then

$$\Delta(I, J) = \begin{cases} 0 & \text{if } h_I = h_J \;, \\ 1 & \text{otherwise} \;. \end{cases}$$

Thus, $\Delta(I, J)$, the difference between cases I and J, is 1 if and only if I and J are orthogonal, i.e., occur in different rows of independent variable B. The variation for categorical responses in the jth category of variable A, $j = 1, \ldots, g$, is given by

$$\xi_j = \binom{n_j}{2}^{-1} \sum_{I<J} \Delta(I, J) \Psi_j(h_I) \Psi_j(h_J) \;,$$

where $\sum_{I<J}$ is the sum over all I and J such that $1 \leq I < J \leq N$ and

$$\Psi_j(h_I) = \begin{cases} 1 & \text{if } h_I \in A_j \;, \\ 0 & \text{otherwise} \;, \end{cases}$$

for $j = 1, \ldots, g$.

The MRPP test statistic is the weighted average of the ξ_j values, $j = 1, \ldots, g$, given by

$$\delta_a = \sum_{j=1}^{g} C_j \xi_j \;,$$

where

$$C_j = \frac{n_j - 1}{N - g} \;, \qquad j = 1, \ldots, g \;,$$

and

$$\sum_{j=1}^{g} C_j = 1 \;.$$

The choice of C_j is dictated by the relationship between t_a and δ_a, but a corresponding test based on

$$C_j = \frac{n_j}{N} \;, \qquad j = 1, \ldots, g \;,$$

simply weighting each treatment group proportional to its size, is a natural alternative since degrees of freedom are meaningless in a permutation context [297, p. 76]. Also, the choice of $v = 2$ or $v = 1$ is irrelevant, since the response measurement scores are coded (0, 1) binary. An efficient computation form for δ_a is given by

$$\delta_a = \frac{N - \sum_{i=1}^{r} \sum_{j=1}^{g} \frac{n_{ij}^2}{n_{.j}}}{N - g} . \tag{7.7}$$

7.2.2 Example Analysis for t_a

Suppose that $N = 64$ individuals are queried as to political party affiliation (variable B) and preference for one of three political candidates (variable A). The results are adapted from Berry and Mielke [25] and given in Fig. 7.2. Variable A consists of $g = 3$ categories: Candidate 1 (a_1), Candidate 2 (a_2), and Candidate 3 (a_3). Variable B consists of $r = 2$ categories: Democrat (b_1) and Republican (b_2). If the null hypothesis posits that candidate preference (A) does not depend on political party affiliation (B), then H_0: $\tau_a = 0$,

$$h_1 = \begin{bmatrix} 1 \\ 0 \end{bmatrix}, \quad h_2 = \begin{bmatrix} 1 \\ 0 \end{bmatrix}, \quad h_3 = \begin{bmatrix} 1 \\ 0 \end{bmatrix}, \quad \ldots, \quad h_{64} = \begin{bmatrix} 0 \\ 1 \end{bmatrix},$$

and

$$\mathbf{H} = \begin{bmatrix} 1 & 1 & 1 & 0 & 0 & 1 & 0 & 0 & \cdots & 0 \\ 0 & 0 & 0 & 1 & 1 & 0 & 1 & 1 & \cdots & 1 \end{bmatrix} .$$

For the frequency data given in Fig. 7.2, $r = 2$, $g = 3$, $v = 1$, $n_{.1} = 52$, $n_{.2} = 7$, $n_{.3} = 5$, $N = n_{.1} + n_{.2} + n_{.3} = 64$, and let

$$C_j = \frac{n_j - 1}{N - g}, \qquad j = 1, \ldots, g ,$$

to correspond to Goodman and Kruskal's t_a test statistic [151]. Because the response measurement scores are coded (0, 1) binary, $v = 2$, employing squared Euclidean

Fig. 7.2 Cross-classification of political party (B) and candidate preference (A)

	A			
B	a_1	a_2	a_3	Total
b_1	5	1	3	9
b_2	47	6	2	55
Total	52	7	5	64

distance between response measurements, and $v = 1$, employing ordinary Euclidean distance between response measurements, yield the same result.

The $\Delta(I, J)$ generalized Minkowski distance-function values are

$$\Delta(1,2) = 0\,, \quad \Delta(1,3) = 0\,, \quad \Delta(1,4) = 1\,, \quad \ldots, \quad \Delta(63,64) = 0\,.$$

Following Eq. (7.2) on p. 368, the $N = 64$ observations given in Fig. 7.2 on p. 374 yield $g = 3$ average distance-function values of

$$\xi_1 = 0.1772\,, \quad \xi_2 = 0.2857\,, \quad \text{and} \quad \xi_3 = 0.60\,.$$

Following Eq. (7.1) on p. 368, the observed value of the MRPP test statistic based on $v = 1$ and treatment-group weights

$$C_j = \frac{n_j - 1}{N - g}\,, \qquad j = 1, 2, 3\,,$$

is

$$\delta_a = \sum_{j=1}^{g} C_j \xi_j = \frac{1}{64 - 3}\big[(52 - 1)(0.1772) + (7 - 1)(0.2857) + (5 - 1)(0.60)\big] = 0.2156\,.$$

Alternatively, following Eq. (7.7) on p. 374, the observed value of the MRPP test statistic is

$$\delta_a = \frac{N - \displaystyle\sum_{i=1}^{r}\sum_{j=1}^{g}\frac{n_{ij}^2}{n_{.j}}}{N - g} = \frac{64 - \left(\dfrac{5^2}{52} + \dfrac{1^2}{7} + \dfrac{3^2}{5} + \dfrac{47^2}{52} + \dfrac{6^2}{7} + \dfrac{2^2}{5}\right)}{64 - 3} = 0.2156\,.$$

Following Eq. (7.6) on p. 372, for the frequency data given in Fig. 7.2, the observed value of Goodman and Kruskal's t_a is

$$t_a = \frac{N \displaystyle\sum_{i=1}^{r} \sum_{j=1}^{g} \frac{n_{ij}^2}{n_{.j}} - \sum_{i=1}^{r} n_{i.}^2}{N^2 - \displaystyle\sum_{i=1}^{r} n_{i.}^2}$$

$$= \frac{64\left(\dfrac{5^2}{52} + \dfrac{1^2}{7} + \dfrac{3^2}{5} + \dfrac{47^2}{52} + \dfrac{6^2}{7} + \dfrac{2^2}{5}\right) - (9^2 + 55^2)}{64^2 - (9^2 + 55^2)} = 0.1497 \, .$$

The computed value of $t_a = 0.1497$ indicates an approximate 15 % reduction in the number of prediction errors, given knowledge of the distribution of political party affiliation (variable B) over knowledge of only the distribution of candidate preference (variable A). The adjusted t_a value of Margolin and Light [263] is

$$t_a(N-1)(r-1) = 0.1497(64-1)(2-1) = 9.4324$$

and, with $(r-1)(g-1) = (2-1)(3-1) = 2$ degrees of freedom, the asymptotic chi-squared probability value of $\chi^2 = 9.4324$ is approximately $P = 0.0089$.

Since there are

$$M = \frac{N!}{\displaystyle\prod_{j=1}^{g} n_{.j}!} = \frac{64!}{52!\ 7!\ 5!} = 2{,}601{,}098{,}044{,}820{,}352$$

possible, equally-likely arrangements of the $N = 64$ observed values given in Fig. 7.2, an exact solution is not practical. If all M possible arrangements of the $N = 64$ observed values given in Fig. 7.2 occur with equal chance, the approximate resampling probability value of $\delta_a = 0.2156$ computed on $L = 1{,}000{,}000$ random arrangements of the observed values with $n_{.1} = 52$, $n_{.2} = 7$, and $n_{.3} = 5$ marginal frequency totals preserved for each arrangement is[1]

$$P(\delta \le \delta_a | H_0) = \frac{\text{number of } \delta \text{ values} \le \delta_a}{L} = \frac{22{,}810}{1{,}000{,}000} = 0.0228 \, .$$

For comparison, the chi-squared approximate probability value under the null hypothesis with $(r-1)(g-1) = (2-1)(3-1) = 2$ degrees of freedom is $P = 0.0089$.

[1]For comparison, the exact probability value is acually $P = 0.0229$.

Following Eq. (7.5) on p. 369, the exact expected value of the M δ values is $\mu_\delta = 0.2455$ and, following Eq. (7.4) on p. 369, the observed chance-corrected measure of effect size is

$$\Re_a = 1 - \frac{\delta_a}{\mu_\delta} = 1 - \frac{0.2156}{0.2455} = +0.1218 \,,$$

indicating approximately 12 % within-group agreement above that expected by chance.

The functional relationships between Goodman and Kruskal's t_a and δ_a are given by

$$t_a = 1 - \frac{N(N-g)\delta_a}{N^2 - \sum_{i=1}^{r} n_{i.}^2} \quad \text{and} \quad \delta_a = (1 - t_a)\frac{N^2 - \sum_{i=1}^{r} n_{i.}^2}{N(N-g)} \,.$$

Thus, for the frequency data given in Fig. 7.2 the observed values of t_a and δ_a are

$$t_a = 1 - \frac{64(64-3)(0.2156)}{64^2 - (9^2 + 55^2)} = 0.1497$$

and

$$\delta_a = (1 - 0.1497)\frac{64^2 - (9^2 + 55^2)}{64(64-3)} = 0.2156 \,.$$

A Reanalysis

The treatment-group weights given by

$$C_j = \frac{n_j - 1}{N - g} \,, \qquad j = 1, \ldots, g \,,$$

are important in associating t_a and δ, but degrees of freedom have no meaning in permutation methods, except for providing analogues to classical tests and measures.

For a reanalysis of the data given in Fig. 7.2, replicated in Fig. 7.3 for convenience, set

$$C_j = \frac{n_j}{N} \,, \qquad j = 1, \ldots, g \,,$$

to simply weight each treatment group proportional to its size, and let $v = 1$ or $v = 2$ as it makes no difference with (0, 1) binary data.

Fig. 7.3 Cross-classification of political party (B) and candidate preference (A)

B	A a_1	a_2	a_3	Total
b_1	5	1	3	9
b_2	47	6	2	55
Total	52	7	5	64

For the frequency data given in Fig. 7.3, $r = 2$, $g = 3$, $n_{.1} = 52$, $n_{.2} = 7$, $n_{.3} = 5$, $N = n_{.1} + n_{.2} + n_{.3} = 64$, and $C_j = n_j/N$, $j = 1, 2, 3$. Following Eq. (7.2) on p. 368, the $N = 64$ observations given in Fig. 7.3 yield $g = 3$ average distance-function values of

$$\xi_1 = 0.1772\,, \quad \xi_2 = 0.2857\,, \quad \text{and} \quad \xi_3 = 0.60\,.$$

Following Eq. (7.1) on p. 368, the observed value of the MRPP test statistic based on $v = 1$ and treatment-group weights

$$C_j = \frac{n_j}{N}\,, \qquad j = 1, 2, 3\,,$$

is

$$\delta_a = \sum_{j=1}^{g} C_j \xi_j = \frac{1}{64}\big[(52)(0.1772) + (7)(0.2857)$$
$$+ (5)(0.60)\big] = 0.2221\,.$$

Since there are still

$$M = \frac{N!}{\prod_{j=1}^{g} n_{.j}!} = \frac{64!}{52!\;7!\;5!} = 2{,}601{,}098{,}044{,}820{,}352$$

possible, equally-likely arrangements of the $N = 64$ observed values given in Fig. 7.3, an exact solution is not practical. If all M possible arrangements of the $N = 64$ observed values given in Fig. 7.3 occur with equal chance, the approximate resampling probability value of $\delta_a = 0.2221$ computed on $L = 1{,}000{,}000$ random arrangements of the observed values with $n_{.1} = 52$, $n_{.2} = 7$, and $n_{.3} = 5$ marginal frequency totals preserved for each arrangement is

$$P\big(\delta \le \delta_a | H_0\big) = \frac{\text{number of } \delta \text{ values} \le \delta_a}{L} = \frac{22{,}921}{1{,}000{,}000} = 0.0229\,.$$

For comparison, the approximate resampling probability value based on $L = 1{,}000{,}000$ and $C_j = (n_j - 1)/(N - g)$ for $i = 1, 2, 3$ is $P = 0.0228$. No comparison is made with the conventional Goodman and Kruskal t_a statistic as t_a is undefined for $C_j = n_j/N, j = 1, \ldots, g$.

Following Eq. (7.5) on p. 369, the exact expected value of the M δ values is $\mu_\delta = 0.2455$ and, following Eq. (7.4) on p. 369, the observed chance-corrected measure of effect size is

$$\Re_a = 1 - \frac{\delta_a}{\mu_\delta} = 1 - \frac{0.2221}{0.2455} = +0.0954 ,$$

indicating approximately 10 % within-group agreement above that expected by chance.

7.2.3 Example Analysis for t_b

If, for illustrative purposes, the null hypothesis posits that political party affiliation (variable B) does not depend on candidate preference (variable A) in this particular election, then the frequency data given in Fig. 7.3 is analyzed over the $r = 2$ rows of variable B instead of the $g = 3$ columns of variable A. The null hypothesis is then, H_0: $\tau_b = 0$,

$$h_1 = \begin{bmatrix} 1 \\ 0 \\ 0 \end{bmatrix}, \quad h_2 = \begin{bmatrix} 1 \\ 0 \\ 0 \end{bmatrix}, \quad h_3 = \begin{bmatrix} 1 \\ 0 \\ 0 \end{bmatrix}, \quad \ldots, \quad h_{64} = \begin{bmatrix} 0 \\ 0 \\ 1 \end{bmatrix},$$

and

$$\mathbf{H} = \begin{bmatrix} 1 & 1 & 1 & 0 & 0 & 0 & 0 & 0 & \cdots & 0 \\ 0 & 0 & 0 & 1 & 0 & 0 & 0 & 0 & \cdots & 0 \\ 0 & 0 & 0 & 0 & 1 & 1 & 1 & 1 & \cdots & 1 \end{bmatrix}.$$

For the frequency data given in Fig. 7.3, $r = 2$, $g = 3$ $v = 1$, $n_{1.} = 9$, $n_{2.} = 55$, $N = n_{1.} + n_{2.} = 64$, and let

$$C_i = \frac{n_i - 1}{N - r} , \qquad i = 1, \ldots, r ,$$

to correspond to Goodman and Kruskal's t_b test statistic [151]. The $\Delta(I, J)$ generalized Minkowski distance-function values are

$$\Delta(1,2) = 0 , \quad \Delta(1,3) = 0 , \quad \Delta(1,4) = 1 , \quad \ldots, \quad \Delta(63,64) = 0 .$$

Following Eq. (7.2) on p. 368, the $N = 64$ observations given in Fig. 7.3 yield $r = 2$ average distance-function values of

$$\xi_1 = 0.6389 \quad \text{and} \quad \xi_2 = 0.2613 \, .$$

Following Eq. (7.1) on p. 368, the observed value of the MRPP test statistic based on $v = 1$ and treatment-group weights

$$C_i = \frac{n_i - 1}{N - r} \, , \qquad i = 1, 2 \, ,$$

is

$$\delta_b = \sum_{i=1}^{r} C_i \xi_i = \frac{1}{64 - 2} \big[(9 - 1)(0.6389) + (55 - 1)(0.2613) \big] = 0.3100 \, .$$

Alternatively,

$$\delta_b = \frac{N - \displaystyle\sum_{j=1}^{g} \sum_{i=1}^{r} \frac{n_{ij}^2}{n_{i.}}}{N - r}$$

$$= \frac{64 - \left(\dfrac{5^2}{9} + \dfrac{1^2}{9} + \dfrac{3^2}{9} + \dfrac{47^2}{55} + \dfrac{6^2}{55} + \dfrac{2^2}{55} \right)}{64 - 2} = 0.3100 \, .$$

Analogous to t_a, Goodman and Kruskal's t_b is given by

$$t_b = \frac{N \displaystyle\sum_{j=1}^{g} \sum_{i=1}^{r} \frac{n_{ij}^2}{n_{i.}} - \sum_{j=1}^{g} n_{.j}^2}{N^2 - \displaystyle\sum_{j=1}^{g} n_{.j}^2} \, . \tag{7.8}$$

Thus, for the frequency data given in Fig. 7.3 on p. 378, the observed value of t_b is

$$t_b = \frac{64 \left(\dfrac{5^2}{9} + \dfrac{1^2}{9} + \dfrac{3^2}{9} + \dfrac{47^2}{55} + \dfrac{6^2}{55} + \dfrac{2^2}{55} \right) - (52^2 + 7^2 + 5^2)}{64^2 - (52^2 + 7^2 + 5^2)} = 0.0667 \, .$$

The computed value of $t_b = 0.0667$ indicates an approximate 7 % reduction in the number of prediction errors, given knowledge of the distribution of candidate preference (variable A) over knowledge of only the distribution of political party affiliation (variable B) . The adjusted t_b value of Margolin and Light [263] is

$$t_b(N-1)(g-1) = 0.0667(64-1)(3-1) = 8.4039$$

and, with $(r-1)(g-1) = (2-1)(3-1) = 2$ degrees of freedom, the asymptotic chi-squared probability value for $\chi^2 = 8.4039$ is $P = 0.0150$.

Since there are

$$M = \frac{N!}{\prod_{i=1}^{r} n_{i.}!} = \frac{64!}{9!\;55!} = 27{,}540{,}584{,}512$$

possible, equally-likely arrangements of the $N = 64$ observed values given in Fig. 7.3, an exact solution is not practical. If all M possible arrangements of the $N = 64$ observed values given in Fig. 7.3 occur with equal chance, the approximate resampling probability value of $\delta_b = 0.3100$ computed on $L = 1{,}000{,}000$ random arrangements of the observed values with $n_{1.} = 9$ and $n_{2.} = 55$ marginal frequency totals preserved for each arrangement is

$$P(\delta \leq \delta_b | H_0) = \frac{\text{number of } \delta \text{ values} \leq \delta_b}{L} = \frac{18{,}910}{1{,}000{,}000} = 0.0189\ .$$

For comparison, the chi-squared approximate probability value under the null hypothesis with $(r-1)(g-1) = (2-1)(3-1) = 2$ degrees of freedom is $P = 0.0150$.

Following Eq. (7.5) on p. 369, the exact expected value of the $M = 27{,}540{,}584{,}512$ δ values is $\mu_\delta = 0.3269$ and, following Eq. (7.4) on p. 369, the observed chance-corrected measure of effect size is

$$\Re_b = 1 - \frac{\delta_b}{\mu_\delta} = 1 - \frac{0.3100}{0.3269} = +0.0516\ ,$$

indicating approximately 5 % within-group agreement above that expected by chance. Analogous to t_a and δ_a, the functional relationships between t_b and δ_b are given by

$$t_b = 1 - \frac{N(N-r)\delta_b}{N^2 - \sum_{j=1}^{g} n_{.j}^2} \quad \text{and} \quad \delta_b = (1-t_b)\frac{N^2 - \sum_{j=1}^{g} n_{.j}^2}{N(N-r)}\ .$$

Thus, for the frequency data given in Fig. 7.3 the observed values of t_b and δ_b are

$$t_b = 1 - \frac{64(64-2)(0.3100)}{64^2 - (52^2 + 7^2 + 5^2)} = 0.0667$$

and

$$\delta_b = (1 - 0.0667)\frac{64^2 - (52^2 + 7^2 + 5^2)}{64(64-2)} = 0.3100 \ .$$

A Reanalysis

For a reanalysis of the frequency data given in Fig. 7.3 on p. 378, set

$$C_i = \frac{n_i}{N} \ , \qquad i = 1, \ldots, r \ ,$$

instead of

$$C_i = \frac{n_i - 1}{N - r} \ , \qquad i = 1, \ldots, r \ ,$$

to simply weight each treatment group proportional to its size, and set v to either 1 or 2 as it makes no difference with (0, 1) binary-coded data.

For the frequency data given in Fig. 7.3, $r = 2$, $g = 3$, $n_{1.} = 9$, $n_{2.} = 55$, $N = n_{1.} + n_{2.} = 64$, and $C_i = n_i/N$, $i = 1, 2$. Following Eq. (7.2) on p. 368, the $N = 64$ observations given in Fig. 7.3 yield $r = 2$ average distance-function values of

$$\xi_1 = 0.6389 \quad \text{and} \quad \xi_2 = 0.2613 \ .$$

Following Eq. (7.1) on p. 368, the observed value of the MRPP test statistic based on $v = 1$ and treatment-group weights

$$C_i = \frac{n_i}{N} \ , \qquad i = 1, 2 \ ,$$

is

$$\delta_b = \sum_{i=1}^{r} C_i \xi_i = \frac{1}{64}\big[(9)(0.6389) + (55)(0.2613)\big] = 0.3144 \ .$$

Since there are still

$$M = \frac{N!}{\prod_{i=1}^{r} n_{i.}!} = \frac{64!}{9!\ 55!} = 27{,}540{,}584{,}512$$

possible, equally-likely arrangements of the $N = 64$ observed values given in Fig. 7.3, an exact solution is not practical. If all M possible arrangements of the $N = 64$ observed values given in Fig. 7.3 occur with equal chance, the approximate resampling probability value of $\delta_b = 0.3144$ computed on $L = 1{,}000{,}000$ random arrangements of the observed values with $n_{1.} = 9$ and $n_{2.} = 55$ marginal frequency totals preserved for each arrangement is

$$P(\delta \leq \delta_a | H_0) = \frac{\text{number of } \delta \text{ values} \leq \delta_a}{L} = \frac{18{,}834}{1{,}000{,}000} = 0.0188 \ .$$

For comparison, the approximate resampling probability value based on $L = 1{,}000{,}000$ and $C_i = (n_i - 1)/(N - r)$ for $i = 1, 2$, and is $P = 0.0189$. No comparison is made with the Goodman and Kruskal t_b statistic as Goodman and Kruskal's t_b is undefined for $C_i = n_i/N$, $i = 1, \ldots, r$.

Following Eq. (7.5) on p. 369, the exact expected value of the M δ values is $\mu_\delta = 0.3269$ and, following Eq. (7.4) on p. 369, the observed chance-corrected measure of effect size is

$$\Re_b = 1 - \frac{\delta_b}{\mu_\delta} = 1 - \frac{0.3144}{0.3269} = +0.0383 \ ,$$

indicating approximately 4 % within-group agreement above that expected by chance.

7.2.4 Goodman–Kruskal's t_a, δ_a, and χ^2

Following the notation given in Fig. 7.1, replicated in Fig. 7.4 for convenience, some interesting simplifications occur for Goodman and Kruskal's t_a and χ^2 when $n_{i.} = N/r$ for $i = 1, \ldots, r$, or when $r = 2$ [297, p. 325]. Consider the frequency data given in Fig. 7.5 for dependent variable A and independent variable B where $r = g = 3$ and $n_{1.} = n_{2.} = n_{3.} = N/r = 10$.

Fig. 7.4 Notation for the cross-classification of two categorical variables, A_j for $j = 1, \ldots, g$ and B_i for $i = 1, \ldots, r$

	A				
B	a_1	a_2	$\cdots$	a_g	Total
b_1	n_{11}	n_{12}	$\cdots$	n_{1g}	$n_{1.}$
b_2	n_{21}	n_{22}	$\cdots$	n_{2g}	$n_{2.}$
$\vdots$	$\vdots$	$\vdots$	$\ddots$	$\vdots$	$\vdots$
b_r	n_{r1}	n_{r2}	$\cdots$	n_{rg}	$n_{r.}$
Total	$n_{.1}$	$n_{.2}$	$\cdots$	$n_{.g}$	N

Fig. 7.5 Example 3×3 contingency table for variables A and B with $n_{1.} = n_{2.} = n_{3.} = 10$

	A			
B	a_1	a_2	a_3	Total
b_1	2	2	6	10
b_2	2	5	3	10
b_3	1	3	6	10
Total	5	10	15	30

For the frequency data given in Fig. 7.5, the observed values of t_a, δ_a, and χ^2 are

$$t_a = \frac{N \displaystyle\sum_{i=1}^{r} \sum_{j=1}^{g} \frac{n_{ij}^2}{n_{.j}} - \sum_{i=1}^{r} n_{i.}^2}{N^2 - \displaystyle\sum_{i=1}^{r} n_{i.}^2}$$

$$= \frac{30(11) - (10^2 + 10^2 + 10^2)}{30^2 - (10^2 + 10^2 + 10^2)} = \frac{30}{600} = 0.05\,, \qquad (7.9)$$

where the value (11) in the numerator of Eq. (7.9) is given by

$$\sum_{i=1}^{r} \sum_{j=1}^{g} \frac{n_{ij}^2}{n_{.j}} = \frac{2^2 + 2^2 + 1^2}{5} + \frac{2^2 + 5^2 + 3^2}{10} + \frac{6^2 + 3^2 + 6^2}{15}$$

$$= 1.8 + 3.8 + 5.4 = 11\,,$$

$$\delta_a = \frac{N - \displaystyle\sum_{i=1}^{r} \sum_{j=1}^{g} \frac{n_{ij}^2}{n_{.j}}}{N - g} = \frac{30 - 11}{30 - 3} = \frac{19}{27} = 0.7037\,,$$

and

$$\chi^2 = N \sum_{i=1}^{r} \sum_{j=1}^{g} \frac{n_{ij}^2}{n_{i.}n_{.j}} - N = 30(1.10) - 30 = 3.00 \,, \tag{7.10}$$

where the value (1.10) in Eq. (7.10) is given by

$$\sum_{i=1}^{r} \sum_{j=1}^{g} \frac{n_{ij}^2}{n_{i.}n_{.j}} = \frac{2^2}{(10)(5)} + \frac{2^2}{(10)(10)} + \cdots + \frac{6^2}{(10)(15)} = 1.10 \,.$$

When $n_{i.} = N/r$ for $i = 1, \ldots, r$, the relationships between t_a and χ^2 simplify to

$$t_a = \frac{\chi^2}{N(r-1)} = \frac{3.00}{30(3-1)} = 0.05$$

and

$$\chi^2 = N(r-1)t_a = 30(3-1)(0.05) = 3.00 \,.$$

Also, for the frequency data given in Fig. 7.5, the observed values of δ_a and t_a are

$$\delta_a = (1 - t_a) \frac{N^2 - \sum_{i=1}^{r} n_{i.}^2}{N(N-g)}$$
$$= (1 - 0.05) \left[\frac{30^2 - (10^2 + 10^2 + 10^2)}{30(30-3)} \right] = (0.95) \left(\frac{600}{810} \right) = 0.7037$$

and

$$t_a = 1 - \frac{N(N-g)\delta_a}{N^2 - \sum_{i=1}^{r} n_{i.}^2} = 1 - \frac{30(30-3)(0.7037)}{30^2 - (10^2 + 10^2 + 10^2)} = 1 - \frac{570}{600} = 0.05 \,,$$

and the observed values of δ_a and χ^2 are

$$\delta_a = \left[1 - \frac{\chi^2}{N(r-1)}\right] \left[\frac{N^2 - \sum_{i=1}^{r} n_{i.}{}^2}{N(N-g)}\right]$$

$$= \left[1 - \frac{3.00}{30(3-1)}\right] \left[\frac{30^2 - (10^2 + 10^2 + 10^2)}{30(30-3)}\right]$$

$$= (0.95)\left(\frac{600}{810}\right) = 0.7037$$

and

$$\chi^2 = N(r-1)\left[1 - \frac{N(N-g)\delta_a}{N^2 - \sum_{i=1}^{r} n_{i.}^2}\right]$$

$$= 30(3-1)\left[1 - \frac{30(30-3)(0.7037)}{30^2 - (10^2 + 10^2 + 10^2)}\right]$$

$$= 60\left(1 - \frac{570}{600}\right) = 3.00\,.$$

By symmetry, analogous results are obtained for t_b, χ^2, and δ_b, when $n_{.j} = N/g$ for $j = 1, \ldots, g$.

Additional simplifications occur for the relationships between t_a and χ^2 when $r = 2$ for any marginal frequency totals [297, p. 325]. Consider the frequency data given in Fig. 7.6 with $r = 2$ rows and $g = 3$ columns. For the frequency data given

Fig. 7.6 Example 2×3 contingency table for variables A and B with $n_{1.} \neq n_{2.}$

	A			
B	a_1	a_2	a_3	Total
b_1	5	3	2	10
b_2	5	5	5	15
Total	10	8	7	25

in Fig. 7.6, t_a, δ_a, and χ^2 are

$$t_a = \frac{N \sum_{i=1}^{r} \sum_{j=1}^{g} \frac{n_{ij}^2}{n_{.j}} - \sum_{i=1}^{r} n_{i.}^2}{N^2 - \sum_{i=1}^{r} n_{i.}^2}$$

$$= \frac{25(13.3929) - (10^2 + 15^2)}{25^2 - (10^2 + 15^2)} = \frac{9.8214}{300} = 0.0327 , \qquad (7.11)$$

where the value (13.3929) in the numerator of Eq. (7.11) is given by

$$\sum_{i=1}^{r} \sum_{j=1}^{g} \frac{n_{ij}^2}{n_{.j}} = \frac{5^2 + 5^2}{10} + \frac{3^2 + 5^2}{8} + \frac{2^2 + 5^2}{7}$$

$$= 5.00 + 4.25 + 4.1429 = 13.3929 ,$$

$$\delta_a = \frac{N - \sum_{i=1}^{r} \sum_{j=1}^{g} \frac{n_{ij}^2}{n_{.j}}}{N - g} = \frac{25 - 13.3929}{25 - 3} = \frac{11.6071}{22} = 0.5276 ,$$

and

$$\chi^2 = N \sum_{i=1}^{r} \sum_{j=1}^{g} \frac{n_{ij}^2}{n_{i.} n_{.j}} - N = 25(1.0327) - 25 = 0.8185 , \qquad (7.12)$$

where the value (1.0327) in Eq. (7.12) is given by

$$\sum_{i=1}^{r} \sum_{j=1}^{g} \frac{n_{ij}^2}{n_{i.} n_{.j}} = \frac{5^2}{(10)(10)} + \frac{3^2}{(10)(8)} + \cdots + \frac{5^2}{(15)(7)} = 1.0327 .$$

When $r = 2$, with any marginal frequency totals for $n_{i.}$, $i = 1, 2$ and $n_{.j}$, $j = 1, \ldots, g$, the relationships between t_a and χ^2 simplify to

$$t_a = \frac{\chi^2}{N} = \frac{0.8185}{25} = 0.0327 \quad \text{and} \quad \chi^2 = N t_a = 25(0.0327) = 0.8185 .$$

Also, for the frequency data given in Fig. 7.6 the observed values of δ_a and t_a are

$$\delta_a = (1 - t_a)\,\frac{N^2 - \sum_{i=1}^{r} n_{i.}^2}{N(N-g)}$$

$$= (1 - 0.0327)\left[\frac{25^2 - (10^2 + 15^2)}{25(25-3)}\right] = (0.9673)\left(\frac{300}{550}\right) = 0.5276$$

and

$$t_a = 1 - \frac{N(N-g)\delta_a}{N^2 - \sum_{i=1}^{r} n_{i.}^2} = 1 - \frac{25(25-3)(0.5276)}{25^2 - (10^2 + 15^2)} = 1 - \frac{290.1786}{300} = 0.0327\,,$$

and the observed values of δ_a and χ^2 are

$$\delta_a = \left[1 - \frac{\chi^2}{N(r-1)}\right]\left[\frac{N^2 - \sum_{i=1}^{r} n_{i.}{}^2}{N(N-g)}\right]$$

$$= \left[1 - \frac{0.8185}{25(2-1)}\right]\left[\frac{25^2 - (10^2 + 15^2)}{25(25-3)}\right]$$

$$= (0.9673)\left(\frac{300}{550}\right) = 0.5276$$

and

$$\chi^2 = N(r-1)\left[1 - \frac{N(N-g)\delta_a}{N^2 - \sum_{i=1}^{r} n_{i.}^2}\right]$$

$$= 25(2-1)\left[1 - \frac{25(25-3)(0.5276)}{25^2 - (10^2 + 15^2)}\right]$$

$$= 25\left(1 - \frac{290.1786}{300}\right) = 0.8185\,.$$

By symmetry, analogous results are obtained for t_b, χ^2, and δ_b when $g = 2$.

Simplifications also occur for the relationships between t_a and χ^2 when $r = g = 2$ for any marginal frequency totals [297, p. 325]. Consider the frequency data given in Fig. 7.7 with $r = g = 2$, where t_a, δ_a, and χ^2 are

$$t_a = \frac{N \sum_{i=1}^{r} \sum_{j=1}^{g} \frac{n_{ij}^2}{n_{.j}} - \sum_{i=1}^{r} n_{i.}^2}{N^2 - \sum_{i=1}^{r} n_{i.}^2}$$

$$= \frac{10(5.2381) - (6^2 + 4^2)}{10^2 - (6^2 + 4^2)} = \frac{0.3810}{48} = 0.0079\,, \qquad (7.13)$$

where the value (5.2381) in the numerator of Eq. (7.13) is given by

$$\sum_{i=1}^{r} \sum_{j=1}^{g} \frac{n_{ij}^2}{n_{.j}} = \frac{4^2 + 3^2}{7} + \frac{2^2 + 1^2}{3} = 3.5714 + 1.6667 = 5.2381\,,$$

$$\delta_a = \frac{N - \sum_{i=1}^{r} \sum_{j=1}^{g} \frac{n_{ij}^2}{n_{.j}}}{N - g} = \frac{10 - 5.2381}{10 - 2} = \frac{4.7619}{8} = 0.5952\,,$$

and

$$\chi^2 = N \sum_{i=1}^{r} \sum_{j=1}^{g} \frac{n_{ij}^2}{n_{i.} n_{.j}} - N = 10(1.0079) - 10 = 0.0794\,, \qquad (7.14)$$

where the value (1.0079) in Eq. (7.14) is given by

$$\sum_{i=1}^{r} \sum_{j=1}^{g} \frac{n_{ij}^2}{n_{i.} n_{.j}} = \frac{4^2}{(6)(7)} + \frac{2^2}{(6)(3)} + \frac{3^2}{(4)(7)} + \frac{1^2}{(4)(3)} = 1.0079\,.$$

Fig. 7.7 Example 2×2 contingency table for variables A and B with $N = 10$ observations

	A		
B	a_1	a_2	Total
b_1	4	2	6
b_2	3	1	4
Total	7	3	10

When $r = g = 2$, with any marginal frequency totals for $n_{i.}$, $i = 1, 2$ and $n_{.j}$, $j = 1, 2$, the relationships between t_a and χ^2 simplify to

$$t_a = \frac{\chi^2}{N} = \frac{0.0794}{10} = 0.0079 \quad \text{and} \quad \chi^2 = Nt_a = 10(0.0079) = 0.0794 \,.$$

Also, for the frequency data given in Fig. 7.7 the observed values of δ_a and t_a are

$$\delta_a = (1 - t_a) \frac{N^2 - \sum_{i=1}^{r} n_{i.}^2}{N(N-g)}$$

$$= (1 - 0.0079) \left[\frac{10^2 - (6^2 + 4^2)}{10(10-2)} \right] = (0.9921)(0.60) = 0.5952$$

and

$$t_a = 1 - \frac{N(N-g)\delta_a}{N^2 - \sum_{i=1}^{r} n_{i.}^2} = 1 - \frac{10(10-2)(0.5952)}{10^2 - (6^2 + 4^2}) = 1 - \frac{47.6190}{48} = 0.0079 \,,$$

and the observed values of δ_a and χ^2 are

$$\delta_a = \left[1 - \frac{\chi^2}{N(r-1)} \right] \left[\frac{N^2 - \sum_{i=1}^{r} n_{i.}{}^2}{N(N-g)} \right]$$

$$= \left[1 - \frac{0.0794}{10(2-1)} \right] \left[\frac{10^2 - (6^2 + 4^2)}{10(10-2)} \right]$$

$$= (0.9921) \left(\frac{48}{80} \right) = 0.5952$$

and

$$\chi^2 = N(r-1)\left[1 - \frac{N(N-g)\delta_a}{N^2 - \sum_{i=1}^{r} n_{i.}^2}\right]$$

$$= 10(2-1)\left[1 - \frac{10(10-2)(0.5952)}{10^2 - (6^2 + 4^2)}\right]$$

$$= 10\left(1 - \frac{47.6190}{48}\right) = 0.0794 \,.$$

By symmetry, analogous results are obtained for t_b, χ^2, and δ_b when $g = 2$. Note that for fourfold contingency tables where $r = g = 2$, as in this case, Goodman and Kruskal's $t_a = t_b = \chi^2/N$, which in turn is equal to Pearson's mean-square contingency coefficient ϕ^2 and also equal to Pearson's product-moment correlation coefficient r^2 if a_1 and b_1 (a_2 and b_2) in Fig. 7.7 are coded as 0 and a_2 and b_2 (a_1 and b_1) are coded as 1.

7.2.5 Fourfold Contingency Tables

Fourfold (2×2) contingency tables are ubiquitous in everyday research; consequently, they deserve special attention—especially with regard to Goodman and Kruskal's t_a and t_b asymmetric measures of nominal association.

Goodman–Kruskal's τ_a Statistic

Consider the 2×2 contingency table given in Fig. 7.8 with $N = 35$ observations. For the frequency data given in Fig. 7.8, $r = 2$, $g = 2$, $v = 1$, $n_{.1} = 19$, $n_{.2} = 16$, $N = n_{.1} + n_{.2} = 35$, and let

$$C_j = \frac{n_j - 1}{N - g} \,, \qquad j = 1, \ldots, g \,,$$

Fig. 7.8 Example 2×2 contingency table for variables A and B with $N = 35$ observations

	A		
B	a_1	a_2	Total
b_1	16	2	18
b_2	3	14	17
Total	19	16	35

to correspond to Goodman and Kruskal's t_a test statistic [151]. Following Eq. (7.2) on p. 368, the $N = 35$ observations given in Fig. 7.8 yield $g = 2$ average distance-function values of

$$\xi_1 = 0.2807 \text{ and } \xi_2 = 0.2333 .$$

Following Eq. (7.1) on p. 368, the observed value of the MRPP test statistic based on $v = 1$ and treatment-group weights

$$C_j = \frac{n_j - 1}{N - g} , \qquad j = 1, 2 ,$$

is

$$\delta_a = \sum_{j=1}^{g} C_j \xi_j = \frac{1}{35 - 2}\big[(19 - 1)(0.2807) + (16 - 1)(0.2333)\big] = 0.2592 .$$

Since there are

$$M = \frac{N!}{\prod_{j=1}^{g} n_{.j}!} = \frac{35!}{19!\ 16!} = 4{,}059{,}928{,}950$$

possible, equally-likely arrangements of the $N = 35$ observed values given in Fig. 7.8, an exact solution is not practical. If all M possible arrangements of the $N = 35$ observed values given in Fig. 7.8 occur with equal chance, the approximate resampling probability value of $\delta_a = 0.2592$ computed on $L = 1{,}000{,}000$ random arrangements of the observed values with $n_{.1} = 19$ and $n_{.2} = 16$ marginal frequency totals preserved for each arrangement is

$$P\big(\delta \leq \delta_a | H_0\big) = \frac{\text{number of } \delta \text{ values} \leq \delta_a}{L} = \frac{0}{1{,}000{,}000} = 0.00 .$$

As explained more completely in Chap. 4, page 187, when M is very large and the probability of an observed δ is very small, resampling permutation procedures often result in zero probability, even with $L = 1{,}000{,}000$ random arrangements of the observed data. Moment-approximation permutation procedures, described briefly in Chap. 1, Sect. 1.2.2, can often provide results in these extreme situations. The moment-approximation of a test statistic requires computation of the exact moments of the test statistic, assuming equally-likely arrangements of the observed response measurement values. Usually, the first three exact moments of δ are used: the exact mean, μ_δ, the exact variance, σ_δ^2, and the exact skewness, γ_δ. The three moments are then used to fit a specified distribution, such as a Pearson type III distribution that approximates the underlying discrete permutation distribution, and provide an approximate probability value. For variable A, a moment-approximation procedure yields $\delta_a = 0.2592$, $\mu_\delta = 0.5143$, $\sigma_\delta^2 = 0.4718 \times 10^{-3}$, $\gamma_\delta = -2.7050$, an observed standardized test statistic of

$$T_a = \frac{\delta_a - \mu_\delta}{\sigma_\delta} = \frac{0.2592 - 0.5143}{0.0217} = -11.7449 ,$$

and a Pearson type III probability value of 0.2107×10^{-4}.[2]

Following Eq. (7.5) on p. 369, the exact expected value of the $M =$ 4,059,928,950 δ values is $\mu_\delta = 0.5143$ and following Eq, (7.4) on p. 369, the observed chance-corrected measure of effect size is

$$\Re_a = 1 - \frac{\delta_a}{\mu_\delta} = 1 - \frac{0.2592}{0.5143} = +0.4961 ,$$

indicating approximately 50 % within-group agreement above that expected by chance.

Following Eq. (7.6), for the frequency data given in Fig. 7.8 on p. 391 the observed value of t_a is

$$t_a = \frac{N \displaystyle\sum_{i=1}^{r} \sum_{j=1}^{g} \frac{n_{ij}^2}{n_{.j}} - \sum_{i=1}^{r} n_{i.}^2}{N^2 - \displaystyle\sum_{i=1}^{r} n_{i.}^2}$$

$$= \frac{35 \left(\dfrac{16^2}{19} + \dfrac{2^2}{16} + \dfrac{3^2}{19} + \dfrac{14^2}{16} \right) - (18^2 + 17^2)}{35^2 - (18^2 + 17^2)} = 0.5109 ,$$

indicating approximately 51 % reduction in the number of predicted errors, given knowledge of the distribution of variable B over knowledge of only the distribution of variable A.

Goodman–Kruskal's t_b Statistic

Now consider Goodman and Kruskal's t_b test statistic. For the frequency data given in Fig. 7.8, replicated in Fig. 7.9 for convenience, $r = 2$, $g = 2$, $v = 1$, $n_{1.} = 18$, $n_{2.} = 17$, $N = n_{1.} + n_{2.} = 35$, and let

$$C_i = \frac{n_i - 1}{N - r} , \qquad i = 1, \ldots, r ,$$

[2]For comparison, the exact probability value is actually $P = 0.2969 \times 10^{-4}$.

Fig. 7.9 Example 2×2 contingency table for variables A and B with $N = 35$ observations

B	A: a_1	a_2	Total
b_1	16	2	18
b_2	3	14	17
Total	19	16	35

to correspond to Goodman and Kruskal's t_b test statistic. Following Eq. (7.2) on p. 368, the $N = 35$ observations given in Fig. 7.9 yield $r = 2$ average distance-function values of

$$\xi_1 = 0.2092 \quad \text{and} \quad \xi_2 = 0.3088 \, .$$

Following Eq. (7.1) on p. 368, the observed value of the MRPP test statistic based on $v = 1$ and treatment-group weights

$$C_i = \frac{n_i - 1}{N - r} \, , \qquad i = 1, 2 \, ,$$

is

$$\delta_b = \sum_{i=1}^{r} C_i \xi_i = \frac{1}{35 - 2}\big[(18 - 1)(0.2092) + (17 - 1)(0.3088)\big] = 0.2576 \, .$$

Since there are

$$M = \frac{N!}{\prod_{i=1}^{r} n_{i.}!} = \frac{35!}{18! \; 17!} = 4{,}537{,}567{,}650$$

possible, equally-likely arrangements of the $N = 35$ observed values with $n_{1.} = 18$ and $n_{2.} = 17$ marginal frequency totals preserved for each arrangement, an exact solution is not practical. If all M possible arrangements of the $N = 35$ observed values given in Fig. 7.9 occur with equal chance, the approximate resampling probability value of $\delta_b = 0.2576$ computed on $L = 1{,}000{,}000$ random arrangements of the observed values with $n_{1.} = 18$ and $n_{2.} = 17$ marginal frequency totals preserved for each arrangement is

$$P\big(\delta \leq \delta_b | H_0\big) = \frac{\text{number of } \delta \text{ values} \leq \delta_b}{L} = \frac{0}{1{,}000{,}000} = 0.00 \, .$$

As previously, when M is very large and the probability of an observed δ is very small, resampling permutation procedures often result in zero probability, even with

$L = 1{,}000{,}000$ random arrangements of the observed data. Moment-approximation permutation procedures can often provide results in these extreme situations. For variable B, a moment-approximation procedure yields $\delta_b = 0.2576$, $\mu_\delta = 0.5109$, $\sigma_\delta^2 = 0.4657{\times}10^{-3}$, $\gamma_\delta = -2.7050$, an observed standardized test statistic of

$$T_b = \frac{\delta_b - \mu_\delta}{\sigma_\delta} = \frac{0.2576 - 0.5109}{0.0216} = -11.7449\ ,$$

and a Pearson type III probability value of $0.2107{\times}10^{-4}$.[3]

Following Eq. (7.5) on p. 369, the exact expected value of the $M = 4{,}059{,}928{,}950$ δ values is $\mu_\delta = 0.5109$ and, following Eq. (7.4) on p. 369, the observed chance-corrected measure of effect size is

$$\Re_b = 1 - \frac{\delta_b}{\mu_\delta} = 1 - \frac{0.2576}{0.5109} = +0.4961\ ,$$

indicating approximately 50 % within-group agreement above that expected by chance.

Following Eq. (7.8) on p. 380, for the frequency data given in Fig. 7.9, the observed value of Goodman and Kruskal's t_b is

$$t_b = \frac{N \displaystyle\sum_{i=1}^{r} \sum_{j=1}^{g} \frac{n_{ij}^2}{n_{i.}} - \sum_{j=1}^{g} n_{.j}^2}{N^2 - \displaystyle\sum_{j=1}^{g} n_{.j}^2}$$

$$= \frac{35\left(\dfrac{16^2}{18} + \dfrac{2^2}{18} + \dfrac{3^2}{17} + \dfrac{14^2}{17}\right) - (19^2 + 16^2)}{35^2 - (19^2 + 16^2)} = 0.5109\ ,$$

indicating approximately 51 % reduction in the number of predicted errors, given knowledge of the distribution of variable B over knowledge of only the distribution of variable A.

Note that for a 2×2 cross-classification table, t_a and t_b yield identical values and are equal to the squared Pearson fourfold point correlation, defined as

$$\phi^2 = \frac{\chi_1^2}{N}\ ,$$

[3]For comparison, the exact probability value is actually $P = 0.2969{\times}10^{-4}$.

Fig. 7.10 Dummy coding of the frequency data given in Fig. 7.9 with a_1 and b_1 coded as 0, and a_2 and b_2 coded as 1

Cell	A	B	Cell	A	B
a_1, b_1	0	0	a_1, b_2	0	1
	0	0		0	1
	0	0		0	1
	0	0			
	0	0	a_2, b_2	1	1
	0	0		1	1
	0	0		1	1
	0	0		1	1
	0	0		1	1
	0	0		1	1
	0	0		1	1
	0	0		1	1
	0	0		1	1
	0	0		1	1
	0	0		1	1
	0	0		1	1
				1	1
a_2, b_1	1	0		1	1
	1	0			

where χ_1^2 is the Pearson chi-squared test statistic for a 2×2 contingency table with one degree of freedom.[4]

For the frequency data given in Fig. 7.9, $t_a = t_b = 0.5109$ and $\chi_1^2 = 17.8808$. Thus,

$$t_a = t_b = \phi^2 = \frac{\chi_1^2}{N} = \frac{17.8808}{35} = 0.5109 \, .$$

Also, it is well known that ϕ^2 is simply a special case of the squared Pearson product-moment correlation coefficient, r^2, for two dichotomous variables. Thus, if a_1 and b_1 in Fig. 7.9 are coded 0 and a_2 and b_2 are coded 1, as in Fig. 7.10, the squared Pearson product-moment correlation coefficient for variables A and B is $r_{ab}^2 = 0.5109$.

7.2.6 Chi-Squared and δ

Since, for a 2×2 contingency table,

$$t_a = t_b = \phi^2 = r^2 = \frac{\chi_1^2}{N} \, ,$$

[4]Note: χ^2 with one degree of freedom is simply a squared normal deviate, i.e., z^2.

then t_a can be defined in terms of δ_a, and vice versa,

$$t_a = 1 - \frac{N(N-g)\delta_a}{N^2 - \sum_{i=1}^{r} n_{i.}^2} \tag{7.15}$$

and

$$\delta_a = (1 - t_a)\frac{N^2 - \sum_{i=1}^{r} n_{i.}^2}{N(N-g)} . \tag{7.16}$$

Consequently, δ_a and χ_1^2 are necessarily related. The relationships between δ_a and χ_1^2 for a 2×2 contingency table are given by

$$\delta_a = \frac{(N - \chi_1^2)\left(N^2 - \sum_{i=1}^{r} n_{i.}^2\right)}{N^2(N-g)} \tag{7.17}$$

and

$$\chi_1^2 = \frac{N\left[N^2 - \sum_{i=1}^{r} n_{i.}^2 - N(N-g)\delta_a\right]}{N^2 - \sum_{i=1}^{r} n_{i.}^2} , \tag{7.18}$$

respectively.

Thus, for the frequency data given in Fig. 7.9, replicated in Fig. 7.11 for convenience, following Eq. (7.17) the observed value of the MRPP test statistic is

$$\delta_a = \frac{(35 - 17.8808)[35^2 - (18^2 + 17^2]}{35^2(35-2)} = \frac{10{,}476.9504}{40{,}425} = 0.2592 \tag{7.19}$$

Fig. 7.11 Example 2×2 contingency table for variables A and B with $N = 35$ observations

	A		
B	a_1	a_2	Total
b_1	16	2	18
b_2	3	14	17
Total	19	16	35

and, following Eq. (7.18), the observed value of χ^2 is

$$\chi_1^2 = \frac{35[35^2 - (18^2 + 17^2) - 35(35 - 2)(0.2592)]}{35^2 - (18^2 + 17^2)}$$

$$= \frac{10,941.8400}{612} = 17.8808 \, . \qquad (7.20)$$

While Eqs. (7.15) to (7.20) are expressed in terms of t_a, the analogue holds for t_b, δ_b, and χ_1^2, since $t_a = t_b$ for a 2×2 contingency table.

7.3 Multiple Binary Choices

Surveys and other methods of data gathering often include questions for which respondents may select any number of categories, i.e., "cafeteria" or "multiple-response" questions. Coombs [79, pp. 295–297, 305–307] and Levine [238] referred to this type of question as a "pick any/r" type question, where "pick any/r" instructs the respondent to choose any or all of r unconstrained categories. For example, subjects may be requested to select every magazine to which they subscribe from a predetermined list, or asked to choose names of close friends from a list of classmates. Usually, it is of interest to determine whether the multiple responses differ among specified groups. However, multiple-response questions are often difficult to analyze as the answers are not independent [4, 385, p. 159]. Current methods used to analyze multiple-response data are limited to assessments of contingency, independence, and the magnitude of predictive association [4, 5, 42, 43, 92, 246, 406]. None of these methods is designed to test for differences among groups within the multiple-response structure.

The MRPP analysis of multiple category choices may be conceptualized as a binary argument problem in which N subjects choose any or all of r presented categories and the responses for each subject are coded 1 if the category is selected and 0 if the category is not selected. The subjects are a priori classified into g distinct groups and the groups are then compared on the multiple responses.

Specifically, let $\Omega = \{\omega_1, \ldots, \omega_N\}$ denote a finite sample of subjects that is representative of a target population, let $x_{I1}, \ldots, x_{Ir}$ denote r binary response measurements for subject ω_I for $I = 1, \ldots, N$, and let $S_1, \ldots, S_g$ designate an exhaustive partitioning of the N subjects into g disjoint treatment groups of sizes $n_1, \ldots, n_g$, where $n_i \geq 2$ for $i = 1, \ldots, g$, and

$$\sum_{i=1}^{g} n_i = N \, .$$

The response of each of the N subjects is a single r-dimensional column vector of size r×1 in which each argument of the response vector is either 0 or 1. The total number of distinct responses for a subject in this context is 2^r. The analysis of the multiple responses depends on the MRPP test statistic given by

$$\delta = \sum_{i=1}^{g} C_i \xi_i ,$$

where, for this application, $C_i = n_i/N$, $i = 1, \ldots, g$, since degrees of freedom are not relevant to permutation methods.

The null hypothesis of no difference in the response structures among the g groups specifies that each of the

$$M = \frac{N!}{\prod_{i=1}^{g} n_i!}$$

possible, equally-likely allocations of the N r-dimensional response measurements to the g treatment groups is equally likely. To illustrate a permutation test for multiple binary choices, consider three example applications in which g groups are compared on r responses and each of the N respondents is allowed to select any one of the 2^r possible arrangements of categorical responses.

7.3.1 Example Analysis 1

Suppose that a class of $N = 19$ elementary school students is assigned three books to read over the summer. Upon their return to school, the students are surveyed as to which of the three books they read during the summer. The $r = 3$ response categories are books A, B, and C, each of the $N = 19$ students is allowed to choose any of the $2^r = 2^3 = 8$ possible response arrangements, and the $g = 2$ groups consist of $n_1 = 8$ girls and $n_2 = 11$ boys. Figure 7.12 lists the $N = 19$ row vectors of observed binary responses from the eight possible response arrangements for each student, where a 1 indicates that a book was read and a 0 indicates that a book was not read. The data are adapted from Mielke and Berry [297, p. 84].

For the binary data listed in Fig. 7.12, $r = 3$, $g = 2$, $v = 1$, $n_1 = 8$, $n_2 = 11$, $N = n_1 + n_2 = 19$, and let

$$C_i = \frac{n_i}{N} , \qquad i = 1, \ldots, g ,$$

simply weighting each treatment group proportional to its size. Following Eq. (7.2) on p. 368, the $N = 19$ binary response measurements listed in Fig. 7.12 yield $g = 2$ average distance-function values of

$$\xi_1 = 0.9598 \quad \text{and} \quad \xi_2 = 0.9463 .$$

Fig. 7.12 Elementary school students example data with $r = 3$, $g = 2$, $n_1 = 8$, $n_2 = 11$, and $N = n_1 + n_2 = 19$

Girls			Boys		
A	*B*	*C*	*A*	*B*	*C*
1	0	0	0	0	1
1	0	1	1	0	1
1	1	0	0	1	1
1	1	0	1	0	1
1	1	0	0	0	0
0	1	1	0	0	1
1	1	1	0	1	1
1	1	0	0	0	1
			0	0	1
			0	0	1
			1	1	0

Following Eq. (7.1) on p. 368, the observed value of the MRPP test statistic based on $v = 1$ and treatment-group weights

$$C_i = \frac{n_i}{N} , \qquad i = 1, 2 ,$$

is

$$\delta_o = \sum_{i=1}^{g} C_i \xi_i = \frac{1}{19}\big[(8)(0.9598) + (11)(0.9463)\big] = 0.9520 .$$

Since there are only

$$M = \frac{N!}{\prod_{i=1}^{g} n_i!} = \frac{19!}{8!\ 11!} = 75{,}582$$

possible, equally-likely arrangements of the $N = 19$ observed values listed in Fig. 7.12, an exact solution is practical.

If all arrangements of the $N = 19$ observed binary values listed in Fig. 7.12 occur with equal chance, the exact probability value of $\delta_o = 0.9520$ computed on the $M = 75{,}582$ possible arrangements of the observed values with $n_1 = 8$ and $n_2 = 11$ binary response measurements preserved for each arrangement is

$$P\big(\delta \le \delta_o | H_0\big) = \frac{\text{number of } \delta \text{ values} \le \delta_o}{M} = \frac{685}{75{,}582} = 0.0091 .$$

Following Eq. (7.5) on p. 369, the exact expected value of the $M = 75{,}582$ δ values is $\mu_\delta = 1.1199$ and, following Eq. (7.4) on p. 369, the observed chance-corrected measure of effect size is

$$\Re_o = 1 - \frac{\delta_o}{\mu_\delta} = 1 - \frac{0.9520}{1.1199} = +0.1499 \,,$$

indicating approximately 15 % within-group agreement above that expected by chance.

7.3.2 Example Analysis 2

Some statistical tests are specifically designed to test for differences among responses, while others are designed to measure similarities among responses. In many studies it is commonplace to design experiments for which a test of differences is requisite, e.g., an F test in a one-way analysis of variance design. To clarify the function of the δ test statistic as a test for differences, consider a modification of the data in Example 1.

Suppose that a class of elementary school students is assigned one of two books to read over the summer, then the students are surveyed as to which of the two books they read during the summer. The $r = 2$ response categories are books A and B, the $g = 2$ treatment groups consist of $n_1 = 12$ girls and $n_2 = 12$ boys, and each of $N = 24$ students may choose any of the $2^r = 2^2 = 4$ possible response arrangements: read only book A, read only book B, read both books A and B, or read neither book A nor book B. In addition, assume that half of the girls read book A only and the other half of the girls read book B only, while half of the boys read both books and the other half of the boys read neither book. Figure 7.13 lists the row vectors of observed binary responses from the four possible response arrangements for each student, where a 1 indicates that a book was read and a 0 indicates that a book was not read. In one sense, the data are the same for the girls and the boys, as there are 12 books read by both the girls and the boys. In another sense, the data are different for the girls and the boys, as the selection of books read by the girls differs from that of the boys.

For the binary data listed in Fig. 7.13, $r = 2$, $g = 2$, $v = 1$, $n_1 = n_2 = 12$, $N = n_1 + n_2 = 24$, and let

$$C_i = \frac{n_i}{N} \,, \qquad i = 1, \ldots, g \,,$$

simply weighting each treatment group proportional to its size. Following Eq. (7.2) on p. 368, the $N = 24$ binary response measurements listed in Fig. 7.13 yield $g = 2$ average distance-function values of

$$\xi_1 = 0.7714 \quad \text{and} \quad \xi_2 = 0.7714 \,.$$

Fig. 7.13 Elementary school students example with $r = g = 2$, $n_1 = n_2 = 12$, and $N = n_1 + n_2 = 24$

Girls		Boys	
A	*B*	*A*	*B*
1	0	1	1
1	0	1	1
1	0	1	1
1	0	1	1
1	0	1	1
1	0	1	1
0	1	0	0
0	1	0	0
0	1	0	0
0	1	0	0
0	1	0	0
0	1	0	0

Following Eq. (7.1) on p. 368, the observed value of the MRPP test statistic based on $v = 1$ and treatment-group weights

$$C_i = \frac{n_i}{N} , \qquad i = 1, 2 ,$$

is

$$\delta_o = \sum_{i=1}^{g} C_i \xi_i = \frac{12}{24}\big(0.7714 + 0.7714\big) = 0.7714 .$$

Since there are only

$$M = \frac{N!}{\prod_{i=1}^{g} n_i!} = \frac{24!}{12!\ 12!} = 2{,}704{,}156$$

possible, equally-likely arrangements of the $N = 24$ observed values listed in Fig. 7.13, an exact solution is possible.

If all arrangements of the $N = 24$ observed binary values listed in Fig. 7.13 occur with equal chance, the exact probability value of $\delta_o = 0.7714$ computed on the $M = 2{,}704{,}156$ possible arrangements of the observed values with $n_1 = n_2 = 12$ binary response measurements preserved for each arrangement is

$$P\big(\delta \le \delta_o | H_0\big) = \frac{\text{number of } \delta \text{ values} \le \delta_o}{M} = \frac{19{,}606}{2{,}704{,}156} = 0.0073 .$$

In this manner, the MRPP test statistic detects the selection differences in reading choices between girls and boys.

Following Eq. (7.5) on p. 369, the exact expected value of the $M = 2{,}704{,}156$ δ values is $\mu_\delta = 0.8907$ and, following Eq. (7.4) on p. 369, the observed chance-corrected measure of effect size is

$$\Re_o = 1 - \frac{\delta_o}{\mu_\delta} = 1 - \frac{0.7714}{0.8907} = +0.1339 ,$$

indicating approximately 13 % within-group agreement above that expected by chance.

7.3.3 Example Analysis 3

Consider a more substantial third example in which $N = 262$ Kansas pig farmers were asked, "What are your primary sources of veterinary information?" The farmers chose as many sources as applied from $r = 5$ response categories: (A) professional consultant, (B) veterinarian, (C) state or local extension service and agents, (D) magazines, and (E) feed companies and representatives. The farmers were also asked their highest attained level of education, providing $g = 5$ educational groups with $n_1 = 88$ high school (HS) graduates, $n_2 = 16$ vocational school (VS) graduates, $n_3 = 31$ 2-year college (2C) graduates, $n_4 = 113$ 4-year college (4C) graduates, and $n_5 = 14$ other (O) graduates. The data are adapted from Bilder et al. [43, p. 1287], and have been extensively analyzed and discussed in an r-dimensional contingency table context by Loughin and Scherer [246], Agresti and Liu [4], Bilder et al. [43], Decady and Thomas [92], and Bilder and Loughin [42]. In this example, the veterinary information data are analyzed in an r-dimensional binary-argument context.

Table 7.1 lists the observed frequencies and row vectors for the $2^r = 2^5 = 32$ possible response arrangements, where a 1 indicates that the source of information was used and a 0 indicates that the information source was not used.

For the data listed in Table 7.1, $r = 5$, $g = 5$, $v = 1$, $n_1 = 88$, $n_2 = 16$, $n_3 = 31$, $n_4 = 113$, $n_5 = 14$, $N = n_1 + n_2 + n_3 + n_4 + n_5 = 262$, and let

$$C_i = \frac{n_i}{N} , \qquad i = 1, \ldots, g ,$$

simply weighting each treatment group proportional to its size. Following Eq. (7.2) on p. 368, the $N = 262$ binary response measurements listed in Table 7.1 yield $g = 5$ average distance-function values of

$$\xi_1 = 1.4168 , \ \xi_2 = 1.3862 , \ \xi_3 = 1.3348 , \ \xi_4 = 1.3017 , \text{ and } \xi_5 = 1.5003 .$$

Following Eq. (7.1) on p. 368, the observed value of the MRPP test statistic based on $v = 1$ and treatment-group weights

$$C_i = \frac{n_i}{N} , \qquad i = 1, \ldots, 5 ,$$

Table 7.1 Frequencies of responses for the veterinary information data from five information sources: (A) professional consultant, (B) veterinarian, (C) state or local extension service or agents, (D) magazines, and (E) feed companies and representatives, and level of education scored as high school (HS), vocational school (VS), 2-year college (2C), 4-year college (4C), and other (O)

	Source of information					Level of education				
Vector	A	B	C	D	E	HS	VS	2C	4C	O
1	0	0	0	0	0	0	0	0	0	0
2	0	0	0	0	1	8	1	6	16	1
3	0	0	0	1	0	17	3	6	30	1
4	0	0	0	1	1	4	1	2	2	1
5	0	0	1	0	0	5	4	2	17	2
6	0	0	1	0	1	1	0	0	1	1
7	0	0	1	1	0	3	0	1	6	1
8	0	0	1	1	1	3	0	0	0	1
9	0	1	0	0	0	9	3	4	11	1
10	0	1	0	0	1	7	0	1	2	0
11	0	1	0	1	0	1	0	0	1	0
12	0	1	0	1	1	2	0	2	0	0
13	0	1	1	0	0	0	0	0	1	1
14	0	1	1	0	1	0	0	0	0	1
15	0	1	1	1	0	1	0	3	3	0
16	0	1	1	1	1	8	2	3	4	0
17	1	0	0	0	0	9	0	0	11	1
18	1	0	0	0	1	0	0	0	0	0
19	1	0	0	1	0	0	0	0	0	1
20	1	0	0	1	1	0	0	0	0	0
21	1	0	1	0	0	0	0	1	0	0
22	1	0	1	0	1	0	0	0	0	0
23	1	0	1	1	0	0	1	0	1	0
24	1	0	1	1	1	0	0	0	0	0
25	1	1	0	0	0	2	0	0	0	0
26	1	1	0	0	1	0	0	0	0	0
27	1	1	0	1	0	0	0	0	0	0
28	1	1	0	1	1	0	0	0	0	0
29	1	1	1	0	0	0	0	0	1	0
30	1	1	1	0	1	0	0	0	0	0
31	1	1	1	1	0	1	1	0	2	0
32	1	1	1	1	1	7	0	0	4	1

is

$$\delta_o = \sum_{i=1}^{g} C_i \xi_i = \frac{1}{262}\big[(88)(1.4168) + (16)(1.3862) + (31)(1.3348) + (113)(1.3017) + (14)(1.5003)\big] = 1.3601\,.$$

Since there are

$$M = \frac{N!}{\prod_{i=1}^{g} n_i!} = \frac{262!}{88!\ 16!\ 31!\ 113!\ 14!} \doteq 4.2196{\times}10^{144}$$

possible, equally-likely arrangements of the $N = 262$ observed values listed in Table 7.1, an exact solution is not possible.

If all M possible arrangements of the $N = 262$ observed binary values listed in Table 7.1 occur with equal chance, the approximate resampling probability value of $\delta_{\text{o}} = 1.3601$ computed on $L = 1{,}000{,}000$ random arrangements of the observed values with $n_1 = 88$, $n_2 = 16$, $n_3 = 31$, $n_4 = 113$, and $n_5 = 14$ binary response measurements preserved for each arrangement is

$$P(\delta \leq \delta_{\text{o}}|H_0) = \frac{\text{number of } \delta \text{ values} \leq \delta_{\text{o}}}{L} = \frac{127{,}263}{1{,}000{,}000} = 0.1273\ .$$

Here, the MRPP δ test statistic detected no definitive differences in sources of veterinary information among the five educational groups.

Following Eq. (7.5) on p. 369, the exact expected value of the M δ values is $\mu_\delta = 1.3663$ and, following Eq. (7.4) on p. 369, the observed chance-corrected measure of effect size is

$$\Re_{\text{o}} = 1 - \frac{\delta_{\text{o}}}{\mu_\delta} = 1 - \frac{1.3601}{1.3663} = +0.0046\ ,$$

indicating very little within-group agreement above that expected by chance.

7.4 Multivariate Measures of Association

A common problem in contemporary data analysis is the measurement of the magnitude of association between a nominal-level independent variable and a dependent variable that may be nominal-, ordinal-, or interval-level. Some representative examples are the measured associations between Religious Affiliation (e.g., Catholic, Jewish, Protestant) and Voting Behavior (e.g., Democrat, Republican, Libertarian, Independent); between Sex (Female, Male) and any attitudinal question that is Likert scaled (e.g., Strongly Agree, Agree, Neutral, Disagree, Strongly Disagree); and between Marital Status (Married, Single, Separated, Divorced, Widowed) and number of days of Work Missed in a year (0, 1, 2, ...).

Additionally, interest may be in the magnitude of association between a nominal-level independent variable and a multivariate dependent variable such as a subject's position in a three-dimensional matrix defined by Occupational Prestige, Income in Dollars, and Years of Education, where the researcher may not want to suffer the loss of information engendered by collapsing the three measurements into a univariate index of socioeconomic status. For a detailed description of problems with collapsing variables into a simple index, see Chap. 6, Sect. 6.1. In this section, a generalized measure of association for nominal independent variables is presented, in which any number and/or combination of interval-, ordinal-, or nominal-level dependent variables can be analyzed.

7.4.1 Interval Dependent Variables

Let $\Omega = \{\omega_1, \ldots, \omega_N\}$ indicate a finite collection of N objects, let $x'_j = (x_{1j}, \ldots, x_{rj})$ denote a transposed vector of r commensurate interval-level response measurements for object ω_j, $j = 1, \ldots, N$, and let $S_1, \ldots, S_g$ designate an exhaustive a priori partitioning of the N objects into g disjoint categories, where $n_i \geq 2$ is the number of objects in category S_i, $i = 1, \ldots, g$. In addition, let

$$\Delta(j,k) = \left(\sum_{i=1}^{r} \left| x_{ij} - x_{ik} \right|^p \right)^{v/p}$$

be a symmetric generalized Minkowski distance-function value of the r response measurements associated with objects ω_j and ω_k. Let

$$\xi_i = \binom{n_i}{2}^{-1} \sum_{j<k} \Delta(j,k)\, \Psi_i(\omega_j)\, \Psi_i(\omega_k)$$

represent the average between-object difference for all objects within category S_i, $i = 1, \ldots, g$, where $\sum_{j<k}$ is the sum over all j and k such that $1 \leq j < k \leq N$, and

$$\Psi_i(\omega_j) = \begin{cases} 1 & \text{if } \omega_j \in S_i \,, \\ 0 & \text{otherwise} \,. \end{cases}$$

Then the average within-category difference, weighted by the number of objects n_i in category i, $i = 1, \ldots, g$, is defined as

$$\delta = \sum_{i=1}^{g} C_i \xi_i \,,$$

where

$$C_i = \frac{n_i}{N} \,, \qquad i = 1, \ldots, g \,,$$

simply weighting each treatment group proportional to its size.

The null hypothesis (H_0) states that equal probabilities are assigned to each of the

$$M = \frac{N!}{\prod_{i=1}^{g} n_i!}$$

possible, equally-likely allocations of the N objects to categories $S_1, \ldots, S_g$. A chance-corrected within-category measure of association is given by

$$\Re = 1 - \frac{\delta}{\mu_\delta} , \tag{7.21}$$

where μ_δ is the arithmetic average of the M δ values calculated on all possible arrangements of the observed response measurement scores given by

$$\mu_\delta = \frac{1}{M} \sum_{i=1}^{M} \delta_i . \tag{7.22}$$

While the measure of association $\Re$ provides a description of the functional relationship between the nominal-level independent variable and the interval-level dependent variable(s), it does not indicate how extreme an observed value of $\Re$ is relative to the M possible values of $\Re$ under the null hypothesis. The probability value associated with an observed value of $\Re$, $\Re_o$, is the probability under the null hypothesis of observing a value of $\Re$ greater than or equal to $\Re_o$. Since μ_δ is invariant under the null hypothesis and δ is a simple linear transformation of $\Re$, i.e.,

$$\delta = \mu_\delta(1 - \Re) ,$$

the exact probability value for $\Re_o$ may be calculated in terms of δ_o and the M possible values of δ, e.g.,

$$P(\Re \geq \Re_o | H_0) = P(\delta \leq \delta_o | H_0) = \frac{\text{number of } \delta \text{ values} \leq \delta_o}{M} .$$

$\Re$ is a chance-corrected measure of association, reflecting the amount of association in excess of what would be expected by chance. $\Re$ attains a maximum value of unity when the association between the nominal independent variable and the interval dependent variable(s) is perfect, i.e., dependent variables scores are identical within each of the g categories of the nominal independent variable. $\Re$ attains a value of zero when the association is equal to chance, i.e., $\mathrm{E}[\Re | H_0] = 0$. Like all chance-corrected measures, $\Re$ occasionally will be slightly negative when the association is less than expected by chance. For a detailed description of chance-corrected measures, see Chap. 2, Sect. 2.2.1.

Example 1
The semantic differential (SD) is a rating scale designed to measure the connotative meaning of objects, events, and concepts. The SD scale measures reactions to stimulus words and concepts in terms of ratings on bipolar scales defined with contrasting adjectives at each end, e.g., good–bad, fast–slow, powerful–powerless [329]. Consider an example application from the semantic differential literature in which it is desired to measure the magnitude of association between Sex (nominal-level independent variable) and scores on three dimensions of the semantic differential (interval-level dependent variables). For this example, let $N = 15$ subjects, with $n_1 = 8$ Females and $n_2 = 7$ Males, and let $r = 3$ dimensions of the semantic differential: Evaluative, Potency, and Activity. The example data are adapted from Berry and Mielke [28, p. 44] and are listed in Fig. 7.14.

For the example data listed in Fig. 7.14, the $N = 15$ observations yield $g = 2$ distance-function values of

$$\xi_1 = 8.9158\times10^{-3} \quad \text{and} \quad \xi_2 = 5.9435\times10^{-3} \ .$$

The observed value of the MRPP test statistic based on $v = 1$ and

$$C_i = \frac{n_i}{N} \ , \qquad i = 1, 2 \ ,$$

is

$$\delta_o = \sum_{i=1}^{g} C_i\xi_i = \frac{1}{15}\Big[(8)(8.9158\times10^{-3}) + (7)(5.9435\times10^{-3})\Big]$$
$$= 7.5287\times10^{-3} \ .$$

Fig. 7.14 Example data with $N = 15$ subjects classified into $g = 2$ categories of a nominal-level independent variable, Female (F) and Male (M), and $r = 3$ dimensions of an interval-level dependent variable: Evaluative (E), Potency (P), and Activity (A)

		Scale		
Subject	Sex	E	P	A
1	F	4.5	5.5	3.9
2	F	2.4	6.0	2.7
3	F	2.7	5.8	3.8
4	F	3.6	6.5	4.5
5	F	4.3	5.6	4.0
6	F	2.5	5.9	2.8
7	F	2.8	5.7	4.0
8	F	3.5	6.4	4.4
9	M	6.4	3.5	6.1
10	M	5.6	4.2	5.5
11	M	5.2	3.1	5.6
12	M	6.2	3.6	6.0
13	M	5.7	4.3	5.7
14	M	5.2	3.0	5.8
15	M	6.1	3.6	6.2

Under the null hypothesis, there are

$$M = \frac{N!}{\prod_{i=1}^{g} n_i!} = \frac{15!}{8!\ 7!} = 6{,}435$$

possible, equally-likely arrangements of the $N = 15$ observed values listed in Fig. 7.14. Following Eq. (7.22), the expected value of the $M = 6{,}435$ δ values is $\mu_\delta = 1.7259\times10^{-2}$ and, following Eq. (7.21), the observed chance-corrected measure of nominal-interval association is

$$\Re_o = 1 - \frac{\delta_o}{\mu_\delta} = 1 - \frac{7.5287\times10^{-3}}{1.7259\times10^{-2}} = +0.5638\ ,$$

indicating approximately 56 % nominal-interval association above that expected by chance.

Since there are only $M = 6{,}435$ possible arrangements of the $N = 15$ observed values listed in Fig. 7.14, an exact solution is possible. If all arrangements of the observed values occur with equal chance, the exact probability value of $\Re_o = +0.5638$ computed on the $M = 6{,}435$ possible arrangements of the observed data with $n_1 = 8$ and $n_2 = 7$ values preserved for each arrangement is

$$P(\Re \geq \Re_o|H_0) = P(\delta \leq \delta_o|H_0) = \frac{\text{number of } \delta \text{ values} \leq \delta_o}{M}$$

$$= \frac{1}{6{,}435} = 1.5540\times10^{-4}\ .$$

7.4.2 Ordinal Dependent Variables

Researchers are often faced with the problem of measuring the magnitude of association between a nominal-level independent variable and ordinal-level dependent variables. Three measures of association have been advanced specifically for a nominal-level independent variable and a single ordinal-level dependent variable: Cureton's rank-biserial correlation coefficient (r_{rb}) [83, 84], Freeman's theta (θ_{ON}) [126], and Crittenden and Montgomery's ν [82], which is a modification of Freeman's θ_{ON} to ensure a proportional-reduction-in-error interpretation.

None of these measures has gained much popularity in the research literature. Cureton's rank-biserial correlation coefficient is defined only for a dichotomous nominal-level variable; consequently, its use is limited. As the sampling distributions of both Freeman's θ_{ON} and Crittenden and Montgomery's ν are unknown, associated tests of significance have not been developed.

Although the focus of this section is on measuring the association between a nominal-level independent variable and ordinal-level dependent variables, it should be noted that Hubert [187] defined θ_{NO}, a modification of Freeman's θ_{ON} for an ordinal independent variable and a nominal dependent variable. In addition, a symmetric version of Freeman's θ_{ON} was independently proposed by Särndal [362], Hubert [187], Agresti [1], and Crittenden and Montgomery [82], which they termed κ, θ_{sym}, $\bar{\delta}$, and I (Iota), respectively, although the sampling distributions remain unknown.

$\Re$ is directly applicable, without modification, to a nominal-level independent variable and any number of ordinal-level dependent variables. Ordinal-level variables, in this context, include the range of dependent variables from (1) fully ranked data where each subject is assigned a unique rank from 1 to N based on the conversion of original scores to ranks, to (2) having N objects associated with a limited number of ordinal categories. The second case differs from the first in that an investigator does not have original data to convert to ranks, but encounters only a crude ordering of the objects into categories, such as low, medium, and high, in the data collection process. In such a case, a simple assignment of ordered values (such as 1, 2, and 3, to low, medium, and high, respectively) to the categories may be used, rather than the values associated with tied ranks.

Example 2

Consider an example application in which it is desired to measure the magnitude of association between Political Affiliation (nominal-level independent variable) and scores on two dimensions of Socioeconomic Status (ordinal-level dependent variables). Let $N = 20$ subjects, with $n_1 = 8$ Democrats and $n_2 = 12$ Republicans, and let $r = 2$ dependent variables where one variable is Years of Education and the other variable is Occupational Prestige, both measured in quintiles. The data are adapted from Berry and Mielke [28, p. 47] and are listed in Fig. 7.15.

For the example data listed in Fig. 7.15, the $N = 20$ observations yield $g = 2$ distance-function values of

$$\xi_1 = 7.9204\times10^{-3} \quad \text{and} \quad \xi_2 = 4.7916\times10^{-3} .$$

The observed value of the MRPP test statistic based on $v = 1$ and

$$C_i = \frac{n_i}{N} , \qquad i = 1, 2 ,$$

is

$$\delta_o = \sum_{i=1}^{g} C_i \xi_i = \frac{1}{20}\Big[(8)(7.9204\times10^{-3}) + (12)(4.7916\times10^{-3})\Big]$$
$$= 6.0431\times10^{-3} .$$

Fig. 7.15 Example data with $N = 20$ subjects classified into $g = 2$ categories of a nominal-level independent variable, Democrat (D) and Republican (R), and $r = 2$ dimensions of an interval-level dependent variable (Education and Prestige)

Subject	Political Affiliation	Socioeconomic Status: Education	Socioeconomic Status: Prestige
1	D	5	3
2	D	4	5
3	D	5	4
4	D	2	3
5	D	2	5
6	D	3	4
7	D	4	2
8	D	2	4
9	R	2	1
10	R	2	1
11	R	1	2
12	R	3	1
13	R	1	2
14	R	2	1
15	R	1	2
16	R	1	1
17	R	3	1
18	R	1	2
19	R	2	3
20	R	3	2

There are

$$M = \frac{N!}{\prod_{i=1}^{g} n_i!} = \frac{20!}{8!\ 12!} = 125{,}970$$

possible, equally-likely arrangements of the $N = 20$ observed values listed in Fig. 7.15. Following Eq. (7.22), the expected value of the $M = 125{,}970$ δ values is $\mu_\delta = 8.3422\times10^{-3}$ and, following Eq. (7.21), the observed chance-corrected measure of nominal-ordinal association is

$$\mathfrak{R}_o = 1 - \frac{\delta_o}{\mu_\delta} = 1 - \frac{6.0431\times10^{-3}}{8.3422\times10^{-3}} = +0.2756\ ,$$

indicating approximately 28 % nominal-interval association above that expected by chance.

Since there are only $M = 125{,}970$ possible arrangements of the $N = 20$ observed values in Fig. 7.15, an exact solution is possible. If all arrangements of the observed values occur with equal chance, the exact probability value of $\Re_o = +0.2756$ computed on the $M = 125{,}970$ possible arrangements of the observed data with $n_1 = 8$ and $n_2 = 12$ preserved for each arrangement is

$$P(\Re \geq \Re_o | H_0) = P(\delta \leq \delta_o | H_0) = \frac{\text{number of } \delta \text{ values} \leq \delta_o}{M}$$

$$= \frac{2}{125{,}970} = 1.5877 \times 10^{-5} .$$

7.4.3 Nominal Dependent Variables

$\Re$ is easily adapted to measure the magnitude of association between a nominal-level independent variable and a nominal-level dependent variable. If the categories of the dependent variable are considered as r dimensions of that variable, then each object can be assigned a binary vector of length r with $r - 1$ values of 0 and a single value of 1 corresponding to the category of the dependent variable into which the object is classified, e.g., for four categories labeled 'A', 'B', 'C', and 'D' and an object that lies in category 'C', the transposed binary vector is $x' = [0\ 0\ 1\ 0]$. An alternative form of nominal data is the result of a question where a subject is asked to "Check all categories that apply." In this case, a vector is constructed in which a value of 1 is assigned to each checked category and a 0 is assigned to each unchecked category, e.g., for four categories labeled 'A', 'B', 'C', and 'D' and a subject who has checked categories 'A' and 'C', the transposed vector is $x' = [1\ 0\ 1\ 0]$. For a detailed description of multiple binary choices and "check all categories that apply," see Chap. 11, Sect. 11.4.2.

Example 3

Consider an example application in which it is desired to measure the magnitude of association between Rural/Urban Residence (nominal-level independent variable) and Marital Status (nominal-level dependent variable). Let $N = 24$ subjects, with $n_1 = 10$ rural residents (R) and $n_2 = 14$ urban residents (U), and let $r = 4$ dimensions of marital status: Single, Married, Widowed, and Divorced (S, M, W, D). The data are adapted from Berry and Mielke [28, p. 48] and are listed in Fig. 7.16.

For the example data listed in Fig. 7.16, the $N = 24$ observations yield $g = 2$ distance-function values of

$$\xi_1 = 4.9841 \times 10^{-3} \quad \text{and} \quad \xi_2 = 1.1814 \times 10^{-2} .$$

The observed value of the MRPP test statistic based on $v = 1$ and

$$C_i = \frac{n_i}{N} , \qquad i = 1, 2 ,$$

Fig. 7.16 Example data with $N = 24$ subjects classified into $g = 2$ categories of a nominal-level independent variable, Rural (R) and Urban (U), and $r = 4$ dimensions of a nominal-level dependent variable: Single (S), Married (M), Widowed (W), and Divorced (D)

		Marital Status			
Subject	Residence	S	M	W	D
1	R	0	0	0	1
2	R	0	1	0	0
3	R	0	1	0	0
4	R	0	1	0	0
5	R	0	1	0	0
6	R	0	1	0	0
7	R	0	1	0	0
8	R	0	1	0	0
9	R	0	1	0	0
10	R	1	0	0	0
11	U	0	0	0	1
12	U	0	0	0	1
13	U	0	0	0	1
14	U	0	1	0	0
15	U	0	1	0	0
16	U	1	0	0	0
17	U	1	0	0	0
18	U	1	0	0	0
19	U	1	0	0	0
20	U	1	0	0	0
21	U	0	0	1	0
22	U	0	0	1	0
23	U	0	0	1	0
24	U	0	0	1	0

is

$$\delta_o = \sum_{i=1}^{g} C_i \xi_i = \frac{1}{20}\Big[(10)(4.9841\times10^{-3}) + (14)(1.1814\times10^{-2})\Big]$$
$$= 8.9683\times10^{-3} .$$

There are

$$M = \frac{N!}{\prod_{i=1}^{g} n_i!} = \frac{24!}{10!\ 14!} = 1{,}961{,}256$$

possible, equally-likely arrangements of the $N = 24$ observed values listed in Fig. 7.16. Following Eq. (7.22), the expected value of the $M = 1{,}961{,}256$ δ values is $\mu_\delta = 1.0445\times10^{-2}$ and, following Eq. (7.21), the observed chance-corrected measure of nominal-ordinal association is

$$\Re_o = 1 - \frac{\delta_o}{\mu_\delta} = 1 - \frac{8.9683\times10^{-3}}{1.0445\times10^{-2}} = +0.1414 \, ,$$

indicating approximately 14 % nominal-nominal association above that expected by chance.

Since there are only $M = 1{,}961{,}256$ possible arrangements of the $N = 24$ observed values in Fig. 7.16, an exact solution is possible. If all arrangements of the observed values occur with equal chance, the exact probability value of $\Re_o = +0.1414$ computed on the $M = 1{,}961{,}256$ possible arrangements of the observed data with $n_1 = 10$ and $n_2 = 14$ preserved for each arrangement is

$$P(\Re \geq \Re_o | H_0) = P(\delta \leq \delta_o | H_0) = \frac{\text{number of } \delta \text{ values} \leq \delta_o}{M}$$

$$= \frac{7{,}192}{1{,}961{,}256} = 3.6670\times10^{-3} \, .$$

7.4.4 Mixed Dependent Variables

A distinctive advantage of the permutation approach to measuring association is the ability to analyze sets of dependent variables that are mixed: interval-, ordinal-, and/or nominal-level response measurements. Each interval- or ordinal-level dependent variable contributes one dimension to the analysis, and each nominal-level dependent variable contributes one dimension for each category of the variable.

Example 4

Consider an example application in which it is desired to measure the magnitude of association between Religious Affiliation (nominal-level independent variable) and Birth Experience, measured as a mixture of three dependent variables: one interval-level, one ordinal-level, and one nominal-level. Let $N = 15$ first-time mothers who have recently given birth, with $n_1 = 4$ Protestant mothers, $n_2 = 5$ Catholic mothers, and $n_3 = 6$ Jewish mothers. In addition, let $r = 5$ dimensions of the birth experience with Hours in Labor constituting the interval-level dependent variable; Birth Weight, measured as Above-normal (1), Normal (2), and Below-normal (3), constituting the ordinal-level dependent variable; and type of Anesthesia (Local, General, and None) constituting the nominal-level dependent variable. One of the $r = 5$ dimensions represents the interval-level dependent variable, one represents the ordinal-level dependent variable, and three (one for each category) represent the nominal-level dependent variable. The data are adapted from Berry and Mielke [28, p. 49] and are listed in Fig. 7.17.

Subject	Religion	Hours in Labor	Birth Weight	Anesthesia Local	Anesthesia General	Anesthesia None
1	P	20	3	0	0	1
2	P	15	3	0	1	0
3	P	10	2	0	0	1
4	P	8	3	0	1	0
5	C	10	3	0	1	0
6	C	8	2	0	1	0
7	C	8	2	0	1	0
8	C	6	1	0	1	0
9	C	5	1	0	0	1
10	J	12	3	1	0	0
11	J	10	2	1	0	0
12	J	5	1	0	1	0
13	J	5	1	1	0	0
14	J	5	1	1	0	0
15	J	4	1	1	0	0

Fig. 7.17 Example data with $N = 24$ subjects classified into $g = 3$ categories of a nominal-level independent variable, Protestant (P), Catholic (C), and Jewish (J), and $r = 5$ dimensions of mixed-level dependent variables: the interval-level variable is Hours in Labor, the ordinal-level variable is Birth Weight, and the nominal-level variable is Anesthesia measured as Local, General, or None

For the example data listed in Fig. 7.17, the $N = 15$ observations yield $g = 3$ distance-function values of

$$\xi_1 = 3.0327\times10^{-2}\ , \quad \xi_2 = 2.0490\times10^{-2}\ , \quad \text{and} \quad \xi_3 = 1.8444\times10^{-2}\ .$$

The observed value of the MRPP test statistic based on $v = 1$ and

$$C_i = \frac{n_i}{N}\ , \qquad i = 1, 2, 3\ ,$$

is

$$\delta_o = \sum_{i=1}^{g} C_i \xi_i = \frac{1}{15}\Big[(4)(3.0327\times10^{-2}) + (5)(2.0490\times10^{-2}) + (6)(1.8444\times10^{-2})\Big] = 2.2295\times10^{-2}\ .$$

There are

$$M = \frac{N!}{\displaystyle\prod_{i=1}^{g} n_i!} = \frac{15!}{4!\ 5!\ 6!} = 630{,}630$$

possible, equally-likely arrangements of the $N = 15$ observed values listed in Fig. 7.17. Following Eq. (7.22), the expected value of the $M = 630{,}630$ δ values is $\mu_\delta = 2.8029 \times 10^{-2}$ and, following Eq. (7.21), the observed chance-corrected measure of nominal-ordinal association is

$$\mathfrak{R}_o = 1 - \frac{\delta_o}{\mu_\delta} = 1 - \frac{2.2295 \times 10^{-2}}{2.8029 \times 10^{-2}} = +0.2046 ,$$

indicating approximately 20 % nominal-mixed association above that expected by chance.

Since there are only $M = 630{,}630$ possible arrangements of the $N = 15$ observed values in Fig. 7.17, an exact solution is possible. If all arrangements of the observed values occur with equal chance, the exact probability value of $\mathfrak{R}_o = +0.2046$ computed on the $M = 630{,}630$ possible arrangements of the observed data with $n_1 = 4$, $n_2 = 5$, and $n_3 = 6$ preserved for each arrangement is

$$P(\mathfrak{R} \geq \mathfrak{R}_o | H_0) = P(\delta \leq \delta_o | H_0) = \frac{\text{number of } \delta \text{ values} \leq \delta_o}{M}$$

$$= \frac{1{,}792}{630{,}630} = 2.8416 \times 10^{-3} .$$

7.5 Relationships Between $\mathfrak{R}$ and Existing Statistics

As is the case with any new statistical method, there exist certain relationships with prevailing methods. It should be noted that the preferred weighting function given by $C_i = n_i/N$ is simply the number of objects in the ith category of the nominal-level independent variable divided by the total number of objects. In the subsequent comparisons with existing methods, the maximum likelihood argument based on the normal distribution dictates that the weighting function must be defined as $C_i = (n_i - 1)/(N - g)$. This alternative weighting function is the number of degrees of freedom associated with the ith category of the nominal-level independent variable divided by the total degrees of freedom over all g categories.

In a permutation context, degrees of freedom are not relevant, as they are a consequence of fitting parameters in a maximum likelihood context. In addition, $v = 1$, which is associated with ordinary Euclidean distances, is replaced by $v = 2$, which also is a consequence of the maximum likelihood argument based on the normal distribution. Since the normal distribution assumption is irrelevant to permutation methods, the use of $v = 2$ is unjustified in a permutation context. Also, squared Euclidean distance with $v = 2$ yields questionable non-metric distance functions.

Finally, it should be noted that $\mathfrak{R}$ is a median-based measure of association when $v = 1$, whereas $\mathfrak{R}$ is a mean-based measure of association when $v = 2$. For clarification, consider the pairwise sum of univariate ($r = 1$) symmetric distance functions given by

$$\sum_{I<J} \Delta(I,J) = \sum_{I<J} \left| x_I - x_J \right|^v ,$$

where $x_1, \ldots, x_N$ are univariate response variables and $\sum_{I<J}$ is the sum over all I and J such that $1 \leq I < J \leq N$. Let $x_{1,N} \leq \cdots \leq x_{N,N}$ denote the order statistics associated with $x_1, \ldots, x_N$. If $v = 1$, then the inequality given by

$$\sum_{I=1}^{N} \left| N - 2I + 1 \right| \left| x_{I,N} - \theta \right| \geq \sum_{I<J} \left| x_I - x_J \right|$$

holds for all θ and equality holds if θ is the median of $x_1, \ldots, x_N$. On the other hand, if $v = 2$, then the inequality given by

$$N \sum_{I=1}^{N} \left(x_I - \theta \right)^2 \geq \sum_{I<J} \left(x_I - x_J \right)^2$$

holds for all θ and equality holds if θ is the mean of $x_1, \ldots, x_N$.

7.5.1 Interval-Level Dependent Variable

It can easily be shown that the permutation version of one-way analysis of variance (ANOVA) is a special case of the permutation approach to nominal-interval association with a single interval-level dependent variable. Specifically, the relationships between $\Re$ and the conventional F-ratio are given by

$$\Re = \frac{(F-1)(g-1)}{F(g-1)+N-g} \quad \text{and} \quad F = \frac{\Re(N-g)+g-1}{(1-\Re)(g-1)} ,$$

when $r = 1$, $v = 2$, and $C_i = (n_i - 1)/(N - g)$. In addition, Kelley's unbiased correlation ratio, ϵ^2, is identical to $\Re$ when $r = 1$, $v = 2$, and $C_i = (n_i - 1)/(N - g)$.[5] Since, in an analysis-of-variance context, ϵ^2 is identical to the adjusted squared correlation coefficient given by

$$\hat{R}^2 = 1 - \frac{(1-R^2)(N-1)}{N-p-1} ,$$

where R^2 is the squared Pearson product-moment correlation coefficient and p denotes the number of predictors, then $\hat{R}^2$, the unbiased estimator of R^2, is also

[5] Technically, Kelley's ϵ^2 is only unbiased under the permutation model of inference.

identical to $\Re$. For a detailed description of Kelley's ϵ^2 measure of effect size, see Chap. 3, Sect. 3.2.4.

Finally, the permutation version of one-way multivariate analysis of variance (MANOVA) is a special case of the permutation approach to nominal-interval association when $r \geq 2$, $v = 2$, $C_i = (n_i - 1)/(N - g)$, and

$$\Delta(j,k) = \left[(\mathbf{x}_j - \mathbf{x}_k)'\hat{\Sigma}^{-1}(\mathbf{x}_j - \mathbf{x}_k)\right]^{v/2} ,$$

where $\hat{\Sigma}$ denotes the $r \times r$ variance–covariance matrix [287].

7.5.2 Ordinal-Level Dependent Variable

It can be shown that the permutation version of the Kruskal–Wallis multi-sample rank-sum test statistic (H) is a special case of the permutation approach to nominal-ordinal association with a single ordinal-level dependent variable. Specifically,

$$\Re = \frac{(H-1)(g-1)}{H(g-1)+N-g} \quad \text{and} \quad H = \frac{\Re(N-g)+g-1}{(1-\Re)(g-1)} ,$$

when $r = 1$, $v = 2$, and $C_i = (n_i - 1)/(N - g)$. For a detailed description of the Kruskal–Wallis rank-sum test, see Chap. 5, Sect. 5.5.

7.5.3 Nominal-Level Dependent Variable

If $C_i = (n_i - 1)/(N - g)$, then

$$\Re = \frac{N-1}{N-g}\left(t - \frac{g-1}{N-1}\right) \quad \text{and} \quad t = \frac{\Re(N-g)+g-1}{N-1} ,$$

where t is Goodman and Kruskal's [151] t statistic associated with g categories of a nominal-level independent variable and r categories of a nominal-level dependent variable [24]. Note that degrees of freedom, from the maximum likelihood approach based on the normal distribution, is an integral component of Goodman and Kruskal's t statistic for cross-classified categorical data. For a detailed description of Goodman and Kruskal's t measure of association, see Sect. 7.2 of this chapter.

7.6 Coda

Chapter 7 utilized the Multi-Response Permutation Procedures (MRPP) developed in Chap. 2 to establish relationships between the test statistics of MRPP, δ and $\Re$, and selected conventional tests and measures designed for the analysis of completely randomized data at the nominal (categorical) level of measurement. Considered in this chapter were Goodman and Kruskal's t_a and t_b asymmetric measures of nominal association, Light and Margolin's categorical analysis of variance, tests to analyze multiple binary choices, and multivariate measures of association.

The relationships between the MRPP test statistic, δ, and Goodman and Kruskal's τ_a and τ_b asymmetrical measures of categorical association were explicated and exact permutation-based probability values were presented for both measures. The exact procedures provide more accurate probability values than those based on the conventional chi-squared procedure. Multiple binary choices are notoriously difficult to analyze. However, the MRPP test statistic solves the analysis problem, providing both an accurate permutation-based probability value and an appropriate chance-corrected measure of effect size.

Chapter 8

Chapter 8 introduces Multivariate Randomized Block Permutation (MRBP) procedures for univariate and multivariate randomized-block response measurement data and establishes the relationships between the MRBP test statistics, δ and $\Re$ developed in Chap. 8, and selected conventional tests and measures designed for the analysis of randomized-block data at the interval level of measurement in Chap. 9, the ordinal level of measurement in Chap. 10, and the nominal level of measurement in Chap. 11.

Randomized Block Data

8

This eighth chapter of *Permutation Statistical Methods* introduces a generalized Minkowski distance function and establishes the foundation for a set of Multivariate Randomized Block Permutation (MRBP) procedures for univariate and multivariate randomized-block data. MRBP procedures were introduced by Mielke and Iyer in 1982 and constitute a class of permutation methods for one or more response measurements among two or more treatments on the same or matched objects [299]. The MRBP procedures presented here provide a synthesizing foundation for a variety of statistical tests and measures that are further developed in Chaps. 9–11 for interval-, ordinal-, and nominal-level response measurements, respectively.

8.1 Multivariate Block Permutation Procedures

Suppose that a number of observed fields are compared to corresponding fields generated by one or more numerical models. Let the observed phenomena and the one or more numerical model predictions of these phenomena be termed "blocks," i.e., the first block might represent the observed phenomena and the remaining $b - 1$ blocks represent additional blocks, such as numerical model predictions of the phenomena for a total of $b \geq 2$ blocks. Also, let $r \geq 1$ denote the number of commensurate response measurements from each phenomenon and let $g \geq 2$ denote the number of phenomena, here called "treatments."

The terms representing "blocks" and "treatments" vary among disciplines. Oftentimes when $g = 2$ treatments and the same objects are represented in each treatment, the design is called a "before-and-after" or "subject-is-own-control" design. When $g = 2$ treatments and matched, but different, objects are represented in each treatment, the design is often called a "matched pairs" design. When $g > 2$ treatments and the same objects are represented in each treatment, the design is sometimes called a "repeated measures" design, and in this case the treatments are often labeled as "trials." Finally, in psychology randomized-block designs are known

K.J. Berry et al., *Permutation Statistical Methods*,
DOI 10.1007/978-3-319-28770-6_8

as "within-subjects" designs to distinguish them from completely randomized or "between-subjects" designs.

Let $x'_{ij} = (x_{1ij}, x_{2ij}, \ldots, x_{rij})$ denote a transposed vector of r response measurement scores associated with the ith treatment and jth block. Then the MRBP test statistic is given by

$$\delta = \left[g \binom{b}{2} \right]^{-1} \sum_{i=1}^{g} \sum_{j<k} \Delta(x_{ij}, x_{ik}) \,, \tag{8.1}$$

where $\sum_{j<k}$ denotes the sum over all j and k such that $1 \le j < k \le b$ and $\Delta(x, y)$ is a symmetric distance-function value of two points $x' = (x_1, x_2, \ldots, x_r)$ and $y' = (y_1, y_2, \ldots, y_r)$ in an r-dimensional Euclidean space. The generalized Minkowski distance function considered here is given by

$$\Delta(x, y) = \left(\sum_{i=1}^{r} |x_i - y_i|^p \right)^{v/p} \,, \tag{8.2}$$

where $p \ge 1$ and $v > 0$. Thus, $p = v = 2$ yields squared Euclidean distance, which is not a metric, and $p = 2$ and $v = 1$ yields ordinary Euclidean distance, which is a metric.[1]

The null hypothesis (H_0) states that the distribution of δ assigns an equal probability to each of the

$$M = (g!)^b$$

possible allocations of the r-dimensional response measurement scores to the g treatment positions within each of the b blocks. Consequently, the collection of r response measurement scores within each block yields g r-dimensional exchangeable random variables under the null hypothesis. The probability value associated with an observed value of δ, say δ_o, is the probability under the null hypothesis (H_0) of observing a value of δ as extreme or more extreme than δ_o. Thus, an exact probability value for δ_o may be expressed as

$$P(\delta \le \delta_o | H_0) = \frac{\text{number of } \delta \text{ values} \le \delta_o}{M} \,.$$

[1]Recall that a distance function is a metric if it satisfies three properties given by (1) $\Delta(x, y) \ge 0$ and $\Delta(x, x) = 0$, i.e., the distance is positive between two different points and is equal to zero from any point to itself; (2) the distance is symmetric: $\Delta(x, y) = \Delta(y, x)$, i.e., the distance between points x and y is the same in either direction; and (3) the triangle inequality is satisfied: $\Delta(x, y) \le \Delta(x, z) + \Delta(z, y)$, i.e., the distance between any two points is the shortest distance along any path.

When M is very large, an approximate probability value for δ may be obtained from a resampling procedure, where

$$P(\delta \leq \delta_o | H_0) = \frac{\text{number of } \delta \text{ values} \leq \delta_o}{L} ,$$

and L denotes the number of randomly sampled test statistic values. Typically, L is set to a large value to ensure accuracy, e.g., $L = 1{,}000{,}000$. When M is very large and P is exceedingly small, a resampling-approximation permutation procedure may produce no δ values equal to or less than δ_o, even with $L = 1{,}000{,}000$, yielding an approximate resampling probability value of $P = 0.00$. In such cases, moment-approximation permutation procedures based on fitting the first three exact moments of the discrete permutation distribution to a Pearson type III distribution provide approximate probability values, as detailed in Chap. 1, Sect. 1.2.2 [284, 299].

As with MRPP, discussed in Chap. 2, a chance-corrected measure of agreement among all b blocks for all g treatments constitutes a universal measure of effect size for all randomized-block designs and is given by

$$\Re = 1 - \frac{\delta}{\mu_\delta} , \tag{8.3}$$

where μ_δ is the arithmetic average of the M δ values calculated on all possible arrangements of the observed response measurement scores given by

$$\mu_\delta = \frac{1}{M} \sum_{i=1}^{M} \delta_i . \tag{8.4}$$

Because μ_δ is a constant under H_0, the permutation distributions of δ and $\Re$ are equivalent, viz.,

$$P(\delta \leq \delta_o | H_0) = P(\Re \geq \Re_o | H_0) ,$$

where

$$\Re_o = 1 - \frac{\delta_o}{\mu_\delta}$$

and δ_o and $\Re_o$ denote the observed values of δ and $\Re$, respectively.

As with the chance-corrected within-group agreement measure presented in Chap. 2, the values of $\Re$ range from negative values to $\Re = +1$ when perfect agreement is achieved, the expected value of $\Re$ is zero under the null hypothesis, and agreement or disagreement is implied by $\Re > 0$ and $\Re < 0$, respectively. While probability values are highly dependent on sample size, this sample-size dependence does not hold for the chance-corrected within-block agreement measure, $\Re$.

8.1.1 Randomized-Block Designs and Alignment

For certain response patterns involving randomized-block designs, the observed test statistic δ_o, as defined in Eq. (8.1) on p. 422, is unable to detect treatment differences [299, pp. 1434–1435]. Such situations occur when the magnitude of the block differences exceeds the magnitude of the treatment differences. For a simple example, consider the univariate response measurement scores listed in Fig. 8.1 with $b = 2$ blocks and $g = 3$ treatments. If $p = 2$ and $v = 1$ in Eq. (8.2) on p. 422, then $\delta_o = 4$ and the random variable δ is also equal to 4 for all permutations of values within blocks; thus, the probability of $\delta_o = 4$ is 1. It is therefore impossible to detect treatment differences.

This problem is rectified by aligning the response measurement scores within each block, a technique initially described by Hodges and Lehmann in 1962 [178]. Alignment is accomplished for the example data in Fig. 8.1 by replacing x_{ij} with $x_{ij} - x_j^*$, where x_j^* is the median of $(x_{ij}, \ldots, x_{gj})$ for $j = 1, \ldots, b$.[2] The observed statistic δ_o is then computed on the aligned data. The median values for Blocks 1 and 2 in Fig. 8.1 are 2 and 6, respectively. If the median value is subtracted from the values in each block, the aligned data are then $y_{111} = 1 - 2 = -1$, $y_{112} = 2 - 2 = 0$, and $y_{113} = 3 - 2 = +1$ for Block 1, and $y_{121} = 5 - 6 = -1$, $y_{122} = 6 - 6 = 0$, and $y_{123} = 7 - 6 = +1$ for Block 2. The median-aligned data are given in Fig. 8.2.

After alignment $\delta_o = 0$ while the random variable δ assumes the values 0.00, 0.67, and 1.33 with respective probability values of 0.1667, 0.3333, and 0.5000, under the null hypothesis (the probability of δ_o is 1/6 after alignment). Note that if $v = 2$ and $r = 1$, the inferential results based on the random variable δ remain unaffected by the alignment.

Fig. 8.1 Example of unaligned data with $g = 3$ treatments, $b = 2$ blocks, and $r = 1$ response measurement

	Treatment		
Block	1	2	3
1	1	2	3
2	5	6	7

Fig. 8.2 Example of aligned data with $g = 3$ treatments, $b = 2$ blocks, and $r = 1$ response measurement

	Treatment		
Block	1	2	3
1	−1	0	+1
2	−1	0	+1

[2] In their 1982 article introducing MRBP, Mielke and Iyer initially suggested using the arithmetic mean instead of the median [299, p. 1435].

8.1.2 Example Univariate MRBP Analysis with $v = 2$

To illustrate an MRBP analysis with univariate response measurement scores and $v = 2$, consider a test of difference between $g = 2$ treatments, where a single response measurement has been obtained from each of $b = 4$ subjects, such as in a matched-pairs experimental design. For this example, there is $r = 1$ response measurement for each subject and $g = 2$ treatments for each of $b = 4$ blocks. The numbers of blocks, treatments, and response measurements are deliberately kept small to simplify the example analysis. The treatments and univariate response measurement scores are listed in Fig. 8.3.

Thus, following Eq. (8.2) on p. 422 for the univariate response measurement scores listed in Fig. 8.3 with $g = 2$, $r = 1$, $b = 4$, $p = 2$, and $v = 2$, the generalized Minkowski distance function yields

$$\Delta(1,2) = \left[\left|(255-171)-(294-202)\right|^2\right]^{2/2} = 64.00\,,$$

$$\Delta(1,3) = \left[\left|(255-171)-(259-247)\right|^2\right]^{2/2} = 5{,}184.00\,,$$

$$\Delta(1,4) = \left[\left|(255-171)-(263-182)\right|^2\right]^{2/2} = 9.00\,,$$

$$\Delta(2,3) = \left[\left|(294-202)-(259-247)\right|^2\right]^{2/2} = 6{,}400.00\,,$$

$$\Delta(2,4) = \left[\left|(294-202)-(263-182)\right|^2\right]^{2/2} = 121.00\,,$$

and

$$\Delta(3,4) = \left[\left|(259-247)-(263-182)\right|^2\right]^{2/2} = 4{,}761.00\,.$$

When $r = 1$ and $g = 2$, Eq. (8.1) on p. 422 reduces to

$$\delta = \binom{b}{2}^{-1} \sum_{j<k} \Delta(x_j - x_k)\,. \tag{8.5}$$

Fig. 8.3 Example univariate data with $g = 2$ treatments, $b = 4$ blocks, and $r = 1$ response measurement

	Treatment	
Block	1	2
1	255	171
2	294	202
3	259	247
4	263	182

Then,

$$\delta = \binom{b}{2}^{-1} \left[\Delta(1,2) + \Delta(1,3) + \Delta(1,4) + \Delta(2,3) + \Delta(2,4) + \Delta(3,4)\right]$$

and the observed value of the MRBP test statistic with $v = 2$ is

$$\delta_o = \binom{4}{2}^{-1} (64.00 + 5{,}184.00 + 9.00 + 6{,}400.00 + 121.00 + 4{,}761.00)$$
$$= \frac{1}{6}(16{,}539.00) = 2{,}756.50\,.$$

Let δ_1 denote the MRBP test statistic for a matched-pairs t test with b blocks, $g = 2$ treatments, and $v = 2$, and let δ_2 denote the MRPP test statistic for a two-sample t test with $g = 2$ treatments, $v = 2$, $n_1 = n_2$, $C_1 = (n_1 - 1)/(N - g)$, and $C_2 = (n_2 - 1)/(N - g)$, where n_1 and n_2 denote the number of objects in treatments 1 and 2, respectively, and $N = n_1 + n_2$. Then the relationship between δ_1 and δ_2 is given by

$$\delta_1 = 2\left(\delta_2 - r_{12}\sqrt{\xi_1\xi_2}\right), \qquad (8.6)$$

where ξ_i, $i = 1, 2$, are the average distance-function values for treatments 1 and 2, respectively, and r_{12} is the Pearson product-moment correlation coefficient calculated on the response measurement scores in treatments 1 and 2. See Chap. 2, Sect. 2.2 for detailed descriptions of ξ_i, $i = 1, 2$, and δ. For the interval-level response measurement scores listed in Fig. 8.3, the sample variances for treatments 1 and 2 are $s_1^2 = 316.9167$ and $s_2^2 = 1{,}125.6667$, $\xi_1 = 2s_1^2 = 2(316.9167) = 633.8333$, $\xi_2 = 2s_2^2 = 2(1{,}125.6667) = 2{,}251.3333$, $r_{12} = +0.0539$, $C_1 = (n_1 - 1)/(N - g) = (4 - 1)/(8 - 1)$, $C_2 = (n_2 - 1)/(N - g) = (4 - 1)/(8 - 2)$, and

$$\delta_2 = \sum_{i=1}^{g} C_i\xi_i = \frac{4-1}{8-2}(633.8333) + \frac{4-1}{8-2}(2{,}251.3333) = 1{,}442.5833\,.$$

Then, following Eq. (8.6),

$$\delta_1 = 2\left(\delta_2 - r_{12}\sqrt{\xi_1\xi_2}\right)$$
$$= 2\left[1{,}442.5833 - 0.0539\sqrt{(633.8333)(2{,}251.3333)}\right]$$
$$= 2{,}756.50\,.$$

The

$$M = (g!)^b = (2!)^4 = 16$$

possible, equally-likely arrangements of the observed univariate response measurement scores described in Fig. 8.3 on p. 425 are listed in Table 8.1 and are ordered by the δ values from lowest to highest.

The observed MRBP test statistic, $\delta_o = 2{,}756.50$, obtained from the original arrangement of the $N = 8$ univariate response measurement scores in Treatments 1 and 2,

$$\{255, 294, 259, 263\} \quad \{171, 202, 247, 182\} ,$$

(Order 1 in Table 8.1) is unusual since 14 of the 16 δ values exceed the observed value of $\delta_o = 2{,}756.50$ and only two values of δ are equal to or less than the observed value.

If all arrangements of the $N = 8$ observed univariate response measurement scores listed in Fig. 8.3 occur with equal chance, the exact probability value of $\delta_o = 2{,}756.50$ computed on the $M = 16$ possible arrangements of the observed response measurement scores with $b = 4$ blocks preserved for each arrangement is

$$P(\delta \leq \delta_o | H_0) = \frac{\text{number of } \delta \text{ values} \leq \delta_o}{M} = \frac{2}{16} = 0.1250 .$$

Table 8.1 Permutations of the observed univariate response measurement scores listed in Fig. 8.3 with values for δ based on $v = 2$ ordered from lowest to highest

Order	Treatment 1	Treatment 2	δ
1	{255, 294, 259, 263}	{171, 202, 247, 182}	2,756.50
2	{171, 202, 247, 182}	{255, 294, 259, 263}	2,756.50
3	{255, 294, 247, 263}	{171, 202, 259, 182}	4,812.50
4	{171, 202, 259, 182}	{255, 294, 247, 263}	4,812.50
5	{171, 202, 247, 263}	{255, 294, 259, 182}	12,908.50
6	{255, 294, 259, 182}	{171, 202, 247, 263}	12,908.50
7	{171, 294, 259, 263}	{255, 202, 247, 182}	13,116.50
8	{255, 202, 247, 182}	{171, 294, 259, 263}	13,116.50
9	{255, 202, 259, 263}	{171, 294, 247, 182}	13,612.50
10	{171, 294, 247, 182}	{255, 202, 259, 263}	13,612.50
11	{171, 202, 259, 263}	{255, 294, 247, 182}	13,668.50
12	{255, 294, 247, 182}	{171, 202, 259, 263}	13,668.50
13	{171, 294, 247, 263}	{255, 202, 259, 182}	13,828.50
14	{255, 202, 259, 182}	{171, 294, 247, 263}	13,828.50
15	{255, 202, 247, 263}	{171, 294, 259, 182}	14,196.50
16	{171, 294, 259, 182}	{255, 202, 247, 263}	14,196.50

For comparison, a conventional matched-pairs t test calculated on the $b = 4$ pairs of response measurement scores listed in Fig. 8.3 yields an observed test statistic of $t_o = +3.6229$. Assuming independence and normality, t is approximately distributed as Student's t under the null hypothesis with $b - 1 = 4 - 1 = 3$ degrees of freedom. Under the null hypothesis, the observed value of $t_o = +3.6229$ yields an approximate two-sided probability value of $P = 0.0362$. Note the large difference between the conventional approximate probability value of $P = 0.0362$ and the exact permutation probability value of $P = 0.1250$. Such discrepancies are common when the number of blocks is small, as in this case with $b = 4$.

The total of the $M = 16$ δ values listed in Table 8.1 is 177,800. Thus, following Eq. (8.4) on p. 423, the exact average value of the $M = 16$ δ values listed in Table 8.1 is

$$\mu_\delta = \frac{1}{M}\sum_{i=1}^{M}\delta_i = \frac{1}{16}(177{,}800) = 11{,}112.50\ .$$

Following Eq. (8.3) on p. 423, the observed chance-corrected measure of effect size is

$$\mathfrak{R}_o = 1 - \frac{\delta_o}{\mu_\delta} = 1 - \frac{2{,}756.50}{11{,}112.50} = +0.7519\ ,$$

indicating approximately 75 % within-block agreement above that expected by chance.[3]

8.1.3 Example Univariate MRBP Analysis with $v = 1$

Because permutation statistical tests are data-dependent, distribution-free, and nonparametric, they require no distributional assumptions and make no estimates of population parameters. Consequently, it is not necessary to set $v = 2$, squaring the response-measurement differences between objects. As with MRPP in Chap. 2, a distance function based on $v = 1$, employing ordinary Euclidean distance between response measurement scores, is an attractive alternative to $v = 2$ as it is a metric, satisfies the triangle inequality, is robust to extreme values, provides an easy-to-understand Euclidean distance between objects, and ensures that the data and analysis spaces are congruent.

[3]The astute reader will have noted that the values of the generalized chance-corrected measure of agreement, $\mathfrak{R}$, are, in general, markedly greater in Chap. 8 than in Chaps. 2–7. Because Chaps. 8–11 analyze randomized-block data, there is less variability to be explained due to the matching of objects or subjects and, therefore, more agreement (less disagreement) between treatments than with the completely randomized designs analyzed in Chaps. 2–7.

To illustrate the computation of the MRBP test statistic with $v = 1$, consider the same finite sample of response measurement scores obtained from the $b = 4$ subjects listed in Fig. 8.3. For these data, there is $r = 1$ response measurement for each subject and $g = 2$ treatments for each of $b = 4$ blocks.

Following Eq. (8.2) on p. 422 for the data listed in Fig. 8.3 with $g = 2$, $r = 1$, $b = 4$, $p = 1$, and $v = 1$, the generalized Minkowski distance function yields

$$\Delta(1,2) = \left[\left| (255 - 171) - (294 - 202) \right|^2 \right]^{1/2} = 8.00 ,$$

$$\Delta(1,3) = \left[\left| (255 - 171) - (259 - 247) \right|^2 \right]^{1/2} = 72.00 ,$$

$$\Delta(1,4) = \left[\left| (255 - 171) - (263 - 182) \right|^2 \right]^{1/2} = 3.00 ,$$

$$\Delta(2,3) = \left[\left| (294 - 202) - (259 - 247) \right|^2 \right]^{1/2} = 80.00 ,$$

$$\Delta(2,4) = \left[\left| (294 - 202) - (263 - 182) \right|^2 \right]^{1/2} = 11.00 ,$$

and

$$\Delta(3,4) = \left[\left| (259 - 247) - (263 - 182) \right|^2 \right]^{1/2} = 69.00 .$$

Then, following Eq. (8.5) on p. 425,

$$\delta = \binom{b}{2}^{-1} \left[\Delta(1,2) + \Delta(1,3) + \Delta(1,4) + \Delta(2,3) + \Delta(2,4) + \Delta(3,4) \right]$$

and the observed value of the MRBP test statistic with $v = 1$ is

$$\delta_{\text{o}} = \binom{4}{2}^{-1} (8.00 + 72.00 + 3.00 + 80.00 + 11.00 + 69.00)$$

$$= \frac{1}{6} (243.00) = 40.50 .$$

The

$$M = (g!)^b = (2!)^4 = 16$$

Table 8.2 Permutations of the observed univariate response measurement scores listed in Fig. 8.3 with values for δ based on $v = 1$ ordered from lowest to highest

Order	Treatment 1	Treatment 2	δ
1	{255, 294, 259, 263}	{171, 202, 247, 182}	40.50
2	{171, 202, 247, 182}	{255, 294, 259, 263}	40.50
3	{255, 294, 247, 263}	{171, 202, 259, 182}	52.50
4	{171, 202, 259, 182}	{255, 294, 247, 263}	52.50
5	{171, 202, 247, 263}	{255, 294, 259, 182}	98.50
6	{255, 294, 259, 182}	{171, 202, 247, 263}	98.50
7	{171, 294, 259, 263}	{255, 202, 247, 182}	99.50
8	{255, 202, 247, 182}	{171, 294, 259, 263}	99.50
9	{255, 202, 259, 263}	{171, 294, 247, 182}	99.50
10	{171, 294, 247, 182}	{255, 202, 259, 263}	99.50
11	{171, 202, 259, 263}	{255, 294, 247, 182}	102.50
12	{255, 294, 247, 182}	{171, 202, 259, 263}	102.50
13	{171, 294, 247, 263}	{255, 202, 259, 182}	103.50
14	{255, 202, 259, 182}	{171, 294, 247, 263}	103.50
15	{255, 202, 247, 263}	{171, 294, 259, 182}	103.50
16	{171, 294, 259, 182}	{255, 202, 247, 263}	103.50

possible, equally-likely arrangements of the observed response measurement scores described in Fig. 8.3 are listed in Table 8.2 and are ordered by the δ values from lowest to highest.

The observed MRBP test statistic, $\delta_o = 40.50$, obtained from the original arrangement of the $N = 8$ univariate response measurement scores in Treatments 1 and 2,

$$\{255, 294, 259, 263\} \quad \{171, 202, 247, 182\} \, ,$$

(Order 1 in Table 8.2) is unusual since 14 of the 16 δ values exceed the observed value of $\delta_o = 40.50$ and only two values of δ are equal to or less than the observed value.

If all arrangements of the $N = 8$ observed univariate response measurement scores listed in Fig. 8.3 occur with equal chance, the exact probability value of $\delta_o = 40.50$ computed on the $M = 16$ possible arrangements of the observed response measurement scores with $b = 4$ blocks preserved for each arrangement is

$$P(\delta \leq \delta_o | H_0) = \frac{\text{number of } \delta \text{ values} \leq \delta_o}{M} = \frac{2}{16} = 0.1250 \, .$$

The fact that both $v = 2$ and $v = 1$ yield the same probability value of $P = 0.1250$ is simply an artifact of the small data set given in Fig. 8.3 and is not, in general, to be expected. No comparison is made with Student's matched-pairs t test as Student's t test is undefined for $v = 1$.

The total of the $M = 16$ δ values listed in Table 8.2 is 1,400. Thus, following Eq. (8.4) on p. 423, the exact average value of the $M = 16$ δ values listed in Table 8.2 is

$$\mu_\delta = \frac{1}{M}\sum_{i=1}^{M}\delta_i = \frac{1}{16}(1{,}400) = 87.50 \ .$$

Following Eq. (8.3) on p. 423, the observed chance-corrected measure of effect size is

$$\Re_{\mathrm{o}} = 1 - \frac{\delta_{\mathrm{o}}}{\mu_\delta} = 1 - \frac{40.50}{87.50} = +0.5371 \ ,$$

indicating approximately 54 % within-block agreement above that expected by chance.

8.1.4 Example Bivariate MRBP Analysis with $v = 2$

In this example, bivariate response measurement scores are used for simplicity to demonstrate a multivariate MRBP analysis. Consider a test of difference between $g = 2$ treatments, where bivariate response measurement scores have been obtained from each of $b = 4$ subjects, such as in a matched-pairs experimental design. For this example, there are $r = 2$ response measurement scores for each subject and $g = 2$ treatments for each of $b = 4$ blocks. The number of blocks, treatments, and response measurement scores are deliberately kept small to simplify the example analysis. The treatments and response measurement scores are listed in Fig. 8.4.

Following Eq. (8.2) on p. 422 for the Treatment 1 response measurement scores listed in Fig. 8.4 with $b = 4$ blocks, $r = 2$ response measurements, $p = 2$, and $v = 2$, the generalized Minkowski distance function yields

$$\Delta(1,2) = \Big[|73-59|^2 + |64-57|^2\Big]^{2/2} = 245.00 \ ,$$

$$\Delta(1,3) = \Big[|73-46|^2 + |64-35|^2\Big]^{2/2} = 1{,}570.00 \ ,$$

Fig. 8.4 Example bivariate response measurement scores with $g = 2$ treatments, $b = 4$ blocks, and $r = 2$ response measurements

	Treatment	
Block	1	2
1	(73, 64)	(23, 47)
2	(59, 57)	(21, 43)
3	(46, 35)	(19, 31)
4	(23, 11)	(16, 28)

$$\Delta(1,4) = \left[|73-23|^2 + |64-11|^2\right]^{2/2} = 5{,}309.00\,,$$

$$\Delta(2,3) = \left[|59-46|^2 + |57-35|^2\right]^{2/2} = 653.00\,,$$

$$\Delta(2,4) = \left[|59-23|^2 + |57-11|^2\right]^{2/2} = 3{,}412.00\,,$$

and

$$\Delta(3,4) = \left[|46-23|^2 + |35-11|^2\right]^{2/2} = 1{,}105.00\,,$$

and for the Treatment 2 response measurement scores listed in Fig. 8.4, the generalized Minkowski distance function yields

$$\Delta(1,2) = \left[|23-21|^2 + |47-43|^2\right]^{2/2} = 20.00\,,$$

$$\Delta(1,3) = \left[|23-19|^2 + |47-31|^2\right]^{2/2} = 272.00\,,$$

$$\Delta(1,4) = \left[|23-16|^2 + |47-28|^2\right]^{2/2} = 410.00\,,$$

$$\Delta(2,3) = \left[|21-19|^2 + |43-31|^2\right]^{2/2} = 148.00\,,$$

$$\Delta(2,4) = \left[|21-16|^2 + |43-28|^2\right]^{2/2} = 250.00\,,$$

and

$$\Delta(3,4) = \left[|19-16|^2 + |31-28|^2\right]^{2/2} = 18.00\,.$$

Then, following Eq. (8.1) on p. 422,

$$\delta = \left[g\binom{b}{2}\right]^{-1}\left[\Delta(1,2) + \Delta(1,3) + \cdots + \Delta(2,4) + \Delta(3,4)\right]$$

and the observed value of the MRBP test statistic with $v = 2$ is

$$\delta_o = \left[2\binom{4}{2}\right]^{-1}(245.00 + 1{,}570.00 + \cdots + 250.00 + 18.00)$$

$$= \frac{1}{12}(13{,}412.00) = 1{,}117.6667\,.$$

In permutation analyses of randomized-block designs it is not always necessary to enumerate all

$$M = (g!)^b$$

possible, equally-likely arrangements of the observed data. It is obvious from a close inspection of Tables 8.1 and 8.2 that half of the arrangements are redundant, yielding duplicate δ values. Considerable savings in computing time can be achieved by eliminating the redundancy and computing only the

$$M = (g!)^{b-1} = (2!)^{4-1} = 8$$

non-redundant arrangements of the observed data.[4]

The $M = 8$ non-redundant, equally-likely arrangements of the observed response measurement scores described in Fig. 8.4 are listed in Table 8.3 and are ordered by the δ values from lowest to highest.

The observed MRBP test statistic, $\delta_o = 1{,}117.6667$, obtained from the original arrangement of the $N = 8$ bivariate response measurement scores in Treatments 1 and 2,

$$\{(73, 64)(59, 57)(46, 35)(23, 11)\} \quad \{(23, 47)(21, 43)(19, 31)(16, 28)\} ,$$

(Order 1 in Table 8.3) is unusual since seven of the eight δ values exceed the observed δ_o value of 1,117.6667 and only one δ value is equal to or less than the observed value.

If all non-redundant arrangements of the $N = 8$ observed bivariate response measurement scores listed in Fig. 8.4 occur with equal chance, the exact probability value of $\delta_o = 1{,}117.6667$ computed on the $M = 8$ arrangements of the observed

Table 8.3 Permutations of the observed bivariate data listed in Fig. 8.4 with values for δ based on $v = 2$ ordered from lowest to highest

Order	Treatment 1	Treatment 2	δ
1	{(73, 64)(59, 57)(46, 35)(23, 11)}	{(23, 47)(21, 43)(19, 31)(16, 28)}	1,117.6667
2	{(73, 64)(59, 57)(46, 35)(16, 28)}	{(23, 47)(21, 43)(19, 31)(23, 11)}	1,152.6667
3	{(73, 64)(59, 57)(19, 31)(16, 28)}	{(23, 47)(21, 43)(46, 35)(23, 11)}	1,549.1667
4	{(73, 64)(59, 57)(19, 31)(23, 11)}	{(23, 47)(21, 43)(46, 35)(16, 28)}	1,554.5000
5	{(73, 64)(21, 43)(46, 35)(23, 11)}	{(23, 47)(59, 57)(19, 31)(16, 28)}	1,659.0000
6	{(73, 64)(21, 43)(46, 35)(16, 28)}	{(23, 47)(59, 57)(19, 31)(23, 11)}	1,684.6667
7	{(73, 64)(21, 43)(19, 31)(16, 28)}	{(23, 47)(59, 57)(46, 35)(23, 11)}	1,720.5000
8	{(73, 64)(21, 43)(19, 31)(23, 11)}	{(23, 47)(59, 57)(46, 35)(16, 28)}	1,735.1667

[4]This was a simplification used as far back as 1933 by Eden and Yates in their randomized-block analysis of Yeoman II wheat shoots [103].

response measurement scores with $b = 4$ blocks preserved for each arrangement is

$$P(\delta \leq \delta_o | H_0) = \frac{\text{number of } \delta \text{ values} \leq \delta_o}{M} = \frac{1}{8} = 0.1250\ .$$

The total of the $M = 8$ δ values listed in Table 8.3 is 12,173.3333. Thus, following Eq. (8.4) on p. 423, the exact average value of the $M = 8$ δ values listed in Table 8.3 is

$$\mu_\delta = \frac{1}{M}\sum_{i=1}^{M} \delta_i = \frac{1}{8}(12{,}173.3333) = 1{,}521.6667\ .$$

Following Eq. (8.3) on p. 423, the observed chance-corrected measure of effect size is

$$\Re_o = 1 - \frac{\delta_o}{\mu_\delta} = 1 - \frac{1{,}117.6667}{1{,}521.6667} = +0.2655\ ,$$

indicating approximately 27 % within-block agreement above that expected by chance.

8.1.5 Example Bivariate MRBP Analysis with $v = 1$

As explained in previous examples, there is no need to square differences when employing permutation tests. To illustrate the computation of MRBP with bivariate response measurement scores and $v = 1$, employing ordinary Euclidean distance instead of squared Euclidean distance between response measurement scores, consider the same finite sample of $b = 4$ subjects listed in Fig. 8.4.

Following Eq. (8.2) on p. 422 for the Treatment 1 response measurement scores listed in Fig. 8.4 with $g = 2$ treatments, $b = 4$ blocks, $r = 2$ response measurements, $p = 2$, and $v = 1$, the generalized Minkowski distance function yields

$$\Delta(1,2) = \left[|73-59|^2 + |64-57|^2\right]^{1/2} = 15.6525\ ,$$

$$\Delta(1,3) = \left[|73-46|^2 + |64-35|^2\right]^{1/2} = 39.6232\ ,$$

$$\Delta(1,4) = \left[|73-23|^2 + |64-11|^2\right]^{1/2} = 72.8629\ ,$$

$$\Delta(2,3) = \left[|59-46|^2 + |57-35|^2\right]^{1/2} = 25.5539\ ,$$

$$\Delta(2,4) = \left[|59-23|^2 + |57-11|^2\right]^{1/2} = 58.4123\ ,$$

and

$$\Delta(3,4) = \left[|46-23|^2 + |35-11|^2\right]^{1/2} = 33.2415\,,$$

and for the Treatment 2 response measurement scores listed in Fig. 8.4, the generalized Minkowski distance function yields

$$\Delta(1,2) = \left[|23-21|^2 + |47-43|^2\right]^{1/2} = 4.4721\,,$$

$$\Delta(1,3) = \left[|23-19|^2 + |47-31|^2\right]^{1/2} = 16.4924\,,$$

$$\Delta(1,4) = \left[|23-16|^2 + |47-28|^2\right]^{1/2} = 20.2485\,,$$

$$\Delta(2,3) = \left[|21-19|^2 + |43-31|^2\right]^{1/2} = 12.1655\,,$$

$$\Delta(2,4) = \left[|21-16|^2 + |43-28|^2\right]^{1/2} = 15.8114\,,$$

and

$$\Delta(3,4) = \left[|19-16|^2 + |31-28|^2\right]^{1/2} = 4.2426\,.$$

Then, following Eq. (8.1) on p. 422,

$$\delta = \left[g\binom{b}{2}\right]^{-1} \left[\Delta(1,2) + \Delta(1,3) + \cdots + \Delta(2,4) + \Delta(3,4)\right]$$

and the observed value of the MRBP test statistic with $v = 1$ is

$$\delta_o = \left[2\binom{4}{2}\right]^{-1} (15.6525 + 39.6232 + \cdots + 15.8114 + 4.2426)$$

$$= \frac{1}{12}(318.7788) = 26.5649\,.$$

The

$$M = (g!)^{b-1} = (2!)^{4-1} = 8$$

Table 8.4 Permutations of the observed bivariate data listed in Fig. 8.4 with values for δ based on $v = 1$ ordered from lowest to highest

Order	Treatment 1	Treatment 2	δ
1	{(73, 64)(59, 57)(46, 35)(23, 11)}	{(23, 47)(21, 43)(19, 31)(16, 28)}	26.5649
2	{(73, 64)(59, 57)(46, 35)(16, 28)}	{(23, 47)(21, 43)(19, 31)(23, 11)}	29.3755
3	{(73, 64)(59, 57)(19, 31)(23, 11)}	{(23, 47)(21, 43)(46, 35)(16, 28)}	33.4871
4	{(73, 64)(59, 57)(19, 31)(16, 28)}	{(23, 47)(21, 43)(46, 35)(23, 11)}	34.0114
5	{(73, 64)(21, 43)(19, 31)(16, 28)}	{(23, 47)(59, 57)(46, 35)(23, 11)}	36.2929
6	{(73, 64)(21, 43)(46, 35)(23, 11)}	{(23, 47)(59, 57)(19, 31)(16, 28)}	36.5032
7	{(73, 64)(21, 43)(19, 31)(23, 11)}	{(23, 47)(59, 57)(46, 35)(16, 28)}	37.3859
8	{(73, 64)(21, 43)(46, 35)(16, 28)}	{(23, 47)(59, 57)(19, 31)(23, 11)}	37.6965

non-redundant, equally-likely arrangements of the observed response measurement scores described in Fig. 8.4 are listed in Table 8.4 and are ordered by the δ values from lowest to highest.

The observed MRBP test statistic, $\delta_o = 26.5649$, obtained from the original arrangement of the $N = 8$ bivariate response measurement scores in Treatments 1 and 2,

$$\{(73, 64)(59, 57)(46, 35)(23, 11)\} \quad \{(23, 47)(21, 43)(19, 31)(16, 28)\} ,$$

(Order 1 in Table 8.4) is unusual since seven of the eight δ values exceed the observed value of $\delta_o = 26.5649$ and only one δ value is equal to or less than the observed value.

If all non-redundant arrangements of the $N = 8$ observed bivariate response measurement scores listed in Fig. 8.4 occur with equal chance, the exact probability value of $\delta_o = 26.5649$ computed on the $M = 8$ arrangements of the observed response measurement scores with $b = 4$ blocks preserved for each arrangement is

$$P(\delta \leq \delta_o | H_0) = \frac{\text{number of } \delta \text{ values} \leq \delta_o}{M} = \frac{1}{8} = 0.1250 .$$

The total of the $M = 8$ δ values listed in Table 8.4 is 271.3176. Thus, following Eq. (8.4) on p. 423, the exact average value of the $M = 8$ δ values listed in Table 8.4 is

$$\mu_\delta = \frac{1}{M} \sum_{i=1}^{M} \delta_i = \frac{1}{8}(271.3176) = 33.9147 .$$

Following Eq. (8.3) on p. 423, the observed chance-corrected measure of effect size is

$$\Re_o = 1 - \frac{\delta_o}{\mu_\delta} = 1 - \frac{26.5649}{33.9147} = +0.2167 ,$$

indicating approximately 22 % within-block agreement above that expected by chance.

8.2 MRBP and Pearson's Product-Moment Correlation

It is not readily apparent that the MRBP test statistic, given by

$$\delta = \left[g \binom{b}{2} \right]^{-1} \sum_{i=1}^{g} \sum_{j<k} \Delta(x_{ij}, x_{ik}) \, ,$$

and the ordinary Pearson product-moment correlation coefficient are closely related when $v = 2$. Let R denote the Pearson product-moment correlation coefficient between two interval-level variables, $(x_{11}, \ldots, x_{g1})$ and $(x_{12}, \ldots, x_{g2})$, given by

$$R = \frac{\text{cov}(x_1, x_2)}{s_1 s_2} \, ,$$

where the covariance of variables x_1 and x_2 is given by

$$\text{cov}(x_1, x_2) = \frac{1}{g-1} \sum_{i=1}^{g} (x_{i1} - \bar{x}_1)(x_{i2} - \bar{x}_2) \, ,$$

and the means and standard deviations are given by

$$\bar{x}_j = \frac{1}{g} \sum_{i=1}^{g} x_{ij} \quad \text{and} \quad s_j = \frac{1}{g-1} \sum_{i=1}^{g} \left(x_{ij} - \bar{x}_j \right)^2 \, ,$$

respectively, for $j = 1, 2$.

If $v = 2$, $b = 2$, and $r = 1$, then the functional relationships between R and δ are given by

$$R = \frac{\mu_\delta - \delta}{2S_1S_2} \quad \text{and} \quad \delta = \mu_\delta - 2RS_1S_2 \, ,$$

where

$$R = \frac{1}{gS_1S_2} \sum_{i=1}^{g} (x_{i1} - \bar{x}_1)(x_{i2} - \bar{x}_2) \, ,$$

$$\mu_\delta = S_1^2 + S_2^2 + (\bar{x}_1 - \bar{x}_2)^2 \, ,$$

$$\bar{x}_j = \frac{1}{g} \sum_{i=1}^{g} x_{ij} \, , \quad \text{and} \quad S_j^2 = \frac{1}{g} \sum_{i=1}^{g} \left(x_{ij} - \bar{x}_j \right)^2 \tag{8.7}$$

for $j = 1, 2$.[5] Thus, R and δ are equivalent under the null hypothesis because $\bar{x}_1$, $\bar{x}_2$, S_1, and S_2 are invariant relative to the M possible permutations of the response measurement scores.

Because R and δ are equivalent under the null hypothesis, the permutation distributions of R and δ are also equivalent when $v = 2$, viz.,

$$P(R \geq R_o | H_0) = P(\delta \leq \delta_o | H_0) = \frac{\text{number of } \delta \text{ values} \leq \delta_o}{M} ,$$

where $M = g!$ and R_o and δ_o denote the observed values of R and δ, respectively. Finally, the functional relationships between R and $\Re$ are given by

$$R = \frac{\Re \mu_\delta}{2S_1S_2} \quad \text{and} \quad \Re = \frac{2RS_1S_2}{\mu_\delta} .$$

8.2.1 Example MRBP Correlation Analysis

To illustrate the relationship between δ and Pearson's R, consider the univariate response measurement scores listed in Fig. 8.5 with $g = 7$ objects, $b = 2$ blocks, $r = 1$ response measurement, and $v = 2$, employing squared Euclidean distance between response measurement scores to correspond to the Pearson product-moment correlation coefficient. For the univariate response measurement scores listed in Fig. 8.5, $\bar{x}_1 = 2.00$, $\bar{x}_2 = 5.00$, $s_1 = 1.00$, $s_2 = 2.00$,

$$\text{cov}(x_1, x_2) = \frac{1}{7-1}(9.00) = 1.50 ,$$

Fig. 8.5 Example data with $g = 7$ objects, $b = 2$ blocks, and $r = 1$ response measurement

Object	x_1	x_2
1	3	8
2	3	6
3	3	5
4	2	6
5	1	5
6	1	3
7	1	2

[5]Note that the summation for S_j^2 in Eq. (8.7) is divided by g and not by $g - 1$, as degrees of freedom are irrelevant to permutation methods.

and the observed Pearson product-moment correlation coefficient is

$$R_o = \frac{\text{cov}(x_1 x_2)}{s_1 s_2} = \frac{1.50}{(1.00)(2.00)} = +0.75 \ .$$

Equivalently, for the response measurement scores listed in Fig. 8.5, $S_1 = 0.9258$, $S_2 = 1.8516$, $\delta_o = 10.7143$, $\mu_\delta = 13.2857$, and

$$R_o = \frac{\mu_\delta - \delta_o}{2S_1 S_2} = \frac{13.2857 - 10.7143}{(2)(0.9258)(1.8516)} = +0.75 \ .$$

Since there are only $M = 7! = 5{,}040$ possible, equally-likely arrangements of the observed response measurement scores listed in Fig. 8.5, an exact solution is feasible. If all arrangements of the $N = 14$ observed response measurement scores listed in Fig. 8.5 occur with equal chance, the exact probability value of $\delta_o = 10.7143$ (or $R_o = +0.75$) computed on the $M = 5{,}040$ possible arrangements of the observed response measurement scores with $b = 2$ blocks preserved for each arrangement is

$$P(\delta \leq \delta_o | H_0) = \frac{\text{number of } \delta \text{ values} \leq \delta_o}{M} = \frac{216}{5{,}040} = 0.0429 \ .$$

For comparison, a conventional test of significance for R is given by

$$t = \left[\frac{(g-2)R^2}{1-R^2} \right]^{1/2}$$

and the observed value of t for $R_o = +0.75$ is

$$t_o = \left[\frac{(7-2)(+0.75)^2}{1-(+0.75)^2} \right]^{1/2} = +2.5355 \ .$$

Assuming independence and normality, t is approximately distributed as Student's t under the null hypothesis with $g - 2 = 7 - 2 = 5$ degrees of freedom. Under the null hypothesis, the observed value of $t_o = +2.5355$ yields an approximate two-sided probability value of $P = 0.0522$.

Also, for the $N = 7$ univariate response measurement scores listed in Fig. 8.5, the exact expected value of the $M = 5{,}040$ δ values is $\mu_\delta = 13.2857$ and following Eq. (8.3) on p. 423, the observed chance-corrected measure of effect size is

$$\Re_o = 1 - \frac{\delta_o}{\mu_\delta} = 1 - \frac{10.7143}{13.2857} = +0.1935 \ ,$$

indicating approximately 19 % within-block agreement above that expected by chance.

Finally, the relationships between the observed values of R_o and $\Re_o$ are

$$R_o = \frac{\Re_o \mu_\delta}{2S_1S_2} = \frac{(+0.1935)(13.2857)}{2(0.9258)(1.8516)} = +0.75$$

and

$$\Re_o = \frac{2R_oS_1S_2}{\mu_\delta} = \frac{2(+0.75)(0.9258)(1.8516)}{13.2857} = +0.1935 \, .$$

Analysis with $v = 1$

Although the Pearson product-moment correlation coefficient is not defined for $v = 1$, it is still possible to analyze the data with δ and $\Re$ based on ordinary Euclidean distances between response measurement scores. For the univariate response measurement scores listed in Fig. 8.5 with $g = 7$, $b = 2$, $r = 1$, and $v = 1$, employing ordinary Euclidean distance between response measurement scores, the observed value of the MRBP test statistic is $\delta_o = 3.00$, the exact expected value of the $M = 5{,}040$ δ values is $\mu_\delta = 3.1224$, and the observed chance-corrected measure of effect size is

$$\Re_o = 1 - \frac{\delta_o}{\mu_\delta} = 1 - \frac{3.00}{3.1224} = +0.0392 \, ,$$

indicating approximately chance within-block agreement. If all arrangements of the $N = 14$ observed response measurement scores listed in Fig. 8.5 occur with equal chance, the exact probability value of $\delta_o = 3.00$ computed on the $M = 5{,}040$ possible arrangements of the observed response measurement scores with $b = 2$ blocks preserved for each arrangement is

$$P(\delta \leq \delta_o | H_0) = \frac{\text{number of } \delta \text{ values} \leq \delta_o}{M} = \frac{216}{5{,}040} = 0.0429 \, ,$$

which is the same as the probability value obtained with $v = 2$.

8.2.2 Permutations of g Response Measurements

If $(x_{11}, \ldots, x_{g1})$ is one of the $g!$ permutations of the observed response measurement scores and $v = 2$, then $\mu_\delta = 2S_1S_2$ and the Pearson product-moment correlation coefficient, R, is equivalent to the chance-corrected within-block measure of effect size, $\Re$, i.e., $R = \Re$, where

$$R = \frac{\text{cov}(x_1, x_2)}{S_1S_2} \quad \text{and} \quad \Re = 1 - \frac{\delta}{\mu_\delta}$$

Fig. 8.6 Example data set with $g = 10$ objects, $b = 2$ blocks, and $r = 1$ response measurement

Object	x_1	x_2
1	21	32
2	27	21
3	32	27
4	35	35
5	43	64
6	47	43
7	50	50
8	58	47
9	64	69
10	69	58

[297, pp. 132–133]. To illustrate the equivalence of Pearson's R and $\Re$ when $(x_{12}, \ldots, x_{g2})$ is a permutation of $(x_{11}, \ldots, x_{g1})$, consider the small data set listed in Fig. 8.6 with $g = 10$ objects, $r = 1$ response measurement, and $b = 2$ blocks. For these data the 10 response measurement scores listed under x_2 in Fig. 8.6 constitute a permutation of the 10 response measurement scores listed under x_1.

For the response measurement scores listed in Fig. 8.6, $\bar{x}_1 = \bar{x}_2 = 44.60$, $s_1 = s_2 = 16.0083$,

$$\text{cov}(x_1, x_2) = \frac{1}{10-1}(1{,}853.4000) = 205.9333 ,$$

and the observed Pearson product-moment correlation coefficient is

$$R_{\text{o}} = \frac{\text{cov}(x_1, x_2)}{s_1 s_2} = \frac{205.9333}{(16.0083)(16.0083)} = +0.8036 .$$

Equivalently, for the response measurement scores listed in Fig. 8.6, the observed value of the MRBP test statistic with $v = 2$ is $\delta_{\text{o}} = 90.60$, the exact expected value of the M δ values is $\mu_\delta = 461.2800$ and, following Eq. (8.3) on p. 423, the observed chance-corrected measure of effect size is

$$\Re_{\text{o}} = 1 - \frac{\delta_{\text{o}}}{\mu_\delta} = 1 - \frac{90.60}{461.2800} = +0.8036 ,$$

indicating approximately 80 % within-block agreement above that expected by chance.

Since there are $M = 10! = 3{,}628{,}800$ possible, equally-likely arrangements of the observed response measurement scores listed in Fig. 8.6, calculation of an exact probability value is prohibitive and an approximate resampling probability value is more practical. For the univariate response measurement scores listed in Fig. 8.6, the approximate resampling probability value of $\delta_{\text{o}} = 90.60$ computed on $L = 1{,}000{,}000$ random arrangements of the observed response measurement scores is

$$P(\delta \leq \delta_o | H_0) = \frac{\text{number of } \delta \text{ values} \leq \delta_o}{L} = \frac{3{,}217}{1{,}000{,}000} = 0.0032 \ .$$

While an exact solution may not be practical, it is not unrealistic, given current computer capabilities. For the response measurement scores listed in Fig. 8.6, the exact probability value of $\delta_o = 90.60$ computed on the $M = 3{,}628{,}800$ possible arrangements of the observed response measurement scores is

$$P(\delta \leq \delta_o | H_0) = \frac{\text{number of } \delta \text{ values} \leq \delta_o}{M} = \frac{11{,}780}{3{,}628{,}800} = 0.0032 \ .$$

Analysis with $v = 1$

Although the Pearson product-moment correlation coefficient is not defined for $v = 1$, it is still possible to analyze the response measurement scores listed in Fig. 8.6 with δ and $\Re$ based on ordinary Euclidean distances between response measurement scores. For the response measurement scores listed in Fig. 8.6 with $g = 10$, $b = 2$, $r = 1$, and $v = 1$, the observed value of δ with $v = 1$ is $\delta_o = 7.40$, the exact expected value of the $M = 3{,}628{,}800$ δ values is $\mu_\delta = 17.40$ and the observed chance-corrected measure of effect size is

$$\Re_o = 1 - \frac{\delta_o}{\mu_\delta} = 1 - \frac{7.40}{17.40} = +0.5747 \ ,$$

indicating approximately 57 % within-block agreement above that expected by chance. If all M possible arrangements of the $N = 20$ observed response measurement scores listed in Fig. 8.6 occur with equal chance, the approximate resampling probability value of $\delta_o = 7.40$ computed on the $L = 1{,}000{,}000$ random arrangements of the observed response measurement scores with $b = 2$ blocks preserved for each arrangement is

$$P(\delta \leq \delta_o | H_0) = \frac{\text{number of } \delta \text{ values} \leq \delta_o}{L} = \frac{5{,}192}{1{,}000{,}000} = 0.0052 \ .$$

For comparison, the exact probability value of $\delta_o = 7.40$ computed on the $M = 3{,}628{,}800$ possible arrangements of the observed response measurement scores is

$$P(\delta \leq \delta_o | H_0) = \frac{\text{number of } \delta \text{ values} \leq \delta_o}{M} = \frac{18{,}669}{3{,}628{,}800} = 0.0051 \ .$$

Finally, it should be noted that if $g = 2$, $r = 1$, $v = 2$, $x_{1j} = -x_{2j} = x_j$, and $|x_j| > 0$ for $j = 1, \ldots, b$, then the test based on δ is equivalent to the permutation version of either the matched-pairs or one-sample t test. When $v = 1$, $\Re$ possesses certain advantages over R; viz., $\Re$ is a measure of chance-corrected agreement rather than a measure of linearity, and second, $\Re$ is much more robust than R since it is based on ordinary Euclidean distances rather than squared Euclidean distances.

8.3 Coda

Chapter 8 provided the foundation for Multivariate Randomized Block Permutation (MRBP) procedures, with special emphasis on the generalized Minkowski distance function, $\Delta(x, y)$, as defined in Eq. (8.2) on p. 422; δ, the weighted mean of the specified distance-function values as defined in Eq. (8.1) on p. 422; and $\Re$, the chance-corrected within-block coefficient of agreement, as defined in Eq. (8.3) on p. 423. Chapters 9, 10, and 11 provide applications of MRBP to randomized-block data at the interval, ordinal, and nominal levels of measurement, respectively.

Chapter 9

Chapter 9 establishes the relationships between the MRBP test statistics, δ and $\Re$, and selected conventional tests and measures designed for the analysis of randomized-block data at the interval level of measurement. Considered in Chap. 9 are Student's t test for matched pairs, Hotelling's multivariate T^2 test for matched pairs, randomized-block analysis of variance, randomized-block multivariate analysis of variance, and Pearson's product-moment correlation coefficient.

Randomized Block Designs: Interval Data

9

This ninth chapter of *Permutation Statistical Methods* utilizes the Multivariate Randomized Block Permutation (MRBP) procedures presented in Chap. 8 to develop the functional relationships between the test statistics of MRBP, δ and $\Re$, and selected conventional tests and measures designed for the analysis of randomized-block data at the interval level of measurement. Included in Chap. 9 are permutation versions of Student's t test for univariate matched-pairs data, Hotelling's T^2 test for multivariate matched-pairs data, randomized-block analysis of variance, randomized-block multivariate analysis of variance, and Pearson's product-moment correlation coefficient.

As detailed in Chap. 8, the structure of the MRBP test statistic, δ, depends on the value of v in the generalized Minkowski distance function given by

$$\Delta(x, y) = \left(\sum_{i=1}^{r} \left| x_i - y_i \right|^p \right)^{v/p} ,$$

where $p \geq 1$ and $v > 0$. The choice of v permits the MRBP test statistic given by

$$\delta = \left[g \binom{b}{2} \right]^{-1} \sum_{i=1}^{g} \sum_{j<k} \Delta(x_{ij}, x_{ik}) \tag{9.1}$$

to be transformed into a variety of tests and measures and provides the flexibility for analyzing univariate and multivariate data at the interval, ordinal, and nominal levels of measurement.

The null hypothesis (H_0) states that the distribution of δ assigns an equal probability to each of the

$$M = (g!)^b$$

K.J. Berry et al., *Permutation Statistical Methods*,
DOI 10.1007/978-3-319-28770-6_9

possible, equally-likely allocations of the g r-dimensional response measurements to the g treatment positions within each of the b blocks. The probability value associated with an observed value of δ, say δ_o, is the probability under the null hypothesis (H_0) of observing a value of δ as extreme or more extreme than δ_o. Thus, an exact probability value for the observed MRBP test statistic, δ_o, may be expressed as

$$P(\delta \leq \delta_o | H_0) = \frac{\text{number of } \delta \text{ values} \leq \delta_o}{M} .$$

As with MRPP, initially described in Chap. 2, a chance-corrected measure of agreement among all b blocks for all g treatments provides an universal measure of effect size for all randomized-block analysis-of-variance designs given by

$$\Re = 1 - \frac{\delta}{\mu_\delta} , \tag{9.2}$$

where μ_δ is the arithmetic average of the M δ values calculated on all possible arrangements of the observed data given by

$$\mu_\delta = \frac{1}{M} \sum_{i=1}^{M} \delta_i . \tag{9.3}$$

Alternatively, for N pairs of values (x_i and y_i for $i = 1, \ldots, N$),

$$\delta = \frac{1}{N} \sum_{i=1}^{N} \left| x_i - y_i \right|^v ,$$

$$\mu_\delta = \frac{1}{N^2} \sum_{i=1}^{N} \sum_{j=1}^{N} \left| x_i - y_j \right|^v ,$$

and

$$\Re = 1 - \frac{\delta}{\mu_\delta} .$$

Permutation analogues of four statistical tests designed for interval-level response measurements are examined in this chapter: (1) Student's matched-pairs t test with univariate interval-level response measurement scores, (2) Hotelling's matched-pairs T^2 test with multivariate interval-level response measurement scores, (3) Fisher's one-way randomized-block analysis of variance with interval-level univariate response measurement scores, and (4) Fisher's one-way randomized-block analysis of variance with interval-level multivariate response measurement scores. The four tests are illustrated and compared with examples analyzed with $v = 2$ and

$v = 1$. Finally, the functional relationship between the Pearson product-moment correlation coefficient, R and δ, is illustrated with $v = 2$.

As with MRPP, discussed more completely in Chap. 2, MRBP procedures are data-dependent, distribution-free, and non-parametric. Therefore, there is no reason to square differences between response measurement scores and $v = 1$ is preferred for all applications of MRBP.

9.1 Permutation Analogue of Student's *t* Test

A research design that calls for a test of differences between two matched treatment groups when univariate ($r = 1$) response measurements have been obtained for each of $b \geq 2$ blocks is commonplace in many fields of research. The conventional approach to such a research design is Student's t test for two matched samples given by

$$t = \frac{\bar{d}}{s_{\bar{d}}} ,$$

where the average difference between the two sets of response measurement scores is given by

$$\bar{d} = \frac{1}{b} \sum_{i=1}^{b} d_i ,$$

$$d_i = x_{1i} - x_{2i} , \qquad i = 1, \ldots, b ,$$

x_{1i} and x_{2i} are univariate response measurement scores for the ith object in treatments 1 and 2, respectively, the sample estimate of the population variance is

$$s_{\bar{d}}^2 = \frac{s_d^2}{b} ,$$

where

$$s_d^2 = \frac{1}{b-1} \sum_{i=1}^{b} \left(d_i - \bar{d}\right)^2 ,$$

and b is the number of objects in each of the two treatment groups.[1] Assuming independence and normality, t is approximately distributed as Student's t under the null

[1]Conventional notation is to use n or N as the number of blocks.

hypothesis with $b-1$ degrees of freedom. Note that $v=2$ yields the permutation version of the classical Fisher–Pitman matched-pairs test where

$$\delta = \frac{2}{t^2+b-1}\sum_{i=1}^{b} d_i^2 \quad \text{and} \quad t = \left(\frac{2}{\delta}\sum_{i=1}^{b} d_i^2 - b + 1\right)^{1/2} \tag{9.4}$$

relate the MRBP test statistic and Student's matched-pairs t test statistic.

If the observed values of δ and t are denoted by δ_o and t_o, respectively, then the exact probability value of δ_o and t_o is given by

$$P(t \geq t_o|H_0) = P(\delta \leq \delta_o|H_0) = \frac{\text{number of } \delta \text{ values} \leq \delta_o}{M} ,$$

where $M = 2^b$ in this application. When b is large, then a method to approximate the probability value is essential. A resampling permutation procedure provides an approximate probability value for δ and is given by

$$P(\delta \leq \delta_o|H_0) = \frac{\text{number of } \delta \text{ values} \leq \delta_o}{L} ,$$

where L is a random sample of all possible arrangements of the $2b$ response measurements. Typically, L is set to a large value to ensure accuracy, e.g., $L =$ 1,000,000. When M is very large and P is exceedingly small, a resampling-approximation permutation procedure may produce no δ values equal to or less than δ_o, even with $L =$ 1,000,000, yielding an approximate resampling probability value of $P = 0.00$. In such cases, moment-approximation permutation procedures based on fitting the first three exact moments of the discrete permutation distribution to a Pearson type III distribution provide approximate probability values, as detailed in Chap. 1, Sect. 1.2.2 [284, 299].

9.1.1 Example 1: $v = 2$

Consider the univariate response measurement scores listed in Fig. 9.1, where $r = 1$ and $b = 9$. When raw scores are defined as differences, as in Fig. 9.1, Eq. (9.1) on p. 445 simplifies to

$$\delta = \binom{b}{2}^{-1} \sum_{j<k} \Delta(x_j, x_k) . \tag{9.5}$$

Employing squared Euclidean distance between response measurement scores, let $v = 2$ to correspond to Student's matched-pairs t test. The data are adapted from Gravetter and Wallnau [153, p. 320].

Fig. 9.1 Example univariate response measurement scores with two treatments, $b = 9$ blocks, and $r = 1$ response measurement

Object	Treatment 1	Treatment 2	d
1	9	7	+2
2	8	7	+1
3	7	3	+4
4	7	8	− 1
5	8	6	+2
6	9	4	+5
7	7	6	+1
8	7	9	− 2
9	8	4	+4

An exact solution is feasible for these data since there are only

$$M = (g!)^b = (2!)^9 = 512$$

possible, equally-likely arrangements of the $b = 9$ objects. Following Eq. (9.5), the observed value of the MRBP test statistic with $v = 2$ is

$$\delta_o = \binom{9}{2}^{-1} |2 - 1|^2 + |2 - 4|^2 + \cdots + |-2 - 4|^2 = \frac{1}{36}(392) = 10.8889\,.$$

If all arrangements of the observed response measurement scores listed in Fig. 9.1 occur with equal chance, the exact probability value of $\delta_o = 10.8889$ computed on the $M = 512$ possible arrangements of the observed response measurement scores with $b = 9$ blocks preserved for each arrangement is

$$P(\delta \le \delta_o | H_0) = \frac{\text{number of } \delta \text{ values} \le \delta_o}{M} = \frac{16}{512} = 0.0781\,.$$

For comparison, a conventional matched-pairs t test on the response measurement scores listed in Fig. 9.1 yields $\bar{d} = +1.7778$, $s_d = 2.3333$, $s_{\bar{d}} = 0.7778$, and an observed matched-pairs t value of $t_o = +2.2857$. Assuming independence and normality, t is approximately distributed as Student's t under the null hypothesis with $b - 1 = 9 - 1 = 8$ degrees of freedom. Under the null hypothesis, the observed value of $t_o = +2.2857$ yields an approximate two-sided probability value of $P = 0.0258$.

Measures of Effect Size

While measures of effect size have been developed for randomized-block analysis-of-variance designs, they are somewhat controversial. One of the problems is whether or not an additive model is appropriate; that is, is it safe to assume that there are no interactions of subjects with treatments? A second problem is deciding whether or not the factor Objects is to be considered random or fixed. For discussions of measures of effect size for randomized-block designs, see articles by Dodd and Schultz [98], Olejnik and Algina [326], Susskind and Howland [391], and Vaughan and Corballis [411].

On the other hand, $\mathfrak{R}$ is a convenient, chance-corrected measure of effect size that is appropriate for all randomized-block designs, is easy to compute, and provides a straightforward interpretation. Moreover, because permutation methods are data-dependent, random or fixed factors are irrelevant to $\mathfrak{R}$, as are additive or non-additive models. Following Eq. (9.3) on p. 446, the exact expected value of the $M = 512$ δ values is $\mu_\delta = 16.00$ and, following Eq. (9.2) on p. 446, the observed chance-corrected measure of effect size is

$$\mathfrak{R}_{\text{o}} = 1 - \frac{\delta_{\text{o}}}{\mu_\delta} = 1 - \frac{10.8889}{16.00} = +0.3194\ ,$$

indicating approximately 32 % within-block agreement above that expected by chance.

Finally, the relationships between the MRBP test statistic and Student's matched-pairs t given in Eq. (9.4) can be confirmed. For the univariate response measurement scores listed in Fig. 9.1 on p. 449, the observed values of δ and t are

$$\delta_{\text{o}} = \frac{2}{t^2 + b - 1} \sum_{i=1}^{b} d_i^2 = \frac{2}{2.2857^2 + 9 - 1}(72) = \frac{144}{13.2244} = 10.8889$$

and

$$t_{\text{o}} = \left(\frac{2}{\delta} \sum_{i=1}^{b} d_i^2 - b + 1 \right)^{1/2} = \left[\frac{2}{10.8889}(72) - 9 + 1 \right]^{1/2}$$

$$= \left(\frac{144}{10.8889} - 8 \right)^{1/2} = +2.2857\ .$$

9.1.2 Example 2: $v = 1$

For a comparison analysis of the response measurement scores listed in Fig. 9.1 on p. 449, set $v = 1$ instead of $v = 2$, employing ordinary Euclidean distance between response measurement scores. For the univariate response measurement scores listed in Fig. 9.1, there are still only

$$M = (g!)^b = (2!)^9 = 512$$

possible, equally-likely arrangements of the $b = 9$ objects. Following Eq. (9.5) on p. 448 the observed value of the MRBP test statistic with $v = 1$ is

$$\delta_o = \binom{9}{2}^{-1} |2-1|^1 + |2-4|^1 + \cdots + |-2-4|^1 = \frac{100}{36} = 2.7778 \,.$$

If all arrangements of the observed response measurement scores listed in Fig. 9.1 occur with equal chance, the exact probability value of $\delta_o = 2.7778$ computed on the $M = 512$ possible arrangements of the observed response measurement scores with $b = 9$ blocks preserved for each arrangement is

$$P(\delta \leq \delta_o | H_0) = \frac{\text{number of } \delta \text{ values} \leq \delta_o}{M} = \frac{16}{512} = 0.0898 \,.$$

For comparison, the exact probability value with $v = 2$ is $P = 0.0781$. No comparison is made with Student's matched-pairs t test as Student's t test is undefined for $v = 1$.

Following Eq. (9.3) on p. 446, the exact expected value of the $M = 512$ δ values is $\mu_\delta = 3.3056$ and, following Eq. (9.2) on p. 446, the observed chance-corrected measure of effect size is

$$\mathfrak{R}_o = 1 - \frac{\delta_o}{\mu_\delta} = 1 - \frac{2.7778}{3.3056} = +0.1597 \,,$$

indicating approximately 16 % within-block agreement above that expected by chance.

9.2 Permutation Analogue of Hotelling's T^2 Test

Oftentimes a research design calls for a test of difference between $g = 2$ matched treatment groups when multivariate ($r \geq 2$) response measurements have been obtained for each of $b \geq 2$ blocks. The conventional approach to such a research design is Hotelling's multivariate T^2 test for two matched samples [181].

Consider that $r \geq 2$ response measurements and $b \geq 2$ subjects are associated with a multivariate pre-treatment and post-treatment matched-pairs permutation test and let $(w_{11j}, \ldots, w_{r1j})$ and $(w_{12j}, \ldots, w_{r2j})$ denote r-dimensional row vectors with elements comprised of the r response measurements on the jth subject from the pre- and post-treatments, respectively, where $j = 1, \ldots, b$. Let

$$x_{1j} = \begin{pmatrix} x_{11j} \\ \vdots \\ x_{r1j} \end{pmatrix} ,$$

where $x_{k1j} = w_{k1j} - w_{k2j}$ for $k = 1, \ldots, r$, be the r-dimensional column vector of differences between pre-treatment and post-treatment response measurement scores for the jth subject, and let $x_{2j} = -x_{1j}$ be the r-dimensional origin reflection of x_{1j} for $j = 1, \ldots, b$. The probability under the null hypothesis is $P(x_{1j}) = P(x_{2j}) = 0.50$ for $j = 1, \ldots, b$. For the multivariate matched-pairs research design, consider the MRBP test statistic given by

$$\delta = \binom{b}{2}^{-1} \sum_{m<n} \Delta(x_{1m}, x_{1n}) \,,$$

where

$$\Delta(x_{1m}, x_{1n}) = \left[(x_{1m}, x_{1n})'(x_{1m}, x_{1n})\right]^{1/2}$$

is the r-dimensional Euclidean distance between the mth and nth subjects' differences, and the sum $\sum_{m<n}$ is over all m and n such that $1 \leq m < n \leq b$.

If the r response measurement scores are in different units, it is necessary that the measurements be made commensurate, i.e., standardized to a common unit of measurement. The replacement of x_{kij} with $x_{kij}^{*} = x_{kij}/\phi_k$, where

$$\phi_k = \sum_{i_1=1}^{2} \sum_{i_2=1}^{2} \sum_{m<n} \left| x_{ki_1m} - x_{ki_2n} \right|$$

for $k = 1, \ldots, r$ and $1 \leq m < n \leq b$ ensures that each response measurement score makes a similar contribution in the r-dimensional Euclidean space since

$$\sum_{i_1=1}^{2} \sum_{i_2=1}^{2} \sum_{m<n} \left| x_{ki_1m}^{*} - x_{ki_2n}^{*} \right| = 1$$

for $k = 1, \ldots, r$. This commensuration is invariant relative to any permutation under the null hypothesis and is termed Euclidean commensuration; see Chap. 3, Sect. 3.4.

Hotelling's multivariate matched-pairs T^2 test statistic is given by

$$T^2 = b\,\bar{x}_1' \, \mathbf{S}_x^{-1} \, \bar{x}_1 \,,$$

where $\mathbf{S}_x$ is an $r \times r$ matrix given by

$$\mathbf{S}_x = \frac{1}{b-1} \sum_{j=1}^{b} (x_{1j} - \bar{x}_1)(x_{1j} - \bar{x}_1)'$$

and

$$\bar{x}_1 = \frac{1}{b} \sum_{j=1}^{b} x_{1j} \; .$$

While the permutation test is applicable to all combinations of r and b, any application of the T^2 test under the null hypothesis with the assumption of multivariate normality requires that $\min(r, b-r) \geq 1$ since the distribution of the adjusted T^2 statistic given by

$$F = \frac{(b-r)T^2}{(b-1)r}$$

is approximately distributed as Snedecor's F with $\nu_1 = r$ and $\nu_2 = b - r$ degrees of freedom. When $v = 2$, the functional relationships between Hotelling's T^2 and the MRBP test statistic are given by

$$T^2 = \frac{r(b-1)^2 \left[2SS_{\text{Total}} - g(b-1)\delta\right]}{g(b-r)(b-1)\delta - 2SS_{\text{Between}}}$$

and

$$\delta = \frac{2\left[r(b-1)^2 SS_{\text{Total}} + T^2 SS_{\text{Between}}\right]}{g(b-1)\left[T^2(b-r) + r(b-1)^2\right]} \, ,$$

where SS_{Between} and SS_{Total} are defined as usual, i.e.,

$$SS_{\text{Between}} = b \sum_{i=1}^{g} \left(\bar{x}_{i.} - \bar{x}_{..}\right)^2 \, ,$$

$$SS_{\text{Total}} = \sum_{i=1}^{g} \sum_{j=1}^{b} \left(x_{ij} - \bar{x}_{..}\right)^2 \, ,$$

$$\bar{x}_{i.} = \frac{1}{b} \sum_{j=1}^{b} x_{ij} \, , \qquad i = 1, \ldots, g \, ,$$

$$\bar{x}_{..} = \frac{1}{bg} \sum_{i=1}^{g} \sum_{j=1}^{b} x_{ij} \, ,$$

and x_{ij} is the univariate response measurement score of the ith object in the jth block.

If the observed values of δ and T^2 are denoted by δ_o and T_o^2, respectively, then the exact probability value of δ_o and T_o^2 is given by

$$P(T^2 \geq T_o^2|H_0) = P(\delta \leq \delta_o|H_0) = \frac{\text{number of } \delta \text{ values} \leq \delta_o}{M} ,$$

where $M = 2^b$ in this application. When b is large (e.g., $2^{30} = 1{,}073{,}741{,}824$), then a method to approximate the probability value is essential. A resampling permutation procedure provides an approximate probability value for δ and is given by

$$P(\delta \leq \delta_o|H_0) = \frac{\text{number of } \delta \text{ values} \leq \delta_o}{L} ,$$

where L is a random sample of all possible arrangements of the $2br$ response measurements.

9.2.1 Example 1: $v = 2$

Consider the following scenario: paired, but randomly arranged, pre-training and post-training writing samples of 11 students were presented blindly to 13 experienced teachers of mathematics and language arts for grading. Each of the 13 judges scored each of the 22 writing samples on a scale from 0 to 10. The data are adapted from Mielke, Berry, and Neidt [304]. The pre- and post-training grades are listed in Tables 9.1 and 9.2, respectively.

The example analysis blocks on the $b = 13$ judges and compares the pre-training and post-training scores of the $r = 11$ students. The analysis evaluates the following

Table 9.1 Pre-training response measurement scores assigned by $b = 13$ judges to writing samples of $r = 11$ students

	Student										
Judge	1	2	3	4	5	6	7	8	9	10	11
1	1	6	1	1	8	1	5	8	6	3	1
2	3	4	6	2	8	3	6	9	9	7	3
3	1	6	2	3	7	3	3	5	5	2	4
4	2	5	5	1	8	2	4	7	8	6	4
5	3	6	5	2	8	2	3	5	9	4	2
6	0	4	3	0	8	0	3	9	7	3	0
7	1	5	0	1	7	0	1	2	8	5	1
8	5	8	4	0	2	0	2	10	2	2	0
9	1	7	2	5	9	2	6	6	9	6	3
10	2	3	2	0	6	1	5	7	5	3	3
11	1	5	2	1	7	1	2	8	8	7	4
12	0	4	1	0	9	0	2	5	3	2	1
13	4	9	2	2	5	3	3	9	8	4	3

Table 9.2 Post-training response measurement scores assigned by $b = 13$ judges to writing samples of $r = 11$ students

	Student										
Judge	1	2	3	4	5	6	7	8	9	10	11
1	9	5	3	1	8	1	7	6	6	4	5
2	8	5	5	2	9	2	6	6	7	5	5
3	5	6	2	3	3	3	6	8	7	5	8
4	7	6	3	2	9	4	5	6	6	4	7
5	8	7	4	2	8	4	7	8	6	3	5
6	6	7	2	0	6	0	5	7	6	5	4
7	5	5	2	1	5	3	5	5	4	0	7
8	4	9	6	0	3	3	10	8	5	3	5
9	9	5	5	7	8	3	8	8	8	7	8
10	4	4	1	0	4	3	4	5	6	6	6
11	6	3	3	2	9	2	9	7	7	5	9
12	6	2	3	1	5	1	6	9	6	5	6
13	9	6	4	4	7	6	9	7	6	7	5

question: Did the course work result in significant pre-training/post-training differences in writing among the students?

While an exact solution is feasible for these data, given that there are only

$$M = (g!)^b = (2!)^{13} = 8{,}192$$

possible, equally-likely arrangements of the $b = 13$ judges, for this example analysis consider a resampling permutation procedure where over-sampling of the M possible arrangements is illustrated.[2] For this analysis, where $r = 11$, $g = 2$, and $b = 13$, let $v = 2$, employing squared Euclidean distance between response measurement scores to correspond to Hotelling's matched-pairs T^2 test. Following Eq. (9.1) on p. 445, the observed value of the MRBP test statistic with $v = 2$ is $\delta_o = 65.9872$. If all M possible arrangements of the observed response measurement scores listed in Tables 9.1 and 9.2 occur with equal chance, the approximate resampling probability value of $\delta_o = 65.9872$ computed on $L = 1{,}000{,}000$ random arrangements of the observed response measurement scores with $b = 13$ blocks preserved for each arrangement is

$$P(\delta \leq \delta_o | H_0) = \frac{\text{number of } \delta \text{ values} \leq \delta_o}{L} = \frac{247}{1{,}000{,}000} = 0.2470 \times 10^{-3} .$$

For comparison, a conventional Hotelling's matched-pairs T^2 test on the response measurement scores listed in Tables 9.1 and 9.2 yields an observed T^2 value of

[2]Over-sampling of the M possible arrangements is quite common in the permutation literature because of its efficiency in certain applications.

$T_o^2 = 766.0821$ and the observed F-ratio value is

$$F_o = \frac{(b-r)T^2}{(b-1)r} = \frac{(13-11)(766.0821)}{(13-1)(11)} = 11.6073\,.$$

Assuming independence, multivariate normality, and homogeneity of variance and covariance, F is approximately distributed as Snedecor's F under the null hypothesis with $\nu_1 = r = 11$ and $\nu_2 = b - r = 13 - 11 = 2$ degrees of freedom. Under the null hypothesis, the observed value of $F_o = 11.6073$ yields an approximate probability value of $P = 0.0819$. Finally, the approximate resampling probability value of $P = 0.2470 \times 10^{-3}$ obtained by over-sampling the data listed in Tables 9.1 and 9.2 may be compared with the exact probability value based on $M = 8{,}192$ given by

$$P(\delta \leq \delta_o | H_0) = \frac{\text{number of } \delta \text{ values} \leq \delta_o}{M} = \frac{2}{8{,}192} = 0.2441 \times 10^{-3}\,.$$

There is a considerable difference between the Hotelling's T^2 probability value of $P = 0.0819$ and the exact probability value of $P = 0.2441 \times 10^{-3}$. The difference is quite possibly due to the violation of assumptions of multivariate normality, and homogeneity of variance and covariance required by Hotelling's T^2, but not required by the permutation test.

Following Eq. (9.3) on p. 446, the exact expected value of the $M = 8{,}192$ δ values is $\mu_\delta = 90.1731$ and, following Eq. (9.2) on p. 446, the observed chance-corrected measure of effect size is

$$\Re_o = 1 - \frac{\delta_o}{\mu_\delta} = 1 - \frac{65.9872}{90.1731} = +0.2682\,,$$

indicating approximately 27 % within-block agreement above that expected by chance.

9.2.2 Example 2: $v = 1$

For a comparison analysis of the multivariate data listed in Tables 9.1 and 9.2, set $v = 1$ instead of $v = 2$, employing ordinary Euclidean distance between response measurement scores, and implement an exact permutation procedure. For the response measurement scores listed in Tables 9.1 and 9.2, the number of possible, equally-likely arrangements is still only

$$M = (g!)^b = (2!)^{13} = 8{,}192\,.$$

Following Eq. (9.1) on p. 445, the observed value of the MRBP test statistic with $v = 1$ is $\delta_o = 0.0928$. If all arrangements of the observed response measurement scores listed in Tables 9.1 and 9.2 occur with equal chance, the exact probability value of $\delta_o = 0.0928$ computed on the $M = 8{,}192$ possible arrangements of the observed response measurement scores with $b = 13$ blocks preserved for each arrangement is

$$P(\delta \leq \delta_o | H_0) = \frac{\text{number of } \delta \text{ values} \leq \delta_o}{M} = \frac{634}{8{,}192} = 0.0774 .$$

In this example there is a considerable difference in probability values, where with $v = 2$, the exact probability value is $P = 0.2441 \times 10^{-3}$, and with $v = 1$, the exact probability value is $P = 0.0774$. The substantial difference in probability values is possibly due to large differences between pre-test and post-test scores that are amplified by squaring the differences with $v = 2$, e.g., Student 1 and Judge 1 with pre- and post-test scores of 1 and 9, respectively; Student 1 and Judge 9 with pre- and post-test scores of 1 and 9, respectively; Student 7 and Judge 8 with pre- and post-test scores of 2 and 10, respectively; and others in Tables 9.1 and 9.2. No comparison is made with Hotelling's T^2 test as Hotelling's T^2 is undefined for $v = 1$.

Following Eq. (9.3) on p. 446, the exact expected value of the $M = 8{,}192$ δ values is $\mu_\delta = 0.1125$ and, following Eq. (9.2) on p. 446, the observed chance-corrected measure of effect size is

$$\Re_o = 1 - \frac{\delta_o}{\mu_\delta} = 1 - \frac{0.0928}{0.1125} = +0.1755 ,$$

indicating approximately 18 % within-block agreement above that expected by chance.

9.3 Permutation Analogue of ANOVA

Consider a research design that calls for a test of differences among $g \geq 3$ matched treatment groups when univariate ($r = 1$) response measurements have been obtained for each of $b \geq 2$ blocks. The conventional approach to such a research design is a randomized-block analysis of variance given by

$$F = \frac{MS_{\text{Between}}}{MS_{\text{Within}}} ,$$

where

$$MS_{\text{Between}} = \frac{1}{g-1} SS_{\text{Between}} ,$$

$$SS_{\text{Between}} = b \sum_{i=1}^{g} (\bar{x}_{i.} - \bar{x}_{..})^2 ,$$

$$MS_{\text{Within}} = \frac{1}{(b-1)(g-1)} SS_{\text{Within}} ,$$

$$SS_{\text{Within}} = SS_{\text{Total}} - SS_{\text{Blocks}} - SS_{\text{Between}} ,$$

$$SS_{\text{Blocks}} = g \sum_{j=1}^{b} (\bar{x}_{.j} - \bar{x}_{..})^2 ,$$

$$SS_{\text{Total}} = \sum_{i=1}^{g} \sum_{j=1}^{b} (x_{ij} - \bar{x}_{..})^2 ,$$

$$\bar{x}_{i.} = \frac{1}{b} \sum_{j=1}^{b} x_{ij} , \qquad i = 1, \ldots, g ,$$

$$\bar{x}_{.j} = \frac{1}{g} \sum_{i=1}^{g} x_{ij} , \qquad j = 1, \ldots, b ,$$

$$\bar{x}_{..} = \frac{1}{bg} \sum_{i=1}^{g} \sum_{j=1}^{b} x_{ij} ,$$

and x_{ij} is the response measurement score of the ith object in the jth block.

Under the null hypothesis, F is approximately distributed as Snedecor's F with $\nu_1 = g - 1$ and $\nu_2 = (b-1)(g-1)$ degrees of freedom. When $v = 2$ the functional relationships between Fisher's F and the MRBP test statistic δ are given by

$$F = \frac{(b-1)\,[2SS_{\text{Total}} - g(b-1)\delta\,]}{g(b-1)\delta - 2SS_{\text{Blocks}}} \tag{9.6}$$

and

$$\delta = \frac{2[FSS_{\text{Blocks}} + (b-1)SS_{\text{Total}}]}{g(b-1)(F+b-1)} , \tag{9.7}$$

where

$$\delta = \left[g \binom{b}{2} \right]^{-1} \sum_{i=1}^{g} \sum_{j<k} \left(x_{ij} - x_{ik} \right)^2 .$$

If the observed values of δ and F are denoted by δ_o and F_o, respectively, the exact probability value of F_o and δ_o is given by

$$P(F \geq F_o|H_0) = P(\delta \leq \delta_o|H_0) = \frac{\text{number of } \delta \text{ values} \leq \delta_o}{M} ,$$

where

$$M = \left(g! \right)^b .$$

When b is large, then a method to approximate the probability value is essential. A resampling permutation procedure provides an approximate probability value for δ and is given by

$$P(\delta \leq \delta_o|H_0) = \frac{\text{number of } \delta \text{ values} \leq \delta_o}{L} ,$$

where L is a random sample of all possible arrangements of the bg response measurements.

9.3.1 Example 1: $v = 2$

Consider a simple one-way randomized-block design with $b = 9$ blocks, $g = 5$ treatments, and $r = 1$ response measurement. For this analysis, let $v = 2$, employing squared Euclidean distance between response measurement scores to correspond to a conventional one-way randomized-block analysis-of-variance F test. Example univariate response measurement scores with $b = 9$, $g = 5$, and $r = 1$ are given in Fig. 9.2.

As exact solution is not feasible for these data since there are

$$M = \left(g! \right)^b = \left(5! \right)^9 = 5{,}159{,}780{,}352{,}000{,}000{,}000$$

possible, equally-likely arrangements of the univariate response measurement scores listed in Fig. 9.2. Therefore, a resampling permutation approach is mandated. Following Eq. (9.1) on p. 445, the observed value of the MRBP test statistic with $v = 2$ is $\delta_o = 35.8556$. If all M possible arrangements of the observed response measurement scores listed in Fig. 9.2 occur with equal chance, the approximate resampling probability value of $\delta_o = 35.8556$ computed on $L = 1{,}000{,}000$ random

Fig. 9.2 Example response measurement scores for a one-way randomized-block analysis of variance with $b = 9$ blocks, $g = 5$ treatments, and $r = 1$ response measurement

	Treatment				
Block	1	2	3	4	5
1	21	22	8	6	6
2	20	19	10	4	4
3	17	15	5	4	5
4	25	30	13	12	17
5	30	27	13	8	6
6	19	27	8	7	4
7	26	16	5	2	5
8	17	18	8	1	5
9	26	24	14	8	9

arrangements of the observed response measurement scores with $b = 9$ blocks preserved for each arrangement is

$$P(\delta \leq \delta_o | H_0) = \frac{\text{number of } \delta \text{ values} \leq \delta_o}{L} = \frac{0}{1{,}000{,}000} = 0.00 ,$$

which is interpreted as less than one in a million.

When M is very large and the probability of an observed δ is very small, resampling permutation procedures often result in zero probability, even with $L = 1{,}000{,}000$. Moment-approximation permutation procedures, described briefly in Chap. 1, Sect. 1.2.2, can often provide results in these extreme situations. The moment-approximation of a test statistic requires computation of the exact moments of the test statistic, assuming equally-likely arrangements of the observed response measurement values. Typically, the first three moments of δ are used: the exact mean, variance, and skewness, denoted by μ_δ, σ_δ^2, and γ_δ, respectively. The three moments are then used to fit a specified distribution, such as a Pearson type III distribution, that approximates the underlying discrete permutation distribution and provide an approximate probability value. For the response measurement scores listed in Fig. 9.2, a moment-approximation procedure yields $\delta_o = 35.8556$, $\mu_\delta = 143.4289$, $\sigma_\delta^2 = 97.1824$, $\gamma_\delta = -1.1613$, an observed standardized test statistic of

$$T_o = \frac{\delta_o - \mu_\delta}{\sigma_\delta} = \frac{35.8556 - 143.4289}{9.8581} = -10.9122 ,$$

and a Pearson type III approximate probability value of $P = 0.8541 \times 10^{-7}$.

For comparison, a conventional one-way randomized-block analysis of variance on the response measurement scores listed in Fig. 9.2 yields an observed value of $F_o = 85.0417$. Assuming independence, normality, and homogeneity of variance and covariance, F is approximately distributed as Snedecor's F with $\nu_1 = g - 1 = 5 - 1 = 4$ and $\nu_2 = (b - 1)(g - 1) = (9 - 1)(5 - 1) = 32$ degrees of freedom. Under the null hypothesis, the observed value of $F_o = 85.0417$ yields an approximate probability value of $P = 1.3944 \times 10^{-16}$.

9.3.2 Homogeneity Assumptions

Typically, one-way randomized-block designs are tested for assumptions of homogeneity of variance and covariance, often called compound symmetry or sphericity.[3] Then, if needed, the degrees of freedom are adjusted to compensate for any discrepancies from the homogeneity assumptions. For the univariate data listed in Fig. 9.2, the symmetric $g \times g$ variance–covariance matrix is

$$\hat{\Sigma} = \begin{bmatrix} 21.0000 & 11.7500 & 9.2500 & 7.8333 & 7.3333 \\ 11.7500 & 28.5000 & 13.7500 & 16.3750 & 13.3750 \\ 9.2500 & 13.7500 & 11.5000 & 8.5833 & 8.2083 \\ 7.8333 & 16.3750 & 8.5833 & 11.6944 & 10.8194 \\ 7.3333 & 13.3750 & 8.2083 & 10.8194 & 16.9444 \end{bmatrix} , \tag{9.8}$$

where the estimated variances are the g elements on the principal diagonal, i.e., 21.0000, 28.5000, ..., 16.9444, and the estimated covariances are the $g(g-1)/2$ elements in each of the upper- and lower-triangular matrices, i.e., 11.7500, 9.2500, ..., 10.8194.

In 1959 Greenhouse and Geisser [154] provided a correction to the numerator and denominator degrees of freedom of the F-ratio that aimed to compensate for a lack of compound symmetry. The Greenhouse and Geisser correction is given by

$$\hat{\varepsilon} = \frac{g^2(A-B)^2}{(g-1)(C-2gD+g^2B^2)} ,$$

where A is the average of the g elements on the principal diagonal of the $\hat{\Sigma}$ variance–covariance matrix given by

$$A = \frac{1}{g} \sum_{i=1}^{g} \hat{S}_{ii} ,$$

B is the average of all g^2 elements in the $\hat{\Sigma}$ variance–covariance matrix given by

$$B = \frac{1}{g^2} \sum_{i=1}^{g} \sum_{j=1}^{g} \hat{S}_{ij} ,$$

[3] As noted by Stevens [386, p. 412], for some time it was thought that the stronger condition, compound symmetry, was necessary in which the population variances and covariances were all required to be equal. However, Huynh and Feldt [190] and Rouanet and Lépine [354] showed that a weaker condition, sphericity, in which only the variances of the differences for all pairs of treatments are required to be equal, was sufficient.

C is the sum of the g^2 squared elements in the $\hat{\Sigma}$ variance–covariance matrix given by

$$C = \sum_{i=1}^{g} \sum_{j=1}^{g} \hat{S}_{ij}^2 ,$$

D is the sum of the squared averages of the elements in each of the g rows (or columns) of the $\hat{\Sigma}$ variance–covariance matrix given by

$$D = \sum_{i=1}^{g} \left(\frac{1}{g} \sum_{j=1}^{g} \hat{S}_{ij} \right)^2 ,$$

and $\hat{S}_{ij}$ for $i,j = 1, \ldots, g$ indicates an estimate of the population variance when $i = j$ and an estimate of the population covariance when $i \neq j$. The maximum value of $\hat{\varepsilon} = 1.00$ is attained when all g estimated variances on the principal diagonal are equal and all $g(g-1)/2$ estimated covariances on the off-diagonals are equal, although it is not required that the estimated covariances be equal to the estimated variances. As shown by Greenhouse and Geisser, the minimum value of $\hat{\varepsilon}$ is given by $1/(g-1)$, which is independent of the elements of the $\hat{\Sigma}$ matrix [154, p. 102].

For the variances and covariances given in the $\hat{\Sigma}$ matrix in Eq. (9.8),

$$A = \frac{1}{5}(21.0000 + 28.5000 + 11.5000 + 11.6944 + 16.9444) = 17.9278 ,$$

$$B = \frac{1}{5^2}(21.0000 + 11.7500 + 9.2500 + \cdots + 16.9444) = 12.1678 ,$$

$$C = 21.0000^2 + 11.7500^2 + 9.2500^2 + \cdots + 16.9444^2 = 4{,}275.3065 ,$$

and

$$\begin{aligned} D = &\left[\frac{1}{5}(21.0000 + 11.7500 + 9.2500 + 7.8333 + 7.3333)\right]^2 \\ &+ \left[\frac{1}{5}(11.7500 + 28.5000 + 13.7500 + 16.3750 + 13.3750)\right]^2 \\ &+ \left[\frac{1}{5}(9.2500 + 13.7500 + 11.5000 + 8.5833 + 8.2083)\right]^2 \\ &+ \left[\frac{1}{5}(7.8333 + 16.3750 + 8.5833 + 11.6944 + 10.8194)\right]^2 \\ &+ \left[\frac{1}{5}(7.3333 + 13.3750 + 8.2083 + 10.8194 + 16.9444)\right]^2 \\ &= 767.3720 . \end{aligned}$$

Then, for the $\hat{\Sigma}$ variance–covariance matrix given in Eq. (9.8), the Greenhouse–Geisser correction for the degrees of freedom is

$$\hat{\varepsilon} = \frac{5^2(17.9278 - 12.1678)^2}{(5-1)[4{,}275.3065 - 2(5)(767.3720) + 5^2(12.1678)^2]} = 0.6845 \; .$$

The adjusted numerator and denominator degrees of freedom for the F-ratio are obtained by multiplying $\hat{\varepsilon}$ by the original degrees of freedom, $\nu_1 = g - 1$ and $\nu_2 = (b-1)(g-1)$. Thus,

$$\nu_1' = \hat{\varepsilon}(g-1) = 0.6845(5-1) = 2.7380$$

and

$$\nu_2' = \hat{\varepsilon}(b-1)(g-1) = 0.6845(9-1)(5-1) = 21.9040$$

replacing $\nu_1 = g - 1 = 5 - 1 = 4$ and $\nu_2 = (b-1)(g-1) = (9-1)(5-1) = 32$ degrees of freedom, respectively. The observed F value of $F_o = 85.0417$ then yields an approximate probability value of $P = 5.7647{\times}10^{-12}$ under the null hypothesis with $\nu_1' = 2.7380$ and $\nu_2' = 21.9040$ degrees of freedom. No adjustment is required for the permutation version of randomized-block designs as permutation tests do not assume homogeneity of variances and covariances; moreover, degrees of freedom are irrelevant to permutation methods.

Following Eq. (9.3) on p. 446, the exact expected value of the M δ values is $\mu_\delta = 143.4289$ and, following Eq. (9.2) on p. 446, the observed chance-corrected measure of effect size is

$$\Re_o = 1 - \frac{\delta_o}{\mu_\delta} = 1 - \frac{35.8556}{143.4289} = +0.7500 \; ,$$

indicating 75 % within-block agreement above that expected by chance.

The relationships between the conventional F-ratio and the MRBP test statistic δ can be demonstrated with the univariate response measurement scores listed in Fig. 9.2 on p. 460. Given $b = 9$ blocks, $g = 5$ groups, $SS_{\text{Blocks}} = 486.7111$, $SS_{\text{Total}} = 3{,}166.3111$, and following Eq. (9.6) on p. 458, the observed value of F in terms of δ is

$$F_o = \frac{(9-1)[2(3{,}166.3111) - 5(9-1)(35.8556)]}{5(9-1)(35.8556) - 2(486.7111)}$$

$$= \frac{39{,}187.2018}{460.7998} = 85.0417$$

and following Eq. (9.7) on p. 458, the observed value of the MRBP test statistic in terms of F is

$$\delta_o = \frac{2[(85.0417)(486.7111) + (9-1)(3{,}166.3111)]}{5(9-1)(85.0417+9-1)}$$

$$= \frac{133{,}442.4584}{3{,}721.6680} = 35.8556 \; .$$

9.3.3 Example 2: $v = 1$

For a comparison analysis of the univariate response measurement scores listed in Fig. 9.2 on p. 460, set $v = 1$ instead of $v = 2$, employing ordinary Euclidean distance between response measurement scores. For the response measurement scores listed in Fig. 9.2, M is still large, i.e.,

$$M = (g!)^b = (5!)^9 = 5{,}159{,}780{,}352{,}000{,}000{,}000 \; ,$$

and resampling is again required. Following Eq. (9.1) on p. 445, the observed value of the MRBP test statistic with $v = 1$ is $\delta_o = 4.7667$. If all M possible arrangements of the observed response measurement scores listed in Fig. 9.2 occur with equal chance, the approximate resampling probability value of $\delta_o = 4.7667$ computed on $L = 1{,}000{,}000$ random arrangements of the observed response measurement scores with $b = 9$ blocks preserved for each arrangement is

$$P(\delta \leq \delta_o | H_0) = \frac{\text{number of } \delta \text{ values} \leq \delta_o}{L} = \frac{0}{1{,}000{,}000} = 0.00 \; .$$

A reanalysis of the data listed in Fig. 9.2 using a moment-approximation permutation procedure yields $\delta_o = 4.7667$, $\mu_\delta = 9.6444$, $\sigma_\delta^2 = 0.1964$, $\gamma_\delta = -1.0674$, an observed standardized test statistic of

$$T_o = \frac{\delta_o - \mu_\delta}{\sigma_\delta} = \frac{4.7667 - 9.6444}{0.4432} = -11.0058 \; ,$$

and a Pearson type III approximate probability value of $P = 0.3232 \times 10^{-7}$. No comparison is made with a conventional randomized-block analysis of variance as the F-ratio is undefined for $v = 1$.

Following Eq. (9.3) on p. 446, the exact expected value of the M δ values is $\mu_\delta = 9.6444$ and, following Eq. (9.2) on p. 446, the observed chance-corrected measure of effect size is

$$\Re_o = 1 - \frac{\delta_o}{\mu_\delta} = 1 - \frac{4.7667}{9.6444} = +0.5058 \; ,$$

indicating approximately 51 % within-block agreement above that expected by chance.

9.4 Permutation Analogue of MANOVA

Occasionally, a research design requires $g \geq 3$ matched treatment groups when multivariate ($r \geq 2$) response measurements are to be obtained for each of $b \geq 2$ blocks. When $v = 2$, this is essentially an extension of Hotelling's matched-pairs T^2 statistic to $g > 2$ treatments.

9.4.1 Example 1: $v = 2$

Consider $b = 3$ subjects (S) that are tested $g = 3$ times (Factor B) and $r = 3$ commensurate response measurements (Factor A) are obtained at each treatment. The data are adapted from Myers and Well [315, p. 260] and are listed in Table 9.3.

The example analysis blocks on the $b = 3$ subjects and compares the $g = 3$ sets of multivariate response measurement scores. For this example analysis, $r = 3$, $b = 3$, $g = 3$, and $v = 2$, employing squared Euclidean distance between response measurement scores to correspond to a multivariate randomized-block analysis. Since there are only

$$M = \left(g!\right)^{b} = \left(3!\right)^{3} = 216$$

possible, equally-likely arrangements of the $b = 3$ subjects, an exact solution is preferable. Following Eq. (9.1) on p. 445, the observed value of the MRBP test statistic with $v = 2$ is $\delta_o = 4.7933$. If all arrangements of the observed response measurement scores listed in Table 9.3 occur with equal chance, the exact probability value of $\delta_o = 4.7933$ computed on the $M = 216$ possible arrangements of the observed response measurement scores with $b = 3$ blocks preserved for each arrangement is

$$P(\delta \leq \delta_o | H_0) = \frac{\text{number of } \delta \text{ values} \leq \delta_o}{M} = \frac{6}{216} = 0.0278 .$$

For comparison, a conventional multivariate randomized-block analysis of variance on the response measurement scores listed in Table 9.3 yields an observed

Table 9.3 Example data for $b = 3$ subjects (S) tested $g = 3$ times (Factor B) with $r = 3$ response measurements (Factor A)

	B_1			B_2			B_3		
	A_1	A_2	A_3	A_1	A_2	A_3	A_1	A_2	A_3
S_1	3.1	2.9	2.4	1.9	2.0	1.7	1.6	1.9	1.5
S_2	5.7	6.8	5.3	4.5	5.7	4.4	4.4	5.3	3.9
S_3	9.7	10.9	8.0	7.4	10.5	6.6	6.9	8.9	6.0

value of $F_o = 22.5488$. Assuming independence, normality, and homogeneity of variance and covariance, F is approximately distributed as Snedecor's F with $\nu_1 = g - 1 = 3 - 1 = 2$ and $\nu_2 = (b-1)(g-1) = (3-1)(3-1) = 4$ degrees of freedom. Under the null hypothesis, the observed value of $F_o = 22.5488$ yields an approximate probability value of $P = 0.0066$.

Following Eq. (9.3) on p. 446, the exact expected value of the $M = 216$ δ values is $\mu_\delta = 44.6659$ and, following Eq. (9.2) on p. 446, the observed chance-corrected measure of effect size is

$$\Re_o = 1 - \frac{\delta_o}{\mu_\delta} = 1 - \frac{4.7933}{44.6659} = +0.8927\,,$$

indicating approximately 89 % within-block agreement above that expected by chance.

9.4.2 Example 2: $v = 1$

For a comparison analysis of the multivariate response measurement scores listed in Table 9.3, set $v = 1$ instead of $v = 2$, employing ordinary Euclidean distance between response measurement scores. Since there are still only

$$M = (g!)^b = (3!)^3 = 216$$

possible, equally-likely arrangements of the $b = 3$ subjects into the $g = 3$ treatment groups, an exact solution is preferable. Following Eq. (9.1) on p. 445, the observed value of the MRBP test statistic with $v = 1$ is $\delta_o = 1.9405$. If all arrangements of the observed response measurement scores listed in Table 9.3 occur with equal chance, the exact probability value of $\delta_o = 1.9405$ computed on the $M = 216$ possible arrangements of the observed response measurement scores with $b = 3$ blocks preserved for each arrangement is

$$P(\delta \le \delta_o | H_0) = \frac{\text{number of } \delta \text{ values} \le \delta_o}{M} = \frac{6}{216} = 0.0278\,.$$

No comparison is made with a conventional multivariate randomized-block analysis of variance as the F-ratio is undefined for $v = 1$.

Following Eq. (9.3) on p. 446, the exact expected value of the $M = 216$ δ values is $\mu_\delta = 5.5414$ and, following Eq. (9.2) on p. 446, the observed chance-corrected measure of effect size is

$$\Re_o = 1 - \frac{\delta_o}{\mu_\delta} = 1 - \frac{1.9405}{5.5414} = +0.6498\,,$$

indicating approximately 65 % within-block agreement above that expected by chance.

9.5 MRBP and Pearson's Product-Moment Correlation

Consider the MRBP test statistic given by

$$\delta = \left[g \binom{b}{2} \right]^{-1} \sum_{i=1}^{g} \sum_{j<k} \Delta(x_{ij}, x_{ik})$$

where g is the number of treatments, b is the number of blocks, and $\Delta(x, y)$ is a generalized Minkowski distance-function value of two points, $x' = (x_1, \ldots, x_r)$ and $y' = (y_1, \ldots, y_r)$ in an r-dimensional Euclidean space given by

$$\Delta(x, y) = \left(\sum_{i=1}^{r} \left| x_i - y_i \right|^p \right)^{v/p} ,$$

where $p \geq 1$ and $v > 0$. As discussed in Chap. 8, Sect. 8.2, when $b = v = 2$, the MRBP test statistic δ is closely related to the ordinary Pearson product-moment correlation coefficient given by

$$R = \frac{\text{cov}(x_1, x_2)}{s_1 s_2} ,$$

where $x_1 = (x_{11}, \ldots, x_{g1}),\ x_2 = (x_{12}, \ldots, x_{g2}),$

$$\text{cov}(x_1, x_2) = \frac{1}{g-1} \sum_{i=1}^{g} (x_{i1} - \bar{x}_1)(x_{i2} - \bar{x}_2) ,$$

$$\bar{x}_j = \frac{1}{g} \sum_{i=1}^{g} x_{ij} , \quad \text{and} \quad s_j = \left[\frac{1}{g-1} \sum_{i=1}^{g} \left(x_{ij} - \bar{x}_j \right)^2 \right]^{1/2}$$

for $j = 1, 2$.

Specifically, the functional relationships between Pearson's R and the MRBP test statistic are given by

$$R = \frac{\mu_\delta - \delta}{2S_1S_2} \quad \text{and} \quad \delta = \mu_\delta - 2RS_1S_2 ,$$

where

$$S_j^2 = \frac{1}{g} \sum_{i=1}^{g} \left(x_{ij} - \bar{x}_j \right)^2 \tag{9.9}$$

for $j = 1, 2$.[4] Furthermore, as noted in Chap. 8, the relationships between R and $\Re$ are given by

$$R = \frac{\Re \mu_\delta}{2S_1 S_2} \quad \text{and} \quad \Re = \frac{2RS_1 S_2}{\mu_\delta} . \tag{9.10}$$

9.5.1 Example 1: $v = 2$

Consider the response measurement scores listed in Fig. 9.3 with $g = 12$ objects, $b = 2$ blocks, and $r = 1$ response measurement. For the response measurement scores listed in Fig. 9.3, $\bar{x}_1 = 43.3333$, $\bar{x}_2 = 42.6667$, $s_1 = 2.1462$, $s_2 = 1.5570$, $\text{cov}(x_1, x_2) = 1.6667$, and the observed value of the Pearson product-moment correlation coefficient is

$$R_o = \frac{\text{cov}(x_1, x_2)}{s_1 s_2} = \frac{1.6667}{(2.1462)(1.5570)} = +0.4988 .$$

Employing squared Euclidean distance between response measurement scores with $v = 2$ to correspond to the Pearson product-moment correlation coefficient with $S_1 = 2.0548$ and $S_2 = 1.4907$, the observed value of the MRBP test statistic with $v = 2$ is $\delta_o = 3.8333$, the exact expected value of the M δ values is $\mu_\delta = 6.8889$ and the observed value of R is, equivalently,

$$R_o = \frac{\mu_\delta - \delta_o}{2S_1 S_2} = \frac{6.8889 - 3.8333}{(2)(2.0548)(1.4907)} = +0.4988 .$$

Fig. 9.3 Example response measurement scores with $g = 12$ objects, $b = 2$ blocks, and $r = 1$ response measurement

Object	x_1	x_2
1	41	41
2	42	41
3	43	41
4	45	41
5	43	42
6	41	42
7	45	43
8	40	43
9	46	44
10	43	44
11	47	45
12	44	45

[4] Note that the summation for S_j^2 in Eq. (9.9) is divided by g and not by $g - 1$, as degrees of freedom are not relevant to permutation methods.

An exact solution is not practical for these data since there are $M = 12! = 479{,}001{,}600$ possible, equally-likely arrangements to be considered. Thus, a resampling approach is mandated. If all M possible arrangements of the observed response measurement scores listed in Fig. 9.3 occur with equal chance, the approximate resampling probability value of $\delta_o = 3.8333$ computed on $L = 1{,}000{,}000$ random arrangements of the observed response measurement scores with $b = 2$ blocks preserved for each arrangement is

$$P(\delta \leq \delta_o | H_0) = \frac{\text{number of } \delta \text{ values} \leq \delta_o}{L} = \frac{55{,}628}{1{,}000{,}000} = 0.0556 .$$

For comparison, a conventional t test of $R = +0.4988$ yields an observed value of

$$t_o = \left[\frac{(g-2)R^2}{1-R^2} \right]^{1/2} = \left[\frac{(12-2)(+0.4988)^2}{1-(+0.4988)^2} \right]^{1/2} = +1.8199 .$$

Assuming independence and normality, t is approximately distributed as Student's t under the null hypothesis with $g - 2 = 12 - 2 = 10$ degrees of freedom. Under the null hypothesis, the observed value of $t_o = +1.8199$ yields an approximate two-sided probability value of $P = 0.0988$.

Finally, for the response measurement scores listed in Fig. 9.3, the observed chance-corrected measure of effect size is

$$\Re_o = 1 - \frac{\delta_o}{\mu_\delta} = 1 - \frac{3.8333}{6.8889} = +0.4435 ,$$

indicating approximately 44 % within-block agreement above that expected by chance. Following Eq. (9.10) on p. 468, the relationships between R and $\Re$ can be confirmed. The observed values of R and $\Re$ are

$$R_o = \frac{\Re_o \mu_\delta}{2 S_1 S_2} = \frac{(+0.4435)(6.8889)}{2(2.0548)(1.4907)} = +0.4988$$

and

$$\Re_o = \frac{2 R_o S_1 S_2}{\mu_\delta} = \frac{2(+0.4988)(2.0548)(1.4907)}{6.8889} = +0.4435 .$$

9.5.2 Example 2: $v = 1$

Although a conventional Pearson product-moment correlation coefficient is undefined for ordinary Euclidean distances between response measurement scores with

$v = 1$, it is still possible to calculate values for the MRBP test statistic, δ, and the chance-corrected measure of effect size, $\Re$, with $v = 1$. For the response measurement scores listed in Fig. 9.3 with $g = 12$, $b = 2$, $r = 1$, and $v = 1$, the observed value for δ is $\delta_o = 1.6667$. Following Eq. (9.3) on p. 446, the exact expected value of the M δ values is $\mu_\delta = 2.1111$ and, following Eq. (9.2) on p. 446, the observed chance-corrected measure of effect size is

$$\Re_o = 1 - \frac{\delta_o}{\mu_\delta} = 1 - \frac{1.6667}{2.1111} = +0.2105 \; ,$$

indicating approximately 21 % within-block agreement above that expected by chance.

Again, an exact solution is not practical for these data since there are $M = 12! = 479{,}001{,}600$ possible, equally-likely arrangements of the observed data to be considered. If all M possible arrangements of the observed response measurement scores listed in Fig. 9.3 occur with equal chance, the approximate resampling probability value of $\delta_o = 1.6667$ based on $v = 1$ and $L = 1{,}000{,}000$ random arrangements of the observed response measurement scores with $b = 2$ blocks preserved for each arrangement is

$$P(\delta \le \delta_o | H_0) = \frac{\text{number of } \delta \text{ values} \le \delta_o}{L} = \frac{169{,}196}{1{,}000{,}000} = 0.1692 \; .$$

9.5.3 Example 3: Permutation Data

As explained in Chap. 8, Sect. 8.2.2, if $(x_{11}, \ldots, x_{g1})$ is one of the $g!$ permutations of the observed response measurement scores and $v = 2$, then the Pearson product-moment correlation coefficient, R, is equivalent to the chance-corrected measure of within-block effect size, $\Re$, where

$$R = \frac{\text{cov}(x_1, x_2)}{s_1 s_2} \quad \text{and} \quad \Re = 1 - \frac{\delta}{\mu_\delta} \; .$$

Consider the small data set listed in Fig. 9.4 with $g = 10$ objects, $b = 2$ blocks, and $r = 1$ response measurement, where the 10 response measurement scores listed under x_2 constitute a permutation of the 10 response measurement scores listed under x_1.

For the response measurement scores listed in Fig. 9.4, $\bar{x}_1 = \bar{x}_2 = 11.00$, $s_1 = s_2 = 4.4721$, $\text{cov}(x_1, x_2) = 16.00$, and the observed Pearson product-moment correlation coefficient is

$$R_o = \frac{\text{cov}(x_1, x_2)}{s_1 s_2} = \frac{16.00}{(4.4721)(4.4721)} = +0.80 \; .$$

Fig. 9.4 Example response measurement scores with $g = 10$ objects, $b = 2$ blocks, and $r = 1$ response measurement

Object	x_1	x_2
1	17	14
2	11	11
3	8	5
4	14	17
5	5	8
6	8	5
7	14	17
8	11	11
9	5	8
10	17	14

Equivalently, for the response measurement scores listed in Fig. 9.4 with $v = 2$, the observed value of the MRBP test statistic is $\delta_o = 7.20$, the exact expected value of the M δ values is $\mu_\delta = 36.00$, and the observed chance-corrected measure of effect size is

$$\Re_o = 1 - \frac{\delta_o}{\mu_\delta} = 1 - \frac{7.20}{36.00} = +0.80 \; .$$

Since there are $M = 10! = 3{,}628{,}800$ possible, equally-likely arrangements of the observed response measurement scores listed in Fig. 9.4, a resampling solution is more practical. If all M possible arrangements of the observed response measurement scores listed in Fig. 9.4 occur with equal chance, the approximate resampling probability value of $\delta_o = 7.20$ computed on $L = 1{,}000{,}000$ random arrangements of the observed response measurement scores with $b = 2$ blocks preserved for each arrangement is

$$P(\delta \leq \delta_o|H_0) = \frac{\text{number of } \delta \text{ values} \leq \delta_o}{L} = \frac{5{,}539}{1{,}000{,}000} = 0.0055 \; .$$

While $M = 3{,}628{,}800$ possible arrangements usually mandates a resampling solution, an exact solution is not out of the question, providing an opportunity to compare exact and resampling probability values. For the response measurement scores listed in Fig. 9.4, the exact probability value of $\delta_o = 7.20$ computed on the $M = 3{,}628{,}800$ possible arrangements of the observed data is

$$P(\delta \leq \delta_o|H_0) = \frac{\text{number of } \delta \text{ values} \leq \delta_o}{M} = \frac{19{,}520}{3{,}628{,}800} = 0.0055 \; .$$

The exact probability value to six decimal places is $P = 0.005538$, the approximate resampling probability value based on $L = 1{,}000{,}000$ is $P = 0.005539$, and the difference is only 0.000001, demonstrating the relative efficiency of resampling-approximation procedures.

For comparison, a conventional t test of $R = +0.80$ yields an observed value of

$$t_o = \left[\frac{(g-2)R^2}{1-R^2}\right]^{1/2} = \left[\frac{(10-2)(+0.80)^2}{1-(+0.80)^2}\right]^{1/2} = +3.7712\,.$$

Assuming independence and normality, t is approximately distributed as Student's t under the null hypothesis with $g - 2 = 10 - 2 = 8$ degrees of freedom. Under the null hypothesis, the observed value of $t_o = +3.7712$ yields an approximate two-sided probability value of $P = 0.0055$.

9.6 Coda

Chapter 9 applied the Multivariate Randomized Block Procedures (MRBP) developed in Chap. 8 to establish relationships between the test statistics of MRBP, δ and $\mathfrak{R}$, and selected conventional tests and measures designed for the analysis of randomized-block data at the interval level of measurement. Considered in this chapter were Student's t test for matched pairs, Hotelling's multivariate T^2 test for matched pairs, randomized-block analysis of variance, and randomized-block multivariate analysis of variance. In addition, the functional relationship between the Pearson product-moment correlation coefficient and δ was detailed with $v = 2$.

Comparisons between $v = 2$, employing squared Euclidean distance between response measurement scores, and $v = 1$, employing ordinary Euclidean distance between response measurement scores, for a variety of statistical tests and measures revealed marked differences between obtained probability values. In this chapter, probability values based on ordinary Euclidean distance were generally greater than those based on squared Euclidean distance between response measurement scores. The permutation-based MRBP test statistic, δ, with $v = 1$ possesses several advantages over conventional statistics based on $v = 2$. The MRBP test statistic is data-dependent, distribution-free, and, with $v = 1$, is robust to extreme response measurement values.

Chapter 10

Chapter 10 establishes the relationships between the MRBP test statistics, δ and $\mathfrak{R}$, and selected conventional tests and measures designed for the analysis of randomized-block data at the ordinal level of measurement. Considered in Chap. 10 are the Wilcoxon signed-ranks test, the sign test, Spearman's rank-order and footrule measures of correlation, Friedman's analysis of variance for ranks, Kendall's coefficient of concordance, Cohen's weighted kappa measure of agreement, Kendall's τ_a and τ_b measures of ordinal association, Yule's Q and Y statistics, and Somers' d_{yx} and d_{xy} asymmetric measures of ordinal association.

Randomized Block Designs: Ordinal Data 10

Chapter 9 of *Permutation Statistical Methods* utilized the Multivariate Randomized Block Permutation (MRBP) procedures presented in Chap. 8 to develop relationships between the test statistics of MRBP, δ and $\Re$, and selected conventional tests and measures designed for the analysis of randomized-block data at the interval level of measurement. This tenth chapter continues the application of the MRBP test statistics to selected conventional tests and measures designed for the analysis of randomized-block data at the ordinal level of measurement. A variety of statistical tests and measures are considered in this chapter, including the Wilcoxon signed-ranks test, the sign test, Spearman's rank-order and footrule measures of correlation, Friedman's analysis of variance for ranks, Kendall's coefficient of concordance, Cohen's weighted kappa measure of agreement, Kendall's τ_a and τ_b measures of ordinal association, Stuart's τ_c statistic, Goodman and Kruskal's γ measure of ordinal association, and Somers' d_{yx} and d_{xy} asymmetric measures of ordinal association.

10.1 Introduction

As detailed in Chap. 8, randomized-block analysis-of-variance designs analyze univariate or multivariate observations on matched objects or subjects. Such designs have been important in academic fields of inquiry ranging from agriculture to zoology. Depending on the field of inquiry, they are variously known as randomized-block, repeated-measures, or within-subjects designs.

Let $x'_{ij} = (x_{1ij}, x_{2ij}, \ldots, x_{rij})$ denote a transposed vector of r response measurements associated with the ith treatment and jth block. Then the MRBP test statistic is given by

$$\delta = \left[g \binom{b}{2} \right]^{-1} \sum_{i=1}^{g} \sum_{j<k} \Delta(x_{ij}, x_{ik}) \ , \tag{10.1}$$

K.J. Berry et al., *Permutation Statistical Methods*,
DOI 10.1007/978-3-319-28770-6_10

where $\sum_{j<k}$ denotes the sum over all j and k such that $1 \le j < k \le b$ and $\Delta(x, y)$ is a symmetric distance-function value of two points $x' = (x_1, x_2, \ldots, x_r)$ and $y' = (y_1, y_2, \ldots, y_r)$ in an r-dimensional Euclidean space. In the context of randomized-block designs, the generalized Minkowski distance function is given by

$$\Delta(x, y) = \left(\sum_{i=1}^{r} |x_i - y_i|^p \right)^{v/p} ,$$

where $p \ge 1$ and $v > 0$.

The null hypothesis (H_0) states that the distribution of δ assigns an equal probability to each of the

$$M = (g!)^b$$

possible, equally-likely allocations of the r-dimensional response measurements to the g treatment positions within each of the b blocks.

An exact probability value for the observed MRBP test statistic, δ_o, may be expressed as

$$P(\delta \le \delta_o | H_0) = \frac{\text{number of } \delta \text{ values} \le \delta_o}{M}$$

and a chance-corrected measure of within-block agreement among all b blocks for all g treatments provides a measure of effect size given by

$$\Re = 1 - \frac{\delta}{\mu_\delta} , \tag{10.2}$$

where μ_δ is the arithmetic average of the M δ values calculated on all possible arrangements of the observed data given by

$$\mu_\delta = \frac{1}{M} \sum_{i=1}^{M} \delta_i . \tag{10.3}$$

When M is very large, an approximate probability value for δ may be obtained from a resampling procedure, where

$$P\big(\delta \le \delta_o | H_0\big) = \frac{\text{number of } \delta \text{ values} \le \delta_o}{L}$$

and L is a random sample of all possible arrangements of the bg response measurements. Typically, L is set to a large value to ensure accuracy, e.g., $L = 1{,}000{,}000$. When M is very large and P is exceedingly small, a resampling-approximation permutation procedure may produce no δ values equal to or less than δ_o, even with

$L = 1{,}000{,}000$, yielding an approximate resampling probability value of $P = 0.00$. In such cases, moment-approximation permutation procedures based on fitting the first three exact moments of the discrete permutation distribution to a Pearson type III distribution provide approximate probability values, as detailed in Chap. 1, Sect. 1.2.2 [284, 299].

10.2 Wilcoxon Signed-Ranks Test

Research designs that posit a test of differences between $g = 2$ matched treatment groups in which univariate ($r = 1$) response measurements have been obtained for each of $b = 2$ blocks are ubiquitous in the statistical literature. When the differences are measured between rank scores, replacing raw scores, the conventional statistic for such research designs is Wilcoxon's signed-ranks test for matched pairs.

In 1945 Frank Wilcoxon published a seminal article on "Individual comparisons by ranking methods" in the initial volume of *Biometrics Bulletin* [429]. Contained within this very brief article of only three pages were two highly innovative rank tests: the rank-sum test for two independent (unpaired) samples and the signed-ranks test for two dependent (paired) samples. The Wilcoxon rank-sum test is described in Chap. 5, Sect. 5.4, as it was designed by Wilcoxon to analyze completely randomized data; the Wilcoxon signed-ranks test is discussed in this chapter as it was originally designed by Wilcoxon to analyze matched-pairs data.[1]

Consider a set of response measurements consisting of N' paired observations, i.e.,

$$\{(x_1, y_1), (x_2, y_2), \ldots, (x_{N'}, y_{N'})\} .$$

The absolute differences between x_i and y_i are given by

$$d_i = |x_i - y_i| , \qquad i = 1, \ldots, N' ,$$

where any $d_i = 0$ is omitted from further consideration and the remaining number of pairs is denoted by N, $N \leq N'$. Next, rank scores $(r_1, \ldots, r_N)$ are assigned to the N pairs according to the relative size of the absolute difference. Finally, to each rank score is affixed the sign of the difference $(R_1, \ldots, R_N)$, indicating those ranked differences that arose from negative values of d_i and those ranked differences that arose from positive values of d_i. Two statistics are defined

$$R^+ = \text{the sum of the positive } R_i \text{ values}$$

[1]For a detailed description of Wilcoxon's signed-ranks test, see a discussion in *A Chronicle of Permutation Statistical Methods* by Berry et al. [41, pp. 137–139].

and

$$R^- = \text{the sum of the negative } R_i \text{ values}$$

for $i = 1, \ldots, N$.

When $g = 2$ and $r = 1$, the MRBP test statistic, as defined in Eq. (10.1) on p. 473, reduces to

$$\delta = \binom{b}{2}^{-1} \sum_{j<k} \Delta(x_j - x_k) \, .$$

Since R^+ and R^- sum to $N(N+1)/2$, it is sufficient to establish the identities relating δ and R^+, which are given by

$$\delta = \frac{N(N+1)(2N+1)}{3(N-1)} - \frac{\left[4R^+ - N(N+1)\right]^2}{2N(N-1)} \tag{10.4}$$

and

$$R^+ = \frac{N(N+1)}{4} + \left\{ \frac{N\left[N(N+1)(2N+1) - 3(N-1)\delta\right]}{24} \right\}^{1/2} . \tag{10.5}$$

For a detailed description of the functional relationships between δ and R^+, see a 1982 article on "An extended class of permutation techniques for matched pairs" by Mielke and Berry [288, p. 1200].

10.2.1 Example 1: $v = 2$

To illustrate the Wilcoxon signed-ranks test, consider the univariate matched-pairs data listed in Fig. 10.1. The raw data represent pounds of head corn per acre and

Fig. 10.1 Example matched-pairs data for the Wilcoxon signed-ranks test with $N = 11$ paired differences and associated signed-ranks [390, p. 24]

Pair	x	y	d	R
1	1,443	1,316	+127	+11
2	2,009	1,903	+106	+10
3	2,011	1,910	+101	+9
4	2,180	2,108	+72	+8
5	1,542	1,612	−70	−7
6	2,122	2,060	+62	+6
7	1,482	1,444	+38	+5
8	1,925	1,971	−36	−4
9	2,463	2,496	−33	−3
10	1,535	1,511	+24	+2
11	1,915	1,935	−20	−1

are adapted from Student's oft-cited 1908 paper on "The probable error of a mean" [390, p. 24].

For the univariate matched-pairs data listed in Fig. 10.1, the observed value of R^+ is $R^+_{\text{o}} = 11 + 10 + 9 + 8 + 6 + 5 + 2 = 51$. For a small sample such as this, tables of exact probability values are available from a variety of sources. Using a standard table published by Siegel and Castellan, the exact probability value of $R^+ = 51$ is given as $P = 0.1230$ [375, pp. 332–334].

It is well established that as $N \to \infty$, R^+ is approximately distributed as $N(0, 1)$ under the null hypothesis with mean and variance given by

$$\mu_{R+} = \frac{N(N+1)}{4}$$

and

$$\sigma^2_{R+} = \frac{N(N+1)(2N+1)}{24} ,$$

respectively. Therefore, the standardized score of R^+ is given by

$$z = \frac{R^+ - \mu_{R+}}{\sigma_{R+}} = \frac{R^+ - \dfrac{N(N+1)}{4}}{\left[\dfrac{N(N+1)(2N+1)}{24}\right]^{1/2}} ,$$

an especially convoluted form expressed in many textbooks. For the example matched-pairs data listed in Fig. 10.1,

$$\mu_{R+} = \frac{N(N+1)}{4} = \frac{11(11+1)}{4} = 33 ,$$

$$\sigma^2_{R+} = \frac{N(N+1)(2N+1)}{24} = \frac{11(11+1)[(2)(11)+1}{24} = 126.50 ,$$

the observed standardized score is

$$z_{\text{o}} = \frac{R^+_{\text{o}} - \mu_{R+}}{\sigma_{R+}} = \frac{51 - 33}{\sqrt{126.5000}} = +1.6004 ,$$

and the two-sided $N(0, 1)$ approximate probability value of $z_{\text{o}} = +1.6004$ under the null hypothesis is $P = 0.1095$. A standard correction for continuity for R^+ provides a closer approximation to the exact probability value of $P = 0.1230$ with $z_{\text{o}} = +1.5559$ and a corrected two-sided $N(0, 1)$ approximate probability value of $P = 0.1197$, under the null hypothesis.

For the univariate matched-pairs data listed in Fig. 10.1, there are only

$$M = (g!)^b = 2^{11} = 2{,}048$$

possible, equally-likely arrangements of the observed data. Therefore, an exact solution is easily accomplished. Employing squared Euclidean distance between the rank scores with $v = 2$ to correspond to Wilcoxon's signed-ranks test and following Eq. (10.1) on p. 473, the observed value of the MRBP test statistic with $v = 2$ is $\delta_o = 77.6363$. If all arrangements of the observed matched-pairs data listed in Fig. 10.1 occur with equal chance, the exact probability value of $\delta_o = 77.6363$ computed on the $M = 2{,}048$ possible arrangements of the observed rank scores with $b = 11$ blocks preserved for each arrangement is

$$P(\delta \leq \delta_o | H_0) = \frac{\text{number of } \delta \text{ values} \leq \delta_o}{M} = \frac{252}{2{,}048} = 0.1230 \, ,$$

which agrees with the tabled value from Siegel and Castellan [375, pp. 332–334]. For comparison, the two-sided $N(0, 1)$ approximate probability value, corrected for continuity, is $P = 0.1197$.

Following Eq. (10.3) on p. 474, the exact expected value of the $M = 2{,}048$ δ values is $\mu_\delta = 92.00$ and, following Eq. (10.2) on p. 474, the observed chance-corrected measure of effect size is

$$\Re_o = 1 - \frac{\delta_o}{\mu_\delta} = 1 - \frac{77.6363}{92.00} = +0.1561 \, ,$$

indicating approximately 16 % within-block agreement above that expected by chance.

Finally, following Eq. (10.4) on p. 476, the relationships between δ and R^+ can be confirmed. For the univariate matched-pairs data listed in Fig. 10.1, the observed value of δ is

$$\delta_o = \frac{N(N+1)(2N+1)}{3(N-1)} - \frac{\left[4R_o^+ - N(N+1)\right]^2}{2N(N-1)}$$

$$= \frac{11(11+1)[2(11)+1]}{3(11-1)} - \frac{[4(51) - 11(11+1)]^2}{2(11)(11-1)}$$

$$= 101.2000 - 23.5636 = 77.6363$$

and following Eq. (10.5) on p. 476, the observed value of R^+ is

$$R_o^+ = \frac{N(N+1)}{4} + \left\{ \frac{N\left[N(N+1)(2N+1) - 3(N-1)\delta_o\right]}{24} \right\}^{1/2}$$

$$= \frac{11(11+1)}{4} + \left(\frac{(11)\{11(11+1)[2(11)+1] - 3(11-1)(77.6363)\}}{24} \right)^{1/2}$$

$$= 33 + 18 = 51 \; .$$

10.2.2 Example 2: $v = 1$

For a comparison analysis of the univariate matched-pairs data listed in Fig. 10.1, set $v = 1$ instead of $v = 2$, employing ordinary Euclidean distance between the rank scores. For the univariate matched-pairs data listed in Fig. 10.1, there are still only

$$M = (g!)^b = 2^{11} = 2{,}048$$

possible, equally-likely arrangements of the observed data. Following Eq. (10.1) on p. 473, the observed value of the MRBP test statistic with $v = 1$ is $\delta_o = 7.4182$. If all arrangements of the observed rank scores listed in Fig. 10.1 occur with equal chance, the exact probability value of $\delta_o = 7.4182$ computed on the $M = 2{,}048$ possible arrangements of the observed rank scores with $b = 11$ blocks preserved for each arrangement is

$$P(\delta \leq \delta_o | H_0) = \frac{\text{number of } \delta \text{ values} \leq \delta_o}{M} = \frac{300}{2{,}048} = 0.1465 \; .$$

For comparison, the exact probability value based on $v = 2$ and $M = 2{,}048$ in Example 1 is $P = 0.1230$. No comparison is made with the conventional Wilcoxon signed-rank test as Wilcoxon's test is undefined for $v = 1$.

Following Eq. (10.3) on p. 474, the exact expected value of the $M = 2{,}048$ δ values is $\mu_\delta = 8.00$ and, following Eq. (10.2) on p. 474, the observed chance-corrected measure of effect size is

$$\Re_o = 1 - \frac{\delta_o}{\mu_\delta} = 1 - \frac{7.4182}{8.00} = +0.0727 \; ,$$

indicating approximately 7 % within-block agreement above that expected by chance.

10.3 Sign Test

The sign test is based solely upon the direction of differences between two sets of response measurements, ignoring the magnitudes of the differences. The sign test is applicable to the case of two matched samples when a researcher simply wishes to establish that $g = 2$ conditions are different.

Like the signed-rank test, consider response measurements consisting of N' paired observations $\{(x_1, y_1), (x_2, y_2), \ldots, (x_{N'}, y_{N'})\}$. The differences are given by

$$d_i = x_i - y_i \ , \qquad i = 1, \ldots, N' \ ,$$

where any $d_i = 0$ is omitted from further consideration and the remaining number of pairs is denoted by N, $N \leq N'$. Next, the magnitudes of the differences are removed and the signs of the differences are preserved, indicating the differences that arose from negative values of d_i and the differences that arose from positive values of d_i. Let $d_i = \pm 1$ for $i = 1, \ldots, N$ and let R^+ denote the number of positive signed values, then the identities relating δ and R^+ are given by

$$\delta = \frac{4R^+(N - R^+)}{N(N-1)} \quad \text{and} \quad R^+ = \frac{N + \sqrt{N^2 - N(N-1)\delta}}{2} \ , \qquad (10.6)$$

as described by Mielke and Berry in 1982 [288].

10.3.1 Example Sign Test

To illustrate the computation of a sign test, consider the univariate matched-pairs data listed in Fig. 10.2. The data are adapted from Siegel and Castellan [375, p. 82]. For the matched-pairs data listed in Fig. 10.2, the observed sum of positive signs is

Fig. 10.2 Example matched-pairs data for the sign test with $N = 14$ paired differences and associated signs

Pair	x	y	d	Sign
1	5	3	+2	+
2	4	3	+1	+
3	6	4	+2	+
4	6	5	+1	+
5	2	3	−1	−
6	5	2	+3	+
7	1	2	−1	−
8	4	3	+1	+
9	5	2	+3	+
10	4	2	+2	+
11	4	5	−1	−
12	7	2	+5	+
13	5	3	+2	+
14	5	1	+4	+

$R_o^+ = 11$ (pairs 1, 2, 3, 4, 6, 8, 9, 10, 12, 13, and 14). The exact probability value of $R_o^+ \geq 11$ is given by the binomial distribution,

$$P(i \geq x|N) = \sum_{i=x}^{N} \binom{N}{i} p^i (1-p)^{N-i} ,$$

where in this example, $N = 14$, $x = R_o^+ = 11$, and $p = 1/2$. Thus,

$$P(i \geq 11|14) = \sum_{i=11}^{14} \binom{14}{i} \left(\frac{1}{2}\right)^i \left(1 - \frac{1}{2}\right)^{14-i}$$

$$= \frac{\binom{14}{11} + \binom{14}{12} + \binom{14}{13} + \binom{14}{14}}{2^{14}} = \frac{364 + 91 + 14 + 1}{16{,}384}$$

$$= \frac{470}{16{,}384} = 0.0287$$

and the exact binomial two-sided probability value is $P = 2(0.0287) = 0.0574$.

For the univariate matched-pairs data listed in Fig. 10.2, there are only

$$M = (g!)^b = 2^{14} = 16{,}384$$

possible, equally-likely arrangements of the observed data. Therefore, an exact solution is feasible. Following Eq. (10.1) on p. 473, the observed value of the MRBP test statistic with $v = 1$ is $\delta_o = 0.7253$.[2] If all arrangements of the observed rank scores listed in Fig. 10.2 occur with equal chance, the exact probability value of $\delta_o = 0.7253$ computed on the $M = 16{,}384$ possible arrangements of the observed rank scores with $b = 14$ blocks preserved for each arrangement is

$$P(\delta \leq \delta_o|H_0) = \frac{\text{number of } \delta \text{ values} \leq \delta_o}{M} = \frac{940}{16{,}384} = 0.0574 .$$

For comparison, the exact binomial two-sided probability value is also $P = 0.0574$.

Following Eq. (10.3) on p. 474, the exact expected value of the $M = 16{,}384$ δ values is $\mu_\delta = 1.00$ and, following Eq. (10.2) on p. 474, the observed chance-corrected measure of effect size is

$$\Re_o = 1 - \frac{\delta_o}{\mu_\delta} = 1 - \frac{0.7253}{1.00} = +0.2747 ,$$

[2]While the values of δ and μ_δ depend on the choice of v, for the sign test $v = 1$ and $v = 2$ yield identical probability values.

indicating approximately 27 % within-block agreement above that expected by chance.

If R^+ denotes the number of positive signs and $d_i = \pm 1$ for $i = 1, \ldots, N$, then following Eq. (10.6) on p. 480, the observed values of δ and R^+ for the matched-pairs data listed in Fig. 10.2 are

$$\delta_o = \frac{4R_o^+(N - R_o^+)}{N(N-1)} = \frac{4(11)(14-11)}{14(14-1)} = \frac{132}{182} = 0.7253$$

and

$$R_o^+ = \frac{N + \sqrt{N^2 - N(N-1)\delta_o}}{2}$$
$$= \frac{14 + \sqrt{14^2 - 14(14-1)0.7253}}{2} = \frac{22}{2} = 11\,.$$

10.4 Spearman's Rank-Order Correlation Coefficient

Consider two rankings of g objects consisting of the first g integers and let x_i and y_i for $i = 1, \ldots, g$ denote the first and second rankings, respectively. A popular measure of correlation between the two rankings is Spearman's rank-order correlation coefficient, ρ, given by

$$\rho = 1 - \frac{\sum_{i=1}^{g} d_i^2}{\dfrac{g(g^2-1)}{6}} = 1 - \frac{6\sum_{i=1}^{g} d_i^2}{g(g^2-1)}\,, \tag{10.7}$$

where $d_i = x_i - y_i$ for $i = 1, \ldots, g$. Rho (ρ) was first developed by Charles Spearman in two articles in 1904 and 1906 that appeared in *American Journal of Psychology* and *British Journal of Psychology*, respectively [381, 382].[3]

Note that the denominator of Spearman's rank-order correlation coefficient, $g(g^2-1)/6$, as given in Eq. (10.7), represents one-half the maximum value of $\sum_{i=1}^{g} d_i^2$ when x_i and y_i for $i = 1, \ldots, g$ both consist of untied rank scores and the y_i rank scores are the exact inverse of the x_i rank scores, i.e., $y_i = g - x_i + 1$ for

[3]Spearman [381] allowed that the general idea of looking at rank differences was first due to Alfred Binet, while his contribution was to work out a formula with the properties of a correlation coefficient [171, pp. 513–514].

$i = 1, \ldots, g$. Thus, Spearman's ρ is a maximum-corrected measure of rank-order correlation.

It is easily confirmed that the denominator of Eq. (10.7), $g(g^2 - 1)/6$, is one-half the maximum value of $\sum_{i=1}^{g} d_i^2$ when x_i and y_i for $i = 1, \ldots, g$ are both untied rank scores and the y_i rank scores are the inverse of the x_i rank scores. For the maximum value of $\sum_{i=1}^{g} d_i^2$, define

$$\sum_{i=1}^{g} d_i^2 = \sum_{i=1}^{g} (x_i - y_i)^2 = \sum_{i=1}^{g} x_i^2 + \sum_{i=1}^{g} y_i^2 - 2 \sum_{i=1}^{g} x_i y_i \, . \qquad (10.8)$$

Since, for g untied rank scores,

$$\sum_{i=1}^{g} x_i^2 = \sum_{i=1}^{g} y_i^2 = \frac{g(g+1)(2g+1)}{6}$$

and, for $y_i = g - x_i + 1$, $i = 1, \ldots, g$,

$$\sum_{i=1}^{g} x_i y_i = \frac{g(g+1)(g+2)}{6} \, ,$$

then substituting into Eq. (10.8) yields

$$\sum_{i=1}^{g} d_i^2 = \frac{2g(g+1)(2g+1)}{6} - \frac{2g(g+1)(g+2)}{6}$$
$$= \frac{2g(g+1)(g-1)}{6} = \frac{g(g^2-1)}{3} \, ,$$

which is twice the value of $g(g^2 - 1)/6$.

Now consider the MRBP test statistic given by

$$\delta = \left[g \binom{b}{2} \right]^{-1} \sum_{i=1}^{g} \sum_{j<k} \Delta(x_{ij}, x_{ik}) \, , \qquad (10.9)$$

where g denotes the number of treatments, b is the number of blocks, and $\Delta(x, y)$ is a generalized Minkowski distance-function value of two points, $x' = (x_1, \ldots, x_r)$ and $y' = (y_1, \ldots, y_r)$ in an r-dimensional Euclidean space given by

$$\Delta(x, y) = \left(\sum_{i=1}^{r} |x_i - y_i|^p \right)^{v/p} \, ,$$

where $p \geq 1$ and $v > 0$.

Also, let a chance-corrected measure of effect size between the two rankings be given by

$$\Re = 1 - \frac{\delta}{\mu_\delta} , \tag{10.10}$$

where μ_δ is the arithmetic average of the $M = g!\ \delta$ values calculated on all possible arrangements of the observed data given by

$$\mu_\delta = \frac{1}{M} \sum_{i=1}^{M} \delta_i . \tag{10.11}$$

When $p = 2$, $v = 2$, and x_i and y_i for $i = 1, \ldots, g$ are rank-order statistics, the MRBP measure of effect size, $\Re$, is identical to the Spearman's rank-order correlation coefficient, ρ, given in Eq. (10.7).

10.4.1 Example: $v = 2$

Consider the rank-correlation data listed in Fig. 10.3 with $g = 8$ objects, $b = 2$ blocks, and $r = 1$ response measurement. For the rank-correlation data listed in Fig. 10.3, the columns labeled x and y contain the observed raw scores, the columns labeled r_x and r_y contain the corresponding rank scores, the column labeled d contains the signed differences between r_x and r_y, and the column labeled d^2 contains the squared rank differences. Following Eq. (10.7) on p. 482, the observed value of Spearman's rank-order correlation coefficient is

$$\rho_o = 1 - \frac{6 \sum_{i=1}^{g} d_i^2}{g(g^2 - 1)} = 1 - \frac{6(18)}{8(8^2 - 1)} = 1 - \frac{108}{504} = +0.7857 . \tag{10.12}$$

Fig. 10.3 Example rank-order correlation data with $g = 8$ objects, $b = 2$ blocks, and $r = 1$ response measurement

Pair	x	y	r_x	r_y	d	d^2
1	72	63	8	7	+1	1
2	46	49	6	6	0	0
3	13	35	2	4	−2	4
4	27	17	4	2	+2	4
5	53	81	7	8	−1	1
6	34	41	5	5	0	0
7	11	26	1	3	−2	4
8	22	15	3	1	+2	4
Total						18

Alternatively, for the rank-correlation data listed in Fig. 10.3, let $v = 2$, employing squared Euclidean distance between the rank scores to correspond to the Spearman's rank-order correlation coefficient [381, 382]. Then, following Eq. (10.9), the observed value of the MRBP test statistic with $v = 2$ is $\delta_o = 2.2500$, following Eq. (10.11) the exact expected value of the $M = 40{,}320$ δ values is $\mu_\delta = 10.50$, and following Eq. (10.10), the observed chance-corrected measure of effect size is

$$\Re_o = 1 - \frac{\delta_o}{\mu_\delta} = 1 - \frac{2.2500}{10.50} = +0.7857 ,$$

indicating approximately 79 % within-block agreement above that expected by chance.

The Spearman's rank-order correlation coefficient, as expressed in Eq. (10.7) on p. 482, is not appropriate for rank-correlation data with tied rank scores, but $\Re$ easily accommodates tied rank scores without any adjustment.[4] Thus, the Spearman's rank-order correlation coefficient is generalized to include tied rank scores and ρ is revealed simply as a special case of $\Re$ when no ties are present. Finally, as shown by Berry and Mielke, Spearman's ρ, like $\Re$, is a chance-corrected measure of agreement since the expected disagreement is $\mathrm{E}[\delta] = \mu_\delta$ [29].

Note that for the rank data listed in Fig. 10.3, $\rho_o = \Re_o = +0.7857$. It is not generally recognized that under special conditions Spearman's maximum-corrected rank-order correlation coefficient, ρ, is also a chance-corrected measure of agreement. When both variable x and variable y consist of ranks from 1 to g with no tied values, or variable x includes tied ranks and variable y is a permutation of variable x, then Spearman's ρ is, paradoxically, both a maximum-corrected measure of correlation and a chance-corrected measure of agreement since any deviation from perfect agreement also counts as a deviation from perfect correlation [223, p. 144]. The Pearson product-moment correlation for interval-level response measurements is also simultaneously a maximum-corrected measure of correlation and a chance-corrected measure of agreement whenever variable y is a permutation of variable x since the standard deviations of variables x and y are necessarily equal [29, 194, p. 7].

Because there are only $M = g! = 8! = 40{,}320$ possible, equally-likely arrangements of the observed data listed in Fig. 10.3, an exact permutation test is feasible. Since $\Re$ is simply a linear transformation of δ, if all arrangements of the observed rank scores listed in Fig. 10.3 occur with equal chance, the exact probability value of $\delta_o = 2.2500$ is identical to the probability of $\Re_o = +0.7857$ computed on the $M = 40{,}320$ possible arrangements of the observed rank scores with $b = 2$ blocks

[4] It is well known that simply calculating Pearson's product-moment correlation coefficient, r_{xy}, on the paired-rank scores provides Spearman's rank-order correlation coefficient and accommodates for any tied rank scores.

preserved for each arrangement. Thus,

$$P(\Re \geq \Re_o|H_0) = P(\delta \leq \delta_o|H_0) = \frac{\text{number of } \delta \text{ values} \leq \delta_o}{M}$$

$$= \frac{563}{40{,}320} = 0.0140\,.$$

10.5 Spearman's Footrule Agreement Measure

The oft-cited 1904 and 1906 articles by Charles Spearman contained two measures of rank-order correlation: the well-known Spearman's rank-order correlation coefficient, ρ, given in Eq. (10.7) on p. 482 and discussed in Sect. 10.4, *vide supra*, and a second, lesser-known, correlation coefficient that Spearman named "the footrule" [381, 382].[5,6]

As with Spearman's rank-order correlation coefficient, consider two rankings of g objects consisting of the first g integers and let x_i and y_i for $i = 1, \ldots, g$ denote the first and second rankings, respectively. Then, Spearman's footrule is given by

$$\mathcal{R} = 1 - \frac{\sum_{i=1}^{g} |x_i - y_i|}{\dfrac{g^2 - 1}{3}} = 1 - \frac{3\sum_{i=1}^{g} |x_i - y_i|}{g^2 - 1}\,. \tag{10.13}$$

Three limitations of the footrule contribute to its lack of use in contemporary research. First, unlike other measures of rank correlation, $\mathcal{R}$ does not norm properly between the limits of -1 and $+1$; second, like Spearman's ρ, $\mathcal{R}$ is limited to fully ranked data and does not accommodate tied rank scores; and third, because of the summation of absolute differences between the rank scores, it is somewhat cumbersome to establish the probability value of an observed value of $\mathcal{R}$, especially when g is small.[7]

[5] Actually, out of some 40 rank coefficients developed by Spearman, a number of measures of rank correlation are presented and discussed in the two Spearman articles, including existing measures such as Pearson's r_{xy}, Yule's Q, Yule's Y, and a number of other suggested new measures.

[6] As noted by Heiser, Spearman included a discussion of his rank-order correlation coefficient in his 1906 paper, only to dismiss it in favor of the footrule [171, p. 514].

[7] Spearman's footrule is one of only a few conventional test statistics based on ordinary Euclidean distances (absolute differences) between values.

10.5.1 Norming and Tied Rank Scores

Spearman's $\mathcal{R}$ attains a maximum value of $+1$ when x_i is identical to y_i for $i = 1, \ldots, g$ and no tied values are present. However, if $y_i = g - x_i + 1$ for $i = 1, \ldots, g$, then $\mathcal{R} = -0.5$ when g is odd and

$$\mathcal{R} = -0.5\left(1 + \frac{3}{g^2 - 1}\right)$$

when g is even [208]. Consequently, $\mathcal{R}$ does not attain a minimum value of -1, except when $g = 2$. Maurice Kendall explicitly pointed to this apparent lack of proper norming as a defect in the footrule [208, p. 33] and Spearman, recognizing that negative values of $\mathcal{R}$ did not represent inverse correlation, actually suggested that "it is better to treat every correlation as positive" [381, pp. 87–88].

Note that, unlike Spearman's rank-order correlation coefficient, the denominator of Spearman's footrule coefficient, $(g^2 - 1)/3$, as given in Eq. (10.13), does not represent one-half the maximum value of $\sum_{i=1}^{g} |x_i - y_i|$ when x_i and y_i for $i = 1, \ldots, g$ are both untied rank scores and the y_i rank scores are the exact inverse of the x_i rank scores, i.e., $y_i = g - x_i + 1$ for $i = 1, \ldots, g$. Thus, Spearman's $\mathcal{R}$ is not a maximum-corrected measure of rank-order correlation and is instead a chance-corrected measure of agreement.

It can easily be shown that Spearman's $\mathcal{R}$ is a chance-corrected measure of agreement and is not, in fact, a conventional measure of correlation, which explains why $\mathcal{R}$ can, on occasion, yield negative values. To show that the expected value of $\sum_{i=1}^{g} |d_i|$ is given by $(g^2 - 1)/3$, let

$$\begin{aligned}
\sum_{i=1}^{g} |d_i| &= \frac{1}{g}\sum_{i=1}^{g}\sum_{j=1}^{g} |i - j| \\
&= \frac{2}{g}\sum_{i=1}^{g-1}\sum_{j=i+1}^{g} (j - i) \\
&= \frac{1}{g}\sum_{i=1}^{g-1}\left[g(g+1) + i^2 - i(2g+1)\right] \\
&= \frac{g(g+1)}{6g}\left[6(g+1) + (2g-1) - 3(2g+1)\right] \\
&= \frac{g^2 - 1}{3} .
\end{aligned}$$

Therefore, Spearman's footrule coefficient,

$$\mathcal{R} = 1 - \frac{\displaystyle\sum_{i=1}^{g} |d_i|}{\dfrac{g^2 - 1}{3}} ,$$

is a chance-corrected measure of agreement when the expected value of $\sum_{i=1}^{g} |d_i|$ is given by $(g^2 - 1)/3$, as it takes the classic form of chance-corrected measures of agreement given by

$$\text{Agreement} = 1 - \frac{\text{Observed disagreement}}{\text{Expected disagreement}}$$

[223, p. 140].[8]

Alternatively, let

$$\delta = \frac{1}{g} \sum_{i=1}^{g} |x_i - y_i| ,$$

$$\mu_\delta = \frac{1}{g^2} \sum_{i=1}^{g} \sum_{j=1}^{g} |x_i - y_j| ,$$

and let

$$\Re = 1 - \frac{\delta}{\mu_\delta}$$

denote a general measure of the relationship between the two sets of rank scores that is not limited to untied rank scores. Let x and y with no subscripts denote $(x_1, \ldots, x_g)$ and $(y_1, \ldots, y_g)$, respectively. If no tied rank scores exist in either x or y, then the expected value of δ is given by

$$\mu_\delta = \frac{1}{g^2} \sum_{i=1}^{g} \sum_{j=1}^{g} |i - j|$$

$$= \frac{2}{g^2} \sum_{i=1}^{g-1} \sum_{j=i+1}^{g} (j - i)$$

[8] Spearman offered a somewhat different derivation of $(g^2 - 1)/3$ in the Appendix to his 1906 paper on the footrule [382, p. 105].

$$= \frac{1}{g^2} \sum_{i=1}^{g-1} \left[g(g+1) + i^2 - i(2g+1) \right]$$

$$= \frac{g(g-1)}{6g^2} \left[6(g+1) + (2g-1) - 3(2g+1) \right]$$

$$= \frac{g^2 - 1}{3g} , \qquad (10.14)$$

[29, pp. 841–842] and the relationships between Spearman's $\mathcal{R}$ and the MRBP test statistic are given by

$$\mathcal{R} = 1 - \frac{3g\delta}{g^2 - 1} \quad \text{and} \quad \delta = \frac{(g^2 - 1)(\mathcal{R} - 1)}{3g} . \qquad (10.15)$$

Finally, Spearman's $\mathcal{R}$ and $\Re$ can be shown to be equivalent. Given

$$\Re = 1 - \frac{\delta}{\mu_\delta} ,$$

then

$$\delta = \mu_\delta (1 - \Re) . \qquad (10.16)$$

Substituting $\mu_\delta(1 - \Re)$ in Eq. (10.16) for δ in

$$\mathcal{R} = 1 - \frac{3g\delta}{g^2 - 1}$$

yields

$$\mathcal{R} = 1 - \frac{3g\mu_\delta(1 - \Re)}{g^2 - 1} .$$

Then substituting $(g^2 - 1)/3g$ in Eq. (10.14) for μ_δ yields

$$\mathcal{R} = 1 - \frac{3g \left(\dfrac{g^2 - 1}{3g} \right) (1 - \Re)}{g^2 - 1} = 1 - \frac{(g^2 - 1)(1 - \Re)}{g^2 - 1} = \Re . \qquad (10.17)$$

Thus, the functional relationships between Spearman's footrule, δ, and $\Re$ are established, the footrule is generalized to include tied rank scores on both x and y, and $\mathcal{R}$ is shown to be equivalent to $\Re$ when no tied rank scores exist. Thus, $\mathcal{R}$, like $\Re$, is a chance-corrected measure of agreement, which explains why a lower limit of -1 is never attained by $\mathcal{R}$, except for the trivial case when $g = 2$. This places

Spearman's footrule into the family of chance-corrected agreement measures that includes such measures as Scott's π coefficient of intercoder agreement [368] and Cohen's κ coefficient of agreement [70].

10.5.2 Probability of Spearman's Footrule

When both the x and y variables consist entirely of untied rank scores from 1 to g and variable y is a permutation of the rank observations in variable x, then methods exist to determine the probability of an observed $\mathcal{R}$ under the null hypothesis that any of the $g!$ orderings of either the x or y values is equally likely. If

$$D = \sum_{i=1}^{g} \left| x_i - y_i \right| = g\delta$$

then, since $\mathcal{R}$ is a linear transformation of D, the probability of an observed value of D is the probability of an observed value of $\mathcal{R}$. Tables of the exact cumulative distribution function of D for $2 \leq g \leq 10$ and approximate probability values based on Monte Carlo methods for $11 \leq g \leq 15$ were given by Ury and Kleinecke in 1979 [408]. In 1988 Franklin extended the work of Ury and Kleinecke, reported the exact cumulative distribution function of D for $11 \leq g \leq 18$, and discussed the rate of convergence to an approximating normal distribution [125]. In 1990 Salama and Quade used Markov-chain properties to obtain the exact cumulative distribution function of D for $4 \leq g \leq 40$ and further investigated approximations to the discrete distribution of D [360].

If either variable x or variable y contains tied values, then the calculation of a probability value becomes more complex. However, because $\mathcal{R} = \Re$ and $\Re$ is merely a linear transformation of δ, the probability of an observed δ is equivalent to the probability of an observed $\mathcal{R}$. Thus, if $\mathcal{R}_o$ and δ_o denote the observed values of $\mathcal{R}$ and δ, respectively, then

$$P(\mathcal{R} \geq \mathcal{R}_o | H_0) = P(\delta \leq \delta_o | H_0) = \frac{\text{number of } \delta \text{ values} \leq \delta_o}{M} \, ,$$

where $M = g!$ in this application.

10.5.3 Example: $v = 1$

Consider the univariate paired-rank data listed in Fig. 10.4 where $g = 8, b = 2, r = 1$, and there are no tied rank scores. Following Eq. (10.13) on p. 486, the observed

Fig. 10.4 Example footrule rank-correlation data with $g = 8$ objects, $b = 2$ blocks, and $r = 1$ response measurement

Pair	x	y	$x-y$	$\lvert x-y\rvert$
1	8	7	+1	1
2	6	6	0	0
3	2	4	−2	2
4	4	2	+2	2
5	7	8	−1	1
6	5	5	0	0
7	1	3	−2	2
8	3	1	+2	2
Total				10

value of Spearman's $\mathcal{R}$ is

$$\mathcal{R}_o = 1 - \frac{3\sum_{i=1}^{g} |x_i - y_i|}{g^2 - 1} = 1 - \frac{3(10)}{8^2 - 1} = 1 - \frac{30}{63} = +0.5238 .$$

Alternatively, for the paired-rank data listed in Fig. 10.4, let $v = 1$, employing ordinary Euclidean distance between the rank scores to correspond to Spearman's footrule [382]. Then, following Eq. (10.9) on p. 483, the observed value of the MRBP test statistic with $v = 1$ is $\delta_o = 1.25$. Following Eq. (10.11) on p. 484, the exact expected value of the $M = 40{,}320$ δ values is $\mu_\delta = 2.6250$ and, following Eq. (10.10) on p. 484, the observed chance-corrected measure of effect size is

$$\Re_o = 1 - \frac{\delta_o}{\mu_\delta} = 1 - \frac{1.25}{2.6250} = +0.5238 ,$$

indicating approximately 52 % within-block agreement above that expected by chance, and establishing the identity between $\mathcal{R}$ and $\Re$ when $v = 1$.

Since there are only $M = 8! = 40{,}320$ possible, equally-likely arrangements of the observed data listed in Fig. 10.4, an exact permutation test is feasible. If all arrangements of the observed rank scores listed in Fig. 10.4 occur with equal chance, the exact probability value of $\delta_o = 1.25$ computed on the $M = 40{,}320$ possible arrangements of the observed rank scores with $b = 2$ blocks preserved for each arrangement is

$$P(\delta \leq \delta_o | H_0) = \frac{\text{number of } \delta \text{ values} \leq \delta_o}{M} = \frac{1{,}248}{40{,}320} = 0.0310 .$$

Finally, the relationships between $\mathcal{R}$ and δ can be confirmed with the univariate paired-rank data listed in Fig. 10.4. Thus, following the expressions in Eq. (10.15) on p. 489, the observed values of Spearman's $\mathcal{R}$ and the MRBP test statistic are

$$\mathcal{R}_o = 1 - \frac{3g\delta_o}{g^2 - 1} = 1 - \frac{(3)(8)(1.25)}{8^2 - 1} = 1 - 0.4762 = +0.5238$$

and

$$\delta_o = \frac{(g^2-1)(1-\mathcal{R}_o)}{3g} = \frac{(8^2-1)(1-0.5238)}{(3)(8)} = \frac{30}{24} = 1.25 \,.$$

10.5.4 Multiple Blocks

Spearman's footrule, as originally presented in his 1904 and 1906 articles, is limited to $g \geq 2$ untied rank scores and $b = 2$ blocks [381, 382]. However, the MRBP measure of effect size, $\Re$, suffers from no such limitations. Spearman's footrule is thus easily generalized to tied or untied rank scores and $b \geq 2$ sets of rankings. Let

$$\delta = \left[g \binom{b}{2} \right]^{-1} \sum_{i=1}^{g} \sum_{r<s} \left| x_{ri} - x_{si} \right|^{v}$$

denote an average distance function based on all $\binom{b}{2}$ possible paired absolute differences among values of the b sets of rankings and let

$$\mu_\delta = \left[g^2 \binom{b}{2} \right]^{-1} \sum_{i=1}^{g} \sum_{j=1}^{g} \sum_{r<s} \left| x_{ri} - x_{sj} \right|^{v}$$

denote the expected value of δ where b is the number of rankings, g is the number of objects, $\sum_{r<s}$ is the sum over all r and s such that $1 \leq r < s \leq g$, and $v = 1$. Then, as previously,

$$\Re = 1 - \frac{\delta}{\mu_\delta} \tag{10.18}$$

is a chance-corrected measure of the agreement among the b sets of rankings that is not limited to untied rank scores. Note that in the case of $b = 2$, Eq. (10.18) reduces to Spearman's 1906 footrule for $b = 2$ blocks as given in Eq. (10.13) on p. 486.

10.5.5 Example Analysis

Consider a generalized footrule analysis where $b = 4$ blocks contain untied rank scores for $g = 8$ objects. The randomized-block data are adapted from Berry and Mielke [30, p. 378] and are given in Fig. 10.5. An exact solution is not possible for these data since there are

$$M = (g!)^b = (8!)^4 = 2{,}642{,}908{,}293{,}365{,}760{,}000$$

Fig. 10.5 Rank scores assigned to $g = 8$ objects by $b = 4$ blocks

	Block			
Object	1	2	3	4
1	6	7	8	8
2	8	5	4	7
3	1	3	6	4
4	2	1	2	2
5	3	2	1	1
6	5	6	7	5
7	4	4	3	3
8	7	8	5	6

possible, equally-likely arrangements of the data listed in Fig. 10.5. Therefore, a resampling permutation approach is mandated. Following Eq. (10.1) on p. 473, the observed value of the MRBP test statistic with $v = 1$ is $\delta_o = 1.4167$. If all M possible arrangements of the observed rank scores listed in Fig. 10.5 occur with equal chance, the approximate resampling probability value of $\delta_o = 1.4167$ computed on $L = 1{,}000{,}000$ random arrangements of the observed rank scores with $b = 4$ blocks preserved for each arrangement is

$$P(\delta \leq \delta_o | H_0) = \frac{\text{number of } \delta \text{ values} \leq \delta_o}{L} = \frac{195}{1{,}000{,}000} = 0.1950 \times 10^{-3} \; .$$

Following Eq. (10.3) on p. 474, the exact expected value of the M δ values is $\mu_\delta = 2.6250$ and, following Eq. (10.2) on p. 474, the observed chance-corrected measure of effect size is

$$\mathfrak{R}_o = 1 - \frac{\delta_o}{\mu_\delta} = 1 - \frac{1.4167}{2.6250} = +0.4603 \; ,$$

indicating approximately 46 % within-block agreement above that expected by chance.

For comparison, consider a more conventional approach to the randomized-block data listed in Fig. 10.5. Although originally designed for continuous data, another measure that accommodates multiple blocks is the intraclass correlation coefficient (ICC), advocated as a measure of agreement by W.S. Robinson in 1957 [350]. While there are at least six different intraclass correlation coefficients, ICC(3, 1) and ICC(2, 1) are the most appropriate for these purposes [373]. ICC(3, 1) treats the blocks as fixed effects, while ICC(2, 1) treats the blocks as random effects. Both measures yield a coefficient of consistency among the scores, indicating the extent to which the blocks are interchangeable.

ICC(2, 1) is the most commonly used intraclass correlation coefficient, where the same b judges rate each of g objects. The b judges are considered to be a random sample from a population of potential raters. Because ICC(2, 1) is used almost exclusively in the literature, it is common in journals to indicate ICC(2, 1) simply as r_I.

For the randomized-block data listed in Fig. 10.5, $r_I = 0.7258$. With $\nu_1 = g - 1 = 8 - 1 = 7$ and $\nu_2 = (b-1)(g-1) = (4-1)(8-1) = 21$ degrees of freedom, $r_I = 0.7258$ yields an F-ratio of $F = 10.26$ with a corresponding approximate probability value of $P = 0.1447{\times}10^{-4}$, under the null hypothesis. Because the data consist of untied rank scores in this example, the sum of squares for blocks is necessarily equal to zero for r_I. Moreover, compared with $\mathfrak{R}$, r_I is designed for continuous data, is a biased estimator of the population ρ_I value, assumes normality, and is based on squared Euclidean differences among rank scores.

10.6 Friedman's Analysis of Variance for Ranks

In 1937 Milton Friedman proposed a balanced randomized-block analysis of variance for ranks in an article titled "The use of ranks to avoid the assumption of normality implicit in the analysis of variance" that appeared in *Journal of the American Statistical Association* [128]. As reflected in the title of the article, the purpose was to avoid the assumption of normality underlying the conventional randomized-block analysis of variance. Friedman called the new procedure "the method of ranks" and designated the associated statistic as χ_r^2 as, he argued, the statistic tends to be distributed according to the usual chi-squared distribution with $g - 1$ degrees of freedom under the null hypothesis that the observed ranking is random [128, p. 676].[9]

Let b denote the number of blocks and g denote the number of objects to be ranked. Then Friedman's statistic is given by

$$\chi_r^2 = \frac{12}{bg(g+1)} \sum_{i=1}^{g} R_i^2 - 3b(g+1) , \tag{10.19}$$

where R_i for $i = 1, \ldots, g$ is the sum of the rank scores for the ith object and there are no tied rank scores [128, p. 679]. A number of statistics are either identical, related, or equivalent to Friedman's χ_r^2. Among these are Kendall and Babington Smith's coefficient of concordance [209] given by[10]

$$W = \frac{12 \displaystyle\sum_{i=1}^{g} R_i^2 - 3gb^2(g+1)}{gb^2(g^2-1)} , \tag{10.20}$$

[9] Although Friedman labeled the statistic as χ_r^2, many textbooks refer to Friedman's analysis of variance for ranks test statistic as T.

[10] The original 1939 article was by Maurice Kendall and Bernard Babington Smith, but the statistic is typically attributed only to Kendall.

the average value of all pairwise Spearman's rank-order correlation coefficients [381, 382] given by

$$\bar{\rho} = \frac{2}{b(b-1)} \sum_{i=1}^{b-1} \sum_{j=i+1}^{b} \rho_{ij} , \tag{10.21}$$

and the Wallis rank-order correlation ratio, given by

$$\eta_r^2 = \frac{\chi_r^2}{b(g-1)} , \tag{10.22}$$

which, as acknowledged by Wallis, is identical to Kendall and Babington Smith's W [414].

The relationships among the various measures are given by

$$\chi_r^2 = Wb(g-1) \quad \text{and} \quad W = \frac{\chi_r^2}{b(g-1)} ,$$

$$\bar{\rho} = \frac{bW - 1}{b-1} \quad \text{and} \quad W = \frac{\bar{\rho}(b-1)+1}{b} ,$$

$$\chi_r^2 = \bar{\rho}(b-1)(g-1) + g - 1 \quad \text{and} \quad \bar{\rho} = \frac{\chi_r^2 - g + 1}{(b-1)(g-1)} ,$$

$$\eta_r^2 = \frac{\chi_r^2}{b(g-1)} \quad \text{and} \quad \chi_r^2 = \eta_r^2 b(g-1) ,$$

$$\eta_r^2 = \frac{\bar{\rho}(b-1)+1}{b} \quad \text{and} \quad \bar{\rho} = \frac{b\eta_r^2 - 1}{b-1} ,$$

and $\eta_r^2 = W$.

The functional relationships between Friedman's χ_r^2 and the MRBP test statistic, when $p = v = 2$, are given by

$$\chi_r^2 = \frac{b[2(g+1)(2g+1) - 3(g+1)^2] - 6(b-1)\delta}{g+1}$$

and

$$\delta = \frac{b[2(g+1)(2g+1) - 3(g+1)^2] - \chi_r^2(g+1)}{6(b-1)} .$$

Alternatively, it can easily be shown that the chance-corrected measure of within-block effect size given by

$$\Re = 1 - \frac{\delta}{\mu_\delta} \tag{10.23}$$

is identical to $\bar{\rho}$ as given in Eq. (10.21). Therefore, the relationships between $\mathfrak{R}$ and Friedman's χ_r^2 are given by

$$\mathfrak{R} = \frac{\chi_r^2 - g + 1}{(b-1)(g-1)} \quad \text{and} \quad \chi_r^2 = \mathfrak{R}(b-1)(g-1) + g - 1 \,,$$

the relationships between $\mathfrak{R}$ and Kendall and Babington Smith's W are given by

$$\mathfrak{R} = \frac{bW - 1}{b-1} \quad \text{and} \quad W = \frac{\mathfrak{R}(b-1)+1}{b} \,,$$

and the relationships between $\mathfrak{R}$ and Wallis's η_r^2 are given by

$$\mathfrak{R} = \frac{b\eta_r^2 - 1}{b-1} \quad \text{and} \quad \eta_r^2 = \frac{\mathfrak{R}(b-1)+1}{b} \,.$$

Note that while χ_r^2, W, $\bar{\rho}$, and η_r^2, as expressed in Eqs. (10.19), (10.20), (10.21), and (10.22), assume no tied rank scores, $\mathfrak{R}$, the chance-corrected measure of effect size as expressed in Eq. (10.23) easily accommodates any number of tied rank scores.[11]

10.6.1 Example 1: $v = 2$

To illustrate the Friedman's analysis of variance for ranks test, and its relationships with other statistics, consider the univariate rank scores listed in Fig. 10.6. The data are adapted from Siegel and Castellan [375, p. 263].

Fig. 10.6 Example data for the Friedman's analysis of variance for ranks with $b = 3$ blocks and $g = 6$ objects

	Block			
Object	1	2	3	R
1	1	1	2	4
2	6	5	3	14
3	3	6	6	15
4	4	4	5	13
5	5	2	4	11
6	2	3	1	6

[11]It should be noted that most textbooks provide elaborate and cumbersome corrections for Friedman's χ_r^2 and Kendall and Babington Smith's W to accommodate tied rank scores.

For the untied rank scores listed in Fig. 10.6, the sum of the squared rank scores is

$$\sum_{i=1}^{g} R_i^2 = 4^2 + 14^2 + 15^2 + 13^2 + 11^2 + 6^2 = 763 ,$$

the observed value of Friedman's χ_r^2 is

$$\chi_r^2 = \frac{12}{bg(g+1)} \sum_{i=1}^{g} R_i^2 - 3b(g+1)$$

$$= \frac{12}{(3)(6)(6+1)} 763 - (3)(3)(6+1) = 9.6667 ,$$

the observed value of Kendall and Babington Smith's W is

$$W = \frac{12 \sum_{i=1}^{g} R_i^2 - 3gb^2(g+1)}{gb^2(g^2-1)}$$

$$= \frac{(12)(763) - (3)(6)(3^2)(6+1)^2}{(6)(3^2)(6^2-1)} = 0.6444 ,$$

and the observed value of the pairwise-average Spearman's rank-order correlation with $\rho_{12} = 0.4286$, $\rho_{13} = 0.3714$, and $\rho_{23} = 0.60$ is

$$\bar{\rho} = \frac{2}{b(b-1)} \sum_{i=1}^{b-1} \sum_{j=i+1}^{b} \rho_{ij} = \frac{2(0.4286 + 0.3714 + 0.60)}{3(3-1)} = 0.4667 .$$

For the randomized-block data listed in Fig. 10.6 with $b = 3$ blocks and $g = 6$ objects, the observed relationships between Friedman's χ_r^2 and Kendall and Babington Smith's W are

$$\chi_r^2 = Wb(g-1) = (0.6444)(3)(6-1) = 9.6667$$

and

$$W = \frac{\chi_r^2}{b(g-1)} = \frac{9.6667}{3(6-1)} = 0.6444 ,$$

the observed relationships between $\bar{\rho}$ and Kendall and Babington Smith's W are

$$\bar{\rho} = \frac{bW-1}{b-1} = \frac{(3)(0.6444)-1}{3-1} = 0.4667$$

and

$$W = \frac{\bar{\rho}(b-1)+1}{b} = \frac{(0.4667)(3-1)+1}{3} = 0.6444 \,,$$

the observed relationships between Friedman's χ_r^2 and $\bar{\rho}$ are

$$\chi_r^2 = \bar{\rho}(b-1)(g-1)+g-1 = (0.4667)(3-1)(6-1)+6-1 = 9.6667$$

and

$$\bar{\rho} = \frac{\chi_r^2 - g + 1}{(b-1)(g-1)} = \frac{9.6667-6+1}{(3-1)(6-1)} = 0.4667 \,,$$

the observed relationships between Wallis's η_r^2 and Friedman's χ_r^2 are

$$\eta_r^2 = \frac{\chi_r^2}{b(g-1)} = \frac{9.6667}{(3)(6-1)} = 0.6444$$

and

$$\chi_r^2 = \eta_r^2 b(g-1) = (0.6444)(3)(6-1) = 9.6667 \,,$$

the observed relationships between Wallis's η_r^2 and $\bar{\rho}$ are

$$\eta_r^2 = \frac{\bar{\rho}(b-1)+1}{b} = \frac{(0.4667)(3-1)+1}{3} = 0.6444$$

and

$$\bar{\rho} = \frac{b\eta_r^2 - 1}{b-1} = \frac{(3)(0.6444)-1}{3-1} = 0.4667 \,,$$

the observed relationships between Friedman's χ_r^2 and $\Re$ are

$$\chi_r^2 = \Re(b-1)(g-1)+g-1 = (0.46667)(3-1)(6-1)+6-1 = 9.6667$$

and

$$\Re = \frac{\chi_r^2 - g + 1}{(b-1)(g-1)} = \frac{9.6667-6+1}{(3-1)(6-1)} = 0.4667 \,,$$

the observed relationships between Kendall and Babington Smith's W and $\Re$ are

$$W = \frac{\Re(b-1)+1}{b} = \frac{(0.4667)(n)+1}{3} = 0.6444$$

and

$$\Re = \frac{bW - 1}{b - 1} = \frac{(3)(0.6444) - 1}{3 - 1} = 0.4667 ,$$

and the observed relationships between η_r^2 and $\Re$ are

$$\eta_r^2 = \frac{\Re(b - 1) + 1}{b} = \frac{(0.4667)(3 - 1) + 1}{3} = 0.6444$$

and

$$\Re = \frac{b\eta_r^2 - 1}{b - 1} = \frac{(3)(0.6444) - 1}{3 - 1} = 0.4667 .$$

Recall that since $\Re$ is simply a linear transformation of δ,

$$P(\Re \geq \Re_o | H_0) = P(\delta \leq \delta_o | H_0) = \frac{\text{number of } \delta \text{ values} \leq \delta_o}{M} .$$

For the rank scores listed in Fig. 10.6 there are

$$M = \left(g!\right)^b = \left(6!\right)^3 = 373{,}248{,}000$$

possible, equally-likely arrangements of the observed data. While an exact solution is perhaps impractical, it is not impossible. For the rank scores listed in Fig. 10.6, let $v = 2$, employing squared Euclidean distance between the rank scores to correspond to Friedman's χ_r^2 statistic [128], then following Eq. (10.1) on p. 473, the observed value of the MRBP test statistic with $v = 2$ is $\delta_o = 3.1111$. If all arrangements of the observed rank scores listed in Fig. 10.6 occur with equal chance, the exact probability value of $\delta_o = 3.1111$ computed on the $M = 373{,}248{,}000$ possible arrangements of the observed rank scores with $b = 3$ blocks preserved for each arrangement is

$$P(\delta \leq \delta_o | H_0) = \frac{\text{number of } \delta \text{ values} \leq \delta_o}{M} = \frac{20{,}913{,}840}{373{,}248{,}000} = 0.0560 .$$

For comparison, the Friedman's test statistic is approximately distributed as chi-squared under the null hypothesis with $g - 1 = 6 - 1 = 5$ degrees of freedom. Under the null hypothesis, the observed value of $\chi_r^2 = 9.6667$ yields an approximate probability value of $P = 0.0853$.

Following Eq. (10.3) on p. 474, the exact expected value of the $M = 373{,}248{,}000$ δ values is $\mu_\delta = 5.8333$ and, following Eq. (10.2) on p. 474, the observed chance-corrected measure of effect size is

$$\Re_o = 1 - \frac{\delta_o}{\mu_\delta} = 1 - \frac{3.1111}{5.8333} = +0.4667 ,$$

indicating approximately 47 % within-block agreement above that expected by chance. Finally, note that when $v = 2$, $\mathfrak{R}$ and $\bar{\rho}$ are equivalent, i.e., $\mathfrak{R} = \bar{\rho} = +0.4667$.

10.6.2 Example 2: $v = 1$

For a comparison analysis of the univariate rank scores listed in Fig. 10.6, set $v = 1$ instead of $v = 2$, employing ordinary Euclidean distance between the rank scores. For the rank scores listed in Fig. 10.6, there are still only

$$M = (g!)^b = (6!)^3 = 373{,}248{,}000$$

possible, equally-likely arrangements of the observed data, an exact solution is feasible. Following Eq. (10.1) on p. 473, the observed value of the MRBP test statistic with $v = 1$ is $\delta_o = 1.4444$. If all arrangements of the observed rank scores listed in Fig. 10.6 occur with equal chance, the exact probability value of $\delta_o = 1.4444$ computed on the $M = 373{,}248{,}000$ possible arrangements of the observed rank scores is

$$P(\delta \leq \delta_o | H_0) = \frac{\text{number of } \delta \text{ values} \leq \delta_o}{M} = \frac{39{,}980{,}161}{373{,}248{,}000} = 0.1071 .$$

For comparison, the exact probability value based on $v = 2$ and $M = 373{,}248{,}000$ in Example 1 is $P = 0.0560$. No comparison is made with the conventional analysis of variance for ranks tests as χ_r^2, W, $\bar{\rho}$, and η_r^2 are undefined for $v = 1$.

Following Eq. (10.3) on p. 474, the exact expected value of the $M = 373{,}248{,}000$ δ values is $\mu_\delta = 1.9444$ and, following Eq. (10.2) on p. 474, the observed chance-corrected measure of effect size is

$$\mathfrak{R}_o = 1 - \frac{\delta_o}{\mu_\delta} = 1 - \frac{1.4444}{1.9444} = +0.2571 ,$$

indicating approximately 26 % within-block agreement above that expected by chance.

10.7 MRBP and the Measurement of Agreement

A number of statistical research problems require the measurement of agreement, rather than association or correlation. Agreement indices measure the extent to which a set of response measurements are identical to another set, i.e., agree, rather than the extent to which one set of response measurements is a linear function of another set of response measurements, i.e., correlated. For details of the important differences between agreement and correlation, see Chap. 4, Sect. 4.1.1.

The usual research situation involving a measure of agreement arises when several judges or raters assign objects to categories, which may be weighted in some manner. Examples include analyzing the extent of agreement between judges on a measure of performance [61], assessing students' learning behaviors [244], appraising the reliability of assigning observations to categories [19, 223], and evaluating the agreement among referees for journal reviews [423]. Other applications include the interchangeability of measures; measurement of the reliability of an instrument, such as a test or scale; measurement of the bias of a concept as applied to observational material by different observers; comparison of observed with theoretically induced values of a variable; and measurement of the degree of homogeneity within groups of observations [350, pp. 17–18].

In 1957 W.S. Robinson published an article in *American Sociological Review* on "The statistical measurement of agreement" [350].[12] In this formative article, Robinson developed the idea of agreement, as contrasted with correlation, and showed that a simple modification of the intraclass correlation coefficient was an appropriate measure of statistical agreement, which he called A, presumably for Agreement [350, p. 20]. He explained that statistical agreement requires that paired values be identical, while correlation requires only that the paired values be linked by some mathematical function [350, p. 19]. Robinson argued that the distinction between agreement and correlation leads to the conclusion that a logically correct estimate of the reliability of a test is given by the intraclass correlation coefficient rather than the Pearsonian (interclass) correlation coefficient and that the concept of agreement, rather than correlation, is the proper basis of reliability theory [350, p. 18]. The 1957 Robinson article, which was quite mathematical, was followed by a more interpretive article in the same journal on "The geometric interpretation of agreement" in 1959 [351].

Currently, the most popular measure of agreement between two judges or raters is the chance-corrected measure of agreement first proposed by Jacob Cohen in 1960 and termed kappa [70].[13] Cohen's kappa measures the agreement between $b = 2$ observers on the assignment of N objects to a set of c discrete unordered categories. In 1968 Cohen proposed a version of kappa that allowed for the weighting of categories [71]. Whereas the original (unweighted) kappa did not distinguish among magnitudes of disagreement, weighted kappa incorporated the magnitude of each disagreement and provided partial credit for disagreements when agreement was not complete. The usual approach is to assign weights to each disagreement pair with larger weights indicating greater disagreement.[14]

In both cases, unweighted and weighted, kappa is equal to 1 when perfect agreement among two of more judges occurs, 0 when agreement is equal to that expected

[12]W.S. Robinson is probably best known for his seminal article on ecological correlations published in the same journal in 1950 [349].

[13]It is interesting that neither of Robinson's articles were cited by Cohen.

[14]Some authors prefer to define kappa in terms of agreement weights, instead of disagreement weights.

under independence, and negative when agreement is less than expected by chance. Because unweighted kappa applies to unordered categories, it is properly included in Chap. 11, Sect. 11.2. Weighted kappa is discussed here as it is typically used for ordered categorical data.

10.7.1 Limitations of Kappa

It should be noted that Cohen's kappa measures of agreement, both unweighted and weighted, are not without detractors and have received considerable criticism over the years. Kappa is well known as a marginal-dependent measure of agreement and is often criticized, based on this dependency; see, for example, articles by Agresti [2], Brennan and Prediger [56], Guggenmoos-Holzmann [159, 160], Maclure and Willett [257], May [268], Thompson and Walter [399], and Zwick [436]. The problem is that there are two sources of disagreement: differences in thresholds and differences in construction of the underlying continuous scale, and it is inherently impossible to represent them by a single number, as noted by Brennan and Hays [55] and Hutchinson [189]. Thus, kappa cannot approximate its maximum value of +1.00 when the marginal frequency distributions in an agreement classification table are not uniform, as noted by von Eye and von Eye [412]. However, kappa will attain its maximum value of +1.00 when the probability for all disagreement cells is zero; consequently, kappa shows no marginal dependency under conditions of perfect agreement [412]. As Brennan and Prediger noted: "It is evident that indiscriminate use of coefficient kappa without modification *may* lead to dramatically incorrect conclusions about the proportion of maximum possible agreement evident in a set of data" [56, p. 698].[15]

Despite this limitation, Cohen's kappa is considered to be the gold standard among agreement coefficients and is interpreted as the proportionate increase in rater agreement above and beyond that expected by chance alone, where chance is defined as the level of agreement expected if the raters had a known base rate for the objects under study and randomly assigned cases corresponding to the base rate; see also an article on the assessment of reliability by Meyer [279]. This definition of chance has been referred to by Brennan and Prediger [56] and Umesh, Peterson, and Sauber [407] as the "fixed marginals" model because the marginal distributions of category assignment are assumed to be known a priori. The problem with the fixed-marginals approach is that it does not give the raters credit for assignments that are independently agreed upon and reflected in the marginal distributions. Thus, as noted by Brennan and Prediger [56], Hanley [166], and Zwick [436], Cohen's kappa statistic penalizes the raters by using the base rate to define the chance agreement level the raters must surpass. For example, consider two raters and two categories, *A* and *B*. If the two raters both feel that the base rate in the population for category *A* is 0.10 and each judge randomly assigns 10 % of the cases to

[15]Emphasis in the original.

category A, then by chance alone the percentage agreement between the two raters is $(0.10)(0.10) + (0.90)(0.90) = 0.82$, and the observed agreement between the two raters must exceed 0.82 for the computed value of kappa to be greater than zero [41, p. 323].

Cohen's kappa is extremely sensitive to the base-rate phenomenon.[16] Because the maximum value that kappa can attain is constrained by differences between the marginal distributions of the two raters, as the base rate moves away from the point of maximum variability a small disagreement between the raters can cause the kappa value to decline dramatically, as noted by Meyer [279]. On the other hand, some researchers have argued that this is appropriate as kappa is a true reliability statistic; see, for example, articles by Bartko [20], Cohen [70], and Shrout et al. [374]. That is to say, as true score variability in the group becomes more restricted, a fixed amount of disagreement plays an increasingly larger role in observed score variability, so calculated reliability coefficients decline in value. In 1960 Cohen noted that it is perfectly reasonable and, in fact, desirable to use a summary agreement measure that is sensitive to both aspects of agreement: item-by-item agreement as reflected in the main diagonal of the agreement matrix, and symmetry between the marginal distributions [70]. Finally, Spitznagel and Helzer protested against even providing base-rate information, arguing that it defeats the purpose of a single measure of reliability [383].

10.7.2 Cohen's Weighted Kappa

For simplicity, consider $N \geq 2$ objects cross-classified by $b = 2$ independent judges into a $c \times c$ contingency table. Let n_{ij}, w_{ij}, $n_{i.}$, and $n_{.j}$ denote the cell frequencies, cell weights, row marginal frequency totals, and column marginal frequency totals, respectively, where

$$n_{i.} = \sum_{j=1}^{c} n_{ij} \; , \quad n_{.j} = \sum_{i=1}^{c} n_{ij} \; , \quad \text{and} \quad N = \sum_{i=1}^{c} \sum_{j=1}^{c} n_{ij} \; .$$

When the c categories for the two judges are similarly arranged, then n_{ii}, $i = 1, \ldots, c$, and n_{ij}, $i \neq j$, denote the agreement and disagreement frequencies, respectively.

Although a variety of weighting schemes have been proposed for weighted kappa, the most popular weighting scheme is quadratic weighting given by $w_{ij} = (i - j)^2$ for $i, j = 1, \ldots, c$, where cell disagreement weights progress geometrically outward from the agreement diagonal, i.e., 0^2, 1^2, 2^2, 3^2, and so on. However, linear weighting is perhaps more intuitive in which $w_{ij} = |i - j|$ for $i, j = 1, \ldots, c$,

[16] For a discussion of the base-rate problem in general, see a 2015 book on *Statistics Done Wrong* by Alex Reinhart [345, pp. 39–47].

where cell disagreement weights progress linearly outward from the agreement diagonal, i.e., 0, 1, 2, 3, and so on. In addition, linear weighting has been shown to have some interesting properties. In 2008 Vanbelle and Albert demonstrated that linear-weighted kappa for $b = 2$ independent judges and $c \geq 3$ ordered categories is equivalent to deriving the weighted kappa coefficient from unweighted kappa values computed on $c - 1$ embedded 2×2 classification tables [410]. In 2009 Mielke and Berry generalized the results of Vanbelle and Albert to $b \geq 2$ independent judges [298].

The weighted kappa test statistic for $b = 2$ judges is defined as

$$\hat{\kappa} = 1 - \frac{\dfrac{1}{N}\displaystyle\sum_{i=1}^{c}\sum_{j=1}^{c} w_{ij} n_{ij}}{\dfrac{1}{N^2}\displaystyle\sum_{i=1}^{c}\sum_{j=1}^{c} w_{ij} n_{i.} n_{.j}} . \tag{10.24}$$

Given a $c{\times}c$ contingency table with N objects cross-classified by the ratings of two independent judges into c ordered categories, an exact permutation test generates all M possible arrangements of the N objects in the c^2 cells, while preserving the total number of objects in each category, i.e., the marginal frequency totals.[17] For each arrangement of cell frequencies with fixed marginal frequency totals, the weighted kappa statistic, $\hat{\kappa}$, and the exact probability, $p(n_{ij}|n_{i.}, n_{.j})$, are calculated, where

$$p(n_{ij}|n_{i.}, n_{.j}) = \frac{\left(\displaystyle\prod_{i=1}^{c} n_{i.}!\right)\left(\displaystyle\prod_{j=1}^{c} n_{.j}!\right)}{N!\displaystyle\prod_{i=1}^{c}\prod_{j=1}^{c} n_{ij}!}$$

is the conventional hypergeometric probability of a $c{\times}c$ contingency table.

Let $\hat{\kappa}_{\text{o}}$ denote the value of the observed weighted kappa statistic and M denote the total number of distinct cell frequency arrangements of the N objects in the $c{\times}c$ contingency table, given fixed marginal frequency totals. Then the exact probability value of $\hat{\kappa}_{\text{o}}$ is given by

$$P = \sum_{k=1}^{M} \Psi(\hat{\kappa}_k)\, p(n_{ij}|n_{i.}, n_{.j}) ,$$

---

[17]While it is straightforward to compute M for 2×2 contingency tables, it is considerably more difficult, and often impossible, to compute M for larger contingency tables. In 1977 Gail and Mantel published exact and approximate methods for determining M consistent with marginal frequency totals in $r{\times}c$ contingency tables [132].

where

$$\Psi(\hat{\kappa}_k) = \begin{cases} 1 & \text{if } \hat{\kappa}_k \geq \hat{\kappa}_o \,, \\ 0 & \text{otherwise} \,. \end{cases}$$

10.7.3 Weighted Kappa Example

Consider a small example data set of $N = 5$ objects classified into $c = 3$ ordered categories by $b = 2$ independent judges. Table 10.1 contains the $c^2 = 9$ cell frequencies. The corresponding linear and quadratic disagreement cell weights are given in parentheses and brackets, respectively. The number of objects and the number of categories are deliberately kept small to simplify the example analysis.

Utilizing linear disagreement weights, given in parentheses in Table 10.1, and following the numerator of Eq. (10.24) on p. 504 with $N = 5$ objects, $c = 3$ categories, $b = 2$ blocks, and $r = 1$ response measurement,

$$\frac{1}{N}\sum_{i=1}^{c}\sum_{j=1}^{c} w_{ij}n_{ij} = \frac{1}{5}\big[(0)(0) + (1)(1) + (2)(0) + (1)(0) + (0)(2)$$
$$+ (1)(0) + (2)(1) + (1)(0) + (0)(1)\big] = \frac{3}{5} = 0.60 \,,$$

and for the denominator of Eq. (10.24),

$$\frac{1}{N^2}\sum_{i=1}^{c}\sum_{j=1}^{c} w_{ij}n_{i.}n_{.j} = \frac{1}{5^2}\big[(0)(1)(1) + (1)(1)(3) + (2)(1)(1)$$
$$+ (1)(2)(1) + (0)(2)(3) + (1)(2)(1) + (2)(2)(1)$$
$$+ (1)(2)(3) + (0)(2)(1)\big] = \frac{19}{25} = 0.76 \,,$$

Table 10.1 Example data for a weighted kappa analysis with $N = 5$ observations, $c = 3$ ordered categories, and $b = 2$ judges

	Judge 2			
Judge 1	Category A	Category B	Category C	Total
Category A	0 (0) [0]	1 (1) [1]	0 (2) [4]	1
Category B	0 (1) [1]	2 (0) [0]	0 (1) [1]	2
Category C	1 (2) [4]	0 (1) [1]	1 (0) [0]	2
Total	1 (2) [4]	3 (2) [4]	1 (2) [4]	5

Note: Linear cell weights are in parentheses and quadratic cell weights are in brackets

then the observed value of weighted $\hat{\kappa}$ with linear weighting is

$$\hat{\kappa}_o = 1 - \frac{\dfrac{1}{N}\displaystyle\sum_{i=1}^{c}\sum_{j=1}^{c} w_{ij}n_{ij}}{\dfrac{1}{N^2}\displaystyle\sum_{i=1}^{c}\sum_{j=1}^{c} w_{ij}n_{i.}n_{.j}} = 1 - \frac{0.60}{0.76} = 0.2105 .$$

There are $M = 8$ possible, equally-likely arrangements of cell frequencies given the observed marginal frequency totals of $\{1, 2, 2\}$ and $\{1, 3, 1\}$ in Table 10.1. The eight arrangements of cell frequencies are listed in Table 10.2, where Table 10.1 of Table 10.2 contains the $N = 5$ observed cell frequencies.

Figure 10.7 lists the computed kappa values and associated hypergeometric probability values for the $M = 8$ tables in Table 10.2, ordered from high to low by the $\hat{\kappa}$ values. As is evident from the kappa and associated probability values listed in Fig. 10.7, the observed value of $\hat{\kappa}_o = +0.2105$ is not unusual as four $\hat{\kappa}$ values are less than $\hat{\kappa}_o = +0.2105$ and four values are equal to or greater than $\hat{\kappa}_o = +0.2105$. Thus, the exact upper-tail probability value of the observed cell configuration is $0.1000 + 0.1000 + 0.1000 + 0.2000 = 0.5000$, i.e., the sum of the hypergeometric probability values associated with values of $\hat{\kappa}_o = +0.2105$ or greater.

Table 10.2 Eight possible arrangements of the cell frequencies in Table 10.1, given fixed marginal frequency totals

Table 1			Table 2			Table 3			Table 4		
0	1	0	1	0	0	1	0	0	0	1	0
0	2	0	0	1	1	0	2	0	1	0	1
1	0	1	0	2	0	0	1	1	0	2	0
Table 5			Table 6			Table 7			Table 8		
0	1	0	0	0	1	0	0	1	0	1	0
1	1	0	0	2	0	1	1	0	0	1	1
0	1	1	1	1	0	0	2	0	1	1	0

Fig. 10.7 Kappa and hypergeometric probability values for the eight 3×3 contingency tables listed in Table 10.2

Table	$\hat{\kappa}$	Probability
3	+0.7368	0.1000
1	+0.2105	0.1000
2	+0.2105	0.1000
5	+0.2105	0.2000
4	−0.3158	0.1000
6	−0.3158	0.1000
7	−0.3158	0.1000
8	−0.3158	0.2000

10.7.4 Relationship of $\mathfrak{R}$ and Cohen's Weighted $\hat{\kappa}$

As previously, consider the MRBP test statistic given by

$$\delta = \left[g \binom{b}{2} \right]^{-1} \sum_{i=1}^{g} \sum_{j<k} \Delta(x_{ij}, x_{ik}) , \tag{10.25}$$

where g is the number of treatments, b is the number of blocks, r is the number of response measurements, $\sum_{j<k}$ denotes the sum over all j and k such that $1 \le j < k \le b$, and $\Delta(x, y)$ is a generalized Minkowski distance function given by

$$\Delta(x, y) = \left(\sum_{i=1}^{r} (x_i - y_i)^p \right)^{v/p}$$

with $p = 2$. If $g = 2$ and $r = 1$, as in this case, the MRBP test statistic given in Eq. (10.25) reduces to

$$\delta = \frac{1}{g} \sum_{i=1}^{g} \Delta(x_{i1}, x_{i2}) , \tag{10.26}$$

where the generalized Minkowski distance function is given by

$$\Delta(x_{i1}, x_{i2}) = \left[\left(x_{i1} - x_{i2} \right)^2 \right]^{v/2} . \tag{10.27}$$

It can easily be shown that the chance-corrected measure of effect size given by

$$\mathfrak{R} = 1 - \frac{\delta}{\mu_\delta} ,$$

with

$$\mu_\delta = \frac{1}{g^2} \sum_{i=1}^{g} \sum_{j=1}^{g} \Delta(x_{i1}, x_{i2}) , \tag{10.28}$$

is equivalent to Cohen's weighted kappa coefficient where $v = 1$ yields $\hat{\kappa}$ with linear weighting and $v = 2$ yields $\hat{\kappa}$ with quadratic weighting.

For comparison, consider a conventional z test calculated on the frequency data listed in Table 10.1 on p. 505. In 1968 Brian S. Everitt developed the exact variance of weighted kappa for two raters that was suitable for any weighting scheme, but found the expression too complicated for routine use [111, 124, p. 323]. In 2005 Mielke, Berry, and Johnston reformulated the exact variance presented by Everitt for two raters into a form conducive to computation and provided an algorithm for

$b = 2$ raters [305]. In 2007 Mielke, Berry, and Johnston extended the exact variance of Everitt to include the classification of N objects by $b \geq 2$ independent raters [306].

Assuming fixed marginal frequency totals and the null hypothesis that the two judges operate independently, the exact expected value of $\hat{\kappa}$ given by

$$\mathrm{E}[\hat{\kappa}] = 0$$

follows from Eq. (10.24) on p. 504 since $\mathrm{E}[n_{ij}] = n_{i.}n_{.j}/N$, and the exact variance of $\hat{\kappa}$ is conveniently given by

$$\sigma_{\hat{\kappa}}^2 = \frac{1}{N-1}\left(\sum_{i=1}^{c}\sum_{j=1}^{c} w_{ij}n_{i.}n_{.j}\right)^{-2}\left[\sum_{i=1}^{c}\sum_{j=1}^{c} w_{ij}^2 n_{i.}(N-n_{i.})\right.$$

$$n_{.j}(N-n_{.j}) - \sum_{i=1}^{c}\sum_{j\neq k} w_{ij}w_{ik}n_{i.}(N-n_{i.})n_{.j}n_{.k} - \sum_{i\neq j}\sum_{k=1}^{c} w_{ik}w_{jk}$$

$$\left. n_{i.}n_{.j}n_{k.}(N-n_{.k}) + \sum_{i\neq k}\sum_{j\neq l} w_{ij}w_{kl}n_{i.}n_{k.}n_{.j}n_{.l}\right] .$$

Let $\hat{\kappa}_{\text{o}}$ denote an observed value of $\hat{\kappa}$. Then the observed standard score is given by

$$z_{\text{o}} = \frac{\hat{\kappa}_{\text{o}} - \mathrm{E}[\hat{\kappa}]}{\sigma_{\hat{\kappa}}} .$$

Since z approaches the $N(0,1)$ distribution as $N \to \infty$ with fixed positive marginal proportions, the approximate probability value under the null hypothesis is given by $P(z \geq z_{\text{o}})$. For the observed frequency data and associated linear cell weights in Table 10.1, $\hat{\kappa}_{\text{o}} = +0.2105$, $\sigma_{\hat{\kappa}}^2 = 0.1342$, $z_{\text{o}} = +0.5747$, and the approximate $N(0,1)$ two-sided probability value is $P = 0.5655$, which approximates the exact probability value of $P = 0.50$. A conventional correction for continuity yields $z_{\text{o}} = +0.7105$ and an approximate $N(0,1)$ two-sided probability value of $P = 0.4774$, which more closely approximates the exact probability value of $P = 0.50$.

Linear Weighting with $v = 1$

Consider the frequency data listed in Table 10.1 on p. 505, but arranged in a randomized-block analysis-of-variance format with $b = 2$ blocks and $g = 5$ observations in each block, as given in Fig. 10.8. Following Eq. (10.27) on p. 507 for the

Fig. 10.8 Example data from Table 10.1 arranged in a block format with $b = 2$ blocks and $g = 5$ observations in each block

	Block	
g	1	2
1	1	2
2	2	2
3	2	2
4	3	1
5	3	3

generalized Minkowski distance function with $p = 2$ and $v = 1$,

$$\Delta(1,1) = \left[(1-2)^2\right]^{1/2} = 1.00\ ,$$
$$\Delta(2,2) = \left[(2-2)^2\right]^{1/2} = 0.00\ ,$$
$$\Delta(3,3) = \left[(2-2)^2\right]^{1/2} = 0.00\ ,$$
$$\Delta(4,4) = \left[(3-1)^2\right]^{1/2} = 2.00\ ,$$

and

$$\Delta(5,5) = \left[(3-3)^2\right]^{1/2} = 0.00\ .$$

Then following Eq. (10.26) on p. 507,

$$\delta = \frac{1}{g}\left[\Delta(1,1) + \Delta(2,2) + \Delta(3,3) + \Delta(4,4) + \Delta(5,5)\right]$$

and the observed value of the MRBP test statistic with $v = 1$ is

$$\delta_{\text{o}} = \frac{1}{5}\left(1.00 + 0.00 + 0.00 + 2.00 + 0.00\right) = \frac{1}{5}(3.00) = 0.60\ .$$

For the observed frequency data listed in Fig. 10.8, there are only

$$M = \left(g!\right)^b = \left(5!\right)^2 = 14{,}400$$

possible, equally-likely arrangements of the observed data, therefore an exact solution is feasible. An exact probability value for $\hat{\kappa}_{\text{o}}$ and δ_{o} may be expressed as

$$P(\kappa \geq \hat{\kappa}_{\text{o}}|H_0) = P(\delta \leq \delta_{\text{o}}|H_0) = \frac{\text{number of } \delta \text{ values} \leq \delta_{\text{o}}}{M}$$

where $\hat{\kappa}_{\text{o}}$ and δ_{o} denote the observed values of Cohen's $\hat{\kappa}$ and δ, respectively.

If all arrangements of the observed data listed in Fig. 10.8 occur with equal chance, the exact probability value of $\delta_{\text{o}} = 0.60$ computed on the $M = 14{,}400$

possible arrangements of the observed data with $b = 2$ blocks preserved for each arrangement is

$$P(\delta \leq \delta_o | H_0) = \frac{\text{number of } \delta \text{ values} \leq \delta_o}{M} = \frac{7{,}200}{14{,}400} = 0.50\,.$$

Following Eq. (10.28) on p. 507 with $v = 1$, the exact expected value of the $M = 14{,}400$ δ values is given by

$$\mu_\delta = \frac{1}{g^2}\big[\Delta(1,1) + \Delta(1,2) + \cdots + \Delta(4,5) + \Delta(5,5)\big]$$

and the exact observed value of μ_δ is

$$\mu_\delta = \frac{1}{5^2}\,(1+1+1+0+2+0+0+\cdots+2+0+1+1+1+2+0)$$
$$= \frac{1}{25}\,(19) = 0.76\,.$$

Then,

$$\hat{\kappa}_o = \mathfrak{R}_o = 1 - \frac{\delta_o}{\mu_\delta} = 1 - \frac{0.60}{0.76} = +0.2105$$

and the identity relating Cohen's $\hat{\kappa}$ with linear weighting and $\mathfrak{R}$ with $v = 1$ is confirmed.

Quadratic Weighting with $v = 2$

Consider again the frequency data listed in Table 10.1 on p. 505, replicated in Fig. 10.9 for convenience, with $N = 5$ objects classified into $c = 3$ unordered categories by $b = 2$ independent judges.

Utilizing quadratic cell disagreement weights, given in brackets in Fig. 10.9, and following the numerator of Eq. (10.24) on p. 504 with $N = 5$ objects, $c = 3$ cate-

Fig. 10.9 Example data for a weighted kappa analysis with $N = 5$ observations, $c = 3$ ordered categories, $b = 2$ judges and quadratic weights in brackets

	Judge 2			
Judge 1	A	B	C	Total
A	0 [0]	1 [1]	0 [4]	1
B	0 [1]	2 [0]	0 [1]	2
C	1 [4]	0 [1]	1 [0]	2
Total	1	3	1	5

gories, $b = 2$ blocks, and $r = 1$ response measurement,

$$\frac{1}{N}\sum_{i=1}^{c}\sum_{j=1}^{c} w_{ij}n_{ij} = \frac{1}{5}\big[(0)(0) + (1)(1) + (4)(0) + (1)(0) + (0)(2) + (1)(0) + (4)(1) + (1)(0) + (0)(1)\big] = \frac{5}{5} = 1.00\;,$$

and for the denominator of Eq. (10.24),

$$\frac{1}{N^2}\sum_{i=1}^{c}\sum_{j=1}^{c} w_{ij}n_{i.}n_{.j} = \frac{1}{5^2}\big[(0)(1)(1) + (1)(1)(3) + (4)(1)(1) + (1)(2)(1) + (0)(2)(3) + (1)(2)(1) + (4)(2)(1) + (1)(2)(3) + (0)(2)(1)\big] = \frac{25}{25} = 1.00\;.$$

Then the observed value of weighted $\hat{\kappa}$ with quadratic weighting is

$$\hat{\kappa}_{\text{o}} = 1 - \frac{\dfrac{1}{N}\displaystyle\sum_{i=1}^{c}\sum_{j=1}^{c} w_{ij}n_{ij}}{\dfrac{1}{N^2}\displaystyle\sum_{i=1}^{c}\sum_{j=1}^{c} w_{ij}n_{i.}n_{.j}} = 1 - \frac{1.00}{1.00} = 0.00\;.$$

Consider again the frequency data listed in Fig. 10.9, but rearranged into a randomized-block analysis-of-variance design with $b = 2$ blocks and $g = 5$ observations in each block, as given in Fig. 10.10.

Following Eq. (10.27) on p. 507 for the generalized Minkowski distance function with $p = v = 2$,

$$\Delta(1,1) = \big[(1-2)^2\big]^{2/2} = 1.00\;,$$
$$\Delta(2,2) = \big[(2-2)^2\big]^{2/2} = 0.00\;,$$

Fig. 10.10 Example data from Fig. 10.9 arranged in a block format with $b = 2$ blocks and $g = 5$ observations in each block

	Block	
g	1	2
1	1	2
2	2	2
3	2	2
4	3	1
5	3	3

$$\Delta(3,3) = \left[(2-2)^2\right]^{2/2} = 0.00\ ,$$
$$\Delta(4,4) = \left[(3-1)^2\right]^{2/2} = 4.00\ ,$$

and

$$\Delta(5,5) = \left[(3-3)^2\right]^{2/2} = 0.00\ .$$

Then following Eq. (10.26) on p. 507,

$$\delta = \frac{1}{g}\left[\Delta(1,1) + \Delta(2,2) + \Delta(3,3) + \Delta(4,4) + \Delta(5,5)\right]$$

and the observed value of the MRBP test statistic with $v = 2$ is

$$\delta_o = \frac{1}{5}\left(1.00 + 0.00 + 0.00 + 4.00 + 0.00\right) = \frac{1}{5}(5.00) = 1.00\ .$$

For the randomized-block data listed in Fig. 10.10, there are only

$$M = (g!)^b = (5!)^2 = 14{,}400$$

possible, equally-likely arrangements of the observed data; therefore, an exact solution is feasible. An exact probability value for $\hat{\kappa}_o$ and δ_o may be expressed as

$$P(\kappa \geq \hat{\kappa}_o | H_0) = P(\delta \leq \delta_o | H_0) = \frac{\text{number of } \delta \text{ values} \leq \delta_o}{M}$$

where $\hat{\kappa}_o$ and δ_o denote the observed values of Cohen's $\hat{\kappa}$ and δ, respectively.

If all arrangements of the observed data listed in Fig. 10.10 occur with equal chance, the exact probability value of $\delta_o = 1.00$ computed on the $M = 14{,}400$ possible arrangements of the observed data with $b = 2$ blocks preserved for each arrangement is

$$P(\delta \leq \delta_o | H_0) = \frac{\text{number of } \delta \text{ values} \leq \delta_o}{M} = \frac{8{,}640}{14{,}400} = 0.60\ .$$

Following Eq. (10.28) on p. 507 with $v = 2$, the exact expected value of the $M = 14{,}400$ δ values is given by

$$\mu_\delta = \frac{1}{g^2}\left[\Delta(1,1) + \Delta(1,2) + \cdots + \Delta(4,5) + \Delta(5,5)\right]$$

and the exact observed value of μ_δ is

$$\mu_\delta = \frac{1}{5^2}\,(1+1+1+0+4+0+0+\cdots+4+0+1+1+1+4+0)$$
$$= \frac{1}{25}\,(25) = 1.00\,.$$

Then,

$$\hat{\kappa}_o = \Re_o = 1 - \frac{\delta_o}{\mu_\delta} = 1 - \frac{1.00}{1.00} = 0.00$$

and the identity relating Cohen's $\hat{\kappa}$ with quadratic weighting and $\Re$ with $v = 2$ is confirmed.

10.7.5 Multiple Judges

While Cohen's $\hat{\kappa}$ is limited to $b = 2$ independent judges, a simple modification to δ and μ_δ generalizes $\Re$, and hence Cohen's $\hat{\kappa}$, to measure agreement among multiple judges [27]. Thus, the MRBP test statistic may be redefined as

$$\delta = \left[g \binom{b}{2} \right]^{-1} \sum_{i=1}^{g} \sum_{s<t} \Delta\left(x_{is}, x_{it}\right)\,, \tag{10.29}$$

where the generalized Minkowski distance function is given by

$$\Delta\left(x_{is}, x_{it}\right) = \left[\sum_{k=1}^{r} \left(x_{isk} - x_{itk}\right)^2 \right]^{1/2}\,,$$

b is the number of judges (i.e., blocks) and $\sum_{s<t}$ is the sum over all s and t such that $1 \le s < t \le b$. The reformulation of the expected value of δ is given by

$$\mu_\delta = \left[g^2 \binom{b}{2} \right]^{-1} \sum_{i=1}^{g} \sum_{j=1}^{g} \sum_{s<t} \Delta\left(x_{is}, x_{jt}\right)\,. \tag{10.30}$$

To illustrate the measurement of agreement for multiple judges, consider the data listed in Fig. 10.11, in which each of $b = 4$ judges is asked to assign $g = 10$ objects to $c = 4$ discrete, mutually exclusive, exhaustive categories, labeled 1, 2, 3, and 4.

Linear Weighting with $v = 1$

For the categorical data listed in Fig. 10.11, $g = 10$, $b = 4$, and $r = 1$. Following Eq. (10.29), the observed value of the MRBP test statistic based on $v = 1$ is

Fig. 10.11 Example data set for multiple judges with categorical data, $g = 10$ objects, $b = 4$ judges, and $r = 1$ response

	Judge			
Object	A	B	C	D
1	3	3	4	3
2	2	1	1	2
3	1	2	1	1
4	4	4	4	3
5	1	2	3	4
6	2	1	3	4
7	1	3	2	4
8	3	2	3	1
9	3	4	3	2
10	2	1	2	2

$\delta_o = 0.9833$, following Eq. (10.30) the exact expected value of δ is $\mu_\delta = 1.2033$, and following Eq. (10.2) on p. 474 the observed chance-corrected measure of effect size is

$$\Re_o = 1 - \frac{\delta_o}{\mu_\delta} = 1 - \frac{0.9833}{1.2033} = +0.1828 \, ,$$

indicating approximately 18 % agreement among the $b = 4$ judges above that expected by chance.

Because there are

$$M = \left(g!\right)^b = \left(10!\right)^4 = 173{,}401{,}213{,}127{,}727{,}513{,}600{,}000{,}000$$

possible, equally-likely arrangements of the observed data listed in Fig. 10.11, an exact permutation solution is not feasible. If all M possible arrangements of the observed data listed in Fig. 10.11 occur with equal chance, the approximate resampling probability value of $\delta_o = 0.9833$ computed on $L = 1{,}000{,}000$ random arrangements of the observed data with $b = 4$ blocks preserved for each arrangement is

$$P(\delta \leq \delta_o | H_0) = \frac{\text{number of } \delta \text{ values} \leq \delta_o}{L} = \frac{47{,}728}{1{,}000{,}000} = 0.0478 \, .$$

Quadratic Weighting with $v = 2$

For the categorical data listed in Fig. 10.11, following Eq. (10.29) the observed value of the MRBP test statistic based on $v = 2$ is $\delta_o = 1.5833$, following Eq. (10.30) the exact expected value of δ is $\mu_\delta = 2.3100$, and following Eq. (10.2) on p. 474 the observed chance-corrected measure of effect size is

$$\Re_o = 1 - \frac{\delta_o}{\mu_\delta} = 1 - \frac{1.5833}{2.3100} = +0.3146 \, ,$$

indicating approximately 31 % agreement among the $b = 4$ judges above that expected by chance.

Because there are still

$$M = (g!)^b = (10!)^4 = 173{,}401{,}213{,}127{,}727{,}513{,}600{,}000{,}000$$

possible, equally-likely arrangements of the observed data listed in Fig. 10.11, an exact permutation solution is not feasible. If all M possible arrangements of the observed data listed in Fig. 10.11 occur with equal chance, the approximate resampling probability value of $\delta_o = 1.5833$ computed on $L = 1{,}000{,}000$ random arrangements of the observed data with $b = 4$ blocks preserved for each arrangement is

$$P(\delta \leq \delta_o | H_0) = \frac{\text{number of } \delta \text{ values} \leq \delta_o}{L} = \frac{19{,}204}{1{,}000{,}000} = 0.0192 \, .$$

10.7.6 An Alternative Approach to Multiple Judges

In this section, an alternative procedure is presented to compute weighted kappa with multiple raters [308]. Although the procedure is appropriate for any number of $c \geq 2$ disjoint ordered categories and $b \geq 2$ judges, the description of the procedure and the examples are limited to three independent judges to simplify presentation.

Consider $b = 3$ judges who independently classify N objects into c disjoint ordered categories. The classification may be conceptualized as $c{\times}c{\times}c$ contingency table with c rows, c columns, and c slices. Let n_{ijk}, R_i, C_j, and S_k denote the cell frequencies and row, column, and slice marginal frequency totals for $i, j, k = 1, \ldots, c$ and let the frequency total be given by

$$N = \sum_{i=1}^{c} \sum_{j=1}^{c} \sum_{k=1}^{c} n_{ijk} \, .$$

Cohen's weighted kappa test statistic for a three-way contingency table is given by

$$\hat{\kappa} = \frac{N^2 \displaystyle\sum_{i=1}^{c} \sum_{i=1}^{c} \sum_{i=1}^{c} w_{ijk} n_{ijk}}{\displaystyle\sum_{i=1}^{c} \sum_{i=1}^{c} \sum_{i=1}^{c} w_{ijk} R_i C_j S_k} \, , \tag{10.31}$$

where w_{ijk} are disagreement weights assigned to each cell for $i, j, k = 1, \ldots, c$. Under the null hypothesis that the judges classify the N objects independently with fixed marginal frequency totals, $\mathrm{E}[\hat{\kappa}] = 0$.

As discussed previously on p. 503, a variety of weighting functions have been proposed for weighted kappa for two judges, where the arbitrary cell weights are denoted as w_{ij} and i and j designate the c categories for each judge [367, p. 246]. Typically, the cell weights are defined such that $w_{ii} = 0$ for $i = 1, \ldots, c$ and the weights are symmetrical, i.e., $w_{ij} = w_{ji}$ for $i, j = 1, \ldots, c$. Examples of weighting systems for two judges include linear weighting where $w_{ij} = |i - j|$, quadratic weighting where $w_{ij} = (i - j)^2$, and unweighted kappa where

$$w_{ij} = \begin{cases} 0 & \text{if } i = j \,, \\ 1 & \text{otherwise} \end{cases}$$

[250, 257].

For three judges, the cell disagreement weights are given by w_{ijk}, where i, j, and k designate the c categories for each judge. Analogously to w_{ij}, w_{ijk} may be defined such that $w_{iii} = 0$ for $i = 1, \ldots, c$ and the weights are symmetrical, i.e., $w_{ijk} = w_{ikj} = w_{jik} = w_{jki} = w_{kij} = w_{kji}$ for $i, j, k = 1, \ldots, c$. Examples of weighting systems for three judges include linear weighting where

$$w_{ijk} = |i - j| + |i - k| + |j - k|$$

and quadratic weighting where

$$w_{ijk} = (i - j)^2 + (i - k)^2 + (j - k)^2 \,.$$

Weighted kappa for three judges reduces to unweighted kappa[18] when

$$w_{ijk} = \begin{cases} 0 & \text{if } i = j = k \,, \\ 1 & \text{otherwise} \,. \end{cases}$$

Given a $c \times c \times c$ contingency table with N objects cross-classified by three independent judges, an exact permutation test involves generating all possible arrangements of the N objects to the c^3 cells, while preserving the marginal frequency totals. For each arrangement of cell frequencies, the weighted kappa

[18] Unweighted kappa for multiple judges is discussed in Chap. 11, Sect. 11.2.1.

statistic, $\hat{\kappa}$, and the exact hypergeometric probability value under the null hypothesis, $P(n_{ijk}|R_i, C_j, S_k)$, are calculated, where

$$P(n_{ijk}|R_i, C_j, S_k) = \frac{\left(\prod_{i=1}^{c} R_i!\right)\left(\prod_{j=1}^{c} C_j!\right)\left(\prod_{k=1}^{c} S_k!\right)}{(N!)^2 \prod_{i=1}^{c}\prod_{j=1}^{c}\prod_{k=1}^{c} n_{ijk}!} \tag{10.32}$$

[290].

If $\hat{\kappa}_o$ denotes the value of the observed weighted kappa test statistic, the exact probability value of $\hat{\kappa}_o$ under the null hypothesis is given by

$$P(\hat{\kappa}_o) = \sum_{l=1}^{M} \Psi_l\left(n_{ijk}|R_i, C_j, S_k\right) ,$$

where

$$\Psi_l\left(n_{ijk}|R_i, C_j, S_k\right) = \begin{cases} P(n_{ijk}|R_i, C_j, S_k) & \text{if } \hat{\kappa} \geq \hat{\kappa}_o , \\ 0 & \text{otherwise} , \end{cases}$$

and M denotes the total number of possible cell frequency arrangements given fixed observed marginal frequency totals. When M is very large, as is typical with multi-way contingency tables, exact tests are impractical and resampling becomes necessary, where a random sample, L, of the M possible arrangements of cell frequencies provides for a comparison of $\hat{\kappa}$ test statistics calculated on the L random tables with the $\hat{\kappa}_o$ test statistic calculated on the observed table.

An efficient resampling algorithm to generate random cell frequency arrangements for multi-way contingency tables with fixed marginal frequency totals was developed by Mielke et al. [307, pp. 19–20]. For a three-way contingency table with r rows, c columns, and s slices, the resampling algorithm is given in 12 simple steps.

Step 1. Construct an $r{\times}c{\times}s$ contingency table from the observed data.
Step 2. Obtain the fixed marginal frequency totals $R_1, \ldots, R_r$, $C_1, \ldots, C_c$, $S_1, \ldots, S_s$, and frequency total N. Set the resampling counter $JL = 0$, and set L equal to the number of samples desired.
Step 3. Set the resampling counter $JL = JL + 1$.
Step 4. Set the marginal frequency counters $JR_i = R_i$ for $i = 1, \ldots, r$; $JC_j = C_j$ for $j = 1, \ldots, c$; $JS_k = S_k$ for $k = 1, \ldots, s$, and $M = N$.
Step 5. Set $n_{ijk} = 0$ for $i = 1, \ldots, r$, $j = 1, \ldots, c$, and $k = 1, \ldots, s$, and set row, column, and slice counters IR, IC, and IS equal to zero.
Step 6. Create cumulative probability distributions PR_i, PC_j, and PS_k from the adjusted marginal frequency totals JR_i, JC_j, and JS_k for $i = 1, \ldots, r$,

$j = 1, \ldots, c$, and $k = 1, \ldots, s$, where

$$PR_1 = JR_1/M \quad \text{and} \quad PR_i = PR_{i-1} + JR_i/M$$

for $i = 1, \ldots, r$,

$$PC_1 = JC_1/M \quad \text{and} \quad PC_j = PC_{j-1} + JC_j/M$$

for $j = 1, \ldots, c$, and

$$PS_1 = JS_1/M \quad \text{and} \quad PS_k = PS_{k-1} + JS_k/M$$

for $k = 1, \ldots, s$.

Step 7. Generate three uniform pseudorandom numbers U_r, U_c, and U_s over $[0, 1)$ and set row, column, and slice indices $i = j = k = 1$, respectively.

Step 8. If $U_r \leq PR_i$, then $IR = i$, $JR_i = JR_i - 1$, and go to Step 9; otherwise, $i = i + 1$ and repeat Step 8.

Step 9. If $U_c \leq PC_j$, then $IC = j$, $JC_j = JC_j - 1$, and go to Step 10; otherwise, $j = j + 1$ and repeat Step 9.

Step 10. If $U_s \leq PS_k$, then $IS = k$, $JS_k = JS_k - 1$, and go to Step 11; otherwise, $k = k + 1$ and repeat Step 10.

Step 11. Set $M = M - 1$ and $n_{IR,IC,IS} = n_{IR,IC,IS} + 1$. If $M > 0$, go to Step 4; otherwise, obtain the required test statistic and calculate the hypergeometric probability value.

Step 12. If $JL < L$, go to Step 3; otherwise, stop.

At the conclusion of Step 11, $\hat{\kappa}$ and the exact probability value as given in Eqs. (10.31) and (10.32), respectively, are obtained for each of the L random three-way contingency tables, given fixed marginal frequency totals. Under the null hypothesis, the resampling approximate probability value for $\hat{\kappa}_o$ is given by

$$P(\hat{\kappa}_o) = \frac{1}{L} \sum_{l=1}^{L} \Psi_l(\hat{\kappa})$$

where

$$\Psi_l(\hat{\kappa}) = \begin{cases} 1 & \text{if } \hat{\kappa} \geq \hat{\kappa}_o\,, \\ 0 & \text{otherwise}\,. \end{cases}$$

The calculation of weighted kappa and the resampling procedure to obtain a probability value for multiple raters can be illustrated with a small example data set. Consider $b = 3$ independent journal reviewers for $N = 93$ submitted manuscripts over a 5-year period. Each reviewer classified each manuscript into one of $c = 3$ disjoint categories: reject, revise and resubmit, or accept. Table 10.3 lists the

Table 10.3 Article recommendations by three independent reviewers for $N = 93$ manuscripts: reject, revise, and accept

		Reviewer 3		
Reviewer 1	Reviewer 2	Reject	Revise	Accept
Reject	Reject	6 (0) [0]	4 (2) [2]	2 (4) [8]
	Revise	3 (2) [2]	5 (2) [2]	4 (4) [6]
	Accept	2 (4) [8]	3 (4) [6]	4 (4) [8]
Revise	Reject	4 (2) [2]	5 (2) [2]	3 (4) [6]
	Revise	5 (2) [2]	8 (0) [0]	4 (2) [2]
	Accept	3 (4) [6]	2 (2) [2]	3 (2) [2]
Accept	Reject	1 (4) [8]	3 (4) [6]	4 (4) [8]
	Revise	3 (4) [6]	2 (2) [2]	2 (2) [2]
	Accept	1 (4) [8]	2 (2) [2]	5 (0) [0]

Note: Linear cell weights are in parentheses and quadratic cell weights are in brackets

c^3 cross-classified observed frequencies and corresponding linear and quadratic weights, where the linear cell weights are given in parentheses and the quadratic cell weights are given in brackets. The frequency data listed in Table 10.3 are adapted from Mielke et al. [308, p. 609].

Linear Weighting

For the observed data listed in Table 10.3 with linear cell disagreement weights, the observed value of $\hat{\kappa}$ is $\hat{\kappa}_o = 0.1000$, indicating 10 % agreement above that expected by chance, and the approximate resampling probability value based on $L = 1{,}000{,}000$ random arrangements of the observed data is

$$P(\hat{\kappa} \geq \hat{\kappa}_o | H_0) = \frac{\text{number of } \hat{\kappa} \text{ values} \geq \delta_o}{L} = \frac{21{,}949}{1{,}000{,}000} = 0.0219 \, .$$

Quadratic Weighting

For the observed data listed in Table 10.3 with quadratic cell disagreement weights, the observed value of $\hat{\kappa}$ is $\hat{\kappa}_o = 0.1036$, indicating approximately 10 % agreement above that expected by chance, and the approximate resampling probability value based on $L = 1{,}000{,}000$ random arrangements of the observed data is

$$P(\hat{\kappa} \geq \hat{\kappa}_o | H_0) = \frac{\text{number of } \hat{\kappa} \text{ values} \geq \delta_o}{L} = \frac{48{,}926}{1{,}000{,}000} = 0.0489 \, .$$

10.8 MRBP and Measures of Ordinal Association

The test statistic S, as defined by Maurice Kendall [207], plays an important role in a variety of statistical measures; it is often expressed as $S = C - D$, where C and D indicate the number of concordant pairs and discordant pairs, respectively, *vide infra*. Consider two ordinal variables that have been cross-classified into an $r{\times}c$ contingency table, where r and c denote the number of rows and columns, respectively. Let $n_{i.}$, $n_{.j}$, and n_{ij} denote the row marginal frequency totals, column marginal frequency totals, and number of objects in the ijth cell, respectively, for $i = 1, \ldots, r$ and $j = 1, \ldots, c$, and let N denote the total number of objects in the $r \times c$ contingency table, i.e.,

$$n_{i.} = \sum_{j=1}^{c} n_{ij} , \quad n_{.j} = \sum_{i=1}^{r} n_{ij} , \quad \text{and} \quad N = \sum_{i=1}^{r} \sum_{j=1}^{c} n_{ij} .$$

If x and y represent the row and column variables, respectively, there are $N(N-1)/2$ pairs of objects in the table that can be partitioned into five mutually exclusive, exhaustive types: concordant pairs, discordant pairs, pairs tied on variable x but differing on variable y, pairs tied on variable y but differing on variable x, and pairs tied on both variable x and variable y.

Concordant pairs (pairs of objects that are ranked in the same order on both variable x and variable y) are given by

$$C = \sum_{i=1}^{r-1} \sum_{j=1}^{c-1} n_{ij} \left(\sum_{k=i+1}^{r} \sum_{l=j+1}^{c} n_{kl} \right) , \tag{10.33}$$

discordant pairs (pairs of objects that are ranked in one order on variable x and the reverse order on variable y) are given by

$$D = \sum_{i=1}^{r-1} \sum_{j=1}^{c-1} n_{i,c-j+1} \left(\sum_{k=i+1}^{r} \sum_{l=1}^{c-j} n_{kl} \right) , \tag{10.34}$$

pairs of objects tied on variable x but differing on variable y are given by

$$T_x = \sum_{i=1}^{r} \sum_{j=1}^{c-1} n_{ij} \left(\sum_{k=j+1}^{c} n_{ik} \right) , \tag{10.35}$$

pairs of objects tied on variable y but differing on variable x are given by

$$T_y = \sum_{j=1}^{c} \sum_{i=1}^{r-1} n_{ij} \left(\sum_{k=i+1}^{r} n_{kj} \right) , \tag{10.36}$$

and pairs of objects tied on both variable x and variable y are given by

$$T_{xy} = \frac{1}{2} \sum_{i=1}^{r} \sum_{j=1}^{c} n_{ij} \left(n_{ij} - 1 \right) . \tag{10.37}$$

Given C, D, T_x, T_y, N, and $S = C - D$, six measures of ordinal association are commonly defined, each having the same numerator, S, but different denominators [39]. The earliest of these measures was Kendall's τ_a [207].[19] Kendall's τ_a is a symmetrical measure of ordinal association that is most suitable when there are no tied pairs and is defined as the simple difference between the proportions of concordant and discordant pairs given by

$$\tau_a = \frac{C}{\frac{N(N-1)}{2}} - \frac{D}{\frac{N(N-1)}{2}} = \frac{2S}{N(N-1)} . \tag{10.38}$$

Kendall's τ_b [207] extends τ_a to measure strong monotonicity in contingency tables when $r = c$. The denominator for τ_b is adjusted for the number of tied pairs for both variable x and variable y; τ_b is given by

$$\tau_b = \frac{S}{\sqrt{(C + D + T_x)(C + D + T_y)}} . \tag{10.39}$$

Stuart's τ_c [389] modifies Kendall's τ_b for contingency tables where $r \neq c$ and is given by

$$\tau_c = \frac{(2m)(S)}{N^2(m-1)} , \tag{10.40}$$

where $m = \min(r, c)$. Goodman and Kruskal's γ [151] is a symmetrical measure of weak monotonicity in which tied pairs of all types are ignored and defined as

$$\gamma = \frac{S}{C + D} . \tag{10.41}$$

Somers' d_{yx} and d_{xy} [380] are asymmetric measures of ordinal association. Unlike the four symmetrical measures, τ_a, τ_b, τ_c, and γ, Somers' d_{yx} and d_{xy} depend on which variable, y or x, is considered to be the dependent variable. If y is the dependent variable, then

$$d_{yx} = \frac{S}{C + D + T_y} , \tag{10.42}$$

[19] Yule's Q for 2×2 contingency tables also has S in the numerator and preceded Kendall's τ_a by some 40 years [434, 435]. While Yule's Q is occasionally prescribed for rank-score data [245, p. 255–256], it was originally designed for categorical data and is therefore described more appropriately in Chap. 6.

and if x is the dependent variable, then

$$d_{xy} = \frac{S}{C + D + T_x} . \tag{10.43}$$

Thus, for both d_{yx} and d_{xy}, when a difference between paired values on the independent variable (i.e., untied pair) is not reflected as a difference between the corresponding paired values on the dependent variable (i.e., tied pair) the denominator of Eqs. (10.42) and (10.43) is increased by T_y or T_x, respectively, and the values of d_{yx} and d_{xy} are diminished accordingly. Finally, it is readily apparent that Kendall's τ_b measure of ordinal association given in Eq. (10.39) is simply the geometric mean of Somers' d_{yx} and d_{xy} given by

$$\tau_b = \sqrt{d_{yx}\, d_{xy}} .$$

10.8.1 Example 1

Consider for this first example, Kendall's τ_a measure of ordinal association, given by

$$\tau_a = \frac{2S}{N(N-1)} .$$

Kendall's τ_a was originally designed to measure the association between two sets of untied rank scores, such as given in Fig. 10.12, where the two sets of rank scores are labeled as x and y. Kendall's τ_a is often described as an alternative to Spearman's rank-order correlation coefficient [221, p. 179]. For the two sets of rankings listed in Fig. 10.12, there are

$$\binom{N}{2} = \frac{N(N-1)}{2} = \frac{8(8-1)}{2} = 28$$

Fig. 10.12 Two sets of $N = 8$ rank scores for Kendall's τ_a measure of ordinal association

	Variable	
Object	x	y
1	1	3
2	3	4
3	2	1
4	4	2
5	5	5
6	7	8
7	8	6
8	6	7

Table 10.4 Paired differences, r_{ij}, s_{ij}, $r_{ij}s_{ij}$, and $|r_{ij} - s_{ij}|$ values for the rank scores listed in Fig. 10.12

Pair	$x_i - x_j$	$y_i - y_j$	r_{ij}	s_{ij}	$r_{ij}s_{ij}$	$\|r_{ij} - s_{ij}\|$
1	$1-3$	$3-4$	-1	-1	$+1$	0
2	$1-2$	$3-1$	-1	$+1$	-1	2
3	$1-4$	$3-2$	-1	$+1$	-1	2
4	$1-5$	$3-5$	-1	-1	$+1$	0
5	$1-7$	$3-8$	-1	-1	$+1$	0
6	$1-8$	$3-6$	-1	-1	$+1$	0
7	$1-6$	$3-7$	-1	-1	$+1$	0
8	$3-2$	$4-1$	$+1$	$+1$	$+1$	0
9	$3-4$	$4-2$	-1	$+1$	-1	2
10	$3-5$	$4-5$	-1	-1	$+1$	0
11	$3-7$	$4-8$	-1	-1	$+1$	0
12	$3-8$	$4-6$	-1	-1	$+1$	0
13	$3-6$	$4-7$	-1	-1	$+1$	0
14	$2-4$	$1-2$	-1	-1	$+1$	0
15	$2-5$	$1-5$	-1	-1	$+1$	0
16	$2-7$	$1-8$	-1	-1	$+1$	0
17	$2-8$	$1-6$	-1	-1	$+1$	0
18	$2-6$	$1-7$	-1	-1	$+1$	0
19	$4-5$	$2-5$	-1	-1	$+1$	0
20	$4-7$	$2-8$	-1	-1	$+1$	0
21	$4-8$	$2-6$	-1	-1	$+1$	0
22	$4-6$	$2-7$	-1	-1	$+1$	0
23	$5-7$	$5-8$	-1	-1	$+1$	0
24	$5-8$	$5-6$	-1	-1	$+1$	0
25	$5-6$	$5-7$	-1	-1	$+1$	0
26	$7-8$	$8-6$	-1	$+1$	-1	2
27	$7-6$	$8-7$	$+1$	$+1$	$+1$	0
28	$8-6$	$6-7$	$+1$	-1	-1	2
Total					$+18$	10

possible pairs, where N denotes the number of paired rankings as listed in the first two columns of Table 10.4.

Because there are no tied rank scores in Fig. 10.12, the $N(N-1)/2$ pairs can be exhaustively divided into just two types: concordant (C) and discordant (D) pairs. To illustrate the calculation of Kendall's S, consider the x and y rank scores for the first pair of objects in Table 10.4: Objects 1 and 2. For variable x calculate $1 - 3 = -2$ and for variable y calculate $3 - 4 = -1$. When the signs agree, either both negative or both positive, as in this case with both signs negative, the pair is considered a concordant pair. Now consider the x and y rank scores for the second pair: Objects 1 and 3. For variable x calculate $1 - 2 = -1$ and for variable y calculate $n = +2$. When the signs disagree, as in this case with one negative sign and one positive sign, the pair is considered a discordant pair.

Given the bivariate rank scores listed in Fig. 10.12, for $i < j$ define

$$r_{ij} = \begin{cases} +1 & \text{if } x_i > x_j \,, \\ 0 & \text{if } x_i = x_j \,, \\ -1 & \text{if } x_i < x_j \,, \end{cases} \quad \text{and} \quad s_{ij} = \begin{cases} +1 & \text{if } y_i > y_j \,, \\ 0 & \text{if } y_i = y_j \,, \\ -1 & \text{if } y_i < y_j \,. \end{cases}$$

Then, following Kendall [207],

$$S = \sum_{i<j} r_{ij} s_{ij}$$

as given in the sixth column of Table 10.4, where there are 23 concordant pairs, each indicated by $+1$ ($C = 23$) and 5 discordant pairs, each indicated by -1 ($D = 5$); therefore, the observed value of Kendall's S is $S_o = C - D = 23 - 5 = +18$. For the rank scores with no tied values listed in Fig. 10.12, the observed value of Kendall's τ_a is

$$\tau_a = \frac{2S_o}{N(N-1)} = \frac{(2)(+18)}{8(8-1)} = +0.6429$$

and, incidentally, because there are no tied rank scores for the data listed in Fig. 10.12, $\tau_a = \tau_b = \tau_c = \gamma = d_{yx} = d_{xy}$.

Now, in a randomized-block analysis-of-variance context, consider the rank scores listed in Fig. 10.12 with $b = 2$ blocks and $g = 8$ univariate measurements for each block, and define the MRBP test statistic

$$\delta = \left[g \binom{b}{2} \right]^{-1} \sum_{i=1}^{g} \sum_{j<k} \Delta(x_{ij}, x_{ik}) \,, \tag{10.44}$$

where $\sum_{j<k}$ denotes the sum over all j and k such that $1 \leq j < k \leq b$ and $\Delta(x, y)$ is a symmetric distance-function value of two points $x' = (x_1, x_2, \ldots, x_r)$ and $y' = (y_1, y_2, \ldots, y_r)$ in an r-dimensional Euclidean space, and the generalized Minkowski distance function is given by

$$\Delta(x, y) = \left(\sum_{i=1}^{r} \left| x_i - y_i \right|^p \right)^{v/p} ,$$

where $p \geq 1$ and $v > 0$.

If $b = 2$, $r = 1$, $p = 2$, $v = 1$, and there are no tied values on either variable x or y ($T_{xy} = 0$), as in this case, then the MRBP test statistic as given in Eq. (10.44) reduces to

$$\delta = \frac{1}{g}\,\Delta(x,y)\;, \tag{10.45}$$

where

$$\Delta(x,y) = \sum_{j<k} \left| r_{ij} - s_{ij} \right| \,.$$

Thus, for the rank scores listed in Fig. 10.12,

$$\Delta(x,y) = 10 \quad \text{and} \quad \delta = \frac{1}{8}\,(10) = 1.25\,.$$

Then it can easily be shown that the functional relationships between the generalized Minkowski distance function $\Delta(x, y)$ and Kendall's S are given by

$$\Delta(x,y) = \frac{g(g-1)}{2} - S \quad \text{and} \quad S = \frac{g(g-1)}{2} - \Delta(x,y)\,.$$

For the rank scores with no tied values listed in Fig. 10.12,

$$\Delta(x,y) = \frac{8(8-1)}{2} - 18 = 10 \quad \text{and} \quad S = \frac{8(8-1)}{2} - 10 = +18\,,$$

as given in the totals for the last two columns of Table 10.4. Also, the relationships between the MRBP test statistic and Kendall's S are given by

$$\delta = \frac{g-1}{2} - \frac{S}{g} \quad \text{and} \quad S = g\left(\frac{g-1}{2} - \delta\right)\,.$$

Thus, for the rank scores with no tied values listed in Table 10.4 on p. 523, the observed values of δ and S are

$$\delta_o = \frac{8-1}{2} - \frac{+18}{8} = 1.25 \quad \text{and} \quad S_o = 8\left(\frac{8-1}{2} - 1.25\right) = +18\,.$$

For the rank scores listed in Fig. 10.12, there are

$$M = (g!)^b = (8!)^2 = 1{,}625{,}702{,}400$$

possible, equally-likely arrangements of the observed data. However, considering variable x fixed, relative to variable y, M can be reduced to

$$M = (g!)^{b-1} = (8!)^{2-1} = 8! = 40{,}320$$

and an exact solution is easily accomplished. Since $g(g-1)/2$ is invariant under permutation,

$$P(S \geq S_o|H_0) = P(\delta \leq \delta_o|H_0) = \frac{\text{number of } \delta \text{ values} \leq \delta_o}{M} ,$$

where S_o and δ_o denote the observed values of Kendall's S and δ, respectively.

For the rank scores listed in Fig. 10.12 on p. 522, following Eq. (10.45), the observed value of the MRBP test statistic with $v = 1$ is

$$\delta_o = \frac{1}{g}\,\Delta(x, y) = \frac{1}{8}\,(10) = 1.25 .$$

If all arrangements of the observed rank scores listed in Fig. 10.12 occur with equal chance, the exact probability value of $\delta_o = 1.25$ computed on the $M = 40{,}324$ possible arrangements of the observed rank scores with $b = 2$ blocks preserved for each arrangement is

$$P(\delta \leq \delta_o|H_0) = \frac{\text{number of } \delta \text{ values} \leq \delta_o}{M} = \frac{1{,}248}{40{,}320} = 0.0310 .$$

For the rank scores listed in Fig. 10.12, following Eq. (10.3) on p. 474, the exact expected value of the $M = 40{,}320$ δ values is

$$\mu_\delta = \frac{1}{M}\sum_{i=1}^{M} \delta_i = \frac{1}{40{,}320}\,(105{,}840) = 2.6250$$

and, following Eq. (10.2) on p. 474, the observed chance-corrected measure of effect size is

$$\Re_o = 1 - \frac{\delta_o}{\mu_\delta} = 1 - \frac{1.25}{2.6250} = +0.5238 ,$$

indicating approximately 52 % within-block agreement above that expected by chance. It should be noted that the relationships between Kendall's τ_a and the MRBP test statistic are given by

$$\tau_a = 1 - \frac{2\delta}{g-1} \quad \text{and} \quad \delta = \frac{(1-\tau_a)(g-1)}{2} .$$

Thus, for the univariate rank scores listed in Fig. 10.12, the observed values of τ_a and δ are

$$\tau_a = 1 - \frac{(2)(1.25)}{8-1} = 1 - 0.3571 = +0.6429$$

and

$$\delta_o = \frac{(1 - 0.6429)(8 - 1)}{2} = \frac{2.5}{2} = 1.25 .$$

Finally, there is an interesting relationship between Kendall's τ_a measure of ordinal association and Spearman's footrule measure of chance-corrected rank agreement $\mathcal{R}$, which has not heretofore been documented. Because $\mathfrak{R}$ and $\mathcal{R}$ yield identical values,[20] the relationships between τ_a and $\mathcal{R}$ are given by

$$\tau_a = 1 + \frac{2(\mathcal{R} - 1)\mu_\delta}{g - 1} \quad \text{and} \quad \mathcal{R} = 1 - \frac{(1 - \tau_a)(g - 1)}{2\mu_\delta} .$$

Thus, for the univariate rank scores listed in Fig. 10.12, the observed values of Kendall's τ_a and Spearman's $\mathcal{R}$ are

$$\tau_a = 1 + \frac{2(0.5238 - 1)(2.6250)}{8 - 1} = +0.6429$$

and

$$\mathcal{R}_o = 1 - \frac{(1 - 0.6429)(8 - 1)}{(2)(2.6250)} = +0.5238 .$$

10.8.2 Example 2

For a second example of measures of ordinal association, consider the small set of univariate rank scores listed in Fig. 10.13 in which tied rank scores on x and y (T_x and T_y, respectively) are introduced.

Fig. 10.13 Two sets of $N = 5$ rank scores with ties for Kendall's τ_a measure of ordinal association

	Variable	
Object	x	y
1	1	2
2	2.5	1
3	2.5	4.5
4	4	4.5
5	5	3

[20]See Eq. (10.17) on p. 489.

Table 10.5 Paired differences, r_{ij}, s_{ij}, $r_{ij}s_{ij}$, and $|r_{ij} - s_{ij}|$ values for the univariate rank scores listed in Fig. 10.13

Pair	$x_i - x_j$	$y_i - y_j$	r_{ij}	s_{ij}	$r_{ij}s_{ij}$	$\|r_{ij} - s_{ij}\|$
1	$1.0 - 2.5$	$2.0 - 1.0$	-1	$+1$	-1	2
2	$1.0 - 2.5$	$2.0 - 4.5$	-1	-1	$+1$	0
3	$1.0 - 4.0$	$2.0 - 4.5$	-1	-1	$+1$	0
4	$1.0 - 5.0$	$2.0 - 3.0$	-1	-1	$+1$	0
5	$2.5 - 2.5$	$1.0 - 4.5$	0	-1	0	1
6	$2.5 - 4.0$	$1.0 - 4.5$	-1	-1	$+1$	0
7	$2.5 - 5.0$	$1.0 - 3.0$	-1	-1	$+1$	0
8	$2.5 - 4.0$	$4.5 - 4.5$	-1	0	0	1
9	$2.5 - 5.0$	$4.5 - 3.0$	-1	$+1$	-1	2
10	$4.0 - 5.0$	$4.5 - 3.0$	-1	$+1$	-1	2
Total					$+2$	8

Table 10.5 lists the ten paired differences, r_{ij}, s_{ij}, $r_{ij}s_{ij}$, and $|r_{ij} - s_{ij}|$ values for the univariate rank scores listed in Fig. 10.13. Following Kendall,

$$S = \sum_{i<j} r_{ij}s_{ij}$$

as given in the sixth column of Table 10.5, where there are five concordant pairs, each indicated by $+1$ ($C = 5$) and 3 discordant pairs, each indicated by -1 ($D = 3$); therefore $S = C - D$ implies that the observed value of S is $S_o = 5 - 3 = +2$. Also, there is one pair of rank scores tied on variable x but not tied on variable y ($T_x = 1$), indicated by a 0 in row 5 of the sixth column and one pair of rank scores tied on variable y but not tied on variable x ($T_y = 1$), indicated by a 0 in row 8 of the sixth column. Then, the observed value of Kendall's τ_a based on $S_o = +2$ is

$$\tau_a = \frac{2S_o}{N(N-1)} = \frac{(2)(+2)}{5(5-1)} = +0.20\ .$$

Now consider the rank scores listed in Fig. 10.13 in a randomized-block analysis-of-variance context with $b = 2$ blocks and $g = 5$ univariate measurements for each block. Then, as shown previously, with $T_{xy} = 0$, the generalized Minkowski distance function is

$$\Delta(x, y) = \frac{g(g-1)}{2} - S_o = \frac{5(5-1)}{2} - 2 = 8\ ,$$

as given in the total for the last column of Table 10.5 and

$$S_o = \frac{g(g-1)}{2} - \Delta(x, y) = \frac{5(5-1)}{2} - 8 = +2\ .$$

Also, the relationships between the MRBP test statistic and Kendall's S are given by

$$\delta = \frac{g-1}{2} - \frac{S}{g} \quad \text{and} \quad S = g\left(\frac{g-1}{2} - \delta\right) .$$

Thus, for the rank scores listed in Table 10.5, the observed values of δ and S are

$$\delta_o = \frac{5-1}{2} - \frac{2}{5} = 1.60 \quad \text{and} \quad S_o = 5\left(\frac{5-1}{2} - 1.60\right) = +2 .$$

For the univariate rank scores listed in Fig. 10.13, there are only

$$M = (g!)^b = (5!)^2 = 14{,}400$$

possible, equally-likely arrangements of the observed data, therefore an exact solution is feasible. As previously,

$$P(S \geq S_o|H_0) = P(\delta \leq \delta_o|H_0) = \frac{\text{number of } \delta \text{ values} \leq \delta_o}{M} ,$$

where S_o and δ_o denote the observed values of Kendall's S and δ, respectively.

For the rank scores listed in Fig. 10.13, following Eq. (10.45) on p. 525 the observed value of the MRBP test statistic with $v = 1$ is

$$\delta_o = \frac{1}{g}\Delta(x, y) = \frac{1}{5}(8) = 1.60 .$$

If all arrangements of the observed rank scores listed in Fig. 10.13 occur with equal chance, the exact probability value of $\delta_o = 1.60$ computed on the $M = 14{,}400$ possible arrangements of the observed rank scores with $b = 2$ blocks preserved for each arrangement is

$$P(\delta \leq \delta_o|H_0) = \frac{\text{number of } \delta \text{ values} \leq \delta_o}{M} = \frac{11{,}030}{14{,}400} = 0.7660 .$$

Following Eq. (10.3) on p. 474, the exact expected value of the $M = 14{,}400$ δ values is

$$\mu_\delta = \frac{1}{M}\sum_{i=1}^{M} \delta_i = \frac{1}{14{,}400}(26{,}332) = 1.8286$$

and, following Eq. (10.2) on p. 474, the observed chance-corrected measure of effect size is

$$\Re_o = 1 - \frac{\delta_o}{\mu_\delta} = 1 - \frac{1.60}{1.8286} = +0.1250 ,$$

indicating approximately 12 % within-block agreement above that expected by chance. It should be noted that $\Re_o = +0.1250$ is equivalent to Spearman's footrule measure $\mathcal{R}$ and, as in Example 1, for the rank scores listed in Fig. 10.13, the observed values of τ_a and the MRBP test statistic δ are

$$\tau_a = 1 - \frac{2\delta_o}{g-1} = 1 - \frac{(2)(1.60)}{5-1} = +0.20$$

and

$$\delta_o = \frac{(1-\tau_a)(g-1)}{2} = \frac{(1-20)(5-1)}{2} = 1.60 \, .$$

Finally, the relationships between the observed values of Kendall's rank-correlation coefficient τ_a and Spearman's footrule measure $\mathcal{R}$ are

$$\tau_a = 1 + \frac{2(\mathcal{R}-1)\mu_\delta}{g-1} = 1 + \frac{2(0.1250-1)(1.8286)}{5-1} = +0.20$$

and

$$\mathcal{R}_o = 1 - \frac{(1-\tau_a)(g-1)}{2\mu_\delta} = 1 - \frac{(1-0.20)(5-1)}{(2)(1.8286)} = 0.1250 \, .$$

10.8.3 Example 3

Whenever two sets of rank scores possess tied scores such that $T_{xy} > 0$, the relationship between Kendall's S and δ must be expressed more completely. Consider the two sets of rank scores listed in Fig. 10.14, where there are multiple tied rank scores. For the rank scores listed in Fig. 10.14, $C = 8$, $D = 2$, $T_x = 1$, $T_y = 2$, and $T_{xy} = 2$. Table 10.6 lists the

$$\frac{N(N-1)}{2} = \frac{6(6-1)}{2} = 15$$

Fig. 10.14 Two sets of rank scores with ties for Kendall's τ_a measure of ordinal association

	Variable	
Object	x	y
1	1.5	2
2	1.5	2
3	3.5	4.5
4	5.5	2
5	3.5	4.5
6	5.5	6

Table 10.6 Paired differences, r_{ij}, s_{ij}, $r_{ij}s_{ij}$, and $|r_{ij} - s_{ij}|$ values for the univariate rank scores listed in Fig. 10.14

Pair	$x_i - x_j$	$y_i - y_j$	r_{ij}	s_{ij}	$r_{ij}s_{ij}$	$\|r_{ij} - s_{ij}\|$	Type
1	1.5 − 1.5	2.0 − 2.0	0	0	0	0	T_{xy}
2	1.5 − 3.5	2.0 − 4.5	−1	−1	+1	0	C
3	1.5 − 5.5	2.0 − 2.0	−1	0	0	1	T_y
4	1.5 − 3.5	2.0 − 4.5	−1	−1	+1	0	C
5	1.5 − 5.5	2.0 − 6.0	−1	−1	+1	0	C
6	1.5 − 3.5	2.0 − 4.5	−1	−1	+1	0	C
7	1.5 − 5.5	2.0 − 2.0	−1	0	0	1	T_y
8	1.5 − 3.5	2.0 − 4.5	−1	−1	+1	0	C
9	1.5 − 5.5	2.0 − 6.0	−1	−1	+1	0	C
10	3.5 − 5.5	4.5 − 2.0	−1	+1	−1	2	D
11	3.5 − 3.5	4.5 − 4.5	0	0	0	0	T_{xy}
12	3.5 − 5.5	4.5 − 6.0	−1	−1	+1	0	C
13	5.5 − 3.5	2.0 − 4.5	+1	−1	−1	2	D
14	5.5 − 5.5	2.0 − 6.0	0	−1	0	1	T_x
15	3.5 − 5.5	4.5 − 6.0	−1	−1	+1	0	C
Total					+6	7	

paired differences, r_{ij}, s_{ij}, $r_{ij}s_{ij}$, and $|r_{ij} - s_{ij}|$ values for the rank scores given in Fig. 10.14.

Following Kendall,

$$\sum_{i<j} r_{ij}s_{ij}$$

is given in the sixth column of Table 10.6, where there are $C = 8$ concordant pairs in rows 2, 4, 5, 6, 8, 9, 12, and 15, indicated by $+1$ values, and $D = 2$ discordant pairs in rows 10 and 13, indicated by -1 values. Values of T_x, T_y, and T_{xy} receive values of 0. Thus,

$$S = \sum_{i<j} r_{ij}s_{ij} = C - D = 8 - 2 = +6$$

and, following Eq. (10.38) on p. 521, Kendall's τ_a statistic is given by

$$\tau_a = \frac{2S}{N(N-1)} = \frac{2(6)}{6(6-1)} = +0.40\ .$$

Now consider the rank scores listed in Fig. 10.14 in a randomized-block analysis-of-variance context with $b = 2$ blocks and $g = 5$ univariate measurements for each block. It is obvious for the $|r_{ij} - s_{ij}|$ column in Table 10.6 that only values of T_x, T_y, and D can receive non-zero values: values of 1 for both T_x and T_y and values of 2 for D. Therefore,

$$\sum_{i<j} |r_{ij} - s_{ij}| = 2D + T_x + T_y = \frac{g(g-1)}{2} - \sum_{i<j} r_{ij}s_{ij} - T_{xy}$$

$$= \frac{g(g-1)}{2} - S - T_{xy} \, .$$

Here, the observed value is $2D + T_x + T_y = (2)(2) + 1 + 2 = 7$ and $g(g-1)/2 - S - T_{xy} = 6(6-1)/2 - 6 - 2 = n - 2 = 7$.

Now, define the generalized Minkowski distance function with $b = 2, r = 1, p = 2$, and $v = 1$,

$$\Delta(x, y) = \frac{g(g-1)}{2} - S \, .$$

Then, substituting $C + D + T_x + T_y + T_{xy}$ for $g(g-1)/2$ and $C - D$ for S,

$$\Delta(x, y) = C + D + T_x + T_y + T_{xy} - (C - D)$$

$$= 2D + T_x + T_y + T_{xy} \, . \qquad (10.46)$$

In this case, $2D + T_x + T_y + T_{xy} = 2(2) + 1 + 2 + 2 = 9$. Finally, the observed MRBP test statistic with $v = 1$ is

$$\delta_o = \frac{1}{g}\Delta(x, y) = \frac{1}{6}(9) = 1.50 \, .$$

Also, the relationships between the MRBP test statistic and Kendall's S are given by

$$\delta = \frac{g-1}{2} - \frac{S}{g} \quad \text{and} \quad S = g\left(\frac{g-1}{2} - \delta\right) .$$

Thus, for the rank scores listed in Fig. 10.14, the observed values of δ and S are

$$\delta_o = \frac{6-1}{2} - \frac{6}{6} = 1.50 \quad \text{and} \quad S_o = 6\left(\frac{6-1}{2} - 1.50\right) = 6 \, .$$

For the rank scores listed in Fig. 10.14, there are only

$$M = (g!)^b = (6!)^2 = 518{,}400$$

possible, equally-likely arrangements of the observed data, therefore an exact solution is possible. As previously,

$$P(S \geq S_o|H_0) = P(\delta \leq \delta_o|H_0) = \frac{\text{number of } \delta \text{ values} \leq \delta_o}{M} ,$$

where S_o and δ_o denote the observed values of Kendall's S and δ, respectively.

If all arrangements of the observed rank scores listed in Fig. 10.14 occur with equal chance, the exact probability value of $\delta_o = 1.50$ computed on the $M = 518{,}400$ possible arrangements of the observed rank scores with $b = 2$ blocks preserved for each arrangement is

$$P(\delta \leq \delta_o|H_0) = \frac{\text{number of } \delta \text{ values} \leq \delta_o}{M} = \frac{69{,}120}{518{,}400} = 0.1333 .$$

Following Eq. (10.3) on p. 474, the exact expected value of the $M = 518{,}400$ δ values is

$$\mu_\delta = \frac{1}{M}\sum_{i=1}^{M} \delta_i = \frac{1}{518{,}400}(1{,}258{,}971) = 2.4286$$

and, following Eq. (10.2) on p. 474, the observed chance-corrected measure of effect size is

$$\Re_o = 1 - \frac{\delta_o}{\mu_\delta} = 1 - \frac{1.50}{2.4286} = +0.3824 ,$$

indicating approximately 38 % within-block agreement above that expected by chance.

As in Examples 1 and 2, $\Re_o$ is equivalent to the observed value of Spearman's footrule measure $\mathcal{R}_o$ and, for the rank scores listed in Fig. 10.14, the relationships between the observed values of τ_a and δ are

$$\tau_a = 1 - \frac{2\delta_o}{g-1} = 1 - \frac{2(1.50)}{6-1} = +0.40$$

and

$$\delta_o = \frac{(1-\tau_a)(g-1)}{2} = \frac{(1-0.40)(6-1)}{2} = 1.50 .$$

Also, the relationships between the observed values of Kendall's τ_a and Spearman's $\mathcal{R}$ are

$$\tau_a = 1 + \frac{2(\mathcal{R}_o - 1)\mu_\delta}{g-1} = 1 + \frac{2(0.3824-1)(2.4286)}{(6-1)} = +0.40$$

and

$$\mathcal{R}_o = 1 - \frac{(1-\tau_a)(g-1)}{2\mu_g} = 1 - \frac{(1-0.40)(6-1)}{2(2.4286)} = +0.3824 .$$

10.8.4 Example 4

For this fourth example of measures of ordinal association, consider the frequency data given in Fig. 10.15, where $N = 20$ bivariate observations have been cross-classified into a 3×3 ordered contingency table. This is a more typical application of measures of ordinal association. For the frequency data given in Fig. 10.15, the number of concordant pairs is

$$C = \sum_{i=1}^{r-1} \sum_{j=1}^{c-1} n_{ij} \left(\sum_{k=i+1}^{r} \sum_{l=j+1}^{c} n_{kl} \right)$$
$$= 6(2+1+1+5) + 2(1+5) + 2(1+5) + 2(5) = 88 ,$$

the number of discordant pairs is

$$D = \sum_{i=1}^{r-1} \sum_{j=1}^{c-1} n_{i,c-j+1} \left(\sum_{k=i+1}^{r} \sum_{l=1}^{c-j} n_{kl} \right)$$
$$= 0(2+2+1+1) + 2(2+1) + 1(1+1) + 2(1) = 10 ,$$

the number of pairs tied on variable x but not tied on variable y is

$$T_x = \sum_{i=1}^{r} \sum_{j=1}^{c-1} n_{ij} \left(\sum_{k=j+1}^{c} n_{ik} \right)$$
$$= 6(2+0) + 2(0) + 2(2+1) + 2(1) + 1(1+5) + 1(5) = 31 ,$$

Fig. 10.15 Example rank-score data for $N = 20$ bivariate observations cross-classified on ordinal variables x and y into a 3×3 contingency table

	y			
x	1	2	3	Total
1	6	2	0	8
2	2	2	1	5
3	1	1	5	7
Total	9	5	6	20

the number of pairs tied on variable y but not tied on variable x is

$$T_y = \sum_{j=1}^{c} \sum_{i=1}^{r-1} n_{ij} \left(\sum_{k=i+1}^{r} n_{kj} \right)$$
$$= 6(2+1) + 2(1) + 2(2+1) + 2(1) + 0(1+5) + 1(5) = 33 ,$$

and the number of pairs tied on both variable x and variable y is

$$T_{xy} = \frac{1}{2} \sum_{i=1}^{r} \sum_{j=1}^{c} n_{ij} \left(n_{ij} - 1 \right)$$
$$= \frac{1}{2} \big[6(6-1) + 2(2-1) + 0(0-1) + 2(2-1) + 2(2-1) + 1(1-1)$$
$$+ 1(1-1) + 1(1-1) + 5(5-1) \big] = 28 .$$

Then, the observed value of Kendall's S is $S_o = C - D = 88 - 10 = +78$ and the observed value of Kendall's τ_a is

$$\tau_a = \frac{2S_o}{N(N-1)} = \frac{2(78)}{20(20-1)} = +0.4105 .$$

Now consider the frequency data given in Fig. 10.15 in a randomized-block analysis-of-variance context with $b = 2$ blocks and $g = 20$ univariate measurements for each block as displayed in Fig. 10.16. Then, as shown in Example 3, the generalized Minkowski distance function with $v = 1$ is

$$\Delta(x, y) = \frac{g(g-1)}{2} - S_o = \frac{20(20-1)}{2} - 78 = 112 .$$

Fig. 10.16 Rank scores assigned to $g = 20$ objects by $b = 2$ blocks

	Block			Block	
Object	1	2	Object	1	2
1	1	1	11	2	2
2	1	1	12	2	2
3	1	1	13	2	3
4	1	1	14	3	1
5	1	1	15	3	2
6	1	1	16	3	3
7	1	2	17	3	3
8	1	2	18	3	3
9	2	1	19	3	3
10	2	1	20	3	3

Alternatively, following Eq. (10.46) on p. 532, the observed value of $\Delta(x, y)$ is given by

$$\Delta(x, y) = 2D + T_x + T_y + T_{xy} = 2(10) + 31 + 33 + 28 = 112 .$$

Following Eq. (10.45) on p. 525, the observed value of the MRBP test statistic δ with $v = 1$ is

$$\delta_o = \frac{1}{g}\,\Delta(x, y) = \frac{1}{20}\,(112) = 5.60 .$$

Also, the relationships between the MRBP test statistic and Kendall's S are given by

$$\delta = \frac{g-1}{2} - \frac{S}{g} \quad \text{and} \quad S = g\left(\frac{g-1}{2} - \delta\right) .$$

Thus, for the randomized-block data listed in Fig. 10.16, the observed values of δ and S are

$$\delta_o = \frac{20-1}{2} - \frac{78}{20} = 5.60 \quad \text{and} \quad S_o = 20\left(\frac{n}{2} - 5.60\right) = +78 .$$

For the randomized-block data listed in Fig. 10.16, there are

$$M = (g!)^{b-1} = (20!)^{2-1} = 2{,}432{,}902{,}008{,}176{,}640{,}000$$

equally-likely arrangements of the observed data, therefore an exact solution is not feasible and a resampling approximation is required, where

$$P(\delta \leq \delta_o|H_0) = \frac{\text{number of } \delta \text{ values} \leq \delta_o}{L}$$

and L is the number of resampled test statistic values. As previously,

$$P(S \geq S_o|H_0) = P(\delta \leq \delta_o|H_0) = \frac{\text{number of } \delta \text{ values} \leq \delta_o}{L} ,$$

where S_o and δ_o denote the observed values of Kendall's S and δ, respectively.

If all M possible arrangements of the frequency data given in Fig. 10.15 occur with equal chance, the approximate resampling probability value of $\delta_o = 5.60$ computed on $L = 1{,}000{,}000$ random arrangements of the observed rank scores with $b = 2$ blocks preserved for each arrangement is

$$P(\delta \leq \delta_o|H_0) = \frac{\text{number of } \delta \text{ values} \leq \delta_o}{L} = \frac{2{,}611}{1{,}000{,}000} = 0.0026 .$$

Following Eq. (10.3) on p. 474, the exact expected value of the M δ values is $\mu_\delta = 93.1006$ and, following Eq. (10.2) on p. 474, the observed chance-corrected measure of effect size is

$$\Re_o = 1 - \frac{\delta_o}{\mu_\delta} = 1 - \frac{5.60}{93.1006} = +0.9399 ,$$

indicating approximately 94 % within-block agreement above that expected by chance.

For the frequency data given in Fig. 10.15, the observed values of τ_a and δ are

$$\tau_a = 1 - \frac{2\delta_o}{g-1} = 1 - \frac{(2)(5.60)}{20-1} = +0.4105$$

and

$$\delta_o = \frac{(1-\tau_a)(g-1)}{2} = \frac{(1-0.4105)(20-1)}{2} = 5.60 .$$

As in Examples 1–3, $\Re_o$ is equivalent to Spearman's footrule measure $\mathcal{R}$ and the observed values of Kendall's rank-correlation coefficient τ_a and Spearman's chance-corrected footrule measure $\mathcal{R}$ are

$$\tau_a = 1 + \frac{2(\mathcal{R}-1)\mu_\delta}{g-1} = 1 + \frac{(2)(0.9399-1)(93.1006)}{20-1} = +0.4105$$

and

$$\mathcal{R}_o = 1 - \frac{(1-\tau_a)(g-1)}{2\mu_\delta} = 1 - \frac{(1-0.4105)(20-1)}{(2)(93.1006)} = +0.9399 .$$

10.9 Selected Measures of Ordinal Association and δ

Considering that the functional relationships between Kendall's S and the MRBP test statistic are given by

$$S = g\left(\frac{g-1}{2} - \delta\right) \quad \text{and} \quad \delta = \frac{g-1}{2} - \frac{S}{g} ,$$

and that S is the common numerator of Kendall's τ_a and τ_b, Stuart's τ_c, Goodman and Kruskal's γ, and Somers' d_{yx} and d_{xy}, then the relationships between δ and τ_a, τ_b, τ_c, γ, d_{yx}, and d_{xy} are easily specified. The seven relationships are illustrated with the frequency data given in Fig. 10.15 on p. 534 where $C = 88$, $D = 10$, $T_x = 31$, $T_y = 33$, $T_{xy} = 28$, the observed value of Kendall's S is $S_o = C - D = 88 - 10 = 78$, and $N = g = 20$.

10.9.1 Kendall's τ_a Statistic and δ

Following Eq. (10.38) on p. 521 for the frequency data given in Fig. 10.15 on p. 534, the observed value of Kendall's τ_a measure of ordinal association is

$$\tau_a = \frac{2S_o}{N(N-1)} = \frac{2(78)}{20(20-1)} = +0.4105 .$$

As stated previously, the functional relationships between Kendall's τ_a, given in Eq. (10.38) on p. 521, and the MRBP test statistic, given in Eq. (10.44) on p. 524, are given by

$$\tau_a = 1 - \frac{2\delta}{g-1} \quad \text{and} \quad \delta = \frac{(1-\tau_a)(g-1)}{2} .$$

Thus, for the frequency data given in Fig. 10.15, the observed values of τ_a and δ are

$$\tau_a = 1 - \frac{2(5.60)}{20-1} = +0.4105 \quad \text{and} \quad \delta_o = \frac{(1-0.4105)(20-1)}{2} = 5.60 .$$

10.9.2 Kendall's τ_b Statistic and δ

Following Eq. (10.39) on p. 521 for the frequency data given in Fig. 10.15 on p. 534, the observed value of Kendall's τ_b measure of ordinal association is

$$\tau_b = \frac{S_o}{\sqrt{(C+D+T_x)(C+D+T_y)}} = \frac{78}{\sqrt{(88+10+31)(88+10+33)}} = +0.60 .$$

The functional relationships between Kendall's τ_b measure of ordinal association, given in Eq. (10.39) on p. 521, and the MRBP test statistic, given in Eq. (10.44) on p. 524, are given by

$$\tau_b = \frac{g\left(\dfrac{g-1}{2} - \delta\right)}{\left[(C+D+T_x)(C+D+T_y)\right]^{1/2}}$$

and

$$\delta = \frac{g(g-1) - 2\tau_b\left[(C+D+T_x)(C+D+T_y)\right]^{1/2}}{2g} .$$

Thus, for the frequency data given in Fig. 10.15, the observed values of τ_b and δ are

$$\tau_b = \frac{20\left(\dfrac{20-1}{2} - 5.60\right)}{\left[(88+10+31)(88+10+33)\right]^{1/2}} = \frac{78}{130} = +0.60$$

and

$$\delta_o = \frac{(20)(20-1)(0.60)\left[(88+10+31)(88+10+33)\right]^{1/2}}{(2)(20)} = \frac{224}{40} = 5.60\ .$$

10.9.3 Stuart's τ_c Statistic and δ

Following Eq. (10.40) on p. 521 for the frequency data given in Fig. 10.15, the observed value of Stuart's τ_c measure of ordinal association is

$$\tau_c = \frac{2mS_o}{N^2(m-1)} = \frac{2(3)(78)}{20^2(3-1)} = +0.5850\ .$$

where $m = \min(r, c) = 3$. The functional relationships between Stuart's τ_c measure of ordinal association, given in Eq. (10.40) on p. 521, and the MRBP test statistic, given in Eq. (10.44) on p. 524, are given by

$$\tau_c = \frac{2m}{m-1}\left[\frac{g\left(\dfrac{g-1}{2} - \delta\right)}{N^2}\right] \quad \text{and} \quad \delta = \frac{m(g-1) - (m-1)(\tau_c)N}{2m}\ ,$$

where $m = \min(r, c)$. Thus, for the frequency data given in Fig. 10.15, the observed values of τ_c and δ are

$$\tau_c = \frac{(2)(3)}{n}\left[\frac{20\left(\dfrac{20-1}{2} - 5.60\right)}{20^2}\right] = \frac{468}{800} = +0.5850$$

and

$$\delta_o = \frac{(3)(20-1) - (20-1)(0.5850)(20)}{(2)(3)} = \frac{33.6}{6} = 5.60\ .$$

10.9.4 Goodman and Kruskal's γ Statistic and δ

Following Eq. (10.41) on p. 521 for the frequency data given in Fig. 10.15 on p. 534, the observed value of Goodman and Kruskal's γ measure of ordinal association is

$$\gamma = \frac{S_{\text{o}}}{C+D} = \frac{78}{88+10} = +0.7959\,.$$

The functional relationships between Goodman and Kruskal's γ measure of ordinal association, given in Eq. (10.41) on p. 521, and the MRBP test statistic, given in Eq. (10.44) on p. 524, are given by

$$\gamma = \frac{g\left(\dfrac{g-1}{2} - \delta\right)}{C+D} \quad \text{and} \quad \delta = \frac{g(g-1) - 2\gamma(C+D)}{2g}\,.$$

Thus, for the frequency data given in Fig. 10.15, the observed values of γ and δ are

$$\gamma_{\text{o}} = \frac{20\left(\dfrac{20-1}{2} - 5.60\right)}{88+10} = \frac{78}{98} = +0.7959$$

and

$$\delta_{\text{o}} = \frac{20(20-1) - (2)(0.7959)(88+10)}{(2)(20)} = \frac{224}{40} = 5.60\,.$$

10.9.5 Somers' d_{yx} Statistic and δ

Following Eq. (10.42) on p. 521 for the frequency data given in Fig. 10.15, the observed value of Somers' d_{yx} measure of ordinal association is

$$d_{yx} = \frac{S_{\text{o}}}{C+D+T_y} = \frac{78}{88+10+33} = +0.5954\,.$$

The functional relationships between Somers' d_{yx} measure of ordinal association, given in Eq. (10.42) on p. 521, and the MRBP test statistic, given in Eq. (10.44) on p. 524, are given by

$$d_{yx} = \frac{g\left(\dfrac{g-1}{2} - \delta\right)}{C+D+T_y} \quad \text{and} \quad \delta = \frac{g(g-1) - 2d_{yx}(C+D+T_y)}{2g}\,.$$

Thus, for the frequency data listed in Fig. 10.15, the observed values of d_{yx} and δ are

$$d_{yx} = \frac{20\left(\dfrac{20-1}{2} - 5.60\right)}{88 + 10 + 33} = \frac{78}{131} = 0.5954$$

and

$$\delta_{\text{o}} = \frac{20(n) - (2)(0.5954)(88 + 10 + 33)}{(2)(20)} = \frac{224}{40} = 5.60 \ .$$

10.9.6 Somers' d_{xy} Statistic and δ

Following Eq. (10.43) on p. 522 for the frequency data listed in Fig. 10.15 on p. 534, Somers' d_{xy} measure of ordinal association is given by

$$d_{xy} = \frac{S_{\text{o}}}{C + D + T_x} = \frac{78}{88 + 10 + 31} = +0.6047 \ .$$

The functional relationships between Somers' d_{xy} measure of ordinal association, given in Eq. (10.43) on p. 522, and the MRBP test statistic, given in Eq. (10.44) on p. 524, are given by

$$d_{xy} = \frac{g\left(\dfrac{g-1}{2} - \delta\right)}{C + D + T_x} \quad \text{and} \quad \delta = \frac{g(g-1) - 2d_{xy}(C + D + T_x)}{2g} \ .$$

Thus, for the frequency data listed in Fig. 10.15, the observed values of d_{xy} and δ are

$$d_{xy} = \frac{20\left(\dfrac{20-1}{2} - 5.60\right)}{88 + 10 + 31} = \frac{78}{129} = +0.6047$$

and

$$\delta_{\text{o}} = \frac{20(n) - (2)(0.6047)(88 + 10 + 31)}{(2)(20)} = \frac{224}{40} = 5.60 \ .$$

10.10 Coda

Chapter 10 utilized the Multivariate Randomized Block Procedures (MRBP) developed in Chap. 8 to establish relationships between the test statistics of MRBP, δ and $\Re$, and selected conventional tests and measures designed for the analysis of randomized-blocks data at the ordinal level of measurement. Considered in this chapter were the Wilcoxon signed-ranks test, the sign test, Spearman's rank-order and footrule measures of rank correlation, Friedman's analysis of variance for ranks, Kendall's coefficient of concordance, Cohen's weighted kappa measure of agreement, Kendall's t_a and t_b measures of ordinal association, Stuart's t_c statistic, Goodman and Kruskal's γ measure of ordinal association, and Somers' d_{xy} and d_{yx} asymmetric measures of ordinal association.

In this chapter, the MRBP test statistic, δ, was shown to replace the various statistics listed above. Moreover, MRBP provides highly accurate probability values, either exact or resampling, without any distributional assumptions. In addition, MRBP is entirely data-dependent. Finally, a universal chance-corrected measure of effect size is provided for each of the listed statistics.

Chapter 11

Chapter 11 establishes the relationships between the MRBP test statistics, δ and $\Re$, and selected conventional tests and measures designed for the analysis of randomized-block data at the nominal level of measurement. Considered in Chap. 11 are Cohen's unweighted κ measure of chance-corrected agreement, McNemar's and Cochran's Q tests for change, Kendall's t_a and Yule's Q and Y measures of categorical association, the odds ratio, Somers' d_{xy} and d_{yx} asymmetric measures of association, Pearson's product-moment correlation coefficient, percentage differences, and chi-squared.

Randomized Block Designs: Nominal Data 11

This last chapter of *Permutation Statistical Methods* utilizes the Multivariate Randomized Block Permutation (MRBP) procedures presented in Chap. 8 to develop relationships between the test statistics of MRBP, δ and $\Re$, and selected conventional tests and measures designed for the analysis of randomized-block data at the nominal (categorical) level of measurement. The statistical evaluation of categorical data is fraught with problems [91], which are detailed more completely in the preface to Chap. 7. Among these are the complex structures that categorical scales attempt to reflect, often with (0, 1) binary coding; simplistic conditional independence assumptions; and concepts unrelated to any realistic notion of measurement error [91].

A variety of statistical tests and measures for categorical data are considered in this chapter, including Cohen's unweighted kappa measure of chance-corrected agreement: McNemar's and Cochran's Q tests for change, Kendall's t_a rank-correlation statistic, Yule's Q and Y measures of association, the odds ratio, Somers' d_{xy} and d_{yx} asymmetric measures of association, Pearson's product-moment correlation coefficient, percentage differences, and Pearson's chi-squared test of independence.

11.1 Introduction

Randomized-block analysis-of-variance designs analyze univariate or multivariate observations on matched objects or subjects. As detailed in Chap. 8, let $x'_{ij} = (x_{1ij}, x_{2ij}, \ldots, x_{rij})$ denote a transposed vector of r response measurements associated with the ith treatment and jth block. Then the MRBP test statistic is given by

$$\delta = \left[g \binom{b}{2} \right]^{-1} \sum_{i=1}^{g} \sum_{j<k} \Delta(x_{ij}, x_{ik}) , \tag{11.1}$$

K.J. Berry et al., *Permutation Statistical Methods*,
DOI 10.1007/978-3-319-28770-6_11

where $\sum_{j<k}$ denotes the sum over all j and k such that $1 \leq j < k \leq b$ and $\Delta(x, y)$ is a symmetric distance-function value of two points $x' = (x_1, x_2, \ldots, x_r)$ and $y' = (y_1, y_2, \ldots, y_r)$ in an r-dimensional Euclidean space. In the context of a randomized-block analysis-of-variance design, the generalized Minkowski distance function is given by

$$\Delta(x, y) = \left(\sum_{i=1}^{r} |x_i - y_i|^p \right)^{v/p} , \tag{11.2}$$

where $p \geq 1$ and $v > 0$.

The null hypothesis (H_0) states that the distribution of δ assigns an equal probability to each of the

$$M = (g!)^b$$

possible, equally-likely allocations of the r-dimensional response measurements to the g treatment positions within each of the b blocks.

An exact probability value for the observed MRBP test statistic, δ_o, may be expressed as

$$P(\delta \leq \delta_o | H_0) = \frac{\text{number of } \delta \text{ values} \leq \delta_o}{M}$$

and a chance-corrected measure of within-block agreement among the b blocks for all g treatments provides a chance-corrected measure of effect size given by

$$\Re = 1 - \frac{\delta}{\mu_\delta} , \tag{11.3}$$

where μ_δ is the arithmetic average of the M δ values calculated on all possible arrangements of the observed data given by

$$\mu_\delta = \frac{1}{M} \sum_{i=1}^{M} \delta_i . \tag{11.4}$$

As previously, when M is large, an approximate probability value for δ may be obtained from a resampling procedure, where

$$P(\delta \leq \delta_o | H_0) = \frac{\text{number of } \delta \text{ values} \leq \delta_o}{L} \tag{11.5}$$

and L is a random sample of all possible arrangements of the bg response measurements. Typically, L is set to a large number to ensure accuracy, e.g., $L = 1{,}000{,}000$.

When M is very large and P is exceedingly small, a resampling-approximation permutation procedure may produce no δ values equal to or less than δ_o, even with $L = 1{,}000{,}000$, yielding an approximate resampling probability value of $P = 0.00$. In such cases, moment-approximation permutation procedures based on fitting the first three exact moments of the discrete permutation distribution to a Pearson type III distribution provide approximate probability values, as detailed in Chap. 1, Sect. 1.2.2 [284, 299].

11.2 Cohen's Kappa Measure of Agreement

As discussed more completely in Chap. 4, Sect. 4.1.1 and Chap. 10, Sect. 10.7, a number of statistical research problems require the measurement of agreement, rather than association or correlation. Measures of agreement describe the extent to which a set of responses are identical with each other, i.e., agree, rather than the extent to which one set of response measurements is a linear function of another set, i.e., correlated.

In 1960, psychologist Jacob Cohen proposed a chance-corrected measure of agreement that he called kappa [70]. Cohen's kappa measure describes the agreement between $b = 2$ observers or judges on the assignment of N objects to a set of c discrete, mutually exclusive categories. In 1968 Cohen proposed an alternative version of kappa that allowed for the weighting of categories [71]. Whereas the original version of (unweighted) kappa, introduced by Cohen in 1960, did not distinguish among magnitudes of disagreement, weighted kappa incorporated the magnitude of each disagreement and provided partial credit for disagreements when agreement was not complete. Weighted kappa is discussed in Chap. 10, Sect. 10.7, as it applies to ordered categories; unweighted kappa is discussed here as it is typically used for unordered categorical data.

Assume that two judges independently classify each of g response measurements into one of c discrete, mutually exclusive, and exhaustive categories, denoted by A, B, and C. The resulting classifications can be displayed in a $c \times c$ cross-classification table, such as Fig. 11.1, with proportions for cell entries, where $p_{..} = 1.00$. In the

Fig. 11.1 Notation for a 3×3 cross-classification with proportions for cell entries

	Judge 2			
Judge 1	A	B	C	Total
A	p_{11}	p_{12}	p_{13}	$p_{1.}$
B	p_{21}	p_{22}	p_{23}	$p_{2.}$
C	p_{31}	p_{32}	p_{33}	$p_{3.}$
Total	$p_{.1}$	$p_{.2}$	$p_{.3}$	$p_{..}$

notation of Fig. 11.1, Cohen's unweighted kappa (hereafter, kappa) is given by

$$\hat{\kappa} = \frac{P_o - P_e}{1 - P_e}, \tag{11.6}$$

where

$$P_o = \sum_{i=1}^{c} p_{ii} \quad \text{and} \quad P_e = \sum_{i=1}^{c} p_{i.}p_{.i} .$$

Thus, P_o is the observed proportion of response measurements on which the two judges agree, P_e is the proportion of response measurements for which agreement is expected by chance, $P_o - P_e$ is the proportion of agreement above that expected by chance, $1 - P_e$ is the maximum possible proportion of agreement above that expected by chance, and $\hat{\kappa}$ is the proportion of agreement between the two judges after chance agreement has been removed [27].

Let $\delta = 1 - P_o$ and $\mu_\delta = 1 - P_e$ denote the observed and expected proportions of disagreement, respectively. Simplification and substitution yield

$$\hat{\kappa} = \frac{P_o - P_e}{1 - P_r} = 1 - \frac{1 - P_o}{1 - P_e} = 1 - \frac{\delta}{\mu_\delta} . \tag{11.7}$$

Thus, Cohen's unweighted $\hat{\kappa}$ may be interpreted as a ratio of measures of disagreement, or distance, between the two judges, where distance is measured by summing a series of zeroes and ones [242]. As expressed in Eq. (11.7), $\hat{\kappa}$ is a specific measure of chance-corrected agreement based on measurement of ordinary Euclidean distances among the classifications of two judges and is identical to $\Re$, the chance-corrected measure of effect size, as given in Eq. (11.3) on p. 544.

An alternative representation of Fig. 11.1 is presented in Fig. 11.2. Figure 11.2 is constructed in the context of a multivariate randomized-block analysis-of-variance design with $g \geq 2$ observations, $b = 2$ blocks (corresponding to the two judges), and the two polytomous variables of Fig. 11.1 represented by a $c \times 1$ vector (x) where the ith element, corresponding to the ith of c categories, is set to $2^{-1/2}$ and where the remaining $c - 1$ elements of x are set to 0. The constant, $2^{-1/2}$, is simply to ensure that the distance between any two vectors will be 0 if the classifications agree and 1

Fig. 11.2 Example data in a multivariate randomized-block representation

	Block	
Object	1	2
1	x_{11}	x_{12}
2	x_{21}	x_{22}
3	x_{31}	x_{32}
⋮	⋮	⋮
g	x_{g1}	x_{g2}

Fig. 11.3 Example data for $g = 5$ objects, $b = 2$ judges, and $c = 3$ categories in a multivariate randomized-block analysis-of-variance representation

	Block	
Object	1	2
1	$\begin{bmatrix} 2^{-1/2} \\ 0 \\ 0 \end{bmatrix}$	$\begin{bmatrix} 0 \\ 2^{-1/2} \\ 0 \end{bmatrix}$
2	$\begin{bmatrix} 0 \\ 2^{-1/2} \\ 0 \end{bmatrix}$	$\begin{bmatrix} 0 \\ 2^{-1/2} \\ 0 \end{bmatrix}$
3	$\begin{bmatrix} 0 \\ 2^{-1/2} \\ 0 \end{bmatrix}$	$\begin{bmatrix} 0 \\ 2^{-1/2} \\ 0 \end{bmatrix}$
4	$\begin{bmatrix} 0 \\ 0 \\ 2^{-1/2} \end{bmatrix}$	$\begin{bmatrix} 2^{-1/2} \\ 0 \\ 0 \end{bmatrix}$
5	$\begin{bmatrix} 0 \\ 0 \\ 2^{-1/2} \end{bmatrix}$	$\begin{bmatrix} 0 \\ 0 \\ 2^{-1/2} \end{bmatrix}$

if the classifications disagree. The choice of the constant is completely arbitrary and any positive value may be chosen. However, the values for δ and μ_δ, but not $\Re$, will be affected by the choice of the constant. Figure 11.3 illustrates Fig. 11.2 with $c \times 1$ vectors for $g = 5$ objects, $b = 2$ blocks, and $c = 3$ categories.

In the alternative representation of the data depicted in Fig. 11.2, δ, the arithmetic average of paired distances between objects, is given by

$$\delta = \frac{1}{g} \sum_{i=1}^{g} \Delta(x_{i1}, x_{i2}) , \tag{11.8}$$

where the generalized Minkowski distance function is given by

$$\Delta(x_{i1}, x_{i2}) = \left[\sum_{k=1}^{c} \left(x_{i1k} - x_{i2k} \right)^p \right]^{v/p} , \tag{11.9}$$

$p \geq 1$, $v > 0$, and x_{isk} denotes the kth element of vector x_{is} with $i = 1, \ldots, g$ (for Objects 1 through g) and $s = 1, 2$ (for Blocks 1 and 2). Then, μ_δ is the expected proportion of disagreement defined as

$$\mu_\delta = \frac{1}{g^2} \sum_{i=1}^{g} \sum_{j=1}^{g} \Delta(x_{i1}, x_{j2}) \tag{11.10}$$

with the generalized Minkowski distance function given by

$$\Delta(x_{i1}, x_{j2}) = \left[\sum_{k=1}^{c} \left(x_{i1k} - x_{j2k} \right)^2 \right]^{1/2} ,$$

where $p = 2$ and $v = 1$, employing ordinary Euclidean distance between elements.

To illustrate the equivalence of the two computation forms, consider the raw data given in Table 10.1 in Chap. 10, replicated for convenience in Fig. 11.4, absent the linear and quadratic weights. Figure 11.3 on p. 547 represents the frequency data given in Fig. 11.4 arranged in a randomized-block format with $g = 5$ objects, $b = 2$ judges, and $c = 3$ discrete categories. For the frequency data given in Fig. 11.4,

$$P_o = \sum_{i=1}^{c} p_{ii} = \frac{0}{5} + \frac{2}{5} + \frac{1}{5} = \frac{3}{5} = 0.60 ,$$

$$P_e = \sum_{i=1}^{c} p_{i.}p_{.i} = \left(\frac{1}{5}\right)\left(\frac{1}{5}\right) + \left(\frac{2}{5}\right)\left(\frac{3}{5}\right) + \left(\frac{2}{5}\right)\left(\frac{1}{5}\right)$$

$$= \frac{1}{25} + \frac{6}{25} + \frac{2}{25} = \frac{9}{25} = 0.36 ,$$

and following Eq. (11.6) on p. 546, the observed value of Cohen's $\hat{\kappa}$ is

$$\hat{\kappa}_o = \frac{P_o - P_e}{1 - P_e} = \frac{0.60 - 0.36}{1.00 - 0.36} = +0.3750 .$$

To illustrate the computation of the Minkowski generalized distance function given in Eq. (11.9) on p. 547, consider $i = 1$, Block 1 of Fig. 11.3, where $\Delta(x_{11}, x_{12})$, following Eq. (11.9) on p. 547 for Objects 1 and 2, is calculated as

$$\Delta(x_{11}, x_{12}) = \left[(2^{-1/2} - 0)^2 + (0 - 2^{-1/2})^2 + (0 - 0)^2 \right]^{1/2} = 1 ,$$

indicating disagreement on the classification of Object 1, i.e., Judge 1 assigned Object 1 to Category A and Judge 2 assigned Object 1 to Category B. Now con-

Fig. 11.4 Example data set with $g = 5$ objects, $b = 2$ judges, and $c = 3$ categories

	Judge 2			
Judge 1	*A*	*B*	*C*	Total
A	0	1	0	1
B	0	2	0	2
C	1	0	1	2
Total	1	3	1	5

sider $i = 2$, Block 2 of Fig. 11.3, where $\Delta(x_{21}, x_{22})$ for Objects 1 and 2 is calculated as

$$\Delta(x_{21}, x_{22}) = \left[(0-0)^2 + (2^{-1/2} - 2^{-1/2})^2 + (0-0)^2\right]^2 = 0 ,$$

indicating agreement on the classification of Object 2, i.e., both Judges assigned Object 2 to Category B.

Then, following Eq. (11.8) on p. 547 with $g = 5$ and $b = 2$, the observed value of the MRBP test statistic based on $v = 1$ is

$$\delta_\text{o} = \frac{1}{g} \sum_{i=1}^{g} \Delta(x_{i1}, x_{i2}) = \frac{1}{5}(1 + 0 + 0 + 1 + 0) = 0.40 ,$$

which is equivalent to $1 - P_\text{o}$, and, following Eq. (11.10) on p. 547, the exact expected value of δ is

$$\mu_\delta = \frac{1}{g^2} \sum_{i=1}^{g} \sum_{j=1}^{g} \Delta(x_{i1}, x_{j2}) = \frac{1}{5^2}(1 + 1 + \cdots + 1 + 0) = 0.64 ,$$

which is equivalent to $1 - P_\text{e}$. Then, following Eq. (11.7) on p. 546,

$$\hat{\kappa}_\text{o} = 1 - \frac{\delta_\text{o}}{\mu_\delta} = 1 - \frac{0.40}{0.64} = +0.3750 .$$

Since $\Re = \hat{\kappa} = +0.3750$ and $\Re$ is simply a linear transformation of δ, a test of significance for δ is a test of significance for $\hat{\kappa}$. Thus, the exact probability value of an observed value of $\hat{\kappa}$, $\hat{\kappa}_\text{o}$, is the exact probability value of an observed value of δ, δ_o, under the null hypothesis, i.e.,

$$P(\hat{\kappa} \geq \hat{\kappa}_\text{o} | H_0) = P(\delta \leq \delta_\text{o} | H_0) = \frac{\text{number of } \delta \text{ values} \leq \delta_\text{o}}{M} ,$$

where

$$M = \left(g!\right)^b$$

in this application. As the $b = 2$ blocks are specified, the randomization associated with a randomized-block analysis-of-variance design is confined to all permutations of the $g = 5$ observations within each block. Under the null hypothesis, each of the M possible arrangements of the observed data occurs with equal probability.

For the data listed in Fig. 11.3, there are only

$$M = \left(g!\right)^b = \left(5!\right)^2 = 14{,}400$$

possible, equally-likely arrangements of the observed data, therefore an exact solution is possible. If all arrangements of the observed randomized-block data given in Fig. 11.3 occur with equal chance, the exact probability value of $\delta_o = 0.40$ computed on the $M = 14{,}400$ possible arrangements of the observed data with $b = 2$ blocks preserved for each arrangement is

$$P(\delta \leq \delta_o | H_0) = \frac{\text{number of } \delta \text{ values} \leq \delta_o}{M} = \frac{2{,}880}{14{,}400} = 0.20 \ .$$

Finally, the observed chance-corrected measure of effect size is

$$\Re_o = 1 - \frac{\delta_o}{\mu_\delta} = 1 - \frac{0.40}{0.64} = +0.3750 \ ,$$

indicating approximately 37 % agreement between the $b = 2$ judges.

11.2.1 Multiple Judges

While Cohen's $\hat{\kappa}$ is limited to $b = 2$ judges, simple modifications to δ and μ_δ generalize $\Re$ to measure agreement among multiple judges [27]. Thus, the MRBP test statistic may be redefined as

$$\delta = \left[g \binom{b}{2} \right]^{-1} \sum_{i=1}^{g} \sum_{s<t} \Delta \left(x_{is}, x_{it} \right) \ , \tag{11.11}$$

where the generalized Minkowski distance function is given by

$$\Delta \left(x_{is}, x_{it} \right) = \left[\sum_{k=1}^{r} \left(x_{isk} - x_{itk} \right)^2 \right]^{1/2} \ ,$$

b is the number of judges (i.e., blocks) and $\sum_{s<t}$ is the sum over all s and t such that $1 \leq s < t \leq b$. The reformulation of the expected value of δ is given by

$$\mu_\delta = \left[g^2 \binom{b}{2} \right]^{-1} \sum_{i=1}^{g} \sum_{j=1}^{g} \sum_{s<t} \Delta \left(x_{is}, x_{jt} \right) \ . \tag{11.12}$$

To illustrate the measurement of agreement for multiple judges, consider the categorical data listed in Fig. 11.5, in which each of $b = 4$ judges (A, B, C, and D) is asked to assign $g = 8$ objects to $c = 4$ discrete, mutually exclusive, exhaustive categories, labeled 1, 2, 3, and 4.

Fig. 11.5 Example data set for multiple judges with categorical data, $g = 8$ objects, $b = 4$ judges, and $r = 1$ response

	Judge			
Object	*A*	*B*	*C*	*D*
1	3	4	3	3
2	4	3	1	4
3	2	1	1	2
4	4	2	3	1
5	1	1	2	1
6	1	3	2	3
7	3	4	4	4
8	2	3	4	3

For the categorical data listed in Fig. 11.5, $g = 8$, $b = 4$, and $r = 1$. Following Eq. (11.11) the observed value of the MRBP test statistic based on $v = 1$ is $\delta_o = 0.9583$, following Eq. (11.12) the exact expected value of δ is $\mu_\delta = 1.2448$, and following Eq. (11.3) on p. 544 the observed chance-corrected measure of effect size is

$$\Re_o = 1 - \frac{\delta_o}{\mu_\delta} = 1 - \frac{0.9583}{1.2448} = +0.2301 ,$$

indicating approximately 23 % agreement among the $b = 4$ judges above that expected by chance.

Because there are

$$M = (g!)^b = (8!)^4 = 2{,}642{,}908{,}293{,}365{,}760{,}000$$

possible, equally-likely arrangements of the observed data listed in Fig. 11.5, an exact permutation solution is not feasible. If all M possible arrangements of the observed randomized-block data listed in Fig. 11.5 occur with equal chance, the approximate resampling probability value of $\delta_o = 0.9583$ computed on $L = 1{,}000{,}000$ random arrangements of the observed data with $b = 4$ blocks preserved for each arrangement is

$$P(\delta \leq \delta_o | H_0) = \frac{\text{number of } \delta \text{ values} \leq \delta_o}{L} = \frac{42{,}514}{1{,}000{,}000} = 0.0425 .$$

Generalization of Cohen's Unweighted $\hat{\kappa}$

The coefficient of effect size, $\Re$, is a generalization of Cohen's unweighted $\hat{\kappa}$ to multiple observers. It preserves the desired qualities of $\hat{\kappa}$ in that it is chance-corrected, Euclidean-based, and applicable to the measurement of reliability. The generalization of Cohen's $\hat{\kappa}$ to multiple observers for categorical data is the special case of $\Re$ when the distance space is restricted to an r-dimensional simplex consisting of r distinct points where the distance between any pair of points is unity and the distance between any two coincident points is zero.

In this context, Cohen's unweighted $\hat{\kappa}$ is the special case of $\Re$ when $b = 2$ and the measure of agreement corresponding to Cochran's Q test is the special case of $\Re$ when $r = 2$. That is, Cochran's Q test involves b judges, g observations, and a two-dimensional simplex. The measure of agreement corresponding to McNemar's test for change is the special case of both Cohen's $\hat{\kappa}$ and the measure of agreement corresponding to Cochran's Q test when $b = r = 2$ [27, 297, pp. 162–163].

11.2.2 An Alternative Approach to Multiple Judges

In this section, an alternative procedure is presented to compute unweighted kappa with multiple judges. Although the procedure is appropriate for any number of $c \geq 2$ disjoint ordered categories and $b \geq 2$ judges, the description of the procedure and the examples are limited to three independent judges to simplify presentation.

Consider $b = 3$ judges who independently classify N objects into c disjoint ordered categories. The classification may be conceptualized as $c{\times}c{\times}c$ contingency table with c rows, c columns, and c slices. Let n_{ijk}, R_i, C_j, and S_k denote the cell frequencies and row, column, and slice marginal frequency totals for $i, j, k = 1, \ldots, c$ and let the frequency total be given by

$$N = \sum_{i=1}^{c} \sum_{j=1}^{c} \sum_{k=1}^{c} n_{ijk} \; .$$

Cohen's unweighted kappa test statistic for a three-way contingency table is given by

$$\hat{\kappa} = \frac{N^2 \displaystyle\sum_{i=1}^{c} \sum_{i=1}^{c} \sum_{i=1}^{c} w_{ijk} n_{ijk}}{\displaystyle\sum_{i=1}^{c} \sum_{i=1}^{c} \sum_{i=1}^{c} w_{ijk} R_i C_j S_k} \; , \tag{11.13}$$

where w_{ijk} are disagreement weights assigned to each cell for $i, j, k = 1, \ldots, c$. "Unweighted" kappa is, in fact, characterized by a specific weighting scheme given by

$$w_{ijk} = \begin{cases} 0 & \text{if } i = j = k \; , \\ 1 & \text{otherwise} \; . \end{cases}$$

Given a $c{\times}c{\times}c$ contingency table with N objects cross-classified by the independent judges, an exact permutation test involves generating all possible arrangements of the N objects to the c^3 cells, while preserving the marginal frequency totals. For each arrangement of cell frequencies, the unweighted kappa statistic, $\hat{\kappa}$, and the exact hypergeometric probability value under the null hypothesis, $P(n_{ijk}|R_i, C_j, S_k)$, are calculated, where

$$P(n_{ijk}|R_i, C_j, S_k) = \frac{\left(\prod_{i=1}^{c} R_i!\right)\left(\prod_{j=1}^{c} C_j!\right)\left(\prod_{k=1}^{c} S_k!\right)}{(N!)^2 \prod_{i=1}^{c}\prod_{j=1}^{c}\prod_{k=1}^{c} n_{ijk}!} \tag{11.14}$$

[290].

If $\hat{\kappa}_o$ denotes the value of the observed unweighted kappa test statistic, the exact probability value of $\hat{\kappa}_o$ under the null hypothesis is given by

$$P(\hat{\kappa}_o) = \sum_{l=1}^{M} \Psi_l\left(n_{ijk}|R_i, C_j, S_k\right) ,$$

where

$$\Psi_l\left(n_{ijk}|R_i, C_j, S_k\right) = \begin{cases} P(n_{ijk}|R_i, C_j, S_k) & \text{if } \hat{\kappa} \geq \hat{\kappa}_o ,\\ 0 & \text{otherwise} ,\end{cases}$$

and M denotes the total number of possible cell frequency arrangements, given fixed observed marginal frequency totals. When M is very large, as is typical with multi-way contingency tables, exact tests are impractical and resampling becomes necessary. In such cases, a random sample of the M possible arrangements of cell frequencies provides a comparison of $\hat{\kappa}$ test statistics calculated on L random multi-way tables with the $\hat{\kappa}_o$ test statistic calculated on the observed multi-way contingency table.

An efficient resampling algorithm to generate random cell frequency arrangements for multi-way contingency tables with fixed marginal frequency totals was developed by Mielke, Berry, and Johnston in 2007 [307, pp. 19–20]. Under the null hypothesis, the approximate resampling probability value for $\hat{\kappa}_o$ is given by

$$P\left(\hat{\kappa}_o\right) = \frac{1}{L}\sum_{l=1}^{L} \Psi_l\left(\hat{\kappa}\right)$$

where

$$\Psi_l\left(\hat{\kappa}\right) = \begin{cases} 1 & \text{if } \hat{\kappa} \geq \hat{\kappa}_o ,\\ 0 & \text{otherwise} .\end{cases}$$

The calculation of unweighted kappa and the resampling procedure to obtain a probability value for multiple judges can be illustrated with a small example data

Fig. 11.6 Classification of $N = 93$ objects by three independent judges into one of three disjoint categories: A, B, or C; disagreement weights are given in parentheses

		Judge 3		
Judge 1	Judge 2	A	B	C
A	A	6 (0)	4 (2)	2 (4)
	B	3 (2)	5 (2)	4 (4)
	C	2 (4)	3 (4)	4 (4)
B	A	4 (2)	5 (2)	3 (4)
	B	5 (2)	8 (0)	4 (2)
	C	3 (4)	2 (2)	3 (2)
C	A	1 (4)	3 (4)	4 (4)
	B	3 (4)	2 (2)	2 (2)
	C	1 (4)	2 (2)	5 (0)

set. Consider $b = 3$ independent judges who classify $N = 93$ objects into one of $c = 3$ disjoint categories: A, B, or C. Figure 11.6 lists the c^3 cross-classified frequencies and corresponding weights, where the cell disagreement weights are given in parentheses. The data are adapted from Mielke et al. [308, p. 609].[1]

For the observed data listed in Fig. 11.6 the observed value of $\hat{\kappa}$ is $\hat{\kappa}_o = 0.1007$, indicating approximately 10 % agreement among the $b = 3$ judges above that expected by chance, and the approximate resampling probability value based on $L = 1{,}000{,}000$ random arrangements of the observed data is

$$P(\hat{\kappa} \geq \hat{\kappa}_o | H_0) = \frac{\text{number of } \hat{\kappa} \text{ values} \geq \delta_o}{L} = \frac{8{,}311}{1{,}000{,}000} = 0.0083 \, .$$

11.3 McNemar's Q Test and δ

In 1947 psychologist Quinn McNemar proposed a test for change that he derived from the matched-pairs t test for proportions [273]. A typical application is to analyze binary responses, coded 0 and 1, at $g = 2$ time periods for each of $b \geq 2$ subjects, such as Success and Failure, Yes and No, Agree and Disagree, or Pro and Con. If the four cells are identified as in Fig. 11.7, then McNemar's test is given by

$$Q = \frac{(B - C)^2}{B + C} \, ,$$

where B and C represent the two cells of change, i.e., Pro to Con and Con to Pro.

Alternatively, McNemar's Q test can be thought of as a chi-squared goodness-of-fit test with one degree of freedom, where the observed frequencies, O_1 and O_2, are B and C, respectively, and the expected frequencies, E_1 and E_2, are given by

[1] These are the same data analyzed in Chap. 10, Sect. 10.7.6 to illustrate weighted kappa with linear and quadratic weighting.

Fig. 11.7 Notation for a 2×2 cross-classification for McNemar's test for change

	Time 2		
Time 1	Pro	Con	Total
Pro	A	B	$A+B$
Con	C	D	$C+D$
Total	$A+C$	$B+D$	N

Fig. 11.8 Example frequency data for McNemar's test for change

	Time 2		
Time 1	Pro	Con	Total
Pro	4	8	12
Con	1	2	3
Total	5	10	15

$E_1 = E_2 = (B + C)/2$, i.e., half the objects are expected to change in one direction (e.g., Pro to Con) and half in the other direction (e.g., Con to Pro), under the null hypothesis of no change from Time 1 to Time 2.

11.3.1 Example Analysis

Consider the frequency data given in Fig. 11.8, where $N = 15$ objects have been recorded as either Pro or Con on a specified issue at Time 1 and again at Time 2. For the frequency data given in Fig. 11.8, the observed value of McNemar's Q test statistic is

$$Q_o = \frac{(B-C)^2}{B+C} = \frac{(8-1)^2}{8+1} = \frac{49}{9} = 5.4444 \ .$$

Alternatively, $O_1 = B = 8.00$, $O_2 = C = 1.00$, $E_1 = E_2 = (O_1 + O_2)/2 = (8 + 1)/2 = 4.50$, and

$$\chi_1^2 = \frac{(O_1 - E_1)^2}{E_1} + \frac{(O_2 - E_2)^2}{E_2}$$
$$= \frac{(8.00 - 4.50)^2}{4.50} + \frac{(1.00 - 4.50)^2}{4.50} = 5.4444 \ .$$

Now consider the data given in Fig. 11.8 in a randomized-block analysis-of-variance context, with $g = 2$ time periods and $b = 15$ blocks. Figure 11.9 displays the frequency data given in Fig. 11.8 in a randomized-block format, where the Pro and Con categories are binary-coded as 0 and 1, respectively.

Fig. 11.9 Data from Fig. 11.8 arranged in a randomized-block format

	Time	
Block	1	2
1	0	0
2	0	0
3	0	0
4	0	0
5	0	1
6	0	1
7	0	1
8	0	1
9	0	1
10	0	1
11	0	1
12	0	1
13	1	0
14	1	1
15	1	1

Define

$$S = \sum_{i=1}^{g} \sum_{j=1}^{b} x_{ij} \quad \text{and} \quad T = \sum_{j=1}^{b} \left(\sum_{i=1}^{g} x_{ij} \right)^2 ,$$

where x_{ij} denotes the cell entry of either 0 or 1 associated with the ith of b rows and the jth of g columns in Fig. 11.9. For the randomized-block binary data listed in Fig. 11.9, the observed values of S and T are

$$S_o = (0+0) + (0+0) + (0+0) + \cdots + (1+0) + (1+1) + (1+1) = 13$$

and

$$T_o = (0+0)^2 + (0+0)^2 + \cdots + (1+0)^2 + (1+1)^2 + (1+1)^2 = 17 .$$

For the binary data listed in Fig. 11.9, there are

$$M = (g!)^b = (2!)^{15} = 32{,}768$$

possible, equally-likely arrangements of the observed data, making an exact solution feasible. Following Eq. (11.1) on p. 543, the observed value of the MRBP test statistic based on $v = 1$ is $\delta_o = 0.4095$. If all arrangements of the observed randomized-block binary data listed in Fig. 11.9 occur with equal chance, the exact probability value of $\delta_o = 0.4095$ computed on the $M = 32{,}768$ possible arrangements of the observed data with $b = 15$ blocks preserved for each arrangement is

$$P(\delta \leq \delta_o|H_0) = \frac{\text{number of } \delta \text{ values} \leq \delta_o}{M} = \frac{510}{32{,}768} = 0.0156\,.$$

For comparison, McNemar's Q is approximately distributed as chi-squared under the null hypothesis with $g - 1 = 2 - 1 = 1$ degree of freedom. Under the null hypothesis, the observed value of $Q_o = 5.4444$ yields an approximate chi-squared probability value of $P = 0.0196$.

Following Eq. (11.4) on p. 544, the exact expected value of the $M = 32{,}768$ δ values is $\mu_\delta = 0.5095$ and, following Eq. (11.3) on p. 544, the observed chance-corrected measure of effect size is

$$\Re_o = 1 - \frac{\delta_o}{\mu_\delta} = 1 - \frac{0.4095}{0.5095} = +0.1963\,,$$

indicating approximately 20 % within-block agreement above that expected by chance.

The functional relationships between McNemar's Q and the MRBP test statistic are given by

$$Q = \frac{S(2b - S) - 2b(b-1)\delta}{2S - T} \quad \text{and} \quad \delta = \frac{S(2b - S) - (2S - T)Q}{2b(b-1)}\,.$$

Thus, for the randomized-block binary data listed in Fig. 11.9, the observed values of McNemar's Q and δ are

$$Q_o = \frac{13[(2)(15) - 13] - [(2)(12) - 17]0.4095}{2(15)(15-1)} = \frac{49}{9} = 5.4444$$

and

$$\delta_o = \frac{13[(2)(15) - 13] - [(2)(13) - 17)]5.4444}{2(15)(15-1)} = \frac{172}{420} = 0.4095\,.$$

11.4 Cochran's Q Test and δ

In 1950 William Cochran published an article in *Biometrika* on "The comparison of percentages in matched samples" [69]. In this brief article, Cochran described a test for equality of matched proportions that is now widely used in educational and psychological research. The Cochran's Q test may be viewed as an extension of the

McNemar [273] test to three or more treatment conditions. For a typical application, suppose that a sample of $b \geq 2$ subjects are observed in a situation wherein each subject performs individually under each of $g \geq 1$ different experimental conditions. The performance is scored as a Success (1) or, otherwise, as a Failure (0). The research question evaluates if the true proportion of successes is constant over the g time periods.

Cochran's Q test for the analysis of g treatment conditions (columns) and N subjects (rows) is given by

$$Q = \frac{(g-1)\left(g \sum_{j=1}^{g} C_j^2 - A^2\right)}{gA - B} , \tag{11.15}$$

where

$$R_i = \sum_{j=1}^{g} x_{ij}$$

is the number of 1s in the ith of N rows,

$$A = \sum_{i=1}^{N} R_i , \quad B = \sum_{i=1}^{N} R_i^2 , \quad C_j = \sum_{i=1}^{N} x_{ij}$$

is the number of 1s in the jth of g columns, and x_{ij} denotes the cell entry of either 0 or 1 associated with the ith of N rows and the jth of g columns. The null hypothesis stipulates that each of the

$$M = \prod_{i=1}^{N} \binom{g}{R_i}$$

distinguishable arrangements of 1s and 0s within each of the N rows occurs with equal probability, given that the values of $R_1, \ldots, R_N$ are fixed [292].

11.4.1 Example Analysis

For an example analysis, consider the binary data listed in Fig. 11.10 consisting of the responses (0 or 1) for $N = 10$ subjects evaluated over $g = 5$ time periods, where a 1 denotes success on a prescribed task and a 0 denotes failure. For the binary data

Fig. 11.10 Successes (1) and failures (0) of $N = 10$ subjects on a series of $g = 5$ time periods

	Time					
Subject	1	2	3	4	5	R_i
1	0	1	1	0	0	2
2	1	0	1	0	1	3
3	0	1	1	0	0	2
4	1	1	0	0	0	2
5	1	0	1	1	0	3
6	0	1	1	0	0	2
7	0	1	0	1	0	2
8	0	0	1	0	0	1
9	0	1	0	1	0	2
10	1	1	1	0	0	3
C_j	4	7	7	3	1	22

listed in Fig. 11.10,

$$A = \sum_{i=1}^{N} R_i = 2 + 3 + 2 + 2 + 3 + 2 + 2 + 1 + 2 + 3 = 22 ,$$

$$B = \sum_{i=1}^{N} R_i^2 = 2^2 + 3^2 + 2^2 + 2^2 + 3^2 + 2^2 + 2^2 + 1^2 + 2^2 + 3^2 = 52 ,$$

$$\sum_{j=1}^{g} C_j^2 = 4^2 + 7^2 + 7^2 + 3^2 + 1^2 = 124 ,$$

and, following Eq. (11.15) on p. 558, the observed value of Cochran's Q is

$$Q_o = \frac{(g-1)\left(g\sum_{j=1}^{g} C_j^2 - A^2\right)}{gA - B} = \frac{(5-1)[5(124) - 22^2]}{5(22) - 52} = 9.3793 .$$

Now consider the binary data listed in Fig. 11.10 in a randomized-block analysis-of-variance context, with $g = 5$ time periods and $b = 10$ blocks. Following Eq. (11.1) on p. 543, the observed value of the MRBP test statistic based on $v = 1$ is $\delta_o = 0.4267$. For the randomized-block binary data listed in Fig. 11.10 where R_i, $i = 1, \ldots, b$, is $\{2, 3, 2, 2, 3, 2, 2, 1, 2, 3\}$, there are

$$M = \prod_{i=1}^{b} \binom{g}{R_i} = \binom{5}{1}^1 \binom{5}{2}^6 \binom{5}{3}^3 = (5)(10^6)(10^3) = 5{,}000{,}000{,}000$$

possible, equally-likely arrangements of the observed data, making an exact solution prohibitive. If all M possible arrangements of the observed randomized-block binary data listed in Fig. 11.10 occur with equal chance, the approximate resampling probability value of $\delta_o = 0.4267$ computed on $L = 1{,}000{,}000$ random arrangements of the observed data with $b = 10$ blocks preserved for each arrangement is

$$P(\delta \leq \delta_o | H_0) = \frac{\text{number of } \delta \text{ values} \leq \delta_o}{L} = \frac{54{,}486}{1{,}000{,}000} = 0.0545 \; .$$

For comparison, Cochran's Q is approximately distributed as chi-squared under the null hypothesis with $g - 1 = 5 - 1 = 4$ degrees of freedom. Under the null hypothesis, the observed value of $Q_o = 9.3793$ yields an approximate chi-squared probability value of $P = 0.0523$.

Following Eq. (11.4) on p. 544, the exact expected value of the M δ values is $\mu_\delta = 0.4960$ and, following Eq. (11.3) on p. 544, the observed chance-corrected measure of effect size is

$$\Re_o = 1 - \frac{\delta_o}{\mu_\delta} = 1 - \frac{0.4267}{0.4960} = +0.1398 \; ,$$

indicating approximately 14 % within-block agreement above that expected by chance.

The functional relationships between Cochran's Q and the MRBP test statistic are given by

$$Q = \frac{(g-1)\left[2A(bg-A) - b(b-1)g^2\delta\right]}{2(gA-B)}$$

and

$$\delta = \frac{2\left[A(bg-A)(g-1) - (gA-B)Q\right]}{b(b-1)(g-1)g^2} \; .$$

Thus, for the randomized-block binary data listed in Fig. 11.10, the observed values of Cochran's Q and δ are

$$Q_o = \frac{(5-1)\{2(22)[(10)(5)-22] - 10(10-1)(5^2)(0.4267)\}}{2[(5)(22)-52]} = 9.3793$$

and

$$\delta_o = \frac{2\{22[(10)(5)-22](5-1) - [(5)(22)-52](9.3793)\}}{10(10-1)(5-1)5^2} = 0.4267 \; .$$

11.4.2 Multiple Binary Responses

Cochran's Q statistic was designed to consider only $r = 1$ binary response for each time period or experimental condition. However, MRBP has no such limitation and can easily analyze $r \geq 1$ binary responses.

Many types of research include multiple-response questions wherein subjects mark all applicable categories. For example, patients may be asked to select from a list and check all illnesses they have had in the past 5 years, employees may be asked to select names of close friends from a list of co-workers, subjects may be asked to select from a list of recreational sites visited in the past year, or frequent flyers may be asked to list the various airlines on which they have flown in the past year. The longitudinal analysis of multiple category choices may be conceptualized as a binary argument problem in which b subjects choose any or all of r presented categories and the responses for each subject are coded 1 if the category is selected and 0 otherwise. The same or matched subjects are assessed at g time periods and the multiple binary responses over the g trials are compared.

The null hypothesis of no differences in the response structures over the g trials specifies that each of the

$$M = (g!)^b$$

possible, equally-likely allocations of the b r-dimensional response measurements to the g trials is equally likely.

To illustrate the analysis of multiple binary responses, consider an example in which b subjects are compared on r binary category choices over g trials [34]. Specifically, consider $b = 12$ subjects who have been diagnosed as clinically depressed. Presented with a list of $r = 14$ symptoms of depression, a clinical psychologist assesses and records any and all symptoms experienced by each subject in the past month. Table 11.1 lists the $r = 14$ response categories from the $2^r = 2^{14} = 16{,}384$ possible response arrangements, where a 1 indicates the symptom was recorded and a 0 indicates that the symptom was not recorded. The data are adapted from Mielke and Berry [297, pp. 140–141].

Table 11.1 provides the baseline values for evaluation of the intervention. After counseling by a clinical psychologist for a period of 6 months, the same subjects are again evaluated and their symptoms recorded. The post-treatment results are given in Table 11.2. Finally, a follow-up evaluation 6 months after the termination of treatment yields the results listed in Table 11.3. The null hypothesis specifies no differences among the binary multiple responses over the $g = 3$ assessments.

For the randomized-block data in listed in Tables 11.1, 11.2, and 11.3, there are

$$M = (g!)^b = (3!)^{12} = 2{,}176{,}782{,}336 \tag{11.16}$$

possible, equally-likely arrangements of the observed data—too many for an exact solution. Therefore, a resampling permutation procedure is mandated. Employing

Table 11.1 Baseline longitudinal multiple binary data recorded on $b = 12$ subjects for $r = 14$ symptoms of depression

	Subject											
Symptom	A	B	C	D	E	F	G	H	I	J	K	L
1	1	1	1	1	1	1	0	1	1	1	1	1
2	0	1	1	0	1	1	1	0	1	1	0	0
3	0	1	0	0	1	1	0	0	1	0	0	1
4	1	1	0	1	0	0	1	1	0	1	1	0
5	1	0	1	1	1	0	1	1	0	1	0	1
6	0	0	0	0	1	0	0	0	0	0	0	1
7	0	0	0	1	0	0	1	0	0	0	0	0
8	0	0	0	0	0	0	0	1	0	0	0	0
9	0	1	1	0	1	1	0	0	0	1	0	0
10	0	1	0	0	1	0	0	0	0	0	0	0
11	1	0	0	0	0	0	0	0	1	0	1	0
12	0	0	1	0	0	0	0	0	0	0	0	0
13	0	1	0	0	1	0	0	0	0	1	0	0
14	1	0	0	1	1	0	0	1	1	0	1	1

Table 11.2 Post-treatment longitudinal multiple binary data recorded on $b = 12$ subjects for $r = 14$ symptoms of depression

	Subject											
Symptom	A	B	C	D	E	F	G	H	I	J	K	L
1	1	0	1	1	1	1	0	1	0	1	1	1
2	0	1	0	0	1	1	0	0	1	1	0	0
3	0	1	0	0	1	1	0	0	1	0	0	1
4	1	1	0	1	0	0	1	1	0	1	1	0
5	1	0	1	1	1	0	1	1	0	1	0	1
6	0	1	0	0	1	0	0	0	1	0	0	1
7	0	0	0	1	0	0	1	0	0	0	0	0
8	0	0	0	0	0	0	0	1	0	0	0	0
9	0	0	1	0	0	1	0	0	0	1	0	0
10	0	1	0	0	1	0	0	0	0	0	1	0
11	1	0	0	0	0	0	0	0	0	0	1	0
12	0	0	1	0	0	0	1	0	0	0	0	0
13	0	1	0	0	1	0	0	0	0	1	0	0
14	1	0	0	1	1	0	0	0	0	0	1	1

ordinary Euclidean distance between response measurements with $v = 1$ and following Eq. (11.1) on p. 543, the observed value of the MRBP test statistic is $\delta_o = 2.4313$. If all M possible arrangements of the observed randomized-block binary data listed in Tables 11.1, 11.2, and 11.3 occur with equal chance, the approximate resampling probability value of $\delta_o = 2.4313$ computed on $L = 1{,}000{,}000$ random arrangements of the observed data with $b = 12$ blocks preserved for each

Table 11.3 Follow-up longitudinal multiple binary data recorded on $b = 12$ subjects for $r = 14$ symptoms of depression

	Subject											
Symptom	A	B	C	D	E	F	G	H	I	J	K	L
1	1	0	1	0	1	1	0	1	0	1	1	0
2	0	0	0	0	1	1	0	0	1	1	0	0
3	0	1	0	0	1	1	0	0	1	0	0	1
4	1	1	0	1	0	0	0	1	0	1	1	0
5	1	0	1	1	1	0	1	1	0	1	0	1
6	0	1	0	0	1	0	0	0	1	0	0	1
7	0	0	0	1	0	0	0	0	0	0	0	0
8	0	0	0	0	0	0	0	1	0	0	0	0
9	0	1	1	0	1	1	0	0	0	1	0	0
10	0	1	0	0	1	0	0	0	0	0	1	0
11	1	0	0	0	0	0	0	0	0	0	1	0
12	0	0	1	0	0	0	1	0	0	0	0	0
13	0	1	0	0	0	0	0	0	0	1	0	0
14	1	0	0	1	1	0	0	1	0	0	1	1

arrangement is

$$P(\delta \leq \delta_o | H_0) = \frac{\text{number of } \delta \text{ values} \leq \delta_o}{L} = \frac{2{,}109}{1{,}000{,}000} = 0.0021 \; .$$

No comparison is made with Cochran's Q test as Q is undefined for $r > 1$.

Following Eq. (11.4) on p. 544, the exact expected value of the M δ values is $\mu_\delta = 2.3533$ and, following Eq. (11.3) on p. 544, the observed chance-corrected measure of effect size is

$$\Re_o = 1 - \frac{\delta_o}{\mu_\delta} = 1 - \frac{2.4313}{2.3533} = -0.0331 \; ,$$

indicating slightly less within-block agreement than expected by chance.

While an exact test is impractical in this case, it is not impossible. Since the ordered responses of one subject may be held fixed relative to the other $b - 1$ subjects, $M = 2{,}176{,}782{,}336$ in Eq. (11.16) can be reduced to

$$M = \left(g!\right)^{b-1} = \left(3!\right)^{12-1} = 362{,}797{,}056$$

equally-likely arrangements of the observed data. If all arrangements of the observed randomized-block binary data listed in Tables 11.1, 11.2, and 11.3 occur with equal chance, the exact probability value of $\delta_o = 2.4313$ computed on the $M = 362{,}797{,}056$ arrangements of the observed data with $b = 12$ blocks preserved for each arrangement is

$$P(\delta \leq \delta_o | H_0) = \frac{\text{number of } \delta \text{ values} \leq \delta_o}{M} = \frac{725{,}594}{362{,}797{,}056} = 0.0020 \; .$$

Note that the approximate resampling probability value of $P = 0.0021$ based on $L = 1{,}000{,}000$ is very close to the exact probability value of $P = 0.0020$ based on $M = 362{,}797{,}056$.[2]

11.5 MRBP and Categorical Fourfold Tables

To illustrate the relationships between the MRBP test statistic and various measures of association for categorical data, consider the small fourfold (2×2) contingency table displayed in Fig. 11.11 with $N = 10$ objects cross-classified by variables x and y, each with two categories coded 0 and 1, respectively. It should be noted that many measures designed for categorical data also serve as measures of association for ordinal data, and vice-versa, when applied to 2×2 contingency tables [344]. Following Eqs. (10.33) to (10.37) in Chap. 10, for the frequency data listed in Fig. 11.11, $C = (3)(2) = 6$ concordant pairs, $D = (1)(4) = 4$ discordant pairs, $T_x = (3)(1) + (4)(2) = 11$ pairs tied on variable x but not tied on variable y, $T_y = (3)(4) + (1)(2) = 14$ pairs tied on variable y but not tied on variable x, $T_{xy} = [(3)(2) + (1)(0) + (4)(3) + (2)(1)]/2 = 10$ pairs tied on both variable x and variable y, $S = C - D$, and the observed value of S is $S_o = C - D = 6 - 4 = +2$.

As explained in Chap. 10, Sect. 10.8.3, whenever two sets of categories possess tied values on both x and y, the doubly tied values represented by T_{xy} must be taken into consideration. Table 11.4 lists a selection of the

$$\frac{N(N-1)}{2} = \frac{10(10-1)}{2} = 45$$

paired differences, r_{ij}, s_{ij}, $r_{ij}s_{ij}$, and $|r_{ij} - s_{ij}|$ values for the frequency data given in Fig. 11.11.[3]

Fig. 11.11 Example data for variables x and y with categories dummy-coded as 0 and 1

	y		
x	0	1	Total
0	3	1	4
1	4	2	6
Total	7	3	10

[2] In general, setting the number of resampled statistics to $L = 1{,}000{,}000$ ensures a minimum of three decimal places of accuracy [195]; see also Chap. 2, Sect. 2.2.

[3] Because of the length of Table 11.4, the listing is abbreviated with less-important pairs 11–15 and 36–40 selectively deleted from the $N(N-1)/2 = 45$ possible pairs.

Table 11.4 Paired differences, r_{ij}, s_{ij}, $r_{ij}s_{ij}$, and $|r_{ij} - s_{ij}|$ values for the univariate rank data listed in Fig. 11.11

Pair	$x_i - x_j$	$y_i - y_j$	r_{ij}	s_{ij}	$r_{ij}s_{ij}$	$\|r_{ij} - s_{ij}\|$	Type
1	0 − 0	0 − 0	0	0	0	0	T_{xy}
2	0 − 0	0 − 0	0	0	0	0	T_{xy}
3	0 − 0	0 − 1	0	−1	0	1	T_x
4	0 − 1	0 − 0	−1	0	0	1	T_y
5	0 − 1	0 − 0	−1	0	0	1	T_y
6	0 − 1	0 − 0	−1	0	0	1	T_y
7	0 − 1	0 − 0	−1	0	0	1	T_y
8	0 − 1	0 − 1	−1	−1	+1	0	C
9	0 − 1	0 − 1	−1	−1	+1	0	C
10	0 − 0	0 − 0	0	0	0	0	T_{xy}
⋮	⋮	⋮	⋮	⋮	⋮	⋮	⋮
16	0 − 1	0 − 1	−1	−1	+1	0	C
17	0 − 1	0 − 1	−1	−1	+1	0	C
18	0 − 0	0 − 1	0	−1	0	1	T_x
19	0 − 1	0 − 0	−1	0	0	1	T_y
20	0 − 1	0 − 0	−1	0	0	1	T_y
21	0 − 1	0 − 0	−1	0	0	1	T_y
22	0 − 1	0 − 0	−1	0	0	1	T_y
23	0 − 1	0 − 1	−1	−1	+1	0	C
24	0 − 1	0 − 1	−1	−1	+1	0	C
25	0 − 1	1 − 0	−1	+1	−1	2	D
26	0 − 1	1 − 0	−1	+1	−1	2	D
27	0 − 1	1 − 0	−1	+1	−1	2	D
28	0 − 1	1 − 0	−1	+1	−1	2	D
29	0 − 1	1 − 1	−1	0	0	1	T_y
30	0 − 1	1 − 1	−1	0	0	1	T_y
31	1 − 1	0 − 0	0	0	0	0	T_{xy}
32	1 − 1	0 − 0	0	0	0	0	T_{xy}
33	1 − 1	0 − 0	0	0	0	0	T_{xy}
34	1 − 1	0 − 1	0	−1	0	1	T_x
35	1 − 1	0 − 1	0	−1	0	1	T_x
⋮	⋮	⋮	⋮	⋮	⋮	⋮	⋮
41	1 − 1	0 − 1	0	−1	0	1	T_x
42	1 − 1	0 − 1	0	−1	0	1	T_x
43	1 − 1	0 − 1	0	−1	0	1	T_x
44	1 − 1	0 − 1	0	−1	0	1	T_x
45	1 − 1	1 − 1	0	0	0	0	T_{xy}
Total					+2	33	

Following Kendall,

$$\sum_{i<j} r_{ij}s_{ij}$$

is given in the sixth column of Table 11.4, where there are $C = 6$ concordant pairs in rows 8, 9, 16, 17, 23, and 24, indicated by $+1$ values, and $D = 4$ discordant pairs in rows 25–28, indicated by -1 values. Tied pairs T_x, T_y, and T_{xy} are each indicated by a value of 0. Thus, the observed value of S is

$$S_o = \sum_{i<j} r_{ij}s_{ij} = C - D = 6 - 4 = +2 .$$

Now consider the categorical data listed in Table 11.4 in a randomized-block analysis-of-variance context with $b = 2$ blocks and $g = 10$ univariate measurements for each block. For the column labeled $|r_{ij} - s_{ij}|$ in Table 11.4, only values of T_x, T_y, and D receive non-zero values: values of 1 for both T_x and T_y and values of 2 for D. Therefore,

$$\sum_{i<j} |r_{ij} - s_{ij}| = 2D + T_x + T_y = \frac{g(g-1)}{2} - \sum_{i<j} r_{ij}s_{ij} - T_{xy}$$

$$= \frac{g(g-1)}{2} - S - T_{xy} .$$

For the frequency data given in Fig. 11.11, the observed value of the generalized Minkowski distance function is

$$\sum_{i<j} |r_{ij} - s_{ij}| = 2D + T_x + T_y = 2(4) + 11 + 14 = 33$$

as described in Table 11.4 and, equivalently,

$$\frac{g(g-1)}{2} - S - T_{xy} = 10(10-1)/2 - 2 - 10 = 33 .$$

Now, define the generalized Minkowski distance function with $b = 2$ and $r = 1$,

$$\Delta(x, y) = \frac{g(g-1)}{2} - S .$$

Then, substituting $C + D + T_x + T_y + T_{xy}$ for $g(g-1)/2$ and $C - D$ for S,

$$\Delta(x, y) = C + D + T_x + T_y + T_{xy} - (C - D) = 2D + T_x + T_y + T_{xy} .$$

For the frequency data given in Fig. 11.11 on p. 564,

$$\Delta(x,y) = 2D + T_x + T_y + T_{xy} = 2(4) + 11 + 14 + 10 = 43\ .$$

With $b = 2$ the observed MRBP test statistic is

$$\delta_o = \frac{1}{g}\Delta(x,y) = \frac{1}{10}(43) = 4.30\ .$$

Also, the relationships between the MRBP test statistic and Kendall's S are given by

$$\delta = \frac{g-1}{2} - \frac{S}{g} \quad \text{and} \quad S = g\left(\frac{g-1}{2} - \delta\right)\ .$$

Thus, for the paired data with no tied values listed in Table 11.4, the observed values of δ and Kendall's S are

$$\delta_o = \frac{10-1}{2} - \frac{2}{10} = 4.30 \quad \text{and} \quad S_o = 10\left(\frac{10-1}{2} - 4.30\right) = +2\ .$$

For the frequency data given in Fig. 11.11, there are

$$M = (g!)^{b-1} = (10!)^{2-1} = 3{,}628{,}800$$

possible, equally-likely arrangements of the observed data. Therefore, the exact expected value of the $M = 3{,}628{,}800$ δ values is

$$\mu_\delta = \frac{1}{M}\sum_{i=1}^{M}\delta_i = \frac{1}{3{,}628{,}800}(16{,}852{,}147) = 4.6440$$

and the observed chance-corrected measure of effect size is

$$\Re_o = 1 - \frac{\delta_o}{\mu_\delta} = 1 - \frac{4.30}{4.6440} = +0.0741\ ,$$

indicating approximately 7 % within-block agreement above that expected by chance.

11.5.1 Kendall's t_a Statistic and δ

As discussed in Chap. 10, Sect. 10.9, Kendall's τ_a is a pair-based measure of ordinal association given by

$$\tau_a = \frac{2S}{N(N-1)}\ ,$$

where S is the number of concordant pairs (C) minus the number of discordant pairs (D). However, τ_a can also be utilized for categorical data in a 2×2 contingency-table format. For the frequency data given in Fig. 11.11, the observed value of τ_a is

$$\tau_a = \frac{2(+2)}{10(10-1)} = \frac{4}{90} = +0.0444 \,.$$

In a randomized-block analysis-of-variance context, the functional relationships between Kendall's τ_a and the MRBP test statistic are given by

$$\tau_a = 1 - \frac{2\delta}{g-1} \quad \text{and} \quad \delta = \frac{(1-g)(\tau_a - 1)}{2} \,.$$

Thus, for the frequency data given in Fig. 11.11, the observed values of Kendall's τ_a and δ are

$$\tau_a = 1 - \frac{2(4.30)}{10-1} = +0.0444 \quad \text{and} \quad \delta_o = \frac{(1-10)(+0.0444-1)}{2} = 4.30 \,.$$

Kendall's τ_a may also be expressed in terms of $\Re$ and μ_δ. Thus, for the frequency data given in Fig. 11.11 on p. 564,

$$\tau_a = 1 + \frac{2(\Re - 1)\mu_\delta}{g-1} = 1 + \frac{2(0.0741)(4.6440)}{10-1} = +0.0444$$

and

$$(1-\Re)\mu_\delta = \frac{g - 1 - \tau_a(g-1)}{2} = \frac{10 - 1 - 0.0444(10-1)}{2} = 4.30 \,.$$

11.5.2 Yule's Q Statistic and δ

In 1912 G. Udny Yule published a paper titled "On the methods of measuring association between two attributes" in *Journal of the Royal Statistical Society* [435]. In this important paper Yule introduced a new statistic for 2×2 contingency tables that he called Q,[4] although he had briefly mentioned Q in a 1900 paper published in *Philosophical Transactions of the Royal Society of London* [434]. Contained within this lengthy paper of 74 pages was strong criticism of the work of Karl Pearson on the analysis of contingency tables. Pearson was greatly offended and a vitriolic response soon followed from Pearson and biometrician David Heron in *Biometrika*; the rejoinder consisted of a remarkable 157 folio pages [336].

[4]The symbol Q was taken from the initial letter of the surname of Lambert Adolphe Jacques Quetelet, the nineteenth century Belgium astronomer, mathematician, statistician, and sociologist [435, p. 586].

Fig. 11.12 Example data for variables x and y with categories dummy-coded as 0 and 1

	y		
x	0	1	Total
0	3	1	4
1	4	2	6
Total	7	3	10

Originally developed for categorical data, Yule's Q is often used for rank data [435]. For a 2×2 contingency table, Yule's Q is identical to Goodman and Kruskal's γ statistic [151]. Yule's Q is given by

$$Q = \frac{C}{C+D} - \frac{D}{C+D} = \frac{C-D}{C+D} = \frac{S}{C+D} ,$$

where C and D denote the number of concordant and discordant pairs, respectively, and $S = C - D$. Thus, for the frequency data given in Fig. 11.11, replicated for convenience in Fig. 11.12, the observed value of Yule's Q is

$$Q_o = \frac{C-D}{C+D} = \frac{(3)(2)-(1)(4)}{(3)(2)+(1)(4)} = \frac{+2}{6+4} = +0.20 .$$

In a randomized-block analysis-of-variance context, the functional relationships between Yule's Q and the MRBP test statistic are given by

$$Q = \frac{g[(g-1)-2\delta]}{2(C+D)} \quad \text{and} \quad \delta = \frac{g(g-1)-2Q(C+D)}{2g} .$$

Thus, for the frequency data given in Fig. 11.12, the observed values for Yule's Q and the MRBP test statistic δ are

$$Q_o = \frac{10[(10-1)-(2)(4.30)]}{2(6+4)} = +0.20$$

and

$$\delta_o = \frac{10(10-1)-(2)(0.20)(6+4)}{(2)(10)} = 4.30 .$$

11.5.3 Yule's Y Statistic and δ

In the 1912 paper where Q was first presented, Yule introduced a second measure of association for 2×2 contingency tables [435, p. 591]. Yule termed the new measure the coefficient of colligation and identified it by the lower-case Greek letter omega,

Fig. 11.13 Notation for a 2×2 cross-classification table

	y		
x	0	1	Total
0	a	b	$a+b$
1	c	d	$c+d$
Total	$a+c$	$b+d$	N

ω, although it is customarily labeled as Yule's Y in the current literature. Given the notation for a 2×2 table in Fig. 11.13, Yule's Y is given by

$$Y = \frac{\sqrt{ad} - \sqrt{bc}}{\sqrt{ad} + \sqrt{bc}} .$$

For the frequency data given in Fig. 11.12, the observed value of Yule's Y is

$$Y_o = \frac{\sqrt{(3)(2)} - \sqrt{(1)(4)}}{\sqrt{(3)(2)} + \sqrt{(1)(4)}} = 0.1010 .$$

In a randomized-block analysis-of-variance context, the functional relationships between Yule's Y and the MRBP test statistic are given by

$$\delta = \frac{1}{2}\left[g - \frac{4Y(ad+bc)}{g(1+Y^2)} - 1\right]$$

and

$$Y = \frac{g(g-2\delta-1)}{2\big(ad+bc\big) + \big[4(ad+bc)^2 - g^2(g-2\delta-1)^2\big]^{1/2}} .$$

Thus, for the frequency data given in Fig. 11.12, the observed value of the MRBP test statistic is

$$\delta_o = \frac{1}{2}\left\{10 - \frac{4(0.1010)[(3)(2)+(1)(4)]}{10(1+0.1010^2)} - 1\right\} = \frac{8.60}{2} = 4.30$$

and the observed value of Y is

$$Y_o = \frac{10[10-(2)(4.30)-1]}{2[(3)(2)+(1)(4)] + \{4[(3)(2)+(1)(4)]^2 - 10^2[10-(2)(4.3)-1]^2\}^{1/2}} = \frac{4.00}{39.5959} = 0.1010 .$$

11.5.4 The Odds Ratio and δ

While useful by itself, the odds ratio has become an important component of more advanced statistical techniques. The natural log (ln) of the odds ratio plays an important role in, for example, both logistic regression and log-linear analysis.

In terms of the pairwise notation of Kendall, the odds ratio may be written as

$$\varphi = \frac{C}{D} ,$$

where C and D denote the number of concordant and discordant pairs, respectively. For the frequency data given in Fig. 11.12, the observed value of the odds ratio is

$$\varphi_{\mathrm{o}} = \frac{C}{D} = \frac{6}{4} = 1.50 .$$

More conventionally, given the notation of Fig. 11.13, the observed value of the odds ratio is given by

$$\varphi_{\mathrm{o}} = \frac{ad}{bc} = \frac{(3)(2)}{(1)(4)} = 1.50 .$$

In a randomized-block analysis-of-variance context and following the notation in Fig. 11.13, the functional relationships between the odds ratio and the MRBP test statistic are given by

$$\varphi = \frac{g(g-1) + 2(ad + bc - g\delta)}{4bc} \quad \text{and} \quad \delta = \frac{g(g-1) + 2(ad + bc) - 4bc\varphi}{2g} .$$

Thus, for the frequency data given in Fig. 11.12, the observed value of the odds ratio is

$$\varphi_{\mathrm{o}} = \frac{10(10-1) + 2[(3)(2) + (1)(4) - (10)(4.3)]}{4(1)(4)} = \frac{24}{16} = 1.50$$

and the observed value of δ is

$$\delta_{\mathrm{o}} = \frac{10(10-1) + 2((3)(2) + (1)(4)) - 4(1)(4)(1.50)}{(2)(10)} = \frac{86}{20} = 4.30 .$$

11.5.5 Relationships Among Q, Y, and φ

Since Yule's Q, Yule's Y, and the odds ratio are all related to the MRBP test statistic δ, they are necessarily related to each other. The relationships between Yule's Q and the odds ratio are given by

$$Q = \frac{\varphi - 1}{\varphi + 1} \quad \text{and} \quad \varphi = \frac{1 + Q}{1 - Q} ,$$

as noted by Yule [435, p. 586].[5] Thus, for the frequency data given in Fig. 11.12, the observed values for Yule's Q and the odds ratio are

$$Q_o = \frac{1.50 - 1}{1.50 + 1} = +0.20 \quad \text{and} \quad \varphi_o = \frac{1 + 0.20}{1 - 0.20} = 1.50 .$$

The relationships between Yule's Q and Yule's Y are given by

$$Q = \frac{2Y}{1 + Y^2} \quad \text{and} \quad Y = \frac{Q}{1 + \sqrt{1 - Q^2}} .$$

For the frequency data given in Fig. 11.12, the observed values for Yule's Q and Yule's Y are

$$Q_o = \frac{2(0.1010)}{1 + 0.1010^2} = 0.20 \quad \text{and} \quad Y_o = \frac{0.20}{1 + \sqrt{1 - 0.20^2}} = 0.1010 .$$

The relationships between Yule's Y and the odds ratio are given by

$$Y = \frac{\sqrt{\varphi} - 1}{\sqrt{\varphi} + 1} \quad \text{and} \quad \varphi = \frac{(Y + 1)^2}{(Y - 1)^2} .$$

For the frequency data given in Fig. 11.12, the observed values for Y and the odds ratio are

$$Y_o = \frac{\sqrt{1.50} - 1}{\sqrt{1.50} + 1} = 0.1010 \quad \text{and} \quad \varphi_o = \frac{(0.1010 + 1)^2}{(0.1010 - 1)^2} = 1.50 .$$

11.5.6 Somers' d_{xy}/d_{yx} and δ

In 1962 sociologist Robert Somers published an article in *American Sociological Review* in which he developed two asymmetric measures of association for ordinal variables, d_{xy} and d_{yx}, given by

$$d_{xy} = \frac{S}{C + D + T_x} \quad \text{and} \quad d_{yx} = \frac{S}{C + D + T_y} ,$$

[5]Because, for a 2×2 contingency table, Yule's Q and Goodman and Kruskal's γ are identical, γ and φ are related in the same manner.

where C and D denote the number of concordant and discordant pairs, respectively, $S = C - D$, and T_x and T_y denote the number of pairs tied on x and y, respectively. Like many other statistics developed for ordinal variables, Somers' measures can be applied to categorical variables when the data are cross-classified into a 2×2 contingency table. Somers' asymmetric measure d_{xy} (d_{yx}) is the y-weak (x-weak) coefficient of monotonicity in which ties on y and x are not taken into account in the denominators of d_{xy} and d_{yx}, respectively [344]. The essential idea is that when there is a difference between pairs in the independent variable that is not reflected by a difference in the dependent variable, the coefficient should be reduced by a compensatory amount. For Somers' d_{xy}, x is the dependent variable and for Somers' d_{yx}, y is the dependent variable.

Recall that for the frequency data given in Fig. 11.12 on p. 569, the number of concordant pairs is $C = 6$, the number of discordant pairs is $D = 4$, the number of pairs tied on variable x but not tied on variable y is $T_x = 11$, the number of pairs tied on variable y but not tied on variable x is $T_y = 14$, the number of pairs tied on both variable x and variable y is $T_{xy} = 10$, and the observed value of S is $S_o = C - D = 6 - 4 = +2$. For the frequency data given in Fig. 11.12, Somers' measures of asymmetric association are

$$d_{xy} = \frac{S}{C + D + T_x} = \frac{+2}{6 + 4 + 11} = +0.0952$$

and

$$d_{yx} = \frac{S}{C + D + T_y} = \frac{+2}{6 + 4 + 14} = +0.0833 \; .$$

In a randomized-block analysis-of-variance context, the functional relationships between the MRBP test statistic and Somers' asymmetric measures of association are given by

$$d_{xy} = \frac{g(g-1) - 2g\delta}{2(C + D + T_x)} \quad \text{and} \quad \delta = \frac{g(g-1) - 2(C + D + T_x)d_{xy}}{2g} \; ;$$

also,

$$d_{yx} = \frac{g(g-1) - 2g\delta}{2(C + D + T_y)} \quad \text{and} \quad \delta = \frac{g(g-1) - 2(C + D + T_y)d_{yx}}{2g} \; .$$

Thus, for the frequency data given in Fig. 11.12 on p. 569, the observed value of Somers' d_{xy} is

$$d_{xy} = \frac{10(10-1) - 2(10)(4.30)}{2(6 + 4 + 11)} = \frac{4}{42} = +0.0952$$

and the observed value of the MRBP test statistic is

$$\delta_{\text{o}} = \frac{10(10-1) - 2(6+4+11)(+0.0952)}{(2)(10)} = \frac{86}{20} = 4.30 \, .$$

Similarly, the observed value of Somers' d_{yx} is

$$d_{yx} = \frac{10(10-1) - 2(10)(4.30)}{2(6+4+14)} = \frac{4}{48} = +0.0833$$

and the observed value of the MRBP test statistic is

$$\delta_{\text{o}} = \frac{10(10-1) - 2(6+4+14)(+0.0833)}{(2)(10)} = \frac{86}{20} = 4.30 \, .$$

11.6 A Reanalysis of the Data

Consider the frequency data given in Fig. 11.12, replicated for convenience in Fig. 11.14 in a raw data format and in Fig. 11.15 in a dummy-coded format. Given the dummy-coded data listed in Fig. 11.15,

$$\sum_{i=1}^{N} x_i = \sum_{i=1}^{N} x_i^2 = 6 \, , \quad \sum_{i=1}^{N} y_i = \sum_{i=1}^{N} y_i^2 = 6 \, , \quad \sum_{i=1}^{N} x_i y_i = 2 \, ,$$

and $N = 10$. The sample means of variables x and y are

$$\bar{x} = \frac{1}{N}\sum_{i=1}^{N} x_i = \frac{1}{10}(6) = 0.60 \quad \text{and} \quad \bar{y} = \frac{1}{N}\sum_{i=1}^{N} y_i = \frac{1}{10}(3) = 0.30$$

Fig. 11.14 Example data for variables x and y with categories dummy-coded as 0 and 1

	y		
x	0	1	Total
0	3	1	4
1	4	2	6
Total	7	3	10

Fig. 11.15 Example dummy-coded frequency data for variables x and y given in Fig. 11.14

Object	Variable x	Variable y
1	0	0
2	0	0
3	0	0
4	0	1
5	1	0
6	1	0
7	1	0
8	1	0
9	1	1
10	1	1

and the sample estimates of the population variances for variables x and y are

$$s_x^2 = \frac{1}{N-1}\left[\sum_{i=1}^{N} x_i^2 - \frac{\left(\sum_{i=1}^{N} x_i\right)^2}{N}\right] = \frac{1}{10-1}\left(6 - \frac{6^2}{10}\right) = 0.2667$$

and

$$s_y^2 = \frac{1}{N-1}\left[\sum_{i=1}^{N} y_i^2 - \frac{\left(\sum_{i=1}^{N} y_i\right)^2}{N}\right] = \frac{1}{10-1}\left(3 - \frac{3^2}{10}\right) = 0.2333\ .$$

Considered as a regression problem, the covariance of variables x and y is

$$\text{cov}(x,y) = \frac{1}{N-1}\left(\sum_{i=1}^{N} x_i y_i - \frac{\sum_{i=1}^{N} x_i \sum_{i=1}^{N} y_i}{N}\right)$$

$$= \frac{1}{10-1}\left(2 - \frac{(6)(3)}{10}\right) = +0.0222\ ,$$

the Pearson product-moment correlation coefficient for variables x and y is

$$r_{xy} = \frac{\text{cov}(x,y)}{\sqrt{s_x^2\, s_y^2}} = \frac{+0.0222}{\sqrt{(0.2667)(0.2333)}} = +0.0891\;,$$

and the unstandardized regression coefficients, b_{xy} and b_{yx}, are

$$b_{xy} = \frac{\text{cov}(x,y)}{s_y^2} = \frac{+0.0222}{0.2333} = +0.0952$$

and

$$b_{yx} = \frac{\text{cov}(x,y)}{s_x^2} = \frac{+0.0222}{0.2777} = +0.0833\;.$$

Now consider the frequency data given in Fig. 11.14 in a randomized-block format, as displayed in Fig. 11.16. For the binary data listed in Fig. 11.16, $g = 10$, $b = 2$, following Eq. (11.2) on p. 544 the observed value of the generalized Minkowski distance function is $\Delta(x,y) = 5.0$, following Eq. (11.1) on p. 543, the observed value of the MRBP test statistic with $v = 2$ is $\delta_o = 0.50$, following Eq. (11.4) on p. 544 the exact expected value of δ is $\mu_\delta = 0.54$, and following Eq. (11.3) on p. 544 the observed chance-corrected measure of effect size is

$$\Re_o = 1 - \frac{\delta_o}{\mu_\delta} = 1 - \frac{0.50}{0.54} = +0.0741\;,$$

indicating approximately 7 % within-block agreement above that expected by chance.

Fig. 11.16 Example randomized-block data for variables x and y given in Fig. 11.14

	Block	
Object	1	2
1	0	0
2	0	0
3	0	0
4	0	1
5	1	0
6	1	0
7	1	0
8	1	0
9	1	1
10	1	1

Note that the values for $\Delta(x, y)$, δ, μ_δ, and $\Re$ differ from those computed on the data listed in Fig. 11.11 on p. 564. In the previous section the values for $\Delta(x, y)$, δ, μ_δ, and $\Re$ were obtained from a comparison of the direction of differences, i.e., the $r_{ij}s_{ij}$ and $|r_{ij} - s_{ij}|$ values. In this section the values are obtained from a comparison of the dummy codes. The two sets of values differ by a factor of $2\delta = 2(4.30) = 8.60$. Thus, the previously obtained value for $\Delta(x, y)$ was 43 and here it is $43/8.60 = 5.00$; the obtained value for δ was 4.30 and here it is $4.30/8.60 = 0.50$; the obtained value for μ_δ was 4.6440 and here it is $4.6440/8.60 = 0.54$. The value for $\Re$ remains unchanged at $+0.0741$ as it is based on the ratio of δ and μ_δ, both of which have been divided by 2δ.

For the randomized-block binary data listed in Fig. 11.16 there are

$$M = (g!)^b = (10!)^2 = 13{,}168{,}189{,}440{,}000 \tag{11.17}$$

possible, equally-likely arrangements of the observed data, making an exact solution impractical. If all M possible arrangements of the observed randomized-block binary data listed in Fig. 11.16 occur with equal chance, the approximate resampling probability value of $\delta_o = 0.50$ computed on $L = 1{,}000{,}000$ random arrangements of the observed data with $b = 2$ blocks preserved for each arrangement is

$$P(\delta \leq \delta_o | H_0) = \frac{\text{number of } \delta \text{ values} \leq \delta_o}{L} = \frac{667{,}214}{1{,}000{,}000} = 0.6672\ .$$

While an exact test is impractical in this case, it is not impossible. Since the ordered responses of one object may be held fixed relative to the other $b - 1$ objects, $M = 13{,}168{,}189{,}440{,}000$ in Eq. (11.17) can be reduced to

$$M = (g!)^{b-1} = (10!)^{2-1} = 3{,}628{,}800$$

possible, equally-likely arrangements of the observed data. If all arrangements of the observed randomized-block binary data listed in Fig. 11.16 occur with equal chance, the exact probability value of $\delta_o = 0.50$ computed on the $M = 3{,}628{,}800$ arrangements of the observed data with $b = 2$ blocks preserved for each arrangement is

$$P(\delta \leq \delta_o | H_0) = \frac{\text{number of } \delta \text{ values} \leq \delta_o}{M} = \frac{2{,}419{,}200}{3{,}628{,}800} = 0.6667\ .$$

Note that the approximate resampling probability value of $P = 0.6672$ based on $L = 1{,}000{,}000$ is very close to the exact probability value of $P = 0.6667$ based on $M = 3{,}628{,}800$.

11.6.1 Pearson's r_{xy} and $\mathfrak{R}$

Because degrees of freedom are unnecessary in a permutation context, define sample standard deviations S_x and S_y as

$$S_x = \left[\frac{1}{g} \sum_{i=1}^{N} (x_i - \bar{x})^2 \right]^{1/2} \quad \text{and} \quad S_y = \left[\frac{1}{g} \sum_{i=1}^{N} (y_i - \bar{y})^2 \right]^{1/2} ,$$

yielding $S_x = 0.4899$ and $S_y = 0.4583$, respectively, for the randomized-block binary data listed in Fig. 11.16. Alternatively,

$$S_x = \left[\frac{(g-1)s_x^2}{g} \right]^{1/2} \quad \text{and} \quad S_y = \left[\frac{(g-1)s_y^2}{g} \right]^{1/2} ,$$

where

$$s_x^2 = \frac{1}{g-1} \sum_{i=1}^{N} (x_i - \bar{x})^2 \quad \text{and} \quad s_y^2 = \frac{1}{g-1} \sum_{i=1}^{N} (y_i - \bar{y})^2 .$$

Then the relationships between Pearson's r_{xy} and $\mathfrak{R}$ are given by

$$r_{xy} = \frac{\mathfrak{R}\mu_\delta}{2S_x S_y} \quad \text{and} \quad \mathfrak{R} = \frac{2r_{xy}S_x S_y}{\mu_\delta}$$

with $v = 2$. For the randomized-block binary data listed in Fig. 11.16, the observed value of r_{xy} is

$$r_{xy} = \frac{(+0.0741)(0.54)}{2(0.4899)(0.4583)} = +0.0891$$

and the observed value of $\mathfrak{R}$ is

$$\mathfrak{R}_o = \frac{2(+0.0891)(0.4899)(0.4583)}{0.54} = +0.0741 .$$

Alternatively, since $\mathfrak{R}\mu_\delta = \mu_\delta - \delta$, r_{xy} can be defined in terms of μ_δ and δ,

$$r_{xy} = \frac{\mu_\delta - \delta}{2S_x S_y} = \frac{0.54 - 0.50}{2(0.4899)(0.4583)} = +0.0891 . \tag{11.18}$$

11.6.2 MRBP and Regression Coefficients

The functional relationships between the unstandardized regression coefficient b_{xy} and the MRBP test statistics δ and μ_δ are given by

$$b_{xy} = \frac{\mu_\delta - \delta_o}{2S_x^2} \quad \text{and} \quad \mu_\delta - \delta = 2b_{xy}S_x^2$$

with $v = 2$. Alternatively, in terms of $\Re$ and μ_δ,

$$b_{xy} = \frac{\Re\mu_\delta}{2S_x^2} \quad \text{and} \quad \Re\mu_\delta = 2b_{xy}S_x^2 \ .$$

Thus, for the randomized-block binary data listed in Fig. 11.16 on p. 576,

$$b_{xy} = \frac{\mu_\delta - \delta}{2S_x^2} = \frac{0.54 - 0.50}{2(0.4899)^2} = \frac{0.04}{0.48} = +0.0833 \ ,$$

$$b_{xy} = \frac{\Re\mu_\delta}{2S_x^2} = \frac{(+0.0741)(0.54)}{2(0.4899)^2} = +0.0833 \ ,$$

$$\mu_\delta - \delta_o = 2b_{xy}S_x^2 = 2(+0.0833)(0.4899)^2 = 0.04 \ ,$$

and

$$\Re\mu_\delta = 2b_{xy}S_x^2 = 2(+0.0833)(0.4899)^2 = 0.04 \ .$$

Similarly, the relationships between the unstandardized regression coefficient b_{yx} and μ_δ and δ are given by

$$b_{yx} = \frac{\mu_\delta - \delta}{2S_y^2} \quad \text{and} \quad \mu_\delta - \delta = 2b_{yx}S_y^2 \ ;$$

or, in terms of $\Re$ and μ_δ,

$$b_{yx} = \frac{\Re\mu_\delta}{2S_y^2} \quad \text{and} \quad \Re\mu_\delta = 2b_{yx}S_y^2 \ .$$

For the binary data listed in Fig. 11.16,

$$b_{yx} = \frac{\mu_\delta - \delta_o}{2S_y^2} = \frac{0.54 - 0.50}{2(0.4583)^2} = \frac{0.04}{0.42} = +0.0952 \ ,$$

$$b_{yx} = \frac{\Re\mu_\delta}{2S_y^2} = \frac{(+0.0741)(0.54)}{2(0.4583)^2} = +0.0952 \ ,$$

$$\mu_\delta - \delta_o = 2b_{yx}S_y^2 = 2(+0.0952)(0.4583)^2 = 0.04 \ ,$$

and

$$\Re\mu_\delta = 2b_{yx}S_y^2 = 2(+0.0952)(0.4583)^2 = 0.04\ .$$

11.6.3 MRBP and Percentage Differences

Simple percentage differences are commonly employed by newspapers and magazines to communicate differences between two groups in an uncomplicated manner to unsophisticated audiences. However, percentage differences have an intimate relationship to randomized-block analysis-of-variance designs, to simple linear regression and correlation, and to the MRBP test statistics, δ and $\Re$.

Consider the frequency data given in Fig. 11.14 on p. 574, replicated for convenience in Fig. 11.17, where the cell entries are expressed as proportions, as in Fig. 11.18.[6] For the proportion data listed in Fig. 11.18, the percentage difference for variable y is given by

$$D_{yx} = |0.4286 - 0.3333| = |0.5714 - 0.0.6667| = +0.0952\ .$$

Computing percentages for variable x yields the proportions listed in Fig. 11.19. For the proportion data listed in Fig. 11.19, the percentage difference for variable x

Fig. 11.17 Example data for variables x and y with categories dummy-coded as 0 and 1

	y		
x	0	1	Total
0	3	1	4
1	4	2	6
Total	7	3	10

Fig. 11.18 Example data from Fig. 11.17 with cell entries expressed as proportions of column totals

	y	
x	0	1
0	0.4286	0.3333
1	0.5714	0.6667
Total	1.0000	1.0000

[6]While generally called "percentage" differences, values are typically calculated and expressed with proportions, as in Fig. 11.18.

Fig. 11.19 Example data from Fig. 11.17 with cell entries expressed as proportions of row totals

	y		
x	0	1	Total
0	0.7500	0.2500	1.0000
1	0.6667	0.3333	1.0000

is given by

$$D_{xy} = |0.7500 - 0.6667| = |0.2500 - 0.3333| = +0.0833 .$$

The functional relationships between percentage differences D_{xy} and D_{yx} and the MRBP statistics δ and $\Re$ are given by

$$D_{xy} = \frac{\mu_\delta - \delta}{2S_x^2} = \frac{\Re\mu_\delta}{2S_x^2} \quad \text{and} \quad \mu_\delta - \delta = 2D_{xy}S_x^2 .$$

Thus, for the frequency data given in Fig. 11.17, the observed value of the percentage difference D_{xy} is

$$D_{xy} = \frac{0.54 - 0.50}{2(0.4899)^2} = \frac{(+0.0741)(0.54)}{2(0.4899)^2} = +0.0833 .$$

and the observed value of $\mu_\delta - \delta$ is

$$\mu_\delta - \delta_o = 2(+0.0833)(0.4899)^2 = 0.04 .$$

Similarly, the functional relationships between percentage difference D_{yx} and $\mu_\delta - \delta$ are

$$D_{yx} = \frac{\mu_\delta - \delta}{2S_y^2} = \frac{\Re\mu_\delta}{2S_y^2} \quad \text{and} \quad \mu_\delta - \delta = 2D_{yx}S_y^2 .$$

For the frequency data given in Fig. 11.17, the observed values of the percentage difference D_{yx} and $\mu_\delta - \delta_o$ are

$$D_{yx} = \frac{0.54 - 0.50}{2(0.4583)^2} = \frac{(+0.0741)(0.54)}{2(0.4583)^2} = +0.0952 .$$

and

$$\mu_\delta - \delta_o = 2(+0.0952)(0.4583)^2 = 0.04 .$$

Note that the percentage differences, the unstandardized regression coefficients, and Somers' two asymmetric measures are all equivalent for a 2×2 contingency table, where for the frequency data given in Fig. 11.17,

$$D_{xy} = b_{xy} = d_{xy} = +0.0833$$

and

$$D_{yx} = b_{yx} = d_{yx} = +0.0952 \ .$$

It is not widely recognized that, given a 2×2 contingency table, a percentage difference is really just the slope of a regression line [45], and that Somers' d_{xy} and d_{yx} measures thereby reduce to simple percentage differences.[7]

It is well known, however, that the Pearson product-moment correlation coefficient is simply the geometric mean of the slopes of two regression lines, i.e.,

$$r_{xy} = \sqrt{b_{xy}b_{yx}} = \sqrt{(+0.0833)(+0.0952)} = 0.0891 \ .$$

Therefore, for a 2×2 contingency table, r_{xy} is also the geometric mean of Somers' two asymmetric coefficients of association, i.e.,

$$r_{xy} = \sqrt{d_{xy}d_{yx}} = \sqrt{(+0.0833)(+0.0952)} = 0.0891 \ ,$$

as well as the geometric mean of the two percentage differences, i.e.,

$$r_{xy} = \sqrt{D_{xy}D_{yx}} = \sqrt{(+0.0833)(+0.0952)} = 0.0891 \ .$$

Finally, since Somers' coefficients are given by

$$d_{xy} = \frac{S}{C+D+T_x} \quad \text{and} \quad d_{yx} = \frac{S}{C+D+T_y} \ ,$$

and Kendall's τ_b coefficient is given by

$$t_b = \frac{S}{\sqrt{(C+D+T_x)(C+D+T_y)}} \ ,$$

then

$$t_b = \sqrt{d_{xy}d_{yx}} = r_{xy}$$

[7] Somers noted that d_{xy} and d_{yx} were equivalent to the corresponding percentage differences in 2×2 contingency tables [380, p. 805].

for a 2×2 contingency table.[8] For the frequency data given in Fig. 11.17,

$$d_{xy} = \frac{+2}{6+4+11} = +0.0952\,, \quad d_{yx} = \frac{+2}{6+4+14} = +0.0833\,,$$

and

$$t_b = \sqrt{(+0.0952)(+0.0833)} = 0.0891 = r_{xy}\,.$$

11.6.4 MRBP and Chi-Squared

It is well known that the chi-squared test of independence with one degree of freedom is related to the Pearson product-moment correlation coefficient when two categories are coded 0 and 1, i.e.,

$$\chi_1^2 = Nr_{xy}^2 \quad \text{and} \quad r_{xy}^2 = \frac{\chi_1^2}{N}\,.$$

It was shown in Eq. (11.18) on p. 578 that

$$r_{xy} = \frac{\mu_\delta - \delta}{2S_xS_y}\,,$$

where

$$S_x = \left[\frac{(g-1)s_x^2}{g}\right]^{1/2} \quad \text{and} \quad S_y = \left[\frac{(g-1)s_y^2}{g}\right]^{1/2}\,.$$

Therefore, with one degree of freedom, the functional relationships between chi-squared and the MRBP statistics, δ and μ_δ, are given by

$$\chi_1^2 = \frac{N(\mu_\delta - \delta)^2}{4S_x^2S_y^2} \quad \text{and} \quad \mu_\delta - \delta = 2S_xS_y\left(\frac{\chi_1^2}{N}\right)^{1/2}\,.$$

For the frequency data given in Fig. 11.17 on p. 580, replicated as Fig. 11.20 for convenience, $\mu_\delta = 0.54$, $\delta = 0.50$,

$$\chi_1^2 = \frac{10(0.54 - 0.50)^2}{4(0.4899)^2(0.4583)^2} = 0.0794\,,$$

[8]Pearson's product-moment correlation coefficient, r_{xy}, for a 2×2 table is more conventionally known as Pearson's ϕ statistic.

Fig. 11.20 Example data for variables x and y with categories dummy-coded as 0 and 1

	y		
x	0	1	Total
0	3	1	4
1	4	2	6
Total	7	3	10

and

$$\mu_\delta - \delta = 2(0.4899)(0.4583)\left(\frac{0.0794}{10}\right)^{1/2} = 0.04\,.$$

Alternatively, defined in terms of μ_δ and $\Re$,

$$\chi_1^2 = \frac{N(\Re\mu_\delta)^2}{4S_x^2S_y^2} = \frac{10[(0.0741)(0.54)]^2}{4(0.4899)^2(0.4583)^2} = 0.0794$$

and

$$\Re\mu_\delta = 2S_xS_y\left(\frac{\chi_1^2}{N}\right)^2 = 2(0.4899)(0.4583)\left(\frac{0.0794}{10}\right)^{1/2} = 0.04\,.$$

11.7 Coda

Chapter 11 utilized the Multivariate Randomized Block Permutation (MRBP) procedures developed in Chap. 8 to establish relationships between the test statistics of MRBP, δ and $\Re$, and selected conventional tests and measures designed for the analysis of randomized-block data at the nominal level of measurement. Considered in this chapter were the relationships between the MRBP test statistic δ and Cohen's unweighted kappa measure of chance-corrected agreement, McNemar's and Cochran's Q tests for change, Kendall's t_a and Yule's Q and Y measures of association, the odds ratio, Somers' d_{xy} and d_{yx} asymmetric measures of association, Pearson's product-moment correlation coefficient, percentage differences, and Pearson's chi-squared test of independence.

Epilogue

As stated in the Preface, the purpose of this book on *Permutation Statistical Methods: An Integrated Approach* is to provide a synthesis of a number of statistical tests and measures which, at first blush, appear unrelated. No attempt is made to provide a synthesis of all statistical methods, but only to derive and illustrate a common model under which a large number of statistical tests and measures can be understood.

Overview

The foundation of the synthesizing model is the generalized Minkowski distance function given by

$$\Delta(x, y) = \left(\sum_{i=1}^{r} |x_i - y_i|^p \right)^{v/p} , \tag{1}$$

where $p \geq 1$ and $v > 0$. When $v = p = 1$, $\Delta(x, y)$ is a city-block metric; when $p = 2$ and $v = 1$, $\Delta(x, y)$ is a Euclidean distance metric; and when $v = p = 2$, $\Delta(x, y)$ is squared Euclidean distance, which is not a metric function.

Derived from the generalized Minkowski distance function are two permutation approaches: Multi-Response Permutation Procedures (MRPP), designed for analyzing completely randomized data, and Multivariate Randomized Block Permutation (MRBP) procedures, designed for analyzing randomized-block data. The generalized Minkowski distance function given in Eq. (1), together with MRPP and MRBP provide for the analysis of completely randomized and randomized-block data, both univariate and multivariate, and utilizing either squared Euclidean distances with $p = v = 2$, or ordinary Euclidean distances with $p = 2$ and $v = 1$.

Both MRPP and MRBP generate two test statistics, δ and $\Re$, providing for a number of statistical tests of differences and measures of association. For MRPP, test statistic δ is the weighted mean of the average distance-function values for all distinct pairs of objects in all treatment groups, and $\Re$ is a chance-corrected within-group measure of effect size. For MRBP, δ is the balanced mean of the

K.J. Berry et al., *Permutation Statistical Methods*,
DOI 10.1007/978-3-319-28770-6

Completely Randomized Experimental Designs							
Interval Level Data Analysis							
Squared Euclidean Distance				Ordinary Euclidean Distance			
Tests of Differences		Measures of Relationships		Tests of Differences		Measures of Relationships	
Univariate analysis	Multivariate analysis	Univariate analysis	Multivariate analysis	Univariate analysis	Multivariate analysis	Univariate analysis	Multivariate analysis
Ordinal Level Data Analysis							
Squared Euclidean Distance				Ordinary Euclidean Distance			
Tests of Differences		Measures of Relationships		Tests of Differences		Measures of Relationships	
Univariate analysis	Multivariate analysis	Univariate analysis	Multivariate analysis	Univariate analysis	Multivariate analysis	Univariate analysis	Multivariate analysis
Nominal Level Data Analysis							
Squared Euclidean Distance				Ordinary Euclidean Distance			
Tests of Differences		Measures of Relationships		Tests of Differences		Measures of Relationships	
Univariate analysis	Multivariate analysis	Univariate analysis	Multivariate analysis	Univariate analysis	Multivariate analysis	Univariate analysis	Multivariate analysis

Fig. 1 Diagram for completely-randomized experimental designs with analysis cells shaded in *gray*

distance-function values for all distinct pairs of objects in all treatment groups, and $\mathfrak{R}$ is a chance-corrected within-blocks measure of effect size. Finally, test statistics δ and $\mathfrak{R}$ are applied to three levels of measurement that are commonly encountered in statistical analyses: interval, ordinal, and nominal.

The resulting five-dimensional structure is composed of (1) completely randomized and randomized-block designs; (2) nominal, ordinal, and interval levels of measurement; (3) ordinary Euclidean and squared Euclidean distance functions; (4) tests of differences and measures of relationships; and (5) univariate and multivariate data structures. Taken together, the five-dimensional structure contains 48 distinct analysis cells, many of which contain new statistical tests and measures. Figure 1 graphically displays the 24 analysis cells for completely randomized experimental designs, shaded in gray, and Fig. 2 graphically displays the 24 analysis cells for randomized-block experimental designs, also shaded in gray.

Many of the new statistics are based on ordinary Euclidean distances ($v = 1$) as most conventional statistics are based on squared Euclidean distances (v=2). However, other new statistics result from generalizing conventional statistics designed for univariate ($r = 1$) data to statistics designed for multivariate ($r \geq 2$) data; for example, multivariate extensions of the Wilcoxon two-sample and the Kruskal–Wallis multi-sample rank-sum tests.

Randomized Block Experimental Designs							
Interval Level Data Analysis							
Squared Euclidean Distance				Ordinary Euclidean Distance			
Tests of Differences		Measures of Relationships		Tests of Differences		Measures of Relationships	
Univariate analysis	Multivariate analysis	Univariate analysis	Multivariate analysis	Univariate analysis	Multivariate analysis	Univariate analysis	Multivariate analysis
Ordinal Level Data Analysis							
Squared Euclidean Distance				Ordinary Euclidean Distance			
Tests of Differences		Measures of Relationships		Tests of Differences		Measures of Relationships	
Univariate analysis	Multivariate analysis	Univariate analysis	Multivariate analysis	Univariate analysis	Multivariate analysis	Univariate analysis	Multivariate analysis
Nominal Level Data Analysis							
Squared Euclidean Distance				Ordinary Euclidean Distance			
Tests of Differences		Measures of Relationships		Tests of Differences		Measures of Relationships	
Univariate analysis	Multivariate analysis	Univariate analysis	Multivariate analysis	Univariate analysis	Multivariate analysis	Univariate analysis	Multivariate analysis

Fig. 2 Diagram for randomized-block experimental designs with analysis cells shaded in *gray*

Permutation Statistical Methods

Throughout the book, emphasis is on permutation statistical methods, both exact and resampling. Permutation methods have a long history with beginnings in the 1920s and 1930s stemming from the early works of R.A. Fisher and E.J.G. Pitman.[1] Permutation methods possess several advantages over conventional statistical methods.

1. Permutation statistical methods are entirely data dependent, in that all of the information required for analysis is contained within the observed data set.
2. Permutation statistical methods do not depend on the assumptions associated with traditional parametric tests, such as normality and homogeneity of variance.
3. Permutation statistical methods provide exact probability values based on the discrete permutation distribution of equally-likely test statistic values, rather than an approximate probability value based on a theoretical distribution, such as a normal, chi-squared, t, or F distribution.
4. Although permutation statistical methods are suitable when a random sample is obtained from a specified population, permutation methods are also appropriate for nonrandom samples, such as are common in everyday research.
5. Permutation statistical methods are appropriate for analyzing entire populations, as permutation methods are not predicated on repeated random sampling from a specified population.

[1]For a comprehensive history of permutation methods, see a 2014 book on *A Chronicle of Permutation Statistical Methods* by Berry et al. [41].

6. Permutation statistical methods can be defined for any selected test statistic. Thus, researchers have the option of using a wide variety of statistics, including the majority of conventional statistics utilized in classical statistical approaches.
7. Permutation statistical methods are ideal for small data sets, when hypothetical distribution functions may provide very poor fits.
8. Appropriate permutation statistical methods are highly resistant to extreme values, such as are common in demographic data, e.g., age at first marriage, income, and so on. Consequently, the need for any data transformation is mitigated in the permutation context and in general is not recommended, e.g., square root, logarithmic, arc cosine, and other transformations, including the conversion of raw scores to ranks.
9. Permutation statistical methods provide data-dependent statistical inferences only to the actual experiment or survey that has been analyzed, and are not dependent on knowledge of a super population.

Summary

Throughout the book a number of conventional statistical tests and measures are described, provided with permutation analogues, illustrated with examples, and, when appropriate, enhanced with multivariate extensions. The organizing scheme of the first half of the book, Chaps. 1–7, dealing with MRPP and the analysis of completely randomized data, is to (1) describe a conventional statistic (e.g., Wilcoxon's two-sample rank-sum test) and illustrate the statistic with an example analysis, (2) present the MRPP permutation analogue of the statistic with $v = 2$ and weighting function $C_i = (n_i - 1)/(N - g)$ for $i = 1, \ldots, g$ to correspond to the conventional test statistic, (3) provide an exact or resampling probability value, depending on the number of possible permutations, and (4) calculate a chance-corrected measure of effect size.

A second example analysis of the same data is then provided with the same weighting function, $C_i = (n_i - 1)/(N - g)$ for $i = 1, \ldots, g$, but with $v = 1$ to illustrate the effect of ordinary Euclidean distance on a conventional statistic. An exact or resampling probability value is calculated and results of the two example analyses are compared. A third example with $v = 1$ and weighting function $C_i = n_i/N$ for $i = 1, \ldots, g$ is constructed, as permutation methods are not dependent on degrees of freedom or squared Euclidean distance. Either an exact or resampling probability value is computed and the three analyses are compared. Finally, when appropriate, the conventional test designed for univariate ($r = 1$) response measurements is generalized to analyze multivariate ($r \geq 2$) response measurements, and the three example analyses are repeated with multivariate data and compared.

The second half of the book, Chaps. 8–11, is devoted to MRBP and the analysis of randomized-block data and follows a similar organizing scheme, with one exception. Since MRBP procedures are balanced, no weighting functions are required. Thus, (1) a conventional statistic is described (e.g., Spearman's rank-order correlation coefficient) and illustrated with an example analysis, (2) the MRBP permutation analogue of the statistic is provided with $v = 2$ to correspond to the conventional test statistic, (3) either an exact or resampling probability value is calculated, depending on the number of possible permutations, and (4) a chance-corrected measure of effect size is provided.

A second example analysis of the same data is then provided with $v = 1$ and the two analyses are compared. Finally, when appropriate, the conventional test designed for univariate ($r = 1$) response measurements is generalized to analyze multivariate ($r \geq 2$) response measurements, and the two example analyses with $v = 2$ and $v = 1$ are repeated with multivariate data and compared.

In this manner, the dimensions of data types (completely-randomized and randomized-block), levels of measurement (nominal, ordinal, and interval), number of dependent variables (univariate and multivariate), type of distance function (squared Euclidean and ordinary Euclidean), and statistical application (tests of differences and measures of association) are explored and investigated. The end result is a large number of new permutation statistical tests and measures based on the generalized Minkowski distance function that have not been presented elsewhere, organized into a systematic framework, and illustrated with numerous examples.

References

1. Agresti, A.: Measures of nominal-ordinal association. J. Am. Stat. Assoc. **76**, 524–529 (1981)
2. Agresti, A.: Categorical Data Analysis, 2nd edn. Wiley, New York (2002)
3. Agresti, A., Finley, B.: Statistical Methods for the Social Sciences. Prentice-Hall, Upper Saddle River (1997)
4. Agresti, A., Liu, I.: Modeling a categorical variable allowing arbitrarily many category choices. Biometrics **55**, 936–943 (1999)
5. Agresti, A., Liu, I.: Strategies for modeling a categorical variable allowing multiple category choies. Sociol. Method Res. **29**, 403–434 (2001)
6. Altman, D.G., Bland, J.M.: Measurement in medicine: the analysis of method comparison studies. Statistician **32**, 307–317 (1983)
7. Anderson, T.W.: An Introduction to Multivariate Statistical Analysis, 2nd edn. Wiley, New York (1984)
8. Anderson, T.W.: Two of Harold Hotelling's contributions to multivariate analysis. Tech. Rep. 40, Stanford University, Stanford (1990)
9. Anderson, D.R., Sweeney, D.J., Williams, T.A.: Introduction to Statistics: Concepts and Applications. West, New York (1994)
10. Ansari, A.R., Bradley, R.A.: Rank-sum tests for dispersion. Ann. Math. Stat. **31**, 1174–1189 (1960)
11. Anscombe, F.J.: Rejection of outliers. Technometrics **2**, 123–147 (1960)
12. Arabie, P.: Was Euclid an unnecessarily sophisticated psychologist? Psychometrika **56**, 567–587 (1991)
13. Arbuckle, J., Aiken, L.S.: A program for Pitman's permutation test for differences in location. Behav. Res. Methods Instrum. **7**, 381 (1975)
14. Author: Resampling Stats User's Guide. Resampling Stats, Arlington (1999)
15. Author: StatXact for Windows. Cytel Software, Cambridge (2000)
16. Bailer, A.J.: Testing variance equality with randomization tests. J. Stat. Comput. Simul. **31**, 1–8 (1989)
17. Bakan, D.: The test of significance in psychological research. Psychol. Bull. **66**, 423–437 (1966)
18. Bakeman, R., Robinson, B.F., Quera, V.: Testing sequential association: estimating exact p values using sampled permutations. Psychol. Methods **1**, 4–15 (1996)
19. Bartko, J.J.: On various intraclass correlation reliability coefficients. Psychol. Bull. **83**, 762–765 (1976)
20. Bartko, J.J.: Measurement and reliability: statistical thinking considerations. Schizophr. Bull. **17**, 483–489 (1991)
21. Bartlett, M.S.: A note on tests of significance in multivariate analysis. Proc. Camb. Philos. Soc. **34**, 33–40 (1939)
22. Bernardin, H.J., Beatty, R.W.: Performance Appraisal: Assessing Human Behavior at Work. Kent, Boston (1984)

K.J. Berry et al., *Permutation Statistical Methods*,
DOI 10.1007/978-3-319-28770-6

23. Berry, K.J., Mielke, P.W.: Moment approximations as an alternative to the *F* test in analysis of variance. Br. J. Math. Stat. Psychol. **36**, 202–206 (1983)
24. Berry, K.J., Mielke, P.W.: An APL function for Radlow and Alf's exact chi-square test. Behav. Res. Methods Instrum. Comput. **17**, 131–132 (1985)
25. Berry, K.J., Mielke, P.W.: Goodman and Kruskal's tau-b statistic: a nonasymptotic test of significance. Sociol. Methods Res. **13**, 543–550 (1985)
26. Berry, K.J., Mielke, P.W.: Subroutines for computing exact chi-square and Fisher's exact probability tests. Educ. Psychol. Meas. **45**, 153–159 (1985)
27. Berry, K.J., Mielke, P.W.: A generalization of Cohen's kappa agreement measure to interval measurement and multiple raters. Educ. Psychol. Meas. **48**, 921–933 (1988)
28. Berry, K.J., Mielke, P.W.: A family of multivariate measures of association for nominal independent variables. Educ. Psychol. Meas. **52**, 41–55 (1992)
29. Berry, K.J., Mielke, P.W.: Spearman's footrule as a measure of agreement. Psychol. Rep. **80**, 839–846 (1997)
30. Berry, K.J., Mielke, P.W.: Extension of Spearman's footrule to multiple rankings. Psychol. Rep. **82**, 376–378 (1998)
31. Berry, K.J., Mielke, P.W.: Least absolute regression residuals: analyses of block designs. Psychol. Rep. **83**, 923–929 (1998)
32. Berry, K.J., Mielke, P.W.: Least sum of absolute deviations regression: distance, leverage, and influence. Percept. Mot. Skills **86**, 1063–1070 (1998)
33. Berry, K.J., Mielke, P.W.: Least sum of Euclidean regression residuals: estimation of effect size. Psychol. Rep. **91**, 955–962 (2002)
34. Berry, K.J., Mielke, P.W.: Longitudinal analysis of data with multiple binary category choices. Psychol. Rep. **93**, 127–131 (2003)
35. Berry, K.J., Martin, T.W., Olson, K.F.: Testing theoretical hypotheses: a PRE statistic. Soc. Forces **53**, 190–196 (1974)
36. Berry, K.J., Martin, T.W., Olson, K.F.: A note on fourfold point correlation. Educ. Psychol. Meas. **34**, 53–56 (1974)
37. Berry, K.J., Mielke, P.W., Iyer, H.K.: Factorial designs and dummy coding. Percept. Mot. Skills **87**, 919–927 (1998)
38. Berry, K.J., Mielke, P.W., Mielke, H.W.: The Fisher–Pitman permutation test: an attractive alternative to the *F* test. Psychol. Rep. **90**, 495–502 (2002)
39. Berry, K.J., Johnston, J.E., Mielke, P.W.: Exact and resampling probability values for measures associated with ordered *R* by *C* contingency tables. Psychol. Rep. **99**, 231–238 (2006)
40. Berry, K.J., Johnston, J.E., Mielke, P.W.: An alternative measure of effect size for Cochran's *Q* test for related proportions. Percept. Mot. Skills **104**, 1236–1242 (2007)
41. Berry, K.J., Johnston, J.E., Mielke, P.W.: A Chronicle of Permutation Statistical Methods: 1920–2000 and Beyond. Springer, Cham (2014)
42. Bilder, C.R., Loughin, T.M.: On the first-order Rao–Scott correction of the Umesh–Loughin–Scherer statistic. Biometrics **57**, 1253–1255 (2001)
43. Bilder, C.R., Loughin, T.M., Nettleton, D.: Multiple marginal independence-testing for pick any/*c* variables. Commun. Stat. Simul. Comput. **29**, 1285–1316 (2000)
44. Biondini, M.E., Mielke, P.W., Berry, K.J.: Data-dependent permutation techniques for the analysis of ecological data. Vegetatio **75**, 161–168 (1988). [The name of the journal was changed to *Plant Ecology* in 1997]
45. Blalock, H.M.: A double standard in measuring degree of association. Am. Sociol. Rev. **28**, 988–989 (1963)
46. Blattberg, R., Sargent, T.: Regression with non-Gaussian stable disturbances. Econometrica **39**, 501–510 (1971)
47. Borgatta, E.F.: My student, the purist: a lament. Soc. Q. **9**, 29–34 (1968)
48. Box, G.E.P.: Science and statistics. J. Am. Stat. Assoc. **71**, 791–799 (1976)
49. Box, J.F.: R. A. Fisher: The Life of a Scientist. Wiley, New York (1978)
50. Box, G.E.P.: An Accidental Statistician: The Life and Memories of George E. P. Box. Wiley, New York (2013). [Inscribed "With a little help from my friend, Judith L. Allen"]

51. Bradbury, I.: Analysis of variance versus randomization: a comparison. Br. J. Math. Stat. Psychol. **40**, 177–187 (1987)
52. Bradley, J.V.: Distribution-free Statistical Tests. Prentice-Hall, Englewood Cliffs (1968)
53. Bradley, J.V.: A common situation conducive to bizarre distribution shapes. Am. Stat. **31**, 147–150 (1977)
54. Brandeau, M.L., Chiu, S.S.: Parametric facility location on a tree network with an L_p norm cost function. Transp. Sci. **22**, 59–69 (1988)
55. Brennan, P.F., Hays, B.J.: The kappa statistic for establishing interrater reliability in the secondary analysis of qualitative clinical data. Res. Nurs. Heal. **15**, 153–158 (1992)
56. Brennan, R.L., Prediger, D.J.: Coefficient kappa: some uses, misuses, and alternatives. Educ. Psychol. Meas. **41**, 687–699 (1981)
57. Brillinger, D.R., Jones, L.V., Tukey, J.W.: The role of statistics in weather resources management. Tech. Rep. II, Weather Modification Advisory Board, United States Department of Commerce, Washington, DC (1978)
58. Bross, I.D.J.: Is there an increased risk? Fed. Proc. **13**, 815–819 (1954)
59. Brown, G.W., Mood, A.M.: On median tests for linear hypotheses. In: Neyman, J. (ed.) Proceedings of the Second Berkeley Symposium on Mathematical Statistics and Probability, vol. II, pp. 159–166. University of California Press, Berkeley (1951)
60. Burr, E.J.: The distribution of Kendall's score S for a pair of tied rankings. Biometrika **47**, 151–171 (1960)
61. Burry-Stock, J.A., Laurie, D.G., Chissom, B.S.: Rater agreement indexes for performance assessment. Educ. Psychol. Meas. **56**, 251–262 (1996)
62. Campbell, M.J., Gardner, M.J.: Calculating confidence intervals for some non-parametric analyses. Br. Med. J. **296**, 1454–1456 (1988)
63. Capraro, R.M., Capraro, M.M.: Treatments of effect sizes and statistical significance tests in textbooks. Educ. Psychol. Meas. **62**, 771–782 (2002)
64. Capraro, R.M., Capraro, M.M.: Exploring the APA fifth edition *Publication Manual's* impact of the analytic preferences of journal editorial board members. Educ. Psychol. Meas. **63**, 554–565 (2003)
65. Carroll, R.M., Nordholm, L.A.: Sampling characteristics of Kelley's ϵ^2 and Hays' $\hat{\omega}^2$. Educ. Psychol. Meas. **35**, 541–554 (1975)
66. Carver, R.P.: The case against statistical significance testing. Harv. Educ. Rev. **48**, 378–399 (1978)
67. Carver, R.P.: The case against statistical significance testing, revisited. J. Exp. Educ. **61**, 287–292 (1993)
68. Chesterton, G.K.: The Complete Father Brown Stories: "The Head of Caesar". Star Books, Vancouver (2003)
69. Cochran, W.G.: The comparison of percentages in matched samples. Biometrika **37**, 256–266 (1950)
70. Cohen, J.: A coefficient of agreement for nominal scales. Educ. Psychol. Meas. **20**, 37–46 (1960)
71. Cohen, J.: Weighted kappa: nominal scale agreement with provision for scaled disagreement or partial credit. Psychol. Bull. **70**, 213–220 (1968)
72. Cohen, J.: Statistical Power Analysis for the Behavioral Sciences. Academic Press, New York (1969)
73. Cohen, J.: Things I have learned (so far). Am. Psychol. **45**, 1304–1312 (1990)
74. Cohen, J.: The earth is round ($p < .05$). Am. Psychol. **49**, 997–1003 (1994)
75. Cohen, J., Cohen, P.: Applied Multiple Regression/Correlation Analysis for the Behavioral Sciences. Erlbaum, Hillsdale (1975)
76. Colwell, D.J., Gillett, J.R.: Spearman versus Kendall. Math. Gaz. **66**, 307–309 (1982)
77. Conover, W.J.: Practical Nonparametric Statistics, 3rd edn. Wiley, New York (1999)

78. Conti, L.H., Musty, R.E.: The effects of delta-9-tetrahydrocannabinol injections to the nucleus accumbens on the locomotor activity of rats. In: Arurell, S., Dewey, W.L., Willette, R.E. (eds.) The Cannabinoids: Chemical, Pharmacologic, and Therapeutic Aspects, pp. 649–655. Academic Press, New York (1984)
79. Coombs, C.H.: A Theory of Data. Wiley, New York (1964)
80. Costner, H.L.: Criteria for measures of association. Am. Sociol. Rev. **30**, 341–353 (1965)
81. Cramér, H.: Mathematical Methods of Statistics. Princeton University Press, Princeton (1946)
82. Crittenden, K.S., Montgomery, A.C.: A system of paired asymmetric measures of association for use with ordinal dependent variables. Soc. Forces **58**, 1178–1194 (1980)
83. Cureton, E.E.: Rank-biserial correlation. Psychometrika **21**, 287–290 (1956)
84. Cureton, E.E.: Rank-biserial correlation when ties are present. Educ. Psychol. Meas. **28**, 77–79 (1968)
85. Curran-Everett, D.: Explorations in statistics: standard deviations and standard errors. Adv. Physiol. Educ. **32**, 203–208 (2008)
86. Daniel, W.W.: Statistical significance versus practical significance. Sci. Educ. **61**, 423–427 (1977)
87. Daniels, H.E.: Rank correlation and population models (with discussion). J. R. Stat. Soc. Ser. B Methodol. **12**, 171–191 (1950)
88. Daniels, H.E.: Note on Durbin and Stuart's formula for $E(r_s)$. J. R. Stat. Soc. Ser. B Methodol. **13**, 310 (1951)
89. Darwin, C.R.: The Effects of Cross and Self Fertilization in the Vegetable Kingdom. John Murray, London (1876)
90. David, F.N.: Review of "Rank Correlation Methods" by M. G. Kendall. Biometrika **37**, 190 (1950)
91. de Mast, J., Akkerhuis, T., Erdmann, T.: The statistical evaluation of categorical measurements: simple scales, but treacherous complexity underneath (2014). [Originally a paper presented at the First Stu Hunter Research Conference in Heemskerk, Netherlands, March, 2013]
92. Decady, Y.R., Thomas, D.R.: A simple test of association for contingency tables with multiple column responses. Biometrics **56**, 893–896 (2000)
93. Diekhoff, G.: Statistics for the Social and Behavioral Sciences: Univariate, Bivariate, Multivariate. Brown, Dubuque (1992)
94. Dielman, T.E.: A comparison of forecasts from least absolute and least squares regression. J. Forecast. **5**, 189–195 (1986)
95. Dielman, T.E.: Corrections to a comparison of forecasts from least absolute and least squares regression. J. Forecast. **8**, 419–420 (1989)
96. Dielman, T.E., Pfaffenberger, R.: Least absolute value regression: necessary sample sizes to use normal theory inference procedures. Decis. Sci. **19**, 734–743 (1988)
97. Dielman, T.E., Rose, E.L.: Forecasting in least absolute value regression with autocorrelated errors: a small-sample study. Int. J. Forecast. **10**, 539–547 (1994)
98. Dodd, D.H., Schultz, R.F.: Computational procedures for estimating magnitude of effects for some analysis of variance designs. Psychol. Bull. **79**, 391–395 (1973)
99. Durbin, J., Stuart, A.: Inversions and rank correlation coefficients. J. R. Stat. Soc. Ser. B Methodol. **13**, 303–309 (1951)
100. Dwass, M.: Modified randomization tests for nonparametric hypotheses. Ann. Math. Stat. **28**, 181–187 (1957)
101. Dwyer, J.H.: Analysis of variance and the magnitude of effect: a general approach. Psychol. Bull. **81**, 731–737 (1974)
102. Dyson, G.: Turing's Cathedral: The Origins of the Digital Universe. Pantheon/Vintage, New York (2012)
103. Eden, T., Yates, F.: On the validity of Fisher's z test when applied to an actual example of non-normal data. J. Agric. Sci. **23**, 6–17 (1933)
104. Edgington, E.S.: Randomization tests. J. Psychol. **57**, 445–449 (1964)

105. Edgington, E.S.: Statistical inference and nonrandom samples. Psychol. Bull. **66**, 485–487 (1966)
106. Edgington, E.S.: Approximate randomization tests. J. Psychol. **72**, 143–149 (1969)
107. Edgington, E.S.: Statistical Inference: The Distribution-Free Approach. McGraw-Hill, New York (1969)
108. Edgington, E.S.: Randomization Tests. Marcel Dekker, New York (1980)
109. Edgington, E.S., Onghena, P.: Randomization Tests, 4th edn. Chapman & Hall/CRC, Boca Raton (2007)
110. Edwards, D.: Exact simulation based inference: a survey, with additions. J. Stat. Comput. Simul. **22**, 307–326 (1985)
111. Everitt, B.S.: Moments of the statistics kappa and weighted kappa. Br. J. Math. Stat. Psychol. **21**, 97–103 (1968)
112. Ezekiel, M.J.B.: Methods of Correlation Analysis. Wiley, New York (1930)
113. Feinstein, A.R.: Clinical biostatistics XXIII: the role of randomization in sampling, testing, allocation, and credulous idolatry (Part 2). Clin. Pharmacol. Ther. **14**, 898–915 (1973)
114. Feinstein, A.R.: Clinical Biostatistics. C.V. Mosby, St. Louis (1977)
115. Ferguson, G.A.: Statistical Analysis in Psychology and Education, 5th edn. McGraw-Hill, New York (1981)
116. Festinger, L.: The significance of differences between means without reference to the frequency distribution function. Psychometrika **11**, 97–105 (1946)
117. Fidler, F., Thompson, B.: Computing correct confidence intervals for ANOVA fixed- and random-effects effect sizes. Educ. Psychol. Meas. **61**, 575–604 (2001)
118. Fisher, R.A.: Statistical Methods for Research Workers. Oliver and Boyd, Edinburgh (1925)
119. Fisher, R.A.: The Design of Experiments. Oliver and Boyd, Edinburgh (1935)
120. Fisher, R.A.: The logic of inductive inference (with discussion). J. R. Stat. Soc. **98**, 39–82 (1935)
121. Fisher, R.A.: Mathematics of a lady tasting tea. In: Newman, J.R. (ed.) The World of Mathematics, vol. III, section VIII, pp. 1512–1521. Simon & Schuster, New York (1956)
122. Fisher, R.A.: The Design of Experiments, 7th edn. Hafner, New York (1960)
123. Fleiss, J.L.: Estimating the magnitude of experimental effects. Psychol. Bull. **72**, 273–276 (1969)
124. Fleiss, J.L., Cohen, J., Everitt, B.S.: Large sample standard errors of kappa and weighted kappa. Psychol. Bull. **72**, 323–327 (1969)
125. Franklin, L.A.: Exact tables of Spearman's footrule for $n = 11(1)18$ with estimate of convergence and errors for the normal approximation. Stat. Probab. Lett. **6**, 399–406 (1988)
126. Freeman, L.C.: Elementary Applied Statistics. Wiley, New York (1965)
127. Frick, R.W.: Interpreting statistical testing: process and propensity, not population and random sampling. Behav. Res. Methods Instrum. Comput. **30**, 527–535 (1998)
128. Friedman, M.: The use of ranks to avoid the assumption of normality implicit in the analysis of variance. J. Am. Stat. Assoc. **32**, 675–701 (1937)
129. Friedman, M.: A comparison of alternative tests of significance for the problem of m rankings. Ann. Math. Stat. **11**, 86–92 (1940)
130. Friedman, H.: Magnitude of experimental effect and a table for its rapid estimation. Psychol. Bull. **70**, 245–251 (1968)
131. Gaebelein, J.W., Soderquist, J.A., Powers, W.A.: A note on the variance explained in the mixed analysis of variance model. Psychol. Bull. **83**, 1110–1112 (1976)
132. Gail, M., Mantel, N.: Counting the number of $r \times c$ contingency tables with fixed margins. J. Am. Stat. Assoc. **72**, 859–862 (1977)
133. Gardner, M.J., Altman, D.G.: Statistics with Confidence: Confidence Intervals and Statistical Guidelines. British Medical Journal, London (1989)
134. Geary, R.C.: Some properties of correlation and regression in a limited universe. Metron **7**, 83–119 (1927)
135. Geary, R.C.: Testing for normality. Biometrika **34**, 209–242 (1947)

136. Gebhard, J., Schmitz, N.: Permutation tests: a revival? I. Optimum properties. Stat. Pap. **39**, 75–85 (1998)
137. Glass, G.V.: Note on rank-buserial correlation. Educ. Psychol. Meas. **26**, 623–631 (1966)
138. Glass, G.V.: Primary, secondary, and meta-analysis of research. Educ. Res. **5**, 3–8 (1976)
139. Glass, G.V.: Statistical Methods in Education and Psychology, 2nd edn. Prentice-Hall, Englewood Cliffs (1984)
140. Glass, G.V., Hakstian, A.R.: Measures of association in comparative experiments: their development and interpretation. Am. Educ. Res. J. **6**, 403–414 (1969)
141. Glass, G.V., Peckham, P.D., Sanders, J.R.: Consequences of failure to meet assumptions underlying the fixed effects analysis of variance and covariance. Rev. Educ. Res. **42**, 237–288 (1972)
142. Glass, G.V., McGraw, B., Smith, M.L.: Meta-Analysis in Social Research: Individual and Neighbourhood Reactions. Sage, Beverly Hills (1981)
143. Golding, S.L.: Flies in the ointment: methodological problems in the analysis of the percentage of variance due to persons and situations. Psychol. Bull. **82**, 278–289 (1975)
144. Good, I.J.: Further comments concerning the lady tasting tea or beer: *P*-values and restricted randomization. J. Stat. Comput. Simul. **40**, 263–267 (1992)
145. Good, P.I.: Permutation, Parametric and Bootstrap Tests of Hypotheses. Springer, New York (1994)
146. Good, P.I.: Permutation Tests: A Practical Guide to Resampling Methods for Testing Hypotheses. Springer, New York (1994)
147. Good, P.I.: Resampling Methods: A Practical Guide to Data Analysis. Birkhäuser, Boston (1999)
148. Good, P.I.: Permutation Tests: A Practical Guide to Resampling Methods for Testing Hypotheses, 2nd edn. Springer, New York (2000)
149. Good, P.I.: Resampling Methods: A Practical Guide to Data Analysis, 2nd edn. Birkhäuser, Boston (2001)
150. Good, P.I.: Extensions of the concept of exchangeability and their applications. J. Mod. Appl. Stat. Methods **1**, 243–247 (2002)
151. Goodman, L.A., Kruskal, W.H.: Measures of association for cross classifications. J. Am. Stat. Assoc. **49**, 732–764 (1954)
152. Goodman, L.A., Kruskal, W.H.: Measures of association for cross classifications, III: approximate sampling theory. J. Am. Stat. Assoc. **58**, 310–364 (1963)
153. Gravetter, F.J., Wallnau, L.B.: Essentials of Statistics for the Behavioral Sciences, 8th edn. Wadsworth, Belmont (2014)
154. Greenhouse, S.W., Geisser, S.: On methods in the analysis of profile data. Psychometrika **24**, 95–112 (1959)
155. Gridgeman, N.T.: The lady tasting tea, and allied topics. J. Am. Stat. Assoc. **54**, 776–783 (1959)
156. Grier, D.A.: Statistical laboratories and the origins of computing. Chance **12**, 14–20 (1999)
157. Grissom, R.J., Kim, J.J.: Effect Sizes for Research: A Broad Practical Approach. Lawrence Erlbaum, Mahwah (2005)
158. Grissom, R.J., Kim, J.J.: Effect Sizes for Research: Univariate and Multivariate Applications. Routledge, New York (2012)
159. Guggenmoos-Holzmann, I.: How reliable are chance-corrected measures of agreement? Stat. Med **12**, 2191–2205 (1993)
160. Guggenmoos-Holzmann, I.: Comment on "Modeling covariate effects in observer agreement studies: the case of nominal scale agreement" by P. Graham. Stat. Med. **14**, 2285–2286 (1995)
161. Guilford, J.P.: Fundamental Statistics in Psychology and Education. McGraw-Hill, New York (1950)
162. Hald, A.: History of Probability and Statistics and Their Applications Before 1750. Wiley, New York (1990)
163. Hald, A.: A History of Mathematical Statistics from 1750 to 1930. Wiley, New York (1998)

164. Haldane, J.B.S., Smith, C.A.B.: A simple exact test for birth-order effect. Ann. Eugen. **14**, 117–124 (1948)
165. Hall, N.S.: R. A. Fisher and his advocacy of randomization. J. Hist. Biol. **40**, 295–325 (2007)
166. Hanley, J.A.: Standard error of the kappa statistic. Psychol. Bull. **102**, 315–321 (1987)
167. Harding, E.F.: An efficient, minimal-storage procedure for calculating the Mann–Whitney U, generalized U and similar distributions. J. R. Stat. Soc.: Ser. C: Appl. Stat. **33**, 1–6 (1984)
168. Hayes, A.F.: Permutation test is not distribution-free: testing H_0: $\rho = 0$. Psychol. Methods **1**, 184–198 (1996)
169. Hays, W.L.: Statistics. Holt, Rinehart and Winston, New York (1963)
170. Hedges, L.V.: Estimation of effect size from a series of independent experiments. Psychol. Bull. **92**, 490–499 (1982)
171. Heiser, W.J.: Geometric representation of association between categories. Psychometrika **69**, 513–545 (2004)
172. Hellman, M.: A study of some etiological factors of malocclusion. Dent. Cosmos **56**, 1017–1032 (1914)
173. Hemelrijk, J.: Note on Wilcoxon's two-sample test when ties are present. Ann. Math. Stat. **23**, 133–135 (1952)
174. Henson, R.K., Smith, A.D.: State of the art in statistical significance and effect size reporting: a review of the APA task force report and current trends. J. Res. Dev. Educ. **33**, 285–296 (2000)
175. Hess, B., Olejnik, S., Huberty, C.J.: The efficacy of two improvement-over-chance effect sizes for two-group univariate comparisons. Educ. Psychol. Meas. **61**, 909–936 (2001)
176. Higgins, J.J.: Introduction to Modern Nonparametric Tests. Brooks/Cole, Pacific Grove (2004)
177. Hitchcock, D.B.: Yates and contingency tables: 75 years later. Electron. J. Hist. Probab. Stat. **5**, 1–14 (2009)
178. Hodges, J.L., Lehmann, E.L.: Rank methods for combination of independent experiments in analysis of variance. Ann. Math. Stat. **33**, 482–497 (1962)
179. Hodges, J.L., Lehmann, E.L.: Estimates of location based on rank tests. Ann. Math. Stat. **34**, 598–611 (1963)
180. Hope, A.C.A.: A simplified Monte Carlo significance test procedure. J. R. Stat. Soc. Ser. B Methodol. **30**, 582–598 (1968)
181. Hotelling, H.: The generalization of student's ratio. Ann. Math. Stat. **2**, 360–378 (1931)
182. Hotelling, H.: A generalized T test and measure of multivariate dispersion. In: Neyman, J. (ed.) Proceedings of the Second Berkeley Symposium on Mathematical Statistics and Probability, vol. II, pp. 23–41. University of California Press, Berkeley (1951)
183. Hotelling, H., Pabst, M.R.: Rank correlation and tests of significance involving no assumption of normality. Ann. Math. Stat. **7**, 29–43 (1936)
184. Howell, D.C.: Statistical Methods for Psychology, 6th edn. Wadsworth, Belmont (2007)
185. Howell, D.C.: Statistical Methods for Psychology, 8th edn. Wadsworth, Belmont (2013)
186. Hubbard, R.: Alphabet soup: Blurring the distinctions between p's and α's in psychological research. Theor. Psychol. **14**, 295–327 (2004)
187. Hubert, L.J.: A note on Freeman's measure of association for relating an ordered to an unordered factor. Psychometrika **39**, 517–520 (1974)
188. Hunter, A.A.: On the validity of measures of association: the nominal-nominal two-by-two case. Am. J. Sociol. **79**, 99–109 (1973)
189. Hutchinson, T.P.: Kappa muddles together two sources of disagreement: Tetrachoric correlation is preferable. Res. Nurs. Health **16**, 313–315 (1993)
190. Huynh, H., Feldt, L.S.: Conditions under which mean square ratios in repeated measurements designs have exact F distributions. J. Am. Stat. Assoc. **65**, 1582–1589 (1970)
191. Irwin, J.O.: Tests of significance for differences between percentages based on small numbers. Metron **12**, 83–94 (1935)
192. Isaacson, W.: The Innovators. Simon & Schuster, New York (2014)

193. Jockel, K.H.: Finite sample properties and asymptotic efficiency of Monte Carlo tests. J. Stat. Comput. Simul. **14**, 336–347 (1986)
194. Johnston, J.E., Berry, K.J., Mielke, P.W.: A measure of effect size for experimental designs with heterogeneous variances. Percept. Mot. Skills **98**, 3–18 (2004)
195. Johnston, J.E., Berry, K.J., Mielke, P.W.: Permutation tests: precision in estimating probability values. Percept. Mot. Skills **105**, 915–920 (2007)
196. Jonckheere, A.R.: A distribution-free *k*-sample test against ordered alternatives. Biometrika **41**, 133–145 (1954)
197. Kahaner, D., Moler, C., Nash, S.: Numerical Methods and Software. Prentice-Hall, Englewood Cliffs (1988)
198. Kaufman, E.H., Taylor, G.D., Mielke, P.W., Berry, K.J.: An algorithm and FORTRAN program for multivariate LAD (ℓ_1 of ℓ_2) regression. Computing **68**, 275–287 (2002)
199. Keller-McNulty, S., Higgins, J.J.: Effect of tail weight and outliers and power and type-I error of robust permutation tests for location. Commun. Stat. Simul. Comput. **16**, 17–35 (1987)
200. Kelley, T.L.: An unbiased correlation ratio measure. Proc. Natl. Acad. Sci. **21**, 554–559 (1935)
201. Kempthorne, O.: The Design and Analysis of Experiments. Wiley, New York (1952)
202. Kempthorne, O.: The randomization theory of experimental inference. J. Am. Stat. Assoc. **50**, 946–967 (1955)
203. Kempthorne, O.: Some aspects of experimental inference. J. Am. Stat. Assoc. **61**, 11–34 (1966)
204. Kempthorne, O.: Why randomize? J. Stat. Plan. Inference **1**, 1–25 (1977)
205. Kendall, M.G.: A new measure of rank correlation. Biometrika **30**, 81–93 (1938)
206. Kendall, M.G.: The treatment of ties in ranking problems. Biometrika **33**, 239–251 (1945)
207. Kendall, M.G.: Rank Correlation Methods. Griffin, London (1948)
208. Kendall, M.G.: Rank Correlation Methods, 3rd edn. Griffin, London (1962)
209. Kendall, M.G., Babington Smith, B.: The problem of *m* rankings. Ann. Math. Stat. **10**, 275–287 (1939)
210. Kendall, M.G., Babington Smith, B.: On the method of paired comparisons. Biometrika **31**, 324–345 (1940)
211. Kendall, M.G., Kendall, S.F.H., Babington Smith, B.: The distribution of Spearman's coefficient of rank correlation in a universe in which all rankings occur an equal number of times. Biometrika **30**, 251–273 (1939)
212. Kennedy, P.E.: Randomization tests in econometrics. J. Bus. Econ. Stat. **13**, 85–94 (1995)
213. Kenny, D.A.: Statistics for the Social and Behavioral Sciences. Little Brown, Boston (1987)
214. Keppel, G.: Design and Analysis: A Researcher's Handbook, 2nd edn. Prentice-Hall, Englewood Cliffs (1982)
215. Keppel, G., Zedeck, S.: Data Analysis for Research Designs: Analysis of Variance and Multiple Regression/Correlation Approaches. Freeman, New York (1989)
216. Kim, M.J., Nelson, C.R., Startz, R.: Mean revision in stock prices? a reappraisal of the empirical evidence. Rev. Econ. Stud. **58**, 515–528 (1991)
217. Kingman, J.F.C.: Uses of exchangeability. Ann. Probab. **6**, 183–197 (1978). [Abraham Wald memorial lecture delivered in Aug 1977 in Seattle, Washington]
218. Kirk, R.E.: Experimental Design: Procedures for the Behavioral Sciences. Brooks/Cole, Belmont (1968)
219. Kirk, R.E.: Practical significance: a concept whose time has come. Educ. Psychol. Meas. **56**, 746–759 (1996)
220. Kirk, R.E.: Effect magnitude: a different focus. J. Stat. Plan. Inference **137**, 1634–1646 (2006). [Keynote address delivered at the 2003 International Conference on Statistics, Combinatorics, and Related Areas, held at the University of Southern Maine]
221. Kraft, C.A., van Eeden, C.: A Nonparametric Introduction to Statistics. Macmillan, New York (1968)
222. Krause, E.F.: Taxicab Geometry. Addison-Wesley, Menlo Park (1975)
223. Krippendorff, K.: Bivariate agreement coefficients for reliability of data. In: Borgatta, E.G. (ed.) Sociological Methodology, pp. 139–150. Jossey-Bass, San Francisco (1970)

224. Kruskal, W.H.: Historical notes on the Wilcoxon unpaired two-sample test. J. Am. Stat. Assoc. **52**, 356–360 (1957)
225. Kruskal, W.H., Wallis, W.A.: Use of ranks in one-criterion variance analysis. J. Am. Stat. Assoc. **47**, 583–621 (1952). [Erratum: J. Am. Stat. Assoc. **48**, 907–911 (1953)]
226. Lachin, J.M.: Statistical properties of randomization in clinical trials. Control. Clin. Trials **9**, 289–311 (1988)
227. LaFleur, B.J., Greevy, R.A.: Introduction to permutation and resampling-based hypothesis tests. J. Clin. Child Adolesc. **38**, 286–294 (2009)
228. Lance, C.E.: More statistical and methodological myths and urban legends. Organ. Res. Methods **14**, 279–286 (2011)
229. Lange, J.: Crime as Destiny: A Study of Criminal Twins. Allen & Unwin, London (1931). [Translated by C. Haldane]
230. Larson, S.C.: The shrinkage of the coefficient of multiple correlation. J. Educ. Psychol. **22**, 45–55 (1931)
231. Larson, R.C., Sadiq, G.: Facility locations with the Manhattan metric in the presence of barriers to travel. Oper. Res. **31**, 652–669 (1983)
232. Lawley, D.N.: A generalization of Fisher's z test. Biometrika **30**, 180–187 (1938)
233. Lawley, D.N.: Corrections to "A generalization of Fisher's z test". Biometrika **30**, 467–469 (1939)
234. Leach, C.: Introduction to Statistics: A Nonparametric Approach for the Social Sciences. Wiley, New York (1979)
235. Lehmann, E.L.: Parametrics vs. nonparametrics: two alternative methodologies. J. Nonparametr. Stat. **21**, 397–405 (2009)
236. Lehmann, E.L.: Fisher, Neyman, and the Creation of Classical Statistics. Springer, New York (2011)
237. Lehmann, E.L., Stein, C.M.: On the theory of some non-parametric hypotheses. Ann. Math. Stat. **20**, 28–45 (1949)
238. Levine, J.H.: Joint-space analysis of "pick-any" data: analysis of choices from an unconstrained set of alternatives. Psychometrika **44**, 85–92 (1979)
239. Levine, T.R., Hullett, C.R.: Eta squared, partial eta squared, and misreporting of effect size in communication research. Hum. Commun. Res. **28**, 612–625 (2002)
240. Levine, T.R., Weber, R., Hullett, C.R., Park, H.S., Massi Lindsey, L.L.: A critical assessment of null hypothesis significance testing in quantitative communication research. Hum. Commun. Res. **34**, 171–187 (2008)
241. Levine, T.R., Weber, R., Park, H.S., Hullett, C.R.: A communication researchers' guide to null hypothesis significance testing and alternatives. Hum. Commun. Res. **34**, 188–209 (2008)
242. Light, R.J.: Measures of response agreement for qualitative data: some generalizations and alternatives. Psychol. Bull. **76**, 365–377 (1971)
243. Light, R.J., Margolin, B.H.: An analysis of variance for categorical data. J. Am. Stat. Assoc. **66**, 534–544 (1971)
244. Linn, R.L., Baker, E.L., Dunbar, S.B.: Complex performance-based assessment: expectations and validation criterion. Educ. Res. **20**, 15–21 (1991)
245. Loether, H.J., McTavish, D.G.: Descriptive and Inferential Statistics: An Introduction, 4th edn. Allyn and Bacon, Boston (1993)
246. Loughin, T.M., Scherer, P.N.: Testing for association in contingency tables with multiple column responses. Biometrics **54**, 630–637 (1998)
247. Ludbrook, J.: Advantages of permutation (randomization) tests in clinical and experimental pharmacology and physiology. Clin. Exp. Pharmacol. Physiol. **21**, 673–686 (1994)
248. Ludbrook, J.: Issues in biomedical statistics: comparing means by computer-intensive tests. Aust. N. Z. J. Surg. **65**, 812–819 (1995)
249. Ludbrook, J.: The Wilcoxon–Mann–Whitney test condemned. Br. J. Surg. **83**, 136–137 (1996)
250. Ludbrook, J.: Statistical techniques for comparing measures and methods of measurement: a critical review. Clin. Exp. Pharmacol. Physiol. **29**, 527–536 (2002)

251. Ludbrook, J.: Outlying observations and missing values: how should they be handled? Clin. Exp. Pharmacol. Physiol. **35**, 670–678 (2008)
252. Ludbrook, J., Dudley, H.A.F.: Issues in biomedical statistics: analyzing 2 × 2 tables of frequencies. Aust. N. Z. J. Surg. **64**, 780–787 (1994)
253. Ludbrook, J., Dudley, H.A.F.: Issues in biomedical statistics: statistical inference. Aust. N. Z. J. Surg. **64**, 630–636 (1994)
254. Ludbrook, J., Dudley, H.A.F.: Why permutation tests are superior to t and F tests in biomedical research. Am. Stat. **52**, 127–132 (1998)
255. Ludbrook, J., Dudley, H.A.F.: Discussion of "Why permutation tests are superior to t and F tests in biomedical research" by J. Ludbrook and H.A.F. Dudley. Am. Stat. **54**, 87 (2000)
256. Lunneborg, C.E.: Data Analysis by Resampling: Concepts and Applications. Duxbury, Pacific Grove (2000)
257. Maclure, M., Willett, W.C.: Misinterpretation and misuse of the kappa statistic. Am. J. Epidemiol. **126**, 161–169 (1987)
258. Manly, B.F.J.: Randomization and Monte Carlo Methods in Biology. Chapman & Hall, London (1991)
259. Manly, B.F.J.: Randomization and Monte Carlo Methods in Biology, 2nd edn. Chapman & Hall, London (1997)
260. Manly, B.F.J.: Randomization, Bootstrap and Monte Carlo Methods in Biology, 3rd edn. Chapman & Hall/CRC, Boca Raton (2007)
261. Manly, B.F.J., Francis, R.I.C.: Analysis of variance by randomization when variances are unequal. Aust. N. Z. J. Stat. **41**, 411–429 (1999)
262. Mann, H.B., Whitney, D.R.: On a test of whether one of two random variables is stochastically larger than the other. Ann. Math. Stat. **18**, 50–60 (1947)
263. Margolin, B.H., Light, R.J.: An analysis of variance for categorical data, II: small sample comparisons with chi square and other competitors. J. Am. Stat. Assoc. **69**, 755–764 (1974)
264. Mathew, T., Nordström, K.: Least squares and least absolute deviation procedures in approximately linear models. Stat. Probab. Lett. **16**, 153–158 (1993)
265. Maxim, P.S.: Quantitative Research Methods in the Social Sciences. Oxford, New York (1999)
266. Maxwell, S.E., Camp, C.J., Arvey, R.D.: Measures of strength of association: a comparative examination. J. Appl. Psychol. **66**, 525–534 (1981)
267. May, R.B., Hunter, M.A.: Some advantages of permutation tests. Can. Psychol. **34**, 401–407 (1993)
268. May, S.M.: Modelling observer agreement: an alternative to kappa. J. Clin. Epidemiol. **47**, 1315–1324 (1994)
269. McCarthy, M.D.: On the application of the z-test to randomized blocks. Ann. Math. Stat. **10**, 337–359 (1939)
270. McGrath, R.E., Meyer, G.J.: When effect sizes disagree: the case of r and d. Psychol. Methods **11**, 386–401 (2006)
271. McHugh, R.B., Mielke, P.W.: Negative variance estimates and statistical dependence in nested sampling. J. Am. Stat. Assoc. **63**, 1000–1003 (1968)
272. McLean, J.E., Ernest, J.M.: The role of statistical significance testing in educational research. J. Health Soc. Behav. **5**, 15–22 (1998)
273. McNemar, Q.: Note on the sampling error of the differences between correlated proportions and percentages. Psychometrika **12**, 153–157 (1947)
274. McQueen, G.: Long-horizon mean-reverting stock priced revisited. J. Financ. Quant. Anal. **27**, 1–17 (1992)
275. Mehta, C.R., Patel, N.R.: Algorithm 643: FEXACT. A FORTRAN subroutine for Fisher's exact test on unordered $r \times c$ contingency tables. ACM Trans. Math. Softw. **12**, 154–161 (1986)
276. Mehta, C.R., Patel, N.R.: A hybrid algorithm for Fisher's exact test in unordered $r \times c$ contingency tables. Commun. Stat. Theory Methods **15**, 387–403 (1986)

277. Mehta, C.R., Patel, N.R., Gray, R.: On computing an exact confidence interval for the common odds ratio in several 2 × 2 contingency tables. J. Am. Stat. Assoc. **80**, 969–973 (1985)
278. Metropolis, N., Ulam, S.: The Monte Carlo method. J. Am. Stat. Assoc. **44**, 335–341 (1949)
279. Meyer, G.J.: Assessing reliability: critical corrections for a critical examination of the Rorschach comprehensive system. Psychol. Assess. **9**, 480–489 (1997)
280. Micceri, T.: The unicorn, the normal curve, and other improbable creatures. Psychol. Bull. **105**, 156–166 (1989)
281. Mielke, P.W.: Asymptotic behavior of two-sample tests based on powers of ranks for detecting scale and location alternatives. J. Am. Stat. Assoc. **67**, 850–854 (1972)
282. Mielke, P.W.: Squared rank test appropriate to weather modification cross-over design. Technometrics **16**, 13–16 (1974)
283. Mielke, P.W.: Convenient beta distribution likelihood techniques for describing and comparing meteorological data. J. Appl. Meterol. **14**, 985–990 (1975)
284. Mielke, P.W.: Meteorological applications of permutation techniques based on distance functions. In: Krishnaiah, P.R., Sen, P.K. (eds.) Handbook of Statistics, vol. IV, pp. 813–830. North-Holland, Amsterdam (1984)
285. Mielke, P.W.: Geometric concerns pertaining to applications of statistical tests in the atmospheric sciences. J. Atmos. Sci. **42**, 1209–1212 (1985)
286. Mielke, P.W.: Non-metric statistical analyses: some metric alternatives. J. Stat. Plan Inference **13**, 377–387 (1986)
287. Mielke, P.W.: The application of multivariate permutation methods based on distance functions in the earth sciences. Earth Sci. Rev. **31**, 55–71 (1991)
288. Mielke, P.W., Berry, K.J.: An extended class of permutation techniques for matched pairs. Commun. Stat. Theory Methods **11**, 1197–1207 (1982)
289. Mielke, P.W., Berry, K.J.: Asymptotic clarifications, generalizations, and concerns regarding an extended class of matched pairs tests based on powers of ranks. Psychometrika **48**, 483–485 (1983)
290. Mielke, P.W., Berry, K.J.: Cumulant methods for analyzing independence of r-way contingency tables and goodness-of-fit frequency data. Biometrika **75**, 790–793 (1988)
291. Mielke, P.W., Berry, K.J.: Permutation tests for common locations among samples with unequal variances. J. Educ. Behav. Stat. **19**, 217–236 (1994)
292. Mielke, P.W., Berry, K.J.: Nonasymptotic inferences based on Cochran's Q test. Percept. Mot. Skill **81**, 319–322 (1995)
293. Mielke, P.W., Berry, K.J.: Permutation-based multivariate regression analysis: the case for least sum of absolute deviations regression. Ann. Oper. Res. **74**, 259–268 (1997)
294. Mielke, P.W., Berry, K.J.: Permutation covariate analyses of residuals based on Euclidean distance. Psychol. Rep. **81**, 795–802 (1997)
295. Mielke, P.W., Berry, K.J.: Euclidean distance based permutation methods in atmospheric science. Data Min. Knowl. Disc. **4**, 7–27 (2000)
296. Mielke, P.W., Berry, K.J.: Data-dependent analyses in psychological research. Psychol. Rep. **91**, 1225–1234 (2002)
297. Mielke, P.W., Berry, K.J.: Permutation Methods: A Distance Function Approach, 2nd edn. Springer, New York (2007)
298. Mielke, P.W., Berry, K.J.: A note on Cohen's weighted kappa coefficient of agreement with linear weights. Stat. Methodol. **6**, 439–446 (2009)
299. Mielke, P.W., Iyer, H.K.: Permutation techniques for analyzing multi-response data from randomized block experiments. Commun. Stat. Theory Methods **11**, 1427–1437 (1982)
300. Mielke, P.W., Berry, K.J., Johnson, E.S.: Multi-response permutation procedures for a priori classifications. Commun. Stat. Theory Methods **5**, 1409–1424 (1976)
301. Mielke, P.W., Berry, K.J., Brier, G.W.: Application of multi-response permutation procedures for examining seasonal changes in monthly mean sea-level pressure patterns. Mon. Weather Rev. **109**, 120–126 (1981)

302. Mielke, H.W., Anderson, J.C., Berry, K.J., Mielke, P.W., Chaney, R.L., Leech, M.: Lead concentrations in inner-city soils as a factor in the child lead problem. Am. J. Public Health **73**, 1366–1369 (1983)
303. Mielke, P.W., Berry, K.J., Landsea, C.W., Gray, W.M.: Artificial skill and validation in meteorological forecasting. Weather Forecast. **11**, 153–169 (1996)
304. Mielke, P.W., Berry, K.J., Neidt, C.O.: A permutation test for multivariate matched-pairs analyses: comparisons with Hotelling's multivariate matched-pairs T^2 test. Psychol. Rep. **78**, 1003–1008 (1996)
305. Mielke, P.W., Berry, K.J., Johnston, J.E.: A FORTRAN program for computing the exact variance of weighted kappa. Percept. Mot. Skill **101**, 468–472 (2005)
306. Mielke, P.W., Berry, K.J., Johnston, J.E.: The exact variance of weighted kappa with multiple raters. Psychol. Rep. **101**, 655–660 (2007)
307. Mielke, P.W., Berry, K.J., Johnston, J.E.: Resampling programs for multiway contingency tables with fixed marginal frequency totals. Psychol. Rep. **101**, 18–24 (2007)
308. Mielke, P.W., Berry, K.J., Johnston, J.E.: Resampling probability values for weighted kappa with multiple raters. Psychol. Rep. **102**, 606–613 (2008)
309. Mielke, P.W., Berry, K.J., Johnston, J.E.: Robustness without rank order statistics. J. Appl. Stat. **38**, 207–214 (2011)
310. Minkowski, H.: Über die positiven quadratishen formen und über kettenbruchähnliche algorithmen. Crelle's J (J. Reine Angew. Math.) **107**, 278–297 (1891). [Also available in H. Minkowski, Gesammelte Abhandlungen, vol. 1, AMS Chelsea, New York, 1967]
311. Mitchell, C., Hartmann, D.P.: A cautionary note on the use of omega squared to evaluate the effectiveness of behavioral treatments. Behav. Assess. **3**, 93–100 (1981)
312. Mood, A.M.: On the asymptotic efficiency of certain nonparametric two-sample tests. Ann. Math. Stat. **25**, 514–522 (1954)
313. Moses, L.E.: Statistical theory and research design. Ann. Rev. Psychol. **7**, 233–258 (1956)
314. Murphy, K.R., Cleveland, J.: Understanding Performance Appraisal: Social, Organizational, and Goal-Based Perspectives. Sage, Thousand Oaks (1995)
315. Myers, J.L., Well, A.D.: Research Design and Statistical Analysis. HarperCollins, New York (1991)
316. Nanda, D.N.: Distribution of the sum of roots of a determinantal equation. Ann. Math. Stat. **21**, 432–439 (1950)
317. Neave, H.R., Worthington, P.L.: Distribution-Free Tests. Unwin Hyman, London (1988)
318. Newson, R.: Parameters behind "nonparametric" statistics: Kendall's tau, Somers' D and median differences. Stata J. **2**, 45–64 (2002)
319. Neyman, J., Pearson, E.S.: On the use and interpretation of certain test criteria for purposes of statistical inference: part I. Biometrika **20A**, 175–240 (1928)
320. Neyman, J., Pearson, E.S.: On the use and interpretation of certain test criteria for purposes of statistical inference: part II. Biometrika **20A**, 263–294 (1928)
321. Nix, T.W., Barnette, J.J.: The data analysis dilemma: Ban or abandon. A review of null hypothesis significance testing. Res. Schools **5**, 3–14 (1998)
322. Nix, T.W., Barnette, J.J.: A review of hypothesis testing revisited: Rejoinder to Thompson, Knapp, and Levin. Res. Schools **5**, 55–57 (1998)
323. O'Boyle, Jr., E., Aguinis, H.: The best and the rest: revisiting the norm of normality of individual performance. Percept. Psychophys. **65**, 79–119 (2012)
324. Okamoto, D.: Letter to the editor: does it work for coffee? Significance **10**, 45–46 (2013)
325. Olds, E.G.: Distribution of sums of squares of rank differences for small numbers of individuals. Ann. Math. Stat. **9**, 133–148 (1938)
326. Olejnik, S., Algina, J.: Measures of effect size for comparative studies: applications, interpretations, and limitations. Contemp. Educ. Psychol. **25**, 241–286 (2000)
327. Olson, C.L.: On choosing a test statistic in multivariate analysis of variance. Psychol. Bull. **83**, 579–586 (1976)
328. Olson, C.L.: Practical considerations in choosing a MANOVA test statistic: a rejoinder to Stevens. Psychol. Bull. **86**, 1350–1352 (1979)

329. Osgood, C.E., Suci, G., Tannenbaum, P.: The Measurement of Meaning. University of Illinois Press, Urbana (1957)
330. Overall, J.E., Spiegel, D.K.: Concerning least squares analysis of experimental data. Psychol. Bull. **72**, 311–322 (1969)
331. Pagano, R.R.: Understanding Statistics in the Behavioral Sciences, 6th edn. Wadsworth, Pacific Grove (2001)
332. Pearson, K.: Contributions to the mathematical theory of evolution. Proc. R. Soc. Lond. **54**, 329–333 (1893)
333. Pearson, K.: Contributions to the mathematical theory of evolution, II. Skew variation in homogeneous material. Philos. Trans. R. Soc. Lond. A **186**, 343–414 (1895)
334. Pearson, K.: Mathematical contributions to the theory of evolution, XIII. On the theory of contingency and its relation to association and normal correlation. In: Drapers' Company Research Memoirs, Biometric Series I, pp. 1–35. Cambridge University Press, Cambridge (1904)
335. Pearson, E.S.: Untitled. Nature **123**, 866–867 (1929). [Review by E.S. Pearson of the second edition of R.A. Fisher's *Statistical Methods for Research Workers*]
336. Pearson, K., Heron, D.: On theories of association. Biometrika **9**, 159–315 (1913)
337. Pfaffenberger, R., Dinkel, J.: Absolute deviations curve-fitting: an alternative to least squares. In: David, H.A. (ed.) Contributions to Survey Sampling and Applied Statistics, pp. 279–294. Academic Press, New York (1978)
338. Picard, R.: Randomization and design: II. In: Feinberg, S.E., Hinkley, D.V. (eds.) R. A. Fisher: An Appreciation, pp. 46–58. Springer, Heidelberg (1980)
339. Pillai, K.C.S.: Some new test criteria in multivariate analysis. Ann. Math. Stat. **26**, 117–121 (1955)
340. Pitman, E.J.G.: Significance tests which may be applied to samples from any populations. Suppl. J. R. Stat. Soc. **4**, 119–130 (1937)
341. Pitman, E.J.G.: Significance tests which may be applied to samples from any populations: II. The correlation coefficient test. Suppl. J. R. Stat. Soc. **4**, 225–232 (1937)
342. Pitman, E.J.G.: Significance tests which may be applied to samples from any populations: III. The analysis of variance test. Biometrika **29**, 322–335 (1938)
343. Randles, R.H., Wolfe, D.A.: Introduction to the Theory of Nonparametric Statistics. Wiley, New York (1979)
344. Raveh, A.: On measures of monotone association. Am. Stat. **40**, 117–123 (1986)
345. Reinhart, A.: Statistics Done Wrong: The Woefully Complete Guide. No Starch Press, San Francisco (2015)
346. Rice, J., White, J.: Norms for smoothing and estimation. SIAM Rev. **6**, 243–256 (1964)
347. Ricketts, C., Berry, J.S.: Teaching statistics through resampling. Teach. Stat. **16**, 41–44 (1994)
348. Roberts, J.K., Henson, R.K.: Correcting for bias in estimating effect sizes. Educ. Psychol. Meas. **62**, 241–253 (2002)
349. Robinson, W.S.: Ecological correlations and the behavior of individuals. Am. Soc. Rev. **15**, 351–357 (1950). [Reprinted in Int. J. Epidemiol. **38**, 337–341 (2009)]
350. Robinson, W.S.: The statistical measurement of agreement. Am. Sociol. Rev. **22**, 17–25 (1957)
351. Robinson, W.S.: The geometric interpretation of agreement. Am. Sociol. Rev. **24**, 338–345 (1959)
352. Rosenberg, B., Carlson, D.: A simple approximation of the sampling distribution of least absolute residuals regression estimates. Commun. Stat. Simul. Comput. **6**, 421–438 (1977)
353. Rosenthal, R., Rosnow, R.L., Rubin, D.B.: Contrasts and Effect Sizes in Behavioral Research: A Correlational Approach. Cambridge University Press, Cambridge (2000)
354. Rouanet, H., Lépine, D.: Comparison between treatments in a repeated measures design: ANOVA and multivariate methods. Br. J. Math. Stat. Psychol. **23**, 147–164 (1970)
355. Rousseeuw, P.J.: Least median of squares regression. J. Am. Stat. Assoc. **79**, 421–438 (1984)
356. Routledge, R.D.: Resolving the conflict over Fisher's exact test. Can. J. Stat. **20**, 201–209 (1992)

357. Roy, S.N.: On a heuristic method of test construction and its use in multivariate analysis. Ann. Math. Stat. **24**, 220–238 (1953)
358. Roy, S.N.: Some Aspects of Multivariate Analysis. Wiley, New York (1957)
359. Saal, F.E., Downey, R.G., Lahey, M.A.: Rating the ratings: assessing the quality of rating data. Psychol. Bull. **88**, 413–428 (1980)
360. Salama, I.A., Quade, D.: A note on Spearman's footrule. Commun. Stat. Simul. Comput. **19**, 591–601 (1990)
361. Salsburg, D.: The Lady Tasting Tea: How Statistics Revolutionized Science in the Twentieth Century. Holt, New York (2001)
362. Särndal, C.E.: A comparative study of association measures. Psychometrika **39**, 165–187 (1974)
363. Satterthwaite, F.E.: An approximate distribution of estimates of variance components. Biom. Bull. **2**, 110–114 (1946)
364. Scheffé, H.: Statistical inference in the non-parametric case. Ann. Math. Stat. **14**, 305–332 (1943)
365. Scheffé, H.: The Analysis of Variance. Wiley, New York (1959)
366. Schmidt, F.L., Johnson, R.H.: Effect of race on peer ratings in an industrial situation. J. Appl. Psychol. **57**, 237–241 (1973)
367. Schuster, C.: A note on the interpretation of weighted kappa and its relations to other rater agreement statistics for metric scales. Educ. Psychol. Meas. **64**, 243–253 (2004)
368. Scott, W.A.: Reliability of content analysis: the case of nominal scale coding. Public Opin. Q. **19**, 321–325 (1955)
369. Senn, S.: Fisher's game with the devil. Stat. Med. **13**, 217–230 (1994). [Publication of a paper presented at the Statisticians in the Pharmaceutical Industry (PSI) annual conference held in Sept 1991 in Bristol, England]
370. Senn, S.: Tea for three: of infusions and inferences and milk in first. Significance **9**, 30–33 (2012)
371. Senn, S.: Response to "Tea break" by S. Springate. Significance **10**, 46 (2013)
372. Sheynin, O.B.: R. J. Boscovich's work on probability. Arch. Hist. Exact Sci. **9**, 306–324 (1973)
373. Shrout, P.E., Fleiss, J.L.: Intraclass correlations: uses in assessing rater relaibility. Psychol. Bull. **86**, 420–428 (1979)
374. Shrout, P.E., Spitzer, R.L., Fleiss, J.L.: Quantification of agreement in psychiatric diagnosis revisited. Arch. Gen. Psychiatry **44**, 172–177 (1987)
375. Siegel, S., Castellan, N.J.: Nonparametric Statistics for the Behavioral Sciences, 2nd edn. McGraw-Hill, New York (1988)
376. Siegel, S., Tukey, J.W.: A nonparametric sum of ranks procedure for relative spread in unpaired samples. J. Am. Stat. Assoc. **55**, 429–445 (1960). [Corrigendum: J. Am. Stat. Assoc. **56**, 1005 (1961)]
377. Siegfried, T.: Odds are, it's wrong. Sci. News **177**, 26–29 (2010)
378. Snedecor, G.W.: Calculation and Interpretation of Analysis of Variance and Covariance. Collegiate Press, Ames (1934)
379. Snyder, P., Lawson, S.: Evaluating results using corrected and uncorrected effect size estimates. J. Exp. Educ. **61**, 334–349 (1993)
380. Somers, R.H.: A new asymmetric measure of association for ordinal variables. Am. Sociol. Rev. **27**, 799–811 (1962)
381. Spearman, C.E.: The proof and measurement of association between two things. Am. J. Psychol. **15**, 72–101 (1904)
382. Spearman, C.E.: 'Footrule' for measuring correlation. Br. J. Psychol. **2**, 89–108 (1906)
383. Spitznagel, E.L., Helzer, J.E.: A proposed solution to the base rate problem in the kappa statistic. Arch. Gen. Psychiatry **42**, 725–728 (1985)
384. Springate, S.: Tea break. Significance **10**, 45–46 (2013)
385. Stark, R., Roberts, I.: Contemporary Social Research Methods. Micro-Case, Bellevue (1996)

386. Stevens, J.P.: Applied Multivariate Statistics for the Social Sciences. Erlbaum, Hillsdale (1986)
387. Stevens, J.P.: Intermediate Statistics: A Modern Approach. Erlbaum, Hillsdale (1990)
388. Still, A.W., White, A.P.: The approximate randomization test as an alternative to the *F* test in analysis of variance. Br. J. Math. Stat. Psychol. **34**, 243–252 (1981)
389. Stuart, A.: The estimation and comparison of strengths of association in contingency tables. Biometrika **40**, 105–110 (1953)
390. "Student": The probable error of a mean. Biometrika **6**, 1–25 (1908). ["Student" is a nom de plume for William Sealy Gosset]
391. Susskind, E.C., Howland, E.W.: Measuring effect magnitude in repeated measures ANOVA designs: implications for gerontological research. J. Gerontol. **35**, 867–876 (1980)
392. Tabachnick, B.G., Fidell, L.S.: Using Multivariate Statistics, 5th edn. Pearson, Boston (2007)
393. Taha, M.A.H.: Rank test for scale parameter for asymmetrical one-sided distributions. Publ. Inst. Stat. Univ. Paris **13**, 169–180 (1964)
394. Taylor, L.D.: Estimation by minimizing the sum of absolute errors. In: Zarembka, P. (ed.) Frontiers in Econometrics, pp. 169–190. Academic Press, New York (1974)
395. Tedin, O.: The influence of systematic plot arrangements upon the estimate of error in field experiments. J. Agric. Sci. **21**, 191–208 (1931)
396. Thompson, D.W.: On Growth and Form: The Complete Revised Edition. Dover, New York (1992)
397. Thompson, W.L.: 402 citations questioning the indiscriminate use of null hypothesis significance tests in observational studies. http://www.warnercnr.colostate.edu/~anderson/thompson1.html (2001). Accessed 18 June 2015
398. Thompson, W.L.: Problems with the hypothesis testing approach. http://www.warnercnr.colostate.edu/~gwhite/fw663/testing.pdf (2001). Accessed 18 June 2015
399. Thompson, W.D., Walter, S.D.: A reappraisal of the kappa coefficient. J. Clin. Epidemiol. **41**, 949–958 (1988)
400. Trafimow, D.: Editorial. Basic Appl. Soc. Psychol. **36**, 1–2 (2014)
401. Trafimow, D., Marks, M.: Editorial. Basic Appl. Soc. Psychol. **37**, 1–2 (2015)
402. Tschuprov, A.A.: Principles of the Mathematical Theory of Correlation. Hodge, London (1939). [Translated by M. Kantorowitsch]
403. Tukey, J.W.: Data analysis and behavioral science (1962). [Unpublished manuscript]
404. Tukey, J.W.: The future of data analysis. Ann. Math. Stat. **33**, 1–67 (1962)
405. Tukey, J.W.: Randomization and re-randomization: the wave of the past in the future. In: Statistics in the Pharmaceutical Industry: Past, Present and Future. Philadelphia Chapter of the American Statistical Association (1988). [Presented at a Symposium in Honor of Joseph L. Ciminera held in June 1988 at Philadelphia, Pennsylvania]
406. Umesh, U.N.: Predicting nominal variable relationships with multiple response. J. Forecast. **14**, 585–596 (1995)
407. Umesh, U.N., Peterson, R.A., Sauber, M.H.: Interjudge agreement and the maximum value of kappa. Educ. Psychol. Meas. **49**, 835–850 (1989)
408. Ury, H.K., Kleinecke, D.C.: Tables of the distribution of Spearman's footrule. J. R. Stat. Soc.: Ser. C: Appl. Stat. **28**, 271–275 (1979)
409. van der Reyden, D.: A simple statistical significance test. Rhod. Agric. J. **49**, 96–104 (1952)
410. Vanbelle, S., Albert, A.: A note on the linearly weighted kappa coefficient for ordinal scales. Stat. Methodol. **6**, 157–163 (2008)
411. Vaughan, G.M., Corballis, M.C.: Beyond tests of significance: estimating strength of effects in selected ANOVA designs. Psychol. Bull. **79**, 391–395 (1969)
412. von Eye, A., von Eye, M.: On the marginal dependency of Cohen's κ. Eur. Pychol. **13**, 305–315 (2008)
413. Wald, A., Wolfowitz, J.: An exact test for randomness in the non-parametric case based on serial correlation. Ann. Math. Stat. **14**, 378–388 (1943)
414. Wallis, W.A.: The correlation ratio for ranked data. J. Am. Stat. Assoc. **34**, 533–538 (1939)
415. Watnik, M.: Early computational statistics. J. Comput. Graph. Stat. **20**, 811–817 (2011)

416. Watterson, I.G.: Nondimensional measures of climate model performance. Int. J. Climatol. **16**, 379–391 (1996)
417. Welch, B.L.: The specification of rules for rejecting too variable a product, with particular reference to an electric lamp problem. Suppl. J. R. Stat. Soc. **3**, 29–48 (1936)
418. Welch, B.L.: On the z-test in randomized blocks and Latin squares. Biometrika **29**, 21–52 (1937)
419. Welch, B.L.: The significance of the difference between two means when the population variances are unequal. Biometrika **29**, 350–362 (1938)
420. Welch, B.L.: On the comparison of several mean values: an alternative approach. Biometrika **38**, 330–336 (1951)
421. Welkowitz, J., Ewen, R.B., Cohen, J.: Introductory Statistics for the Behavioral Sciences, 5th edn. Harcourt Brace, Orlando (2000)
422. Wherry, R.J.: A new formula for predicting the shrinkage of the coefficient of multiple correlation. Ann. Math. Stat. **2**, 440–457 (1931)
423. Whitehurst, G.J.: Interrater agreement for journal manuscript reviews. Am. Psychol. **39**, 22–28 (1984)
424. Whitfield, J.W.: Rank correlation between two variables, one of which is ranked, the other dichotomous. Biometrika **34**, 292–296 (1947)
425. Wickens, T.D.: Multiway Contingency Tables Analysis for the Social Sciences. Erlbaum, Hillsdale (1989)
426. Wilcox, R.R.: Statistics for the Social Sciences. Academic Press, San Diego (1996)
427. Wilcox, R.R.: Applying Contemporary Statistical Techniques. Academic Press, San Diego (2003)
428. Wilcox, R.R., Muska, J.: Measuring effect size: a non-parametric analgue of $\hat{\omega}^2$. Br. J. Math. Stat. Psychol. **52**, 93–110 (1999)
429. Wilcoxon, F.: Individual comparisons by ranking methods. Biom. Bull. **1**, 80–83 (1945)
430. Wilkinson, L.: Statistical methods in psychology journals: guidelines and explanations. Am. Psychol. **54**, 594–604 (1999)
431. Wilks, S.S.: Certain generalizations in the analysis of variance. Biometrika **24**, 471–494 (1932)
432. Wilson, H.G.: Least squares versus minimum absolute deviations estimation in linear models. Decis. Sci. **9**, 322–325 (1978)
433. Yates, F.: Contingency tables involving small numbers and the χ^2 test. Suppl. J. R. Stat. Soc. **1**, 217–235 (1934)
434. Yule, G.U.: On the association of attributes in statistics: with illustrations from the material childhood society. Philos. Trans. R. Soc. Lond. **194**, 257–319 (1900)
435. Yule, G.U.: On the methods of measuring association between two attributes. J. R. Stat. Soc. **75**, 579–652 (1912). [Originally a paper read before the Royal Statistical Society on 23 April 1912]
436. Zwick, R.: Another look at interrater agreement. Psychol. Bull. **103**, 374–378 (1988)

Author Index

An "n" following a page number indicates an entry contained within a footnote on that page, an *italic* number indicates an entry in a figure or table heading, and a page number in Roman type indicates a textural reference.

K.J. Berry et al., *Permutation Statistical Methods*,
DOI 10.1007/978-3-319-28770-6

Subject Index

K.J. Berry et al., *Permutation Statistical Methods*,
DOI 10.1007/978-3-319-28770-6

MIX
Papier aus verantwortungsvollen Quellen
Paper from responsible sources
FSC® C105338

If you have any concerns about our products,
you can contact us on
ProductSafety@springernature.com

In case Publisher is established outside the EU,
the EU authorized representative is:
Springer Nature Customer Service Center GmbH
Europaplatz 3, 69115 Heidelberg, Germany

Printed by Libri Plureos GmbH
in Hamburg, Germany